Tropical Products Institute

Proceedings of the Conference on animal feeds of tropical and subtropical origin

**Held at the
London School of Pharmacy,
Brunswick Square
London WC1N 1AX
1st-5th April 1974**

Tropical Products Institute
56/62 Gray's Inn Road,
London WC1X 8LU

1975

Ministry of Overseas Development

The Institute wishes to acknowledge financial assistance from the British Council, the Commonwealth
Foundation and the International Association of Fish
Meal Manufacturers for travelling expenses and
maintenance of some of the delegates.

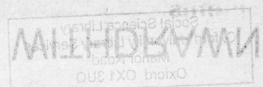

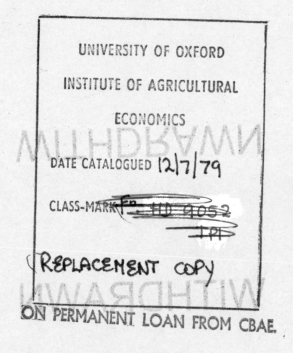
Price £4.05 excluding postage

Tropical Products Institute
ISBN: 0 85954 031 6

Contents

Seventh Session: Contamination of animal feeds by pests

Eighth Session: Use of concentrate feeds in developing countries

Concluding Session

Programme Committee

Dr N. R. Jones	*Chairman*	Tropical Products Institute, 56–62 Gray's Inn Road, London WC1X 8LU
Dr J. Nabney		
Mr D. Halliday	*Programme Secretary*	
Mr B. Wills Dr C. L. Green/Dr J. B. Davies }	*Joint organising Secretaries*	
Mr A. A. Bindloss		Unilever Ltd., St Bridget's House, Bridewell Place, London EC4P 4BP
Mr A. D. Bird		Director General, Compound Animal Feed Manufacturers', National Association (CAFMNA), 58 Southwark Bridge Road, London SE1 0AS
Professor D. Lewis		School of Agriculture, University of Nottingham, Sutton Bonington, Loughborough, Leicestershire LE12 5RD
Dr R. Roberts		J. Bibby Agriculture Ltd., Richmond House, 1 Rumford Place, Liverpool L3 9QQ
Dr A. J. Smith		Centre for Tropical Veterinary Medicine, Easter Bush, Roslin, Midlothian, Scotland

Editor	Mr D Halliday
Production Editors	Mr B. Wills and Miss B. A. Dines

Introductory remarks

by the Chairman, Dr P. C. Spensley,
Director, Tropical Products Institute

Mr Price, Ladies and Gentlemen.

It gives me great pleasure on behalf of my colleagues at the Tropical Products Institute and myself, to welcome you all, very warmly, to this International Conference on Animals Feeds of Tropical and Subtropical Origin. And may I convey greetings especially to those participants who have travelled from overseas to be with us. We are delighted to see you all and to have such a good number of countries represented — 47 is the number, I believe.

The organisation of international conferences on particular commodities or groups of commodities is now quite a regular feature of the Tropical Products Institute's programme and they all have a basic theme running through them. The subject matter is obviously very different from one conference to the next. Clearly there will be little common ground between a meeting on spices, which was the subject of our last conference, and the one we are starting today. But in each we have had the basic objective of bringing together the producer in the less-developed parts of the world with his problems, and the user in the more-developed countries, with his particular requirements and difficulties. Certainly with so many of you coming from overseas it promises very well for establishing this dialogue that we believe to be valuable.

There can, I think, be little doubt about the importance of the subject matter of our meeting this week. With the supply and demand situation for animal feeds and feed ingredients as it is today, the problems we can usefully discuss are perhaps even more significant and urgent now than they were two years ago, when we started to plan the conference. I believe, incidentally, that this may be the first international conference which has attempted a broad view of the whole range of animal feed materials which arise particularly in tropical and subtropical regions.

This year, we in TPI are celebrating our 80th anniversary — quite a respectable age for a scientific establishment, even in this country. For the last nine of these years, TPI has operated as a part of the Ministry of Overseas Development and so as a direct element in British official aid. This has been a challenging and rewarding period in the Institute's history. When I chaired the opening session of the Spice Conference in 1972, I mentioned that the Ministry of Overseas Development had been absorbed into the Foreign and Commonwealth Office and re-named the Overseas Development Administration. Today, I can inform you that we are back where we came in and that with the recent change of Government in Britain, our parent body is once again becoming a separate Ministry of Overseas Development.

This then is my cue to introduce to you Mr. William Price who, just in the last few weeks, has been appointed Parliamentary Under-Secretary to the Minister of Overseas Development, in the new Government. Mr Price is a journalist by profession and has been a member of Parliament since 1966. His constituency is Rugby in the Midlands. He is, I know, deeply concerned about the less-advantaged and the handicapped sections of society and has more than just an official interest in the needs and problems of the Third World.

It is therefore with the greatest pleasure that I welcome him on this occasion and ask him if he would honour us with an address to open this conference.

Opening Address

by Mr William Price, MP.
Parliamentary Secretary, Ministry of Overseas Development

Mr Chairman, Ladies and Gentlemen.

It gives me great pleasure to welcome you to this international conference on 'Animal Feeds of Tropical and Subtropical Origin'. I welcome especially those participants from overseas and hope that they will enjoy their stay in London.

This conference is one of a series organised biennially by the Tropical Products Institute, which is one of the scientific establishments of my ministry. This series of conferences has covered a number of aspects of the very varied work of the Institute. Previous subjects have been 'The Oil Palm', 'Essential Oils Production in Developing Countries', 'Tropical and Subtropical Fruits' and 'Spices'.

The subject of the present conference was chosen because it is now apparent that the supply of animal feeds of the right quality at the right price is one of the major constraints to the increase of supplies of animal protein at prices which the less affluent can afford. This problem is particularly acute in the poorer countries and the recent sharp rise in the world market price of animal products associated with an even greater increase in the price of feedingstuffs has made the situation much worse. We in the more-developed countries of the Western World have suffered from this but our problems in this regard are of course much less than those of the less advanced countries.

The greatest unrealised potential for increasing animal production lies, in fact, in the less-developed countries of the tropics and sub-tropics. These contain most of the world's livestock population, but efficiency of production of meat and milk is very much less than in the developed world.

There are many reasons for this related to such factors as animal health, genetic potential, marketing, etc. but undoubtedly the availability and quality of animal feed is one of the most important reasons for this low productivity.

During this conference you will be hearing papers dealing with the problems of low productivity of livestock in the less developed countries and suggesting ways in which matters could be improved. One of the most important aspects of this will be the greater use for animal feed of tropical and subtropical food crops such as cassava, bananas and sugar cane. Also there are waste materials to be used arising for instance from the processing of economic crops such as coffee, and even from animal slaughter itself. Such use can also help with problems of environmental pollution in some cases. We should also not forget the dry roughages such as cereal straws which together with grazing on natural pastures largely sustain most of the livestock in the less developed countries. The more efficient use of such roughages could make a very important contribution towards improving livestock production in the less developed countries.

Of course not all animals feeds produced in the less developed countries of the tropics and sub-tropics are used locally, and large quantities of, for example, oilseed cakes and meals are exported to Europe and Japan. This provides valuable foreign exchange to the producing countries and enables the developed countries to meet the ever-increasing demand for meat by their populations. It is to the mutual advantage of both groups that such materials be of the maximum possible quality so as to enable the exporter to sell at the best possible price and the importer to be able to make the most effective use of them. The development of local animal production industries will lead to a gradual decline in this trade unless supplies are increased.

Higher quality standards in locally produced feeds will also, of course, be of benefit in improving the efficiency of local animal production. The development of industries for the production of mixed feeds in the less developed countries is an important element in improving the efficiency and level of animal production, especially pigs and poultry. You will hear several papers referring to this subject. Also you will hear the importance of maintaining appropriate minimum standards of quality for mixed feeds. Accurate information on the composition and nutritive value of feed materials is also of great importance for such industries.

12   One of the most important functions of an international conference of this type is to enable as wide a range of people as possible to meet together to discuss problems of common interest. I hope, therefore, that you will make the most of your opportunities in this regard.

I welcome you all once again and wish you every success in your deliberations.

First Session

Monday 1st April
Morning

Chairman
Dr P. C. Spensley
Director
Tropical Products Institute

World animal production and feed supplies

B. L. Nestel

International Development Research Centre, Latin American Office,
Bogota, Colombia

Summary

During the decade of the 1960's the Gross Domestic Product (GDP) of a number of developing countries met the UN Development Decade growth target of 5% per annum. However, agricultural production grew at this rate during the decade in only 13 out of 93 countries studied. Within the agricultural sector livestock production grew at 2.7% per annum, a figure only marginally greater than the population growth of 2.6% per annum and one well below the growth in demand for livestock products.

This slow growth of the livestock sector has not only constrained overall economic development in many countries where livestock are a major natural resource but has also led to a supply deficit which has led to a substantial and widespread rise in prices for livestock products, especially meat. This situation has been accentuated by poor harvests which have limited the amount of grain available for livestock feeds.

A major part of the world's livestock are found in tropical areas of the third world. These livestock often suffer from fluctuating feed supplies and their productivity is well below that of stock in developed countries. Nevertheless many industrial and export crops are produced in the tropics and their by-products, which often have a feed potential, are frequently wasted. Excellent prospects also exist for utilising better the extra photosynthetic potential of the tropics by producing higher yields and quality in grassland and by making use of surpluses of crops, such as sugar cane and cassava, specifically for livestock feed.

In the long run the growth of animal production in the tropics (and indeed in the world at large) is likely to depend upon the use of by-products and new tropical feeds, coupled with the stratification of the livestock industry, rather than on the development of concentrate feeding practices based on traditional temperate zone feeds.

Résumé

La production animale mondiale et les provisions de pâtures

Pendant la décade de 1960, la Production Domestique Brute (PDB) d'un nombre de pays en développement a rempli l'objectif de croissance de 5% par an pour la Décade de Développement des NU. Cependant, la production agricole pendant la décade s'accrut dans cette proportion seulement dans 13 de 93 pays étudiés. Dans le secteur agricole, la production du bétail s'accrut par 2.7% par an, un chiffre seulement en marge de la croissance de la population de 2.6% par an et qui est bien au-dessous de demandes des produits du bétail.

Cette croissance lente du secteur bétail n'a pas forcée seulement le développement global dans des nombreux pays où le bétail est une ressource naturelle majeure, mais elle a menée aussi à un déficit des provisions qui par conséquent a conduit à une augmentation substantielle et répandue des prix pour les produits du bétail, surtout de la viande. Cette situation a été accentuée par des récoltes mauvaises qui ont limitées la quantité disponible de céréales pour la nourriture du bétail.

Une partie majeure du bétail mondial se trouve dans les régions tropicales du troisième monde. Ce bétail souffre souvent à cause de provisions fluctuantes de fourrage et leur productivité est bien au-dessous du stock dans les pays développés. Néanmoins, beaucoup de récoltes industrielles et pour l'exportation sont produites dans les régions tropicales et leurs sous-produits qui ont souvent un potentiel nutritif, sont fréquemment gaspillés. Il y a aussi des perspectives excellentes pour une meilleure utilisation du potentiel photosynthétique supplémentaire des tropiques, en produisant des plus grandes récoltes et une meilleure qualité de pâturages et en utilisant les excédents des récoltes comme la canne à sucre et le manioc, spécifiquement pour la nourriture du bétail.

A la longue, la croissance de la production d'animaux dans les régions tropicales (et à vrai dire à travers le monde), dépendra probablement de l'usage des sous-produits et des nouvelles pâtures tropicales couplé avec la stratification de l'industrie du bétail plutôt que du développement de l'usage des pâtures concentrées, basées sur les nourritures traditionnelles utilisées dans les zones tempérées.

16 Resumen

La producción animal mundial y los suministros de piensos

Durante la década de los años sesenta, el Producto Nacional Bruto (PNB) de cierto número de países en vías de desarrollo alcanzó la meta de la Década de Desarrollo de las NU del 5% anual. Sin embargo, solamente en 13 de los 93 países estudiados aumentó la producción agrícola en esta proporción durante la década. Dentro del sector agrícola, la producción ganadera creció el 2,7% por año, una cifra ligeramente mayor que el aumento de población del 2,6% anual y muy por debajo del incremento de la demanda de productos de la ganadería.

Este lento crecimiento del sector ganadero no solamente ha limitado el desarrollo económico general en muchos países donde la ganadería es un recurso natural principal, sino que ha conducido a un déficit de suministros que ha originado un alza sustancial extensa de los productos ganaderos, especialmente de la carne. Esta situación se ha acentuado por cosechas pobres, que han limitado la cantidad de cereales disponible para piensos del ganado.

Una parte importante de la ganadería del mundo se encuentra en áreas tropicales del tercer mundo. Esta ganadería sufre frecuentemente las fluctuaciones de los suministros de piensos y su productividad está muy por debajo de la del ganado en los países desarrollados. Sin embargo, muchas cosechas industriales y de exportación se producen en los trópicos, y sus productos secundarios, que a veces tienen potencial como pienso, se desperdician frecuentemente.

Existen, también, perspectivas excelentes para utilizar mejor el potencial fotosintético extraordinario de los trópicos produciendo rendimientos y calidades más altos en las praderas y haciendo uso de excedentes de cosechas, tales como la caña de azúcar y el cazabe, específicamente para pienso del ganado.

A largo plaza, el aumento de la producción animal en los trópicos (y, realmente, en el mundo en general) es probable que dependa del uso de productos secundarios y nuevos piensos tropicales, unido a la estratificación de la industria ganadera, más bien que del desarrollo de prácticas de piensos concentrados basadas en piensos de las zonas templades tradicionales.

The world's livestock producers are having difficulty in satisfying the current growth rate in demand for animal products. Indeed, in terms of market demand, livestock products, especially meat, appear to have some of the most promising growth prospects amongst all tropical agricultural commodities. The potential for growth in animal production is even greater if the nutritional desirability of raising the per capita intake of animal protein, which is currently only 5–10 g/day in the poorest countries, to a figure a little nearer to the 75 g/caput/day average of the populations of the developed countries, is taken into consideration.

Much of this potential lies in the less developed countries (LDCs), which contain 58% of the world's agricultural land, 70% of its cattle, 63% of its sheep and 60% of its pigs, but produce only 21% of its milk, 34% of its beef, 50% of its mutton, and 37% of its pork. In terms of productivity per head of stock, beef and veal ranges from 93 kg in North America to 4 kg in the Far East (Table 1); for pigs, comparable figures are 98 kg in North America and 16 kg in Latin America; for poultry, North America produces 12 times as much meat per head of bird as the Far East; and for milk, 30 times as much as Africa.

Whilst the biological limits of animal productivity in North America and Europe have certainly not been reached, the scope for both genetic and nutritional improvement appears to be much less in these regions than in the LDCs. For example, wastage from disease in Europe is currently only a fraction of that in some semi-arid areas where over 40% of the calves born usually die before weaning.

In the developing countries, livestock production has grown traditionally at about 1.5% per annum, but at present, with population growth often approaching 3%/a and GDP growth frequently exceeding 5%, the growth in domestic demand for livestock products often exceeds 6%/a.

In 1972 per capita agricultural production in the LDCs was actually lower than it was in the early 1960s (Table 2). For most animal products, human

TABLE 1. *Regional production of some livestock products per head of animal population*

| | Kg/product/head livestock population | | |
	Beef/Veal	*Mutton/Lamb*	*Pork*
USA and Canada	93	11	98
Latin America	36	3	16
Africa	15	3	29
Near East	17	4	12
Far East	4	4	24

Source: FAO *Production yearbook.*

TABLE 2. *Per capita agricultural production, 1963–72 (1961–65 = 100)*

	Developed countries	*Developing countries*
1963	99	101
1964	103	101
1965	102	99
1966	106	97
1967	108	100
1968	111	101
1969	109	102
1970	110	103
1971	113	103
1972	112	99

Source: FAO *Indices of agricultural production.*

population growth in the period 1961/63–1969/71 outstripped the growth of animal production; only in the case of pork and poultry did animal production grow at a faster rate than the human population (Table 3).

TABLE 3. *Annual average rate of growth of production of main livestock commodities in developing regions, 1961/63–1969/71 (%/a)*

	Africa South of Sahara	Far East	Latin America	Near East and Mahgreb	Total
Beef & veal	2.8	2.6	2.2	2.6	2.3
Mutton & lamb	1.5	1.3	1.6	2.5	1.7
Pig meat	4.3	4.5	3.7	-7.7	6.1
Poultry meat	4.6	6.2	6.7	5.7	6.1
Milk	2.5	1.5	3.2	3.2	2.2
Total	2.7	2.3	2.8	3.2	2.7
Human population growth	2.5	2.6	2.9	2.7	2.6

Source: FAO *Mon. Bull. agric. Econ. Stat.*, April 1973.

However, because of the inequitable distribution of protein supplies within the population, these figures suggest that in both relative and absolute terms the average animal protein intake of the lower income groups in the LDCs declined during the 1960s.

This situation appears to have been accentuated because people in the developed countries – and the upper socio-economic groups in the developing countries – have been consuming a disproportionate amount of meat. For example, in the USA per capita beef consumption rose from 25 to 52 kg between 1940 and 1972; the same trend can be seen in Europe and Japan. Indeed the high income-elasticity of demand for beef, at most income levels in most societies, and the inability of the industry to meet this demand, have put prices under increasing pressure and this has had the effect of making beef even less accessible to low-income consumers.

Much of the increase in production to meet the growing demand for livestock products in the wealthier countries has come from the increased use of concentrate feeds (Table 4). Thus although cereals are the most important food staple and provide, for

example, two thirds of the average human calorie intake in the Far East, as well as a large part of the protein supply for man in many countries, about 300 m/t of the world's grain production, in addition to about 60 m/t of milling offals, are currently used as animal feed.

In the developed countries, in the period 1969/71, per capita usage of coarse grains (used mainly for animal feed) averaged 387 kg whereas in the developing countries total use of all grains (used almost entirely for human food) averaged only 178 kg/caput. In the USA and Canada per capita utilisation of grain is now approaching 1 t/a of which only 70 kg are consumed directly through bakery-type products; the major part of the rest is consumed through the plant-animal food chain.

The same general trend in increased grain use has also been apparent in some LDCs where economic prosperity has both increased the demand for meat and provided the foreign exchange for feed imports. This occurred in Israel and Taiwan (Table 5) some years ago and has taken place more recently in countries such as Trinidad and Malaysia.

In Brazil, which not only has a booming economy but is fortunate enough to be able to produce its own feed supplies, soya bean use in the poultry industry increased from 60,000 t in 1962 to 600,000 t in 1973, in which year national requirements of balanced concentrates for poultry were 3 m/t or double the 1972 level. However, few LDCs have either the land resources available to maintain such a growth rate in feed production from domestic resources or the foreign exchange to import large quantities of feed.

As a result of this situation an increasing proportion of the world's grain and oilseed production is being utilised for feeding animals to meet the demand for animal products in affluent nations. For example, oilcake consumption for animal feeds rose from 15.4 m to 31.2 m/t in the developed countries during the period 1955–68 while at the same time consumption in the LDCs, not all of which was for feed, rose only from 4.5 m to 6.2 m/t.

World trade in feed products increased 83% between 1960 and 1967 and at current prices was worth around US $ 5 billion a year in 1967. This situation has accentuated an existing trend for basic food

TABLE 4. *Compound feed production in the EEC from 1955 to 1970 and percentage of increase ('000 t)*

	Belgium inc. Luxembourg	France	Germany	Italy	Netherlands	Total EEC
1955	993	1,270	1,968	380	2,900	7,511
1960	1,550	2,220	3,578	800	4,600	12,746
1965	2,527	4,544	6,594	2,600	5,625	21,290
1970	4,282	6,475	9,727	3,633	7,851	31,968
Increase (%)						
1955–70	331	410	394	856	171	326
Annual growth (%)						
1961–65	10.3	15.5	13.0	20.3	4.1	10.8
1965–70	9.0	4.9	8.3	12.9	6.8	8.6

Source: EEC statistics

TABLE 5. *Imports of feedgrains, soya beans and fishmeal into Taiwan, 1961–71 ('000 t)*

Year	Corn	Feed wheat	Barley	Sorghum	Total Feedgrains	Soya beans	Fishmeal
1962	2	67	10	12	91	62	4
1963	6	97	6	8	117	168	3
1964	9	70	10	5	94	182	3
1965	56	94	7	0	157	182	6
1966	65	72	0	2	139	165	6
1967	134	74	27	0	235	351	15
1968	361	112	53	2	528	385	16
1969	388	176	93	1	658	472	32
1970	602	151	239	4	996	618	35
1971	554	129	321	29	1,033	525	45

Source: Trade of China, Joint-Commission on Rural Reconstruction.

commodity prices to increase, and a phenomenal price rise has recently taken place. A number of developing countries have cashed in on this situation, sometimes at the expense of maintaining their own national protein supplies. Indeed the growth of beef exports to the USA from some Central American countries had led to a decline in domestic per capita protein intakes in the exporting countries.

Thus, the food problem of the Third World is now no longer solely one associated with population and food supplies but is also being strongly influenced by rising affluence, a trend that will surely be accentuated by the enormous new purchasing power of the oil-producing countries, in almost all of which current animal protein intakes are low.

FAO's Indicative World Plan provided some idea of the magnitude of the increase in concentrate feeds that might be required to provide a marginal increase in per capita animal consumption by 1985 (Table 6). To

TABLE 6. *Estimated demand for grains for feed use, 1962 (actual) and 1985 (projected) ('000,000 tons)*

	1962	1985
Developed countries	202	320
64 developing countries	17	48–68
Centrally planned countries (excl. China)	52	126
Total	271	494–514

Source: FAO *Indicative World Plan* (1969).

meet this limited goal, more than 200 m/t of additional cereals will be required for feed during a period in which the population of the LDCs is projected to grow by 1.1 billion. Unless this grain can be made available or alternate feed sources can be found, per capita availability of animal protein in the LDCs could well decline in the next decade.

However, before grain can be fed to livestock in many LDCs it will be necessary to increase supplies for human food. Tremendous efforts have been made to do this in recent years and in some countries cereal yields have been raised dramatically. This undoubtedly had had a major impact on food and feed supplies and on indicating the potential of new technology in the LDCs. Unfortunately the availability of the inputs necessary to apply the new technology varies a great

deal and there is a tendency for them to be more accessible to better-off farmers and to those with the best land. As a result, the so-called 'green revolution' has sometimes accentuated the nutritional and economic problems related to income differentials and in doing so has highlighted the complex relationships between the introduction of new technology and the bringing about of socio-economic progress to lower-income rural people.

These relationships are even more complex when we move into the livestock sector in which the stock represent a multi-purpose component of a complex farming system and may serve as sources of capital, hides, skins, hair, wool, manure, meat, milk and draught power. There is often a delicate inter-relationship between these functions, especially in ruminants, and disturbing this delicate balance by giving undue emphasis solely to meat or milk production may easily decrease overall productivity. In the case of pigs and poultry the transfer of technology is less difficult to bring about and, provided that appropriate price relationships and marketing channels exist, modernisation of these industries can occur very rapidly when appropriate feed is available.

In large parts of Asia, animal draught power is essential if crop planting is to be completed within the time limits imposed by climate. Draught animals may be required to work, perhaps, only 30 to 40 days/a, although they may use up 25–30% of the total annual farm energy production.

Given this situation, low animal productivity may represent the optimum use of a farmer's total resources. His total farming system is often a sophisticated equilibrium between man, his draught animals and his crop and, for example, the introduction of an extra rice crop a year, which may mean a higher yield and income from rice. It may not necessarily benefit total income if the extra crop can be planted only by burning the stubble, thereby depriving the draught animals of their major energy source.

A further impediment to rapid growth in livestock production is the small subsistence farmer with one or two multiple-purpose animals. Such farmers may not have the feed or capital resources to increase their stock numbers and may prefer the security of a cow producing 1 litre of milk a day for domestic consumption throughout a period of 20 months to an animal producing 5 litres daily for 10 months, but which is more susceptible to disease and whose milk has to be transported in the hot sun to a maybe distant and unreliable marketing outlet. To such a farmer the maintenance of stable production with limited risks may be more important than maximising income by introducing a new risk element.

Although this is perhaps not the place to discuss population strategies, there can be little hope for meeting animal protein requirements unless and until the rate of growth of world population can be checked. Every year this is increasing by about 75 m with 80% of the increase taking place in the LDCs. To meet current food intake levels a century hence implies a 10-fold or greater increase in agricultural production if population growth rates continue at their current level. Enormous investments will be needed to bring more land into production. In many countries extra land is just not available and in others we know little of the long-term effects on the environment of large scale forest or jungle clearance; large areas of good arable land are steadily being lost through urbanisation to house the expanding population; incredible quantities of water and energy will be required for future domestic, industrial and agricultural use.

Because of the increasing pressure on land for crop production, livestock are being pushed more and more on to marginal land and as a result both production and numbers may well decline, as has already occurred in parts of Turkey, Korea and West Africa. Given this situation, as well as the rising costs of both the labour and the physical inputs needed for intensive grassland production, I suggest that in the future we may expect a change in the pattern of ruminant meat production with more emphasis being placed on rearing from lands unsuited for food production, with stock produced in this way being finished either on by-products and/or on new types of feed which are more effective converters of solar energy than traditional cereals.

Since a large part of the world's ruminant population, rangelands and unutilised by-products are found in the tropics, where in any case there is a greater intensity of solar radiation, this is the area from which greater production must come, even though passing range grassland through cattle can be a wasteful process (Love, 1970; MacFadyen, 1964). However, until an alternative system of utilising rangeland and its incident

solar energy can be found, extensive cattle grazing would appear to be the most effective use.

Many traditional rangeland systems are already under pressure from a variety of factors such as lowered stock and human mortality, encroachment of crop lands and the availability of consumer goods. Earlier attempts to control range use, especially communal grazing, have persistently failed, but evidence is now accruing, especially in Africa, that the development of a monetary economy is inducing stock owners to voluntarily reduce grazing pressure and to conserve rangeland by selling off stock. This change is usually closely related to the development of a more modern marketing system associated with a stratified pattern of production.

The restructuring of the cattle industry in this way has come about quite rapidly in Kenya where three types of enterprises have evolved: rangeland rearing, growing out on better grassland, and feedlot finishing (Creek et al., 1973). As a result both the quality of national meat production has been improved and higher prices have had their impact even on the nomadic producers.

The Kenya system has used corn as the primary source of its feedlot energy, although in the long run its success will probably need to depend on an energy source that is not competitive with human food. However, historically the range of alternatives has never been very large and even today most concentrated rations fed in the tropics are largely based on formulations used by large European and North American compounders. Only in the last few years does a serious effort appear to have been made to utilise tropical by-products on a large scale and to produce tropical crops of high biological efficiency specifically for use as livestock feeds.

Indeed in terms of utilising solar energy rather than producing specific commodities, the potential of tropical agriculture has never been extensively exploited. In tropical regions the higher light saturation values and higher maximum rates of photosynthesis provide higher efficiencies of energy conversion and, taken in conjunction with the greater annual energy input and the absence of limitations due to low temperature, enable much higher dry-matter yields to be attained than are feasible in temperate agriculture.

For example, whereas a cereal yield of 5 t/ha represents a conversion of only about 0.6% of the annual input of light energy into grain (Cooper, 1970), recorded annual yields of both dry-matter production and solar energy conversion from tropical crops such as *Pennisetum purpureum,* sugar cane and cassava, are often multiples of the temperate data.

Progress reports on the use of some by-products such as coffee pulp and surplus bananas and some tropical crops of high biological efficiency such as sugar cane (molasses) and cassava are being presented here this week. These reports do not cover all the opportunities available for developing new feeds in the tropics. In Senegal a diet containing 40–50% groundnut hulls is being used with undecorticated cotton seed, sorghum bran and molasses to obtain daily gains of 1 kg in

TABLE 7. *Maximum yields of some tropical staples in selected experiment stations*

Crop	t/ha/a	Cal/ha/day x 10³
Cassava	71	250
Maize	20	200
Sweet potato	65	180
Rice	26	176
Sorghum*	16–18	140–158
Wheat	12	110
Banana	39	80

Source: de Vries *et al.,* 1967.
* Data updated in the light of recent research.

cattle. In the Caribbean the product of an interesting new process for derinding sugar cane forms the main component of a ration giving gains of a similar magnitude (Donefer *et al.,* 1973). Although the use of a number of these new diets may not yet be economic it could well become so in the future, within the dynamic framework which surround feed:meat price ratios and the future prospects for these two commodities.

Both cassava and sugar cane are used to produce human food and have received relatively limited attention as animal feeds until recently, although Preston (1971) has pointed out that the readily available energy per hectare from molasses may be up to 4 times that of cereals, and de Vries *et al.* (1967) have stressed that cassava, too, appears to outproduce cereals and to have a high biological efficiency, perhaps because of the fact that its yield is not constrained by the many problems associated with flowering.

There is currently a trade of 2 m/t a year in dried cassava from the Far East to the EEC where it is substituted for cereals, because the Common Agricultural Policy allows its importation at a low tariff, although in energy terms its value is similar to maize. Recent projections (Phillips, 1974) indicate a market potential for cassava in the EEC of between 4.5 m and 9 m/t by 1980, in addition to a great deal of interest from Japan and a vast untapped potential existing in the cassava-producing developing countries themselves.

At present, feed compounders rarely use more than 15% dried cassava in their rations although there is now ample evidence (Nestel, 1973) to show that, providing the ration is adequately textured and supplemented, principally with methionine, up to 60% dried cassava can be used in poultry and swine diets.

Cassava-based rations, like those of many other tropical crops and grasses, are generally low in protein and require supplementation. Currently much of such supplementation is from oilcakes of which over 30 m/t a year go into animal feeds. Oilcake production is expected to continue to grow as the human demand for vegetable oils increases. However, in the future there may be a need to rely more on non-protein nitrogen, such as biuret or urea, whose use in livestock feeds has been growing rapidly in recent years.

A number of efforts are also being made to develop the use of single-cell protein for use as both food and feed. Amongst the substrates currently under study are carobs, methanol, spent bisulphite liquor from pulp mills, molasses, gas oil, animal feedlot manure and cassava (Nestel, 1973; PAG, 1973). Some of these projects have already reached the pilot or commercial stage.

In view of the rising price of bag nitrogen there would also seem to be a real need to give more emphasis to the role of the pasture legume, especially in the tropics. Tropical pasture legumes are not widely used outside of Australia, a fact that is probably related to the cheap price of bag nitrogen in recent years and to the relative ease of managing grass pastures compared with grass legume ones. Perhaps in the long term we may hope to see the enrichment of pastures resulting from current research in the field of protoplast fusion, with the production of grasses which have their own nitrogen-fixing rhizobia.

In referring to animal protein requirements in the future we may also need to give some thought to the type of animal that will be most productive on the feed available.

Some years ago (Hentges & Howes, 1973) it was shown that Zebu cattle digested crude protein and other nutrients more efficiently and consumed more feed DM on low protein diets than did European cattle, whereas on standard nutritionally adequate diets there was no difference between the two races. More recently Bressani (1974) has shown that native (criollo) pigs in Guatemala outperformed Duroc-Jerseys on 6.5% but not on 14% protein rations. This kind of genotype-environmental interaction assumes considerable importance when we look at the vast numbers of criollo pigs and Zebu-type cattle owned by peasant farmers and when we explore the potential for increased production through better feeding of these animals.

Many past efforts at livestock improvement in the tropics have failed through trying to introduce the adoption of skilled and capital-intensive Western technology. Bressani's Criollo pigs on a 6.5% protein high coffee pulp ration certainly do not look as nice as Large Whites on a balanced diet, but they are better than the average local Criollo pig, are well adapted to their environment, grow well on a rather cheap food and do not involve a big cash outlay in a society that lacks money.

There are two other sources of animal production which I wish to mention briefly — wildlife and fish. At the risk of being controversial, because it is a highly emotive subject, I am going to express the opinion that neither domestication of species such as the eland, nor game ranching, will make any significant impact on world animal production in the future. Both concepts are attractive to the biologist and the conservationist but I do not see either proposal making a real impact on future meat supplies because, in the case of domestication, the resource base is so small and the time involved too long and, with respect to game cropping, the problem of marketing is so complex.

In contrast, a real growth centre for animal production — if that is the correct expression — in the future would seem to be aquaculture. Inland waters do not usually compete with cropland, they are largely unexploited, food conversion can be as high as 1 kg fish per kg of dry food and per ha yields in excess of 3 tonnes have been recorded. In addition, apart from its direct use as human food, one third of the world's fish catch (mainly marine) is currently processed into fish meal for use as animal feed.

Unless population growth can be drastically curtailed and better use can be made of resources such as atmospheric and waste nitrogen and of incident solar energy, the world of the future may find itself short of both carbohydrate and protein. The greatest potential for long-term growth in animal production appears to lie in a better utilisation of tropical lands through the improved use of the ruminant for producing young stock on land unsuitable for crop production, coupled with the fattening of these stock on a combination of feed produced from crops of high biological efficiency (such as sugar cane and cassava) and tropical crop by-products, with protein supplies being increasingly derived from sources such as non-protein nitrogen, single-cell protein and by-products of inland fisheries.

If all this sounds too reactionary to those of you accustomed primarily to European agriculture, perhaps I may refer you to later papers in which the authors describe how some of these changes are actually taking place.

References

Bressani, R. (1974) in *Annual Report for 1973*. Guatemala City: Instituto de Nutricion de Centro America, America y Panama (INCAP).

Cooper, J. P. (1970) Potential production and energy conversion in temperate and tropical grasses. *Herb. Abstr.* **40**(1), 1–15.

Creek, M. J., Destro, D., Miles, D. G., Redfern, D. M., Robb, J., Schleifer, E. W. & Squire, H. A. (1973) A case study in the transfer of technology, the UNDP/ FAO Feedlot project in Kenya. *Proc. 3rd Wld Conf. Anim. Prod., Melbourne* Vol. **2**, paper 3(a) 29.

de Vries, C. A., Ferwerda, J. D. & Flach, M. (1967) Choice of food crops in relation to actual and potential production in the tropics. *Neth. J. agric. Sci.* **15**, 241–248.

Donefer, E., James, L. A., & Laurie, C. K. (1973) Use of sugarcane derived feedstuffs for livestock, *Proc. 3rd Wld Conf. on Anim. Prod., Melbourne* Vol. **3**, paper 5(c) 27.

Hentges, J. F. Jr. & Howes, J. R. (1963) Digestibility of feed by Brahmans and Herefords in *Crossbreeding beef cattle*, pp. 148–152. Eds. T. J. Gunha, M. Koger, & A. C. Warnick. Gainesville: University of Florida Press.

Love, R. M. (1970) The rangelands of the Western U.S. *Sci. Am.* **222**(2), 88–97.

MacFadyen, A. (1964) Energy flow in ecosystems and its exploitation by grazing, in *Grazing in terrestial and marine environments,* pp. 3–20. Ed. Crisp. Oxford: Blackwell.

Nestel, B. L. (1973) Current trends in world cassava research. *Proc. 3rd int. Symp. trop. Roots Tubers, Ibadan.*

Phillips, T. P. (1974) *Cassava: a study of its utilisation and potential markets.* Ottawa: I.D.R.C.

Preston, T. R. (1971) The use of urea in high molasses diets for milk and beef production in the humid tropics in *Report of an ad hoc consultation on the value of non-protein nitrogen for ruminants consuming poor herbages,* pp. 101–113. Rome: FAO.

Protein Advisory Group of the United Nations System (PAG) (1973) *Rep. 3rd Mtg PAG Ad Hoc Wking Gp on Single-Cell Protein.* New York: UN. *Rep.* 10017.

Factors limiting the production of animal products in the tropics, with particular reference to animal feeds

W. J. A. Payne
Consultant
A. J. Smith
Centre for Tropical Veterinary Medicine, Roslin, Midlothian, Scotland

Summary

The major factors affecting the production of domestic animals and hence animal products in the tropics are listed and discussed. These factors include the direct and indirect effects of climatic environment, the genetic merit of the available livestock; the system of animal feeding; the incidence of animal diseases, the skill, motivation and cultural attitudes of farmers in the community; the efficacy of local research and extension services; the availability of credit; the existence of processing and marketing facilities; price structure and policy; and the priority given to the industry by local governments.

The production of animal feeds is also governed by many factors of which the two most important ones are soil type and climate. Because the possibilities and problems of animal feed production are legion, discussion has been limited to those associated with the humid tropics. Within the humid tropics, food for animals may be obtained from natural and planted forage, planted field and tree crops, by-products of field and tree crops and by-products from non-crop materials. The factors which limit the production of animal feeds from each of these sources are detailed.

Finally an assessment is made of the potential availability of animal feeds in one area of the humid tropics, namely Southern Thailand.

Résumé

Les facteurs limitatifs de la production des produits pour les animaux sous les tropiques surtout au sujet de la nourriture pour les animaux

Les facteurs majeurs affectant la production d'animaux domestiques et ainsi des produits d'animaux sous les tropiques sont enregistrés et discutés. Ces facteurs incluent les effets directs et indirects de l'environnement climatique, le mérite génétique du bétail disponible, le système de nourriture des animaux, l'incidence des maladies des animaux, l'habileté, les résultats et les attitudes culturelles des fermiers dans la communauté, l'efficacité des recherches locales et des services supplémentaires, la disponibilité de crédit, l'existence des facilités de traitement et commercialisation, la structure des prix et le système et la priorité donnée aux industries par les gouvernements locaux.

La production des nourritures pour les animaux est dirigée aussi par des nombreux facteurs desquels les plus importants sont la qualité de la terre et le climat. Etant donné que les possibilités et les problèmes de la production de nourritures pour animaux sont nombreuses, les discussions se sont limitées à celles asociées avec les régions tropicales humides. Dans les régions tropicales humides, la nourriture pour les animaux peut être obtenue de fourrages naturels et plantés, de champs plantés et de cueillettes d'arbres, les sous-produits des champs d'arbres fruitiers et les sous-produits de matériels sans appartenance aux récoltes. Les facteurs qui limitent la production de nourritures pour les animaux de chacune de ces sources, sont expliqués en détail. Finalement, on fait une évaluation de la disponibilité potentielle de nourritures pour les animaux dans une région tropicale humide nommément le Sud de la Thailande.

Resumen

Factores que limitan le producción de productos animales en los trópicos, con referencia particular a los piensos para animales

Se reseñan y discuten los factores principales que afectan a la producción de animales domésticos y, por consiguiente, de productos animales en los trópicos. Estos factores incluyen los efectos directos e indirectos del medio ambiente climático; el mérito genético del ganado disponible; el sistema de alimentación animal; la incidencia de enfermedades animales; la destreza, estímulo y actitudes culturales de los granjeros de la comunidad; la eficacia de la investigación local y de los servicios de extensión; la disponibilidad de crédito; la existencia de medios de fabricación y comercialización; estructura y política de precios; y la prioridad dada a la industria por los gobiernon locales.

La producción de piensos para animales está, también, regulada por muchos factores, de los cuales los dos más importantes son el tipo de suelo y el clima. A causa de que las posibilidades y problemas de los piensos animales son enormes, la discusión se ha limitado a los asociados con los trópicos húmedos. Dentro de los trópicos húmedos, el alimento para los animales puede obtenerse de forrajes naturales y sembrados; de cosechas de campos sembrados y de árboles plantados; de productos secundarios de las cosechas de campos y de los árboles; y productos secundarios de materiales no cosechados. Se detallan los factores que limitan la producción de piensos para animales de cada una de estas procedencias.

Finalmente, se hace una valoración de la disponibilidad potencial de piensos para animales en una zona de los trópicos húmedos, es decir, Thailandia del Sur.

Introduction

The domestic farm animal population of the world is extremely large and the number of cattle, pigs, sheep, goats, buffaloes and camels approximately equals the number of people in the world (Table 1). Of this animal population one half of the cattle and buffalo, one quarter of the sheep, more than one half of the goats and one half of the pigs are kept in the tropics. These animals are generally very unproductive. For example, in Latin America beef cattle take three, four or more years to reach slaughter weight and only 40 calves per 100 breeding cows are produced each year. In India lactation yields from indigenous cows of 500 kg would be regarded as normal. In Africa, indigenous pigs might be expected to produce one litter a year consisting of three to five piglets. These figures compare very unfavourably with standards achieved in the developed regions of the world (Table 2).

TABLE 1. *Domestic farm animal population of the world*

	Cattle	Sheep	Goats	Buffaloes	Camels	Pigs
Europe	122.6	127.5	12.6	0.3	–	139.4
USSR	99.1	137.9	5.4	0.5	0.2	67.4
North and Central America	172.0	26.8	13.6	–	–	95.2
South America	203.3	123.5	28.0	0.1	–	83.3
Asia	288.9	210.4	146.5	92.9	4.1	49.1
China	63.2	71.0	57.5	29.4	–	223.0
Africa	158.4	140.4	119.4	2.1	10.0	6.9
	33.7	237.2	0.2	–	–	3.5
Near East	42.1	149.9	55.2	3.8	4.8	0.2
World	1,141.2	1,074.7	383.0	125.3	14.6	667.7

Source: FAO

TABLE 2. *Estimates of the productivity of domestic farm animals in developed and developing countries*

	Class of livestock		
	Sheep	Cow/Steer	Pig
Number of offspring per year			
Developed countries	1.5	0.9	18
Developing countries	0.8	0.4	5
Growth rate (kg liveweight/d)			
Developed countries	0.11	1.13	0.23
Developing countries	0.05	0.23	0.05
Milk yield (l/lactation)			
Developed countries	–	3637	–
Developing countries	–	455	–

Source: Wilson (1968) and others

Why are the domestic farm animals that are kept in the tropics so unproductive? What are the factors that limit production and how important is the quantity and quality of the available animal feeds?

The factors that limit animal production in the tropics can be conveniently grouped in seven categories.

(1) The direct and indirect effect of climate.
(2) The genetic merit of the available livestock.
(3) The nutritional level that is feasible and economic.
(4) Animal health.
(5) The level of management as affected by social attitudes.
(6) The availability of processing and marketing facilities.
(7) The availability of credit.

We will now discuss sequentially the relative importance of these seven categories and then in Part II discuss how animal feeds are produced and how their availability can be increased. Finally in Part III we will use a case study to highlight the potential availability of animal feeds using data from Southern Thailand.

Part I. The major factors that affect the production of domestic animals in the tropics

(1) *The direct and indirect effects of climate*

The main climatic elements that exert a direct influence on animal production are ambient temperature, radiation, humidity and day-length, and it is the free-grazing ruminant animal that is most affected. To a greater or less degree non-ruminants, such as pigs and poultry, can usually be economically protected from intense climatic stress by the provision of adequate housing and other methods of modifying the climate.

The overall effects of direct climatic stress on free-grazing ruminants can radically reduce milk production, and slow down the attainment of physical and sexual maturity, thus increasing the time intervals between birth and mature weight and between generations. Where climatic stress is very severe there is a decrease in reproductive rate and increased mortality. On the other hand the more or less equal periods of day and night that occurs in the tropics favours year round breeding by livestock such as goats and sheep, that are seasonal breeders in the mid-latitude regions.

Topography, particularly altitude, can exert a marked effect on climate and in many tropical countries there are montane areas whose climate does not usually cause undue stress in livestock.

The major indirect effect of the climatic environment is on the feed supply and this will be dealt with in a later section. Other indirect effects are that the climate may provide an optimal environment for the proliferation of animal diseases and parasites and that it may increase the cost of processing and marketing animal products.

(2) *The genetic merit of the available livestock*

The primary problem is that although indigenous livestock breeds in the tropics may be well adapted to their respective environments, part of this adaptation has probably been achieved by a natural selection against productivity. Both high rates of milk production and rapid growth rate exacerbate the effects of climatic stress by increasing the metabolic heat output of the animal.

In this situation, the animal husbandman who has improved environmental conditions, feeding levels, management and animal health and who seeks to improve the genetic merit of his livestock to take advantage of these improvements, has a choice of pursuing several different breeding policies. He may select for productivity in indigenous stock, upgrade indigenous stock by the introduction of exotic males or by the importation of the semen, introduce a crisscross breeding system using exotic and indigenous males or introduce exotic stock and attempt to select for adaptation. There is a further policy that could be pursued and that is to ameliorate the climatic stress to such an extent that exotic stock of high merit can be used. This latter policy is often economically untenable.

Experience to date suggests that the most suitable breeding policy to adopt will vary from country to country and from region to region within countries and will depend upon many factors. These include the type of indigenous livestock available, the agricultural system prevailing, the managerial ability of the local farmers and the size and the type of the market for livestock products. For example, in some regions of Kenya and in the island of Bali in Indonesia pure breeding of the Boran and Bali breeds, respectively, should be encouraged, whilst in many part of mainland Southeast Asia and in the north of South America cross-breeding between indigenous and imported exotic cattle should be the preferred breeding policy.

In general some form of crossbreeding using indigenous and exotic cattle is likely to be the most successful breeding system (Table 3), but it is of little use to

TABLE 3. *The effect of crossbreeding on the productivity of indigenous cattle*

Milk production (lactation yield, in kg)			
Exotic	Indigenous	Crossbred	Author and Country
2,454 (Friesian)	–	2,274 (Friesian/local Egyptian cattle)	Abdel Ghani and Fahomy (1966) (Egypt).
2,550 (Friesian)	840 (White Fulani)	1,688 (Friesian/ Fulani)	Knudsen and Schael (1970) (Nigeria)
	1,674 (Sahiwal)	2,024 (Holstein/ Sahiwal)	Singh and Desai (1964) (India)
1,823 (Holstein)	–	1,400 (Holstein/ Criollo)	McDowell (1972)

upgrade indigenous cattle if the managerial abilities of the local farmers are not upgraded simultaneously. If crossbreeding is to continue beyond the first and second generations it will of course also be necessary to safeguard the continued existence of purebred indigenous cattle by encouraging the selection of the most productive indigenous stock in special bull breeding herds.

(3) *The nutritional level that is feasible and economic*

Livestock require an adequate level of nutrition if they are to express their inherent productive abilities, but in the tropics it is often uneconomic or impossible to feed at the level necessary for achievement of maximum productivity.

There are five major sources of animal feed in the tropics; natural forage, planted forage, field crops and the by-products of field crops, forage grown in association with tree crops and the by-products of tree crops, and by-product feed from other non-crop sources.

In the arid tropics, where a long dry season is the norm, the availability of natural forage and its quality is extremely seasonal (Figure 1). The area of planted forage is usually very limited and seasonal. Field crop and by-product feeds are usually in short supply. Stocking rates must therefore be limited to carrying

Figure 1

Seasonal changes in the yield and quality of *Hyparrhenia* grassland in Matabeleland (Rhodesia)

Average month end herbage analysis on a DM basis 1961–1968

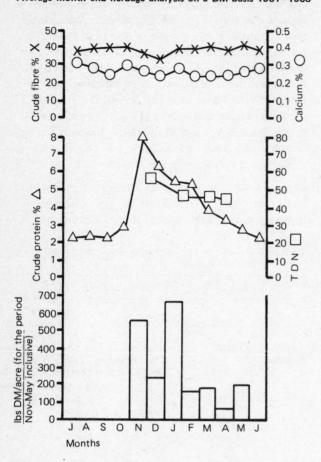

Source: Bembridge (1970)

equatorial tropics is that of ascertaining the most economic methods of feeding livestock, on the very large quantities of forage and by-product feeds that could be grown.

(4) *Animal health*

Although veterinarians have achieved considerable success in eliminating or reducing the incidence of some of the major endemic diseases, such as rinderpest and contagious bovine pleuropneumonia in many parts of the tropical world, animal health problems are still acute in most tropical countries. These problems exist despite the fact that indigenous livestock often demonstrate considerable tolerance of some tropical diseases and parasites. For example the N'dama cattle of West Africa exhibit tolerance to trypanosomiasis as do Bali cattle and most Southeast Asian water buffaloes to liver fluke. Animal health problems are particularly serious in the equatorial climatic regions where the potential for production of animal feeds is greatest and consequently some special consideration should be accorded to animal health investigations in these regions. If exotic breeds are to be introduced, it will certainly be necessary to give animal health measures some priority in development plans.

(5) *The level of management as affected by social attitudes*

Management is influenced by the cultural attitudes of the community in which the farmer lives, his educational level, his degree of skill and the advice and support that he receives from his country's extension service. The ability of the latter depends to a considerable extent on what factual information can be made available by local research services.

Cultural attitudes are possibly of more importance in livestock production than in any other sector of agriculture. It is, for example, extremely difficult to improve the genetic merit of cattle in countries such as India, where the slaughter of culled animals is taboo. Similarly it is difficult to improve the productivity of livestock in countries where the populations place more importance on the quantity rather than quality of their livestock. Almost everywhere in the tropical world larger livestock, such as cattle and water buffaloes, are raised not just for direct economic motives, but also on account of the prestige that ownership confers. These animals are used as a readily cashable investment, for use in community and family ceremonies such as birth, death and marriage, or as a convenient form of easily divisable, inheritable property.

Although many owners of tropical livestock, particularly the nomadic peoples, are excellent animal husbandmen, the majority are illiterate, and resistant to and suspicious of change.

The major instrument of change in most countries is the government extension service and although there are some dedicated livestock extension officers too, many are ill-treated, under-motivated urban dwellers

capacity of the land in the dry season because it is usually uneconomic to conserve forage on any scale or to feed relatively expensive crop or by-product concentrate feeds to breeding or growing stock. Although the availability of natural and planted forage can still be very seasonal in the monsoonal 'wet and dry' tropics, the dry season is comparatively short, crop and by-products feeds are more freely available and some conservation of food may be economic. In the humid tropics climatic conditions are very favourable for the growth of all forms of forage and far higher yields of dry matter per unit area of land can be achieved than is possible in the mid-latitude regions (Table 4). In addition livestock can utilise forage grown under tree crops, such as coconuts, and very large quantities of by-product feeds are potentially available. The major nutritional problem in the

TABLE 4. *Annual yield of dry matter from intensively managed forage in different climatic zones*

Forage	Location	Climate	Dry matter kg/ha
Napier	Puerto Rico	Humid tropical	42,600[1]
Guinea	Puerto Rico	Humid tropical	34,750[1]
Pangola	Puerto Rico	Humid tropical	33,620[1]
Para	Puerto Rico	Humid tropical	30,270[1]
Mixed	New Zealand	Warm tropical	25,220[2]

Sources: Vicente-Chandler *et al.* (1964), Brougham (1961).

who are psychologically hostile to the attitudes of the rural community from which they are probably one or two generations removed. Furthermore, even if the local research organisations are investigating the practical problems of livestock husbandry, the information channels between research and extension are often clogged, so that extension officers remain ignorant of important research findings. Given this situation the initiation of change becomes a major problem.

(6) *The availability of processing and marketing facilities*

All too often strenuous efforts are made to increase productivity on the farm whilst little attention is paid to the availability of processing facilities, the price structure and the marketing system. Even if many livestock producers are motivated by some non-economic factors they are usually still responsive to price changes and certainly require, though seldom obtain, an efficient marketing service.

In almost all countries the processing and marketing facilities, if intelligently manipulated, can be used to create new opportunities for traditional livestock industries. For example, in arid regions the organisation of feedlots to finish cattle purchased by a marketing organisation could assist in the reduction of excessive stocking rates on grazing land. In the humid tropics a breeding-marketing organisation could agist growing cattle to smallholders to utilise forage and by-product feeds that would otherwise be wasted.

(7) *Availability of credit*

All livestock farmers, large or small, nomads or specialised producers are likely to require some form of credit if they are going to expand or to radically change their operations. Only too often credit is available only to the large, specialised, land-owning producers and the credit needs of small producers and large producers who own livestock but not land, are totally neglected. Traditional attitudes to the granting of credit essentially based on using freehold title to land as security for loans, must be changed if the credit requirements of an expanding and changing livestock industry are to be satisfied.

In the next part of this paper we will deal with the major factors that limit the production of animal feeds in the tropics.

Part II. Major factors affecting the production of animal feeds in the humid tropics

(1) *Introduction*

Although many factors such as topography, soil fertility, climate, the density of the human and live-stock populations, the type of agricultural system practised and the degree of sophistication of the local economy including agro-industries, affect the production of animal feeds; the primary factors are soil fertility and climate. It is obvious that the problems of live-stock feed production in arid lands must be different from those in the humid regions. Because it is impossible to consider all factors in all tropical climates in one paper, discussion in this paper will be limited to the problems and possibilities in the humid tropics.

The natural vegetation of the major part of the humid tropics (Figure 2) was rain forest, but the activities of man have long since replaced a large proportion of the aboriginal forest by field crops, fruit orchards, tree crop plantations, secondary bush and open grasslands. Only in the mountainous areas of Southeast Asia, the

Figure 2
World distribution of tropical rain forests

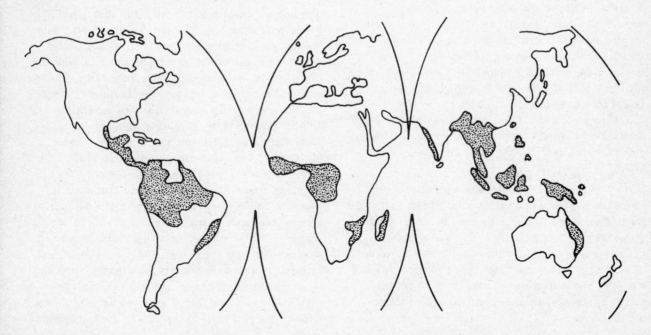

Congo basin in Africa and the Amazon basin in South America are large areas of aboriginal rain forest still intact and even the future vegetational stability of the Amazon basin is now in doubt. Apart from the rain forests, there are very considerable areas of seasonally flooded pastures, perennial swamps and montane forests and grasslands in many humid tropical countries. Within the humid tropics food for animals may be obtained from natural and planted forage, planted field and tree crops, by-products of field and tree crops and by-products of non crop materials.

(2) *Natural forage*

In the humid tropics natural forage can be subdivided into several categories. These comprise roadside, footpath, river or canal bank and field fringe forage in cropping areas, natural pastures under tree crops; forage in accessible secondary bush lands; seasonally flooded grasslands, the fringes and islands or hummocks within perennial swamps; montane pastures and almost pure stands of *Imperta spp* and other grasses that have replaced degraded rain forest.

Roadside, footpath, river or canal bank and field fringe forage can be of considerable importance in arable areas. Once the crops are planted livestock are tethered or herded where there is suitable grazing, or fed cut forage. After the harvest, livestock are usually allowed to graze the stubbles, and when these are eaten out the animals may graze in bush areas, where these still exist, or be returned to the fringe forage. In South and Southeast Asia, on humid tropical islands and in the humid regions of Africa, these grazing practices are almost universal. The practice of cutting and carrying forage to the animal is not widespread but it is carried out in a number of countries by peoples as diverse as the Javanese and Balinese in Indonesia and the Chagga who live on the slopes of Kilimanjaro in Tanzania. This supply of forage may be augmented by the planting of forage trees along roads, footpaths, river and canal banks and on the field bunds in rice growing areas. In the islands of Madura and Bali in Indonesia there are very extensive plantings of the legume tree, *Sesbania grandiflora,* whose branches are out for browze during the dry season. Indeed, the best example of what can be achieved by the utilisation of this type of forage occurs in the island of Madura. This island, with a total land area of 4,497 km² has no planted forage except for the Sesbania trees and approximately 25,000 ha of degraded forest supports a total ruminant livestock population of approximately 8,000 buffaloes, 560,000 cattle and 150,000 sheep and goats.

Natural forage under tree crops such as coconuts, rubber, oil-palm and fruit varies widely in species composition, yield and quality. In Southeast Asia where such forage is frequently not fully utilised the dominant grass species is often *Imperata cylindrica* but where it is well grazed the sward usually consists of a mixture of *Axonopus spp., Paspalum spp., Mimosa spp.,* and *Desmodium spp.* Neither type of grazing is particularly productive but they provide a palatable and useful feed.

Forage in accessible secondary bush areas varies even more than that under tree crops in species composition yield and quality, though within an ecological zone more or less the same species are present as exist under tree crops.

Seasonally flooded grasslands can provide excellent forage for a part of the year. They are very extensive in South America, in some localised areas in Africa such as around Lake Chad and in some countries in Southeast Asia, such as Indonesia. A very palatable and nutritious grass species found in the seasonally flooded areas of northern South America and Indonesia is *Leersia hexandra.* The drier margins and islands within perennial swamps often produce the same type of forage as is found in the seasonally flooded grasslands.

Montane forage in South America is to some extent utilised by llamoids, sheep and cattle, whilst in Africa it is utilised by wild game, cattle, camels, sheep and goats. The smaller areas found in Southeast Asia are not well used.

Some of the largest areas of natural forage available in the tropics, especially in the Amazon and Orinoco basins in South America and in Southeast Asia, exist as a direct result of man's mismanagement of the humid rain forest. Where forest land is cleared for cropping and them mismanaged, soil fertility decreases very rapidly and the land is invaded by low fertility grasses such as *Imperata cylindrica*. The resulting pastures are frequently burnt to provide regrowth feed for ruminant livestock. These practices prevent the regeneration of the forest which is replaced by a low-fertility, unproductive, fire climax grassland. Millions of hectares of humid tropical forest have already been reduced to this status in Southeast Asia and the area is increasing annually. The same factors are now operating on a very large scale in the Amazon basin in South America, with the difference that the fire climax often includes the babassu palm (*Orbignya spp*).

(3) *Planted forage*

The humid tropics possess a potential advantage in forage production that is now being investigated but has not yet been properly exploited. It is probable that more forage can be grown per unit area, with less effort and at less unit cost in this than in any other environment. Forage will grow continuously through-out the year and although intensity of growth may still be seasonal, with appropriate inputs the production of dry matter per unit area can be very high (Table 4). In initial investigations on the production of forage, attention has been concentrated on the use of one or a small number of species. It is possible that even higher total yields of dry matter per unit area, and certainly nutritionally superior yields of forage, could be obtained by using several forage species. The most suitable type of species to use will vary from region to region. It is particularly important, if nitrogenous fertilizers are not used, that legumes should be introduced into the forage mixture, as these improve the yields and quality of the accompanying grasses and increase the total protein content of the animal's nutrient intake. Two of the most useful legumes to use are *Stylosanthes guyanensis* and *Centrosema pubescens*.

The production potential of planted forage in the humid tropics can only be fully exploited by ruminant livestock, but the most economical methods of accomplishing this end have still to be formulated.

If labour is cheap, forage can be cut and fed to livestock managed indoors, but where this practice is uneconomic, forage must be grazed. Emphasis should be placed on total animal production per unit area of land and not on the production of the individual animal, and on the grazing of the forage when the nutrient content is high. This means that forage must always be grazed when it is young and this aim can only be achieved by heavy stocking and by the very rapid rotation of livestock around the grazings. If humid tropical pastures are properly managed it has been found that the intake of digestible energy is the first limiting factor in the proper nutrition of milking cattle. At present, under-, not overstocking is only too often the norm in grazing management in humid tropical environments. Livestock are invariably grazing old forage that has a low nutritional value. Because the land is under grazed the major part of the dry matter goes up in smoke at the annual fires instead of into the bellies of livestock.

(4) *Feeds from field and tree crops*

Some field crops can be directly fed to livestock, or alternatively if they are processed to produce human foods they yield valuable by-product feeds. Other field crops yield only by-product feeds. Tree crops also produce by-product feeds but in addition livestock production can sometimes be integrated with tree crops by planting improved forage species under the trees.

Field crops that can be directly used as animal feeds may be classified into three major groups: cereal crops such as rice, maize, sorghum and millet; root crops such as sweet potato, cassava, yam or dalo; and grain legumes such as soya beans and groundnuts. Man directly competes with livestock in the consumption of these foods and a decision as to whether they should be used for livestock feeding depends upon whether the supply is in surplus to man's needs and on the relative prices of these feeds and animal products. In general it will only be economic to feed them to non-ruminant livestock such as pigs and poultry, whose efficiency in converting them into animal products is superior to that of ruminant livestock.

Almost all field crops grown in the humid tropics yield some by-product that is useful as a livestock feed. Major by-product feeds derived from field crops together with their extraction rate are listed in Table 5. In addition the cereals provide straw and the grain legumes haulms. These are useful and in some cases are the only roughage feeds available to livestock for part of the year. Sugar cane also provides bagasse,

Crop	By-product feed	By-product feed as % original crop: by weight*
Castor	Castor meal	50
Cotton	Cottonseed meal	47
Maize	Corn bran	10
	Corn germ meal	19
Peanut (unshelled)	Groundnut meal	43
Rice (paddy)	Rice bran	10
	Broken rice	4
Sesame	Sesame meal	80
Soya bean	Soya bean	78
Sugar cane	Green tops	7
	Molasses	3

*These data refer to crops processed in the tropics and not exported and processed overseas.

that is now being pelleted and widely used as a roughage component of concentrate rations used for fattening beef animals.

The major by-product feeds derived from tree crops together with their extraction rates are shown in Table 6. In addition, banana stems may be used for ruminant livestock or pig feeding and surplus or spoilt fruit may be fed to pigs.

Ruminant livestock can be grazed under tall and hybrid coconuts, though bovines might damage dwarf coconuts. Coconut plantations provide a particularly attractive environment for grazing cattle, buffaloes and sheep and integration of livestock with the crop increases total production and return per unit area of land and reduces weeding costs. The growth of forage for ruminant animals under the trees does, however, increase competition between the coconuts and the ground cover plants both for water and for other nutrients. Consequently livestock should not be kept in coconut plantations in regions where the total rainfall is less than 1,500 to 1,700 mm and/or there is a long dry season. Additional fertilisers also need to be applied to the pasture under the nuts and on heavy soils livestock should not be grazed during the wet season as they may consolidate the soil and cause an adverse effect on coconut yield.

In the past the possibility of using forage available in rubber plantations has been almost entirely neglected. The inherent difficulties are that the availability of forage varies as the rubber trees age, livestock are required that will not damage the young trees or knock the collection cups off the mature trees, and that no effort has yet been made to select suitable forage plants for the rubber plantation environment. These difficulties can be overcome. Livestock particularly sheep could be agisted to plantations in different numbers at different times.

The situation with regard to integrating livestock with oil-palm production is somewhat similar to that of integrating livestock with rubber production.

TABLE 6. *Extraction rates for by-product feeds derived from tree crops grown in the humid tropics*

Crop	By-product feed	By-product feed as % original crop: by weight
Babassu nuts (kernels)	Babassu kernel meal	35
Cocoa beans	Cocoa shell meal[1]	11
Coconuts	Coconut meal	
	hydraulic press	34
	expeller press	35
	primitive press	42
Illipe nuts (kernels)	Illipe nut meal	60
Oil Palm (fresh fruit bunches)	Sludge (dried)[2]	3
	Palm kernel meal	22
Rubber seeds	Rubber seed meal	50
Sago (trunks)	Coarse sawdust[3]	60
	Crude wet sago[4]	40
	Unrefined sago flour[4]	21
	Sago refuse[5]	19

Note: 1. This meal contains theobromine and care should be exercised in feeding.
2. Sludge has not yet been used commercially as a feed but appears to have possibilities.
3. This is the coarse material scraped from the trunk and may be fed to ruminants.
4. These feeds may be used for non-ruminants.
5. This feed is only suitable for ruminants.

(5) By-product feeds from non-crop sources

Major sources are the fishing industry and integrated abattoirs. Fishing industries are being developed in most humid tropical countries and in many locally produced fishmeal is gradually become available. The establishment of integrated abattoirs that could produce bonemeal, meat and bonemeal, meatmeal and blood meal is difficult in most humid tropical countries, except where there are very large cities. Consumer demand is for fresh and not for chilled meat so that abattoirs must be sited close to populated areas. Almost all offals are sold for direct consumption. The opportunity does exist for the production of bone- and bloodmeals but not meatmeals, where the daily throughput of an abbatoir can economically justify the installation of the necessary plant.

Part III. Potential availability of animal feeds in Southern Thailand. A case study

The Southern Thailand region includes 14 provinces situated between the narrow neck of the Kra peninsula in the north and the Malaysian border in the south; ie approximately between latitudes 6 and 11°N and longitudes 98 and 107°E. The total area is approximately 73,000 km² (Table 7).

The climate is humid tropical but not equatorial. There is little variation in ambient temperature throughout the year: mean minimum and maximum monthly temperatures being 24.8 and 28.3°C and 26.6 and 28.7°C at Chumpon in the north and at Songkhla in the south, respectively. The humidity is almost always high. Total rainfall varies from a minimum of 1,800 mm per annum on the coast to perhaps twice that amount in the interior. Forage growth may cease for as long as three months on the west coast or for two months on the east coast and in some limited areas of the interior, but may hardly cease at all in other areas.

TABLE 7. *Gross land utilisation in South Thailand in 1973*

Category	Area (km²)	% total area
Primary forest	22,100	30
Secondary forest[1]	4,500	6
Limestone areas	1,800	2
Rubber	18,000	25
Coconuts	900	1
Oil palm[2]	<100	trace
Fruit[3]	300	trace
Rice	9,400	13
Swamps[4]	5,000	7
Miscellaneous[5]	9,500	13
Other[6]	1,500	3
Total	73,000	

Notes: 1. This occurs principally on the margins of agricultural land.
2. This is a new crop and only a small area has so far been planted.
3. This area is probably underestimated.
4. Coastal and inland swamps included.
5. A major proportion of this land is under secondary scrub or lalang (*Imperata cylindrica*).
6. This includes roadside and village grazings, etc.

Source: Regional Planning Study (1973).

Soils are varied, many being of a lateritic type and several districts are characterised by very scenic limestone outcrops. Apart from coastal and inland swamps and a few small areas of savanna the indigenous vegetation was originally almost entirely tropical rain forest. The major proportion has long since been replaced by tree and field crops, so that today only 30% of the total area is still clothed in rain forest (Table 7).

The only practical method of assessing the productivity of forage resources is to measure and compare them in terms of an animal unit. In this study the Thai livestock unit or 'Thai lsu' is used. This is the cattle biomass equivalent to the average liveweight of a mature indigenous Thai bull, ie 300 kg. Other classes of cattle and types of ruminant used in South Thailand have been given equivalent values according

TABLE 8. *Estimated carrying capacities of the various types of forage and stubbles available in South Thailand*

Category	Carrying capacity (Thai lsu/ha)
Secondary forest	0.20
Roadside and village grazing	1.00
Common grazing land	1.00
Limestone areas	0.50
Swamps	
Rubber plantations (legume mixture)	1.00
Coconut plantations (legume mixture)	1.30
Rice and other stubbles	0.40
Unused land (mainly *Imperata cylindrica*)	0.20
Planted pasture (legume mixture)	2.00

Source: Regional Planning Study (1973)

to their weight for age. The productivities of the various types of forage available in South Thailand have then been expressed in terms of carrying capacity (Table 8). Using data from Tables 7 and 8 it has been

calculated that without any improvement in forage yield, present forage resources in South Thailand could carry 1.40 m Thai livestock units. The present total population of buffaloes, cattle, sheep and goats is only equivalent to 0.84 m Thai livestock units.

No field crop at present makes any major direct contribution to animal feed resources. Very limited areas of maize and sorghum are grown and small quantities of these cereals may be fed to non-ruminant livestock. Limited quantities of fresh tapioca roots are also fed to pigs.

The major crop by-product feeds are those obtained from rice. Rice straw is an important roughage feed for cattle and buffaloes and broken rice and rice bran are major constituents of pig and poultry rations. Estimates of the potential contribution of crops to the supply of by-product feeds are given in Table 9.

The present and the potential contributions of the principal tree crops to the supply of by-product feeds

TABLE 9. *Estimates of the potential contribution of field crops to the supply of by-product feeds in South Thailand*[1]

Crop	By-product feed	Area in South Thailand (ha)	Estimated yield (t)
Maize	Stover	18,738	_[2]
Mung	haulm	7,300	_[2]
Peanuts	haulm	5,800	12,000
	groundnut meal[3]	–	3,200
Rice	straw	542,723	1,266,900
	broken rice	–	40,500
	rice bran	–	101,000
Sesame	sesame meal[3]	60	20
Soya beans	haulm	38	76
	soya bean meal	–	30
Sorghum	straw	1,423	3,500[4]
Sugarcane	green tops[5]	4,600	5,000
Sweet potato	haulm	11,000	4,000
	unsaleable tubers		3,300

Notes: 1. These figures should only be taken as very rough approximations.
2. Not known.
3. The assumption is that the total crop is crushed for oil.
4. Only the leaves are normally consumed by livestock.
5. There is no molasses available as sugarcane is used for local consumption.

Source: Regional Planning Study (1973)

are given in Table 10. One by-product that is not fully utilised at present is rubber seed meal, which has a feeding value approximately equivalent to that of coconut meal. The present production of coconut meal is not fully used in Thailand, as more than half of the crop is exported.

The major by-product feed not produced from crops is fishmeal, it being estimated that approximately 44,000 t were manufactured in 1973 (1). The quality of most of this fishmeal is poor as it contains less then 55% crude protein. Exports are at present limited to 20% of total production in order to conserve the product for use in the local livestock feed mixing industry. In addition considerable quantities of raw scrap fish are sold directly to duck farmers, especially in the province of Surat Thani. There is no abattoir by-product industry.

TABLE 10. *Estimates of the potential contribution of major tree crops to the supply of by-product feeds in South Thailand*

Crop	By-product feed	Estimated production (t)	
		Present	Potential
Rubber	Rubber seed meal	750	20,000
Coconuts	Coconut meal	16,000[1]	>16,000
Oil palm	Sludge	nil	9,900[2]
	Palm kernel meal	nil	1,800

Notes: 1. Approximately half this total is exported.
2. Sludge has potential as a feed but is not yet used on a commercial basis.

Source: Regional Planning Study (1973)

To summarise the feed situation in South Thailand forage supplies are at present underused especially in the coconut and rubber plantations. Rice is the only crop that provides any quantity of by-product feeds, limited quantities of protein concentrates are produced as by-products of tree crops, and there is a very considerable production of fishmeal.

It is considered that South Thailand could support a far larger ruminant livestock population than exists at present and that the potential for further increases in forage production is very considerable. Providing the production of maize could be considerably increased there are sufficient supplies of plant and fish protein concentrate feeds at present to produce the raw materials for an indigenous feed mixing industry that could provide the concentrate feeds needed for the expansion of the non-ruminant live-stock population.

References

Abdel-Ghani, W. & Fahony, S. K. (1966). Effect of service period on milk yield and lactation period in Fresian cattle and their crosses in *U.A.R. Agricultural Research Review. Cairo,* **44** (4) 31–36.

Bembridge, T. J. (1970). Influence of two levels of supplementation feeding and these stocking rates on the infections performance of beef cows. *Rhod. Agric. J.* **67**, 139–143.

Brougham, R. W. (1961). Symposium on grazing management. *New Zealand Soc. of Ani. Prod.* **21**, 33.

FAO (1971). Production Yearbook Vol. **25**.

Knudsen, P. B., & Schael, A. S. (1970). The Vom herd; a study of the performance of a mixed Friesian/Zebu herd in a tropical environment. *Trop. Agric. Trinidad* **47**, 189–203.

McDowell, R. E. (1972). Improvement of livestock production in warm climates. Wlf Freeman and Co., San Francisco.

Regional Planning Study (1973). Physical environment and progress of sector studies. *South Thailand Regional Planning Study Songkhla, Thailand.*

Singh, R. A., and Desai, R. N. (1964). Genetic study on relative efficiency of milk production in crossbreds (Holstein x Sahiwal) as compared to purebred sahiwal cattle. *Indian Vet. J.* **41**, 169–174.

Vicente-Chandler, J. R., Caro Costies, R. W., Pearson, F., Abrune, Figorella, J., & Silva, S. (1964). The Intensive Management of Tropical Forages in Puerto Rico, University of Puerto Rico, Bull **187**.

Wilson, P. N. (1968). Biological ceilings and economic efficiency for the production of animal protein *Chemistry and Industry* 6th July, 899.

Some aspects of animal production in the Philippines

K. A. Alim

Faculty of Agriculture, Shatby, Alexandria, Egypt

Summary

Consumption of animal protein in the Philippines is considered to be inadequate. Demand for poultry products and pork is almost entirely met by local production, but about 20% of requirements for other meats are imported. Only 2% of supplies of milk and milk products are produced locally the balance being imported.

It is considered that there is considerable potential for increasing milk and beef production from both carabao and cattle, and the ways of achieving this are discussed. Problems associated with the operations of the feed milling industry in the Philippines, which has grown with pig and poultry production, are discussed.

Résumé

Quelques aspects de la production animale dans les Philippines

La consommation de protéines animales dans les Philippines est considérée inadéquate. La demande pour des produits de volaille et de porc est presque entièrement remplie par la production locale, mais environ 20% des besoins pour d'autres viandes sont importés. Seulement 2% des provisions de lait et de produits laitiers sont produites localement, le reste étant importé.

On considère qu'il existe un potentiel considérable pour l'augmentation de la production de lait et de viande de buffles et du bétail et on discute les moyens de l'obtenir. On discute aussi les problèmes associés avec les opérations de l'industrie du moulage des aliments dans les Philippines, qui s'est développée avec la production des cochons et de la volaille.

Resumen

Algunos aspectos de la producción animal en las Filipinas

Se considera que el consumo de proteínas animales en las Filipinas es inadecuado. La demanda de productos avícolas y cerdo se satisface casi completamente por la producción local, pero se importa alrededor del 20% de las demandas de otras carnes. Se produce localmente el 2% de los suministros de leche y de productos lácteos, siendo importado el resto.

Se estima que hay un considerable potencial para aumentar la producción de leche y de carne de vaca, tanto de carabao como de ganado vacuno, y se discuten los medios de lograrlo. Se discuten los problemas asociados con las operaciones de la industria de la molienda de piensos en las Filipinas, que se ha desarrollado con la producción de cerdos y de aves.

The Philippines consist of more than 7,100 islands with a total area of some 299,400 km^2, the islands of Luzon and Mindanao constituting a high proportion of the total area. The country is rich in agricultural resources and Mindanao has vast areas of undeveloped land. The population of the Philippines today is approximately 40m and the birth rate is about 3.50%

The home supply of animal protein is inadequate. It has been shown that the 1968 per capita consumption of protein of animal origin was only 19.20g per day of which fish and marine products contributed 9.39g. Taking 50g as the minimum daily protein requirement from animal origin recommended for the Filipino in the report, 19.20g represented only 38.40% of the assessed requirement.

The major economic farm animals in the country are carabao (Malayan buffalo), cattle, pigs and poultry. The pig and poultry industries are increasing in size and making a good contribution to human protein needs but much of the grain and protein feeds needed for them are imported.

The country is rich in grasslands both on the hillsides and on the plains. In Luzon pasture is found mostly on the hillsides whereas in Mindanao, pasture is on both the hillsides and the plains. The pastures are mainly native grasses and these show considerable differences in carrying capacity. However, cattle are normally raised on the hillsides and carabao used in farm operations are mainly found in the lowlands.

The availability of pasture feed depends to a large extent on rainfall. In the north and central islands there are distinct wet and dry seasons. During the wet season water is abundant, water holes are filled and there is ample feed, but in the very dry years cattle on hillsides have to be moved to areas where feed and water are available or marketed prematurely to prevent starvation and death. In these situations, some stored forage in the form of hay or silage is required to bridge the nutritional gaps caused by the dry season. Such a practice may result in more production per animal and per acre.

The animal production situation

In 1968, the total meat supply was 562.790 t of which 97.54% was locally produced. The total population of swine for that year was estimated at 8.82 m head with a total pork production of 335.530 t which supplied about 59% of the effective demand (that is the ability to purchase), or 99.7% of the pork consumed. Most of the pigs going to the abattoir are local types which are raised on small-holdings.

The poultry population in 1968 was 93.31 m and poultry meat production was 88.470 t, representing 99.89% of total supply. In that year 15.70% of total meat consumption came from poultry.

Egg production totalled 95.840 t providing for a per capita consumption of 2.57 kg/a. The poultry in the country may be grouped into two categories: (a) the native chickens which comprise the majority of the poultry population, and (b) the modern or commercial stock. There is a widespread shifting from home flocks to large-scale projects for commercial production of eggs and broilers.

In the same year the country had 4.58 m head of carabao, 1.99 m head of cattle, 624,000 head of goats, 282,000 head of horses and a few thousand sheep. These produced 85.050 t of carabeef (carabao meat), 34.720 t of beef, and 19.120 t of other meats (chevon, mutton, horse meat), or a total production of only 138.890 t. This production accounted for only 24.67% of the total meat yield and was far from adequate in meeting the effective

demand for these meats. Approximately 20% of the supply requirement was filled up by importation of beef products mostly in the form of corned beef and frozen beef.

The dairy situation was completely different: only about 2% (9,000 t) of the total supply of milk and milk products were produced locally and the remaining 98% were imported. The cost of the importations amounted to about US $41 m.

The possibilities for expansion

In making reasonable assessments of the practical possibilities for the expansion of animal production it is important to bear in mind that poultry and pigs may compete with human beings for food while carabao and cattle can provide food for humans from raw materials which would otherwise be non-utilisable. The distinction to some extent is not absolute for hens and pigs to a greater extent can eat roughages. Carabao or cattle of course can be fed intensively on grains. The meat producing ruminants like cattle and carabao may not be very efficient converters of feed comparatively, but their value obviously is that they provide good food from cheap roughages without competing directly against human beings.

Some of the measures that could be taken in the development of carabao and cattle production and the problems facing the feed milling industry in the country are outlined as follows:

The carabao

The carabao is a tropical beef breed and is used in agricultural operations, but under some conditions the importance of breeds mainly for beef and production should not be under estimated. Evidently the genetic potential of the carabao for milk production can be improved, though such improvement from within can be made only slowly. It is also possible to increase the milk production of the carabao cows by improving their existing level of feeding and management. As might be expected feeding and management determine the actual or immediate level of production of an individual animal.

The calving interval of the carabao cows is unnecessary long. According to local information, the carabao cows calve only every 2 to 3 years. One of the factors which could contribute greatly to the increase in meat and milk production of the carabao is the improvement of reproductive performance or fertility. The number of carabao in the country is below optimum, and there has been a ban on carabao slaughter since 1953 for the purpose of increasing carabao numbers. With a decreasing ratio of carabao numbers to population, the consumer will be provided with an increasing supply per capita of palatable nutritious food for a decreasing portion of his expendable income. A high level of reproductive performance also means regular calvings of carabao cows at the most suitable time of the year. Because of this anything that can be done to obtain maximum fertility in the carabao cows is of the greatest economic importance. Efforts for development should take into account that this improvement is probably almost entirely under the control of management and that the carabao is owned by farmers in the barrios (villages) and is kept in small groups under a wide variety of conditions.

Cattle

There are various types of tropical and temperate cattle breeds in the country but the majority are local types. The cattle owners prefer the American Brahman breed. The temperate cattle breeds do not seem to do well even at an altitude of approximately 1,000 ft. It would appear that lack of suitable pasture is another problem facing the temperate cattle breeds in the Philippines. There is much crossbreeding carried out.

The cattle herds are large in size. In some of the herds there appears to be a need to increase the offtake of beef stock, increase slaughter weight and reduce time to maturity of animals. Although better breeding and better management are important factors for development, perhaps the greatest potential for increased production lies in better feeding by pasture improvement (at present, the country does not produce enough cereals to satisfy the demands of the people). Improved pastures greatly increase rate of gain and branding percentages (that is the calves branded in relation to the number of cows joined with the bulls in the previous year). It is usual in beef cattle breeding to strive for maximum economic branding percentages, and adequate fertilisation and improved varieties of forages are essential for this practice. Other improvement work takes the form of disease control, the application of a system of pasture management, grazing control, the conservation of fodder and improvement of methods of herd management.

The feed milling industry

The progress made in the poultry and swine production could not have been reached without the developments made in the feed milling industry. However, some difficulties are encountered in the industry and one of the problems is the importation of the necessary machinery and their spare parts. Additionally, the country is still not self-supporting in plant products which constitute the basic parts of the commercial feeds produced and other essential supplements and additives which must be imported. Sometimes locally produced ingredients also create important procurement problems. For example, the production of maize in the country is seasonal and the feed millers have to store enough of the crop to cover the requirements between seasons. This involves warehousing and the protection of maize from destruction by weevils and rodents during storage.

Another feed ingredient with procurement problems is rice bran. It appears that good rice bran is a suitable feed ingredient of all mashes for livestock and poultry. But as it is in short supply, the dealers adulterate it with finely ground rice hulls lowering the nutritive value. Finally in the majority of feed companies in the country there are problems of shortage of trained staff which also need to be overcome.

Discussion

Dr Babatunde: I refer to Table 7 of Dr Nestel's paper and would like to ask him if the data are based on yields of dry matter, and the means of calculating yields of cassava in terms of annual production per unit area when it is very difficult for cassava specie to yield within one year. I would also like the information on soil fertility levels and the countries which were involved.

Dr Nestel: Yields are not expressed on a dry matter basis. Recent work indicates that the age of harvesting cassava is related to moisture stress. Cassava can be harvested in a number of countries at ages as early as 8 or 9 months and these figures are converted to a 12 month basis. The figures quoted for cassava in Table 7 relate to Java but it is not recalled whether the cassava was harvested after 12 or 14/15 months. The objective of Table 7 was to show the potential of maximum yields recorded in the literature from experimental stations. The point about soil fertility is taken; it certainly would not be possible to produce 71 t of fresh cassava per ha on many tropical soils. The point was that of the utilisation of solar energy and photosynthetic potential taking the best yields from any source. The figures quoted in Table 7 originated from a range of countries, including Java (cassava), Mexico (maize) and the Philippines (triplecropped rice).

Dr Babatunde: I refer to Table 2 of the paper by Payne and Smith, and would like to know if the figures quoted are for local breeds only, and if so why data for exotic breeds managed in developing countries are not also included. Given good management and an appropriate plane of nutrition exotic animals perform well in developing country environments although perhaps not as well as in the developed countries.

Dr Smith: The figures quoted in Table 2 are purely averages and do not relate necessarily to exotic or indigenous stock alone. For example the low calving percentages of cows in developing countries are due more to poor nutrition than the breed, and it is quite possible to increase this to 0.9% by correct management. In my own experiments I have been able to get indigenous pigs to produce 15 to 16 piglets per annum by improved management. In the case of dairy cattle, it is possible to increase milk production from zebu cattle to around 3200 litres (700 gallons) per annum. Importation of exotic stock would improve this if the nutrition were correct. It is not a question of indigenous or exotic breeds but a combination of factors ranging from management, through nutrition to breeding.

Mr Blair-Rains: For 60 years the developed countries have been importing oilseeds from developing countries, and in effect taking protein and phosphorus from them. Now with our greater purchasing power we have turned our attention to obtaining animal protein from the developing countries, and they are responding to this demand by establishing ranches and feedlots. The export of meat will inevitably mean that less will be available for the ordinary people of developing countries, where diets are often already deficient in protein. I am concerned at this trend and would invite the comments of Dr Nestel on this matter.

Dr Nestel: Essentially it is a question of a political decision of better nutrition for the domestic population versus foreign exchange for industrial growth. Obviously one would like to see people better fed on humanitarian grounds but many politicians will choose foreign exchange instead. Although I personally have much sympathy with the views expressed by Mr Blair-Rains decisions of this nature must be taken nationally and it is not for outsiders to intervene.

Dr Smith: This is not just an international problem. I have seen cases of kwashiorkor in a village in Southern India which had an extremely productive poultry unit, because the eggs were being sold to buy other things.

38   *Mr Sunkwa-Mills:* There are several important factors contributing to low production of animals in the tropics which do not appear to be covered in the paper by Payne and Smith. In particular I would like to mention those of the education and training of personnel handling stock and problems associated with traditional system of land tenure which may make it difficult to acquire land for development purposes.

Dr Smith: These factors are mentioned in the paper but due to limited time they were not covered in my presentation. I agree that they are very important. I mentioned the difficulty of obtaining credit for livestock development. Obviously if the developer does not own the land there will be difficulties in obtaining credit.

Second Session

World trade in feed materials

Monday 1st April
Afternoon

Chairman
Mr A. D. Bird
Director General
CAFMNA
London

World cereal supplies and trade and usage for animal feeds

E. M. Low

Home Grown Cereals Authority, London

Summary

World cereal markets have been subject to cyclical movements with periods of high prices being followed by perhaps longer periods of low prices, surpluses and national stock piling. The year 1973 saw a major upturn in demand, dramatic price increases and dispersals of national stocks. So great has the supply and demand balance changed that the question has arisen — has a new era of shortage started, in which the present high prices will be considered as normal? The argument put forward for the change is that there has been a major upturn in demand, which will not be balanced by increases in supply. This question is vital to the livestock producers of developing countries. The expansion of meat production in these countries depends in part upon a movement from grazing to supplementary feeding types of husbandry. The extent to which this change will be possible depends upon the availability of cereals or other vegetable products for livestock feeding.

Cereal production on a world scale has been expanding rapidly. During the 1960's it increased by approximately 50%. Expansion has been greater for wheat and maize than for the other cereals and the increases, attributable more to yield improvements than to acreage changes, have been greater in the developed than in the developing countries. Trade in cereals over the same period has also increased rapidly, but the quantities involved are still small in relation to total world cereal production. By and large, the expansion in trade has been to satisfy human needs in the developing countries and the needs of livestock product industries in the developed countries.

The paper is concerned with present developments and trends in cereal supplies, trade and usage for animal feeds that can be observed in the statistics that are available, and with the identification of the forces that have been operating in the past and are likely to influence developments in the future. The analysis leads to a discussion of the extent to which cereals, either imported or home-grown, might be available to contribute to the development of animal production in the developing countries.

Résumé

Les provisions mondiales de céréales, le commerce et l'utilisation pour les pâtures des animaux

Les marchés mondiaux de céréales ont été sujets à des mouvements cycliques avec des périodes de prix élevés, suivies par des périodes peut-être plus longues de prix bas, de surplus et d'entassement des stocks nationaux. L'an 1973 a vu une augmentation majeure des demandes, une croissance dramatique des prix et une dispersion des stocks nationaux. L'équilibre entre les provisions et les demandes changea d'une telle manière que la question qui se posait était si une nouvelle ère de déficit avait commencée, dans laquelle les prix élevés courants seront considérés comme normaux? L'argument avancé pour ce changement est qu'il y a eu une augmentation majeure de demandes qui ne sera pas équilibrée par les croissances des provisions. Cette question est vitale pour les producteurs de bétail dans les pays en développement. L'expansion de la production de viande dans ces pays dépend en partie du changement de la paissance aux types de nourriture supplémentaire de ferme. En quelle mesure l'étendue de ce changement sera possible dépend de la disponibilité de céréales ou d'autres produits végétaux pour la nutrition du bétail.

La production de céréales sur une échelle mondiale s'est amplifiée rapidement. Durant la période autour de 1960, elle augmenta d'environ 50%. L'expansion a été plus grande pour le blé et le maïs que pour les autres céréales et les augmentations attribuables d'avantage aux améliorations des récoltes qu'aux changements de superficies cultivées ont été plus grandes dans les pays développés que dans les pays en développement. Le commerce de céréales dans la même période a également augmenté rapidement, mais les quantités impliquées sont encore réduites en rapport avec la production mondiale totale de céréales. Plus ou moins, l'expansion du commerce a été pour

42   satis faire les besoins humains dans les pays en développement et les besoins pour les industries des produits pour le bétail dans les pays développés.

Le rapport s'occupe avec les développements présents et les tendances concernant les provisions de céréales, le commerce et l'utilisation pour les pâtures des animaux qu'on observe dans les statistiques disponibles et avec l'identification des forces qui ont opérées dans le passé et qui probablement vont influencer le développement dans l'avenir. L'analyse mène à une discussion pour évaluer jusqu'à quel point les céréales, soit importées soit cultivées sur place, pourraient être disponibles pour contribuer au développement de la production du bétail dans les pays en développement.

Resumen

Los suministros de cereales mundiales y el comercio y uso para piensos de los animales

Los mercados de cereales mundiales han estado sujetos a movimientos cílicos con períodos de precios altos que han sido seguidos por, quizá, más largos períodos de precios bajos, excedentes y acumulación de las existencias nacionales. Tanto ha cambiado la balanza de la oferta y la demanda, que ha surgido la interrogación — ¿ha comenzado una nueva era de escasez, en la que los precios altos actuales serán considerados como normales? El razonamiento presentado para explicar el cambio es que ha habido un importante aumento en la demanda, que no será equilibrado por los aumentos en el suministro. Esta cuestión es vital para los productores de ganados de los países en vías de desarrollo. La expansión de la producción de carne en esos países depende en parte de un cambio de pastos a tipos de piensos suplementarios de producción agrícola. La extensión en que se pueda producir este cambio dependerá de la disponibilidad de cereales o de ostros productos vegetales para alimento del ganado.

La producción de cereales a escala mundial ha estado aumentando rápidamente. Durante los años 1960 aumentó el 50% aproximadamente. La expansión ha sido mayor para el trigo y el maiz que para los otros cereales y los aumentos, atributibles más a mejora de producción que a cambios en la superficie cultivada, han sido más grandes en los países desarrollados que en los en vías de desarrollo. También ha aumentado rápidamente el comercio de cereales durante el mismo período, pero el volumen de las operaciones es todavía pequeño en relación con la producción de cereales mundial total. Por todos conceptos, la expansión en el comercio ha sido para satisfacer las necesidades humanas en los países en vías de desarrollo y las necesidades de las industrias de productos de la ganadería en los países desarrollados.

Este documento trata de la evolución y tendencias actuales en los suministros, comercio y uso de cereales para piensos de animales que pueden observarse en las estadísticas de que se dispone, y de la identificación de las fuerzas que han estado operando en el pasado y que es probable que influyan en la evolución en el futuro. El análisis conduce a una discusión de la magnitud de los cereales de que pudiera disponerse, ya sean importados o de producción nacional, para contribuir al progreso de la producción animal en los países en vías de desarrollo.

Introduction

World cereal markets have been subject to cyclical movements with periods of high prices being followed by perhaps longer periods of low prices, surpluses and national stock piling. The year 1973 saw a major upturn in demand, dramatic price increases and dispersals of national stocks. So great has the supply and demand balance changed that the question has arisen — has a new era of shortage started, in which the present high prices will be considered as normal? The argument put forward for the change is that there has been a major upturn in demand, which will not be balanced by increases in supply. This question is vital to the livestock producers of developing countries. The expansion of meat production in these countries depends in part upon a movement from grazing to supplementary feeding types of husbandry. The extent to which this change will be possible depends upon the availability of cereals or other vegetable products for livestock feeding.

The paper is concerned with present developments and trends in cereal supplies, trade and usage for animal feeds that can be observed in the statistics that are available, and with the identification of the forces that have been operating in the past and are likely to influence developments in the future. The analysis shows that there are so many influences at work that it is not possible to reach firm conclusions on the extend to which cereals, either imported or home-grown, might be available or contribute to the development of animal production in the developing countries. It is hoped however that the analysis helps to clarify the issues involved.

General considerations

A large proportion of people in the world is underfed. Many are under-fed only in relation to their own expectations of what their diet should be, but a very large number are in absolute terms under-nourished. This state exists not because the world does not contain the resources needed for the production of sufficient food but because the political and economic regime allows under-utilisation of resources in some regions and shortages in others.

Cereals may be consumed either as primary or derived food products. In this context a primary food product is defined as one obtained directly from a food-producing activity and the major primary foods are vegetable products, fish, game and products derived from animals, reared on feeds that cannot be eaten by humans; derived foods are those obtained from animals which are fed on feeds that can equally well be consumed by humans. Because animals are unable to convert vegetable products to provide the same amount of energy, protein, etc. as was contained in the original vegetable products, there is a food loss involved in the production of these derived foods. The inevitable consequence is that derived food products are expensive.

Diets change as peoples become more wealthy. Past experience indicates that as peoples become more wealthy they increase their consumption of basic foods. When basic nutritional requirements are satisfied expenditure becomes diversified into more nutritional and palatable foods, clothing, consumer durables and luxury items. The actual patterns of expenditure that have emerged have varied with regional circumstances, and with the availability and relative prices of competing products. The more expensive primary and therefore derived food products are in relation to consumer durables, clothes, etc., the more will demand be directed towards primary foods and away from derived livestock products.

The price system in a free market has a function of equating supply to demand. In the short term it acts as an instrument for rationing the supplies available among the people most anxious and able to buy. In the longer term it entices production to the level where the marginal producer is just encouraged to produce and the marginal buyer is just prepared to buy.

The world's requirement for cereals may be defined by population and by man's nutritional needs but neither demand nor supply can be readily forecast. Future demand depends on the range of and the attractiveness of substitute products that may be available; future supply depends on technological developments and these as a rule cannot be anticipated.

Disturbances to the cereal market since 1972

Cereal prices, which had been stable during the 1960's, increased rapidly from the autumn of 1972. The extent of the change is illustrated in Table 1.

The increases in world cereal prices can be attributed to an improbable combination and succession of events. Cereal harvests in 1972 were generally poor and the USSR in particular experienced an unusually bad harvest. The following rice harvest led to serious shortfalls in many parts of Asia and at the same time shortages and high prices developed in other agricultural commodities, particularly protein foods. Consumption of cereals continued to rise and the normal demands of cereal importing countries were augmented by heavy orders from the USSR and China. The increased demand for exports and reductions in world stocks created appropriate conditions for rapid price increases. Finally the changing approach of the Arab countries to their oil marketing policies created further inflationary pressures.

In 1973/74 world production of cereals, in both wheat and coarse grains, is assessed to have been much higher than in the previous year, but exportable supplies were lower because of the very low level of stocks at the beginning of the season. Demand through the season has been highly stimulated by unfavourable crops in parts of Latin America, the Near and the Far East and by the increasing requirement for feed grains in Europe and Japan. Prices have continued at high levels. It must be

TABLE 1. *Price of main grains 1962 − 1974 (pounds sterling per long ton), imported into the United Kingdom, C.I.F., U.K. basis (Commonwealth Secretariat, 1974; Home Grown Cereals Authority, 1974)*

	American wheat (14% protein Dark northern spring	American maize (US No 3 Yellow corn)	Sorghum (US No 2 Milo)	Canadian barley (No 2 Feed)	Home-grown grain Feeding barley
1962/63	27.9[1]	19.6	17.9	24.4	19.7[2]
1965/66	26.5[1]	23.4	20.5	24.5	21.8[2]
Aug 71	31.3	22.7	26.3	21.0	21.3
Aug 72	31.9	26.0	26.75	26.0	24.8
Sept 72	38.5	27.9	30.30	–	24.6
Sept 73	94.0	56.6	54.30	57.0(3)	45.6
Feb 74	114.0	65.0	66.50	64.25[3]	60.0
Mar 74	99.5	66.0	64.50	63.50[3]	61.0

1. No 2 Dark hard winter at 14% protein   2. All UK barley   3. EEC barley

recognised however, that the present high price of cereals is the product not only of a shift in the supply/demand balance, which it is argued below is probably temporary, but also the result of world inflation and exceptional increases in fuel prices and freight charges. A reversal of the effect on price of inflation is improbable, but the world cereal supply/demand balance is likely to change and freight rates should fall when the present intense activity in shipbuilding has its effect on world shipping capacity.

Production trends

World production of cereals (excluding rice) in 1971 was a little under 1,000 m/t. Over a third of this is accounted for by wheat, almost one third by maize and the remainder by barley, oats, rye, millet and sorghum. Wheat is primarily grown for human consumption but it is used for animal feed and for this reason warrants attention with the feedgrains. Rice is not considered in this paper for it is used almost exclusively for human food.

The pattern of world cereal production in 1971 is set out in Table 2. No individual year can be accepted as being completely normal, but 1971 is certainly the most recent year that can be accepted as a basis for discussion. Table 2 is based on FAO statistics. The estimates of production relate to crops harvested during 1971.

Of the world's cereals in 1971 44% were produced in the developed countries: countries accounting for some 20% of the world's population. The centrally planned economies, containing 31% of the world's population produced 35% of the world's cereals, and the developing countries, with nearly half of the world's population, produced only 21% of its cereals.

Not only the quantities but the types of cereals produced varied between regions. In relation to their populations, the developed countries produced disproportionately large quantities of wheat, barley, oats, corn and sorghum; the centrally planned economies produced large quantities of wheat, barley, oats, rye and millet; the developing countries produced less than their proportionate share of all cereals except sorghums.

Differences in cereal production between regions were partly accounted for by the areas of land devoted to cereal production, but the greater part of the differences were attributable to yields. In 1971, for example, wheat yields in North America and Western Europe were 21.7 and 29.7 q/ha respectively, whereas in Africa and Asia yields were only 9.6 and 11.0 q/ha.

Similar differences between the developed and developing countries were evident for barley and maize. Climate and soil conditions clearly influence yields,

TABLE 2. *Cereal production 1971 (million tonnes and %)*

	Wheat		Barley Oats Rye		Corn		Millet Sorghum		All Grains		Population 1970 m	
		%		%		%		%		%		%
Developed Market Economies												
North America	58.4	17	43.4	18	146.2	48	22.2	22	270.2	27		
W. Europe	56.7	16	61.0	23	25.6	8	0.4		143.7	14		
Oceania	8.8	2	4.6	1	0.3		1.4	2	15.1	2		
Other	2.3	1	2.4	1	8.6	3	0.6		13.9	1		
Total	126.2	44	111.4	43	180.7	59	24.6	24	442.9	44	725	20
Developing Market Economies												
Africa	5.4	2	4.8	1	11.1	4	15.0	16	36.3	4		
Latin America	11.9	3	2.5	1	39.2	13	8.5	9	62.1	6		
Near East	23.0	6	7.8	3	4.2	1	4.6	4	39.6	4		
Far East	31.0	9	5.0	1	13.8	4	17.9	19	67.7	7		
Total	71.3	20	20.1	6	68.3	22	46.0	48	205.7	21	1765	49
Centrally Planned Economies												
Asia	32.9	9	21.5	6	32.1	11	23.5	26	110.0	11		
USSR and E. Europe	123.4	35	87.6	45	24.5	8	2.2	2	237.7	24		
Total	156.3	44	109.1	51	56.6	19	25.7	28	347.7	35	1145	31
World	353.8	100	240.6	100	305.6	100	96.2	100	996.2		3632	

but much of the difference observed has generally been judged to relate to qualities of seed used, fertiliser applications and to husbandry practices.

The increase in production over the past decade or so on a world scale has been rapid. Between 1960 and 1973 output both of wheat and of the coarse grains increased by about 50%. Amongst the individual coarse grains the production of rye decreased, small or medium increases occurred for oats and sorghum, but increases of rather more than 50% were recorded for barley and maize.

Rates of change in levels of production for the major cereals in the developing and developed regions of the world are shown in Table 3. The measures used are average annual percentage changes in production and yields for the period 1965–1972. The table illustrates the differing trends found through the world. In North America and Europe, regions well supplied with wheat, the trend has been towards greater output of feed grains. In the developing countries in contrast, the emphasis has been rather on expanding the production of wheat.

For wheat the increase in production on a world scale has been almost entirely attributable to changes in yields. Since the early 1960's, wheat yields have increased annually by about 3% per year. Regional differences have been apparent. In North America and Western Europe acreages fell back but yields increased with the result that total production increased. In Africa and Asia on the other hand both harvested areas and yields increased and production rose by approximately 6% per annum in Africa and 5% in Asia.

For coarse grains the increase in world production has come about both through increases in cropping area and through improvements in yields. Yield increases for all coarse grains taken together averaged just under 3% per year. The regional pattern of change in coarse grain production has not been the same as for wheat. The developed countries, with expanding livestock industries, have continued to expand their production. In Western Europe the most notable development has been the increase in maize growing. In North America expansion has come from both barley and maize.

In the developing countries changes from year to year have tended to be much greater, but there has been an underlying upward trend in feed grain pro-

duction with barley as the expansion crop in Africa and maize in Asia. In almost all areas, however, both harvested areas and yields have increased.

Table 3 points to an improving cereal situation in all major regions with increases in grain production greater than population change. The absolute quantity of cereals per head however, is still low in many areas.

International trade in cereals

Cereals are commodities in which the majority of countries are self sufficient. Taken together the quantity of cereals entering into international trade in 1971 was approximately 11% of the total quantity produced. Wheat enters into international trade to a greater extent than feedgrains and in 1971 world wheat exports were about 15% of total world production. Even so relatively small changes in world production lead to proportionately large changes in the quantities available on the world's market and therefore on world market prices.

World trade in cereals over the past 10–15 years has been increasing by about 3% for wheat and 6% for coarse grains. The increase in the trade in wheat has been stimulated by increasing demand for human food and much of the increased trade has been directed to the developing countries and very large quantities have been traded under food aid programmes such as the US PL480 or the Food Aid Convention. The increase in the trade in coarse grains has been to satisfy the increasing needs of the livestock industries in the developed countries, notably in Japan and Western Europe.

The grain entering into international trade comes from a small number of exporting countries. In the case of wheat Canada, the USA, Australia and France are the principal exporting countries, and they account for some 80% of the total. Argentina is a regular and substantial wheat exporter but surplus supplies from other developing countries are very small. Similarly for coarse grains the number of countries supplying the world market is small. Argentina and Thailand are the only developing countries that play an important role as exporters of feedgrains.

TABLE 3. *Cereal production and yields and human population in selected developed and developing regions*

	Wheat		*Barley*		*Maize*		*Population*
	Prod	*Yields*	*Prod*	*Yields*	*Prod*	*Yields*	
N America	1.0	2.9	5.9	1.6	4.0	3.2	1.3
W Europe	2.1	3.2	2.9	1.3	9.8	5.6	1.0
L America	-0.2	-0.6	2.9	0.3	3.0	0.6	2.9
Africa	6.1	3.2	6.7	3.0	1.9	0.7	3.0
Asia	4.9	2.9	-0.4	0.6	6.6	4.7	2.0

Average Percentage Change per year 1965-1972

Based on data in USDA reports and UN Statistical Yearbook.

Future trends have not been and almost certainly cannot be satisfactorily forecast but the following considerations are likely to be important.

Demand

(1) Without a decrease in the birth rate in the developing countries, world population growth will accelerate; as the developing regions become, in absolute terms, more important the high birth rate in these regions will become more significant as a pressure on world food resources.

(2) Demand for cereals will increase not only because of population increase but with improvements in living standards. Very large numbers of people in the developing countries are underfed and a much larger number, given greater wealth, would replace part of their present cereal diet with livestock products. Increased supplies of cereals will be required to satisfy both types of demand. Through the world as a whole potential demand is enormous. In Asia alone, it is estimated, an extra 100 m t of cereals will be required by 1980 to feed the extra population that will then exist and to raise the diet of those who are at present under-nourished to an acceptable nutritional standard. Heavy demands for cereals can also be expected to result from increasing demand for livestock products. Asia is the region with the most serious food deficit but all the developing countries consume less than is consumed in the advanced countries. As national wealth increases in the developing countries an increase in demand for food on a very large scale must therefore be expected.

(3) World opinion appears to be changing towards an attitude which does not accept hunger and famine as inevitable. Although hunger and food shortage is still common in many parts of the world, the supply of subsidised food to less well off regions and free food for famine relief has become a regular feature and an example of international co-operation. Moreover an increasing number of countries, the communist countries included, now react to shortage by buying on the world market rather than by internal economies.

(4) Although there are strong forces tending to increase the demand for cereals on the world market, it must be recognised that much of the development in the international trade in cereals has taken place during an era of relatively low cereal prices. Not only have cereal prices increased less than the prices of most other commodities and manufactured products but in recent years one third or more of the wheat sold in international trade has been subsidised in one way or another.

(5) If cereal prices were to stay at the current levels there can be no doubt that normal demand on a world scale would be cut back. Moreover with grains at high prices and with severely restricted stocks it is doubtful if the exporting countries would be as generous with subsidies and free grain as they were when burdened with stocks for which no profitable market could be found.

Supply

Despite periodic shortages the present century has been a period of adequate cereal supplies. Technological advances have more than balanced increases in effective demand and cereal prices have acted as a brake on increased production. The potential for increased cereal production although not easily quantifiable must be great.

The decline in cereal acreages in the USA during the 1960's is perhaps the best known evidence that there is scope for expansion in cereal production in the developed countries. A recent USDA study (1973) designed to measure potential food production in the USA concluded that by the mid 1980's with favourable prices, no restrictions on land use, availability of supplies of inputs such as seed, fertilizer, machinery and labour, and with normal growing weather, wheat production could be increased by nearly 50%, feedgrains by over 50% and soya beans by 33%. A considerable part of these increases it was calculated would result from extra acres but the greater part would be due to improvements in yields.

A similar if not greater potential must exist in the centrally planned and in the developing economies. In many of these regions land is available for further agricultural expansion and in others greater productivity could be achieved by improved cropping patterns, fertiliser use, by seed improvement, by expansion of irrigation facilities and by other types of improvement in husbandry techniques. It cannot be argued that crop yields all over the world can be raised to the levels achieved in the regions which at present record the highest yields, for climate and other environmental conditions are not equal. There can be no doubt however, that cereal productivity in many parts of the world is far below the potential for these regions. Africa, with about 10% of the world's population uses about 3% of the world's fertilisers and Asia with over half of the world's population uses only 10% of the world's fertilisers. Cereal production in the developing countries has improved in recent years more than has the production of many other crops, but even so productivity is in the majority of countries below the objectives set in the FAO Indicative World Plan for Agricultural Development.

This failure to meet the objectives set can be attributed to shortage or misuse of suitable seeds, lack of fertilisers and pesticides, delays in the development of irrigation projects and difficulties in bringing about necessary changes in systems of land tenure, agricultural credit and education and extension services. The 'Green Revolution' has not led to the production increases that had been hoped for, but this failure does not cast any real doubt on the capacity of developing countries in the coming years to increase their cereal productions.

Availability of cereals for livestock production in developing countries

The analysis of the potential supply of cereals in the world suggests that provided the trend in demand evident in recent years does not change to a marked degree, cereals could easily become cheaper relative to other commodities than they are at present. In this case, cereals could form an ideal base for the development of livestock industries. The analysis of the factors affecting demand however, point to a very large requirement for cereals as a primary food product. If adequate nutritional standards are to be achieved for all the people in the world, it is unlikely that there will be surpluses upon which livestock production can develop.

The actual outcome on a world scale however will depend upon both economic and political forces. The economic forces are, on the supply side the extent to which cereal production will be increased by price incentives and by changes in production techniques, and on the demand side, by the extent to which effective demand will be increased by growing populations and increasing standards of living. The political forces are the pressures on individual countries to increase national standards of living and the conflicting desires to minimise famine and hunger in the less favoured parts of the world.

In past years, national desires to increase local prosperity have been the dominant political force and on the economic front advances in technology have allowed production to increase without the incentive of substantially high prices.

The expansion in cereal production in recent years throughout the world has in general been directed towards the satisfaction of growing local demand. Exports have increased but excluding the rapid increases of the past two years which were achieved not by extra production in the exporting countries, but by depletion of stocks, the increase in exports has been much smaller proportionately than the increasing production.

This trend towards self-sufficiency in cereals is likely to continue unless there is a rapid development of world finance for the feeding of peoples in regions of shortage. Further development of systems of relief for local catastrophes is almost certain but there is little support for policies involving continuous aid. It is recognised that the problem of chronic food shortage cannot be solved until population increases in these areas are brought within bounds. The development of depressed regions must come from within and the aid philosophy that is currently gaining ground is that assistance to developing regions must be in the form of materials and know-how which will help people in these regions to help themselves.

Countries that produce cereals surplus to human requirements will have a feed base for a livestock industry. Those with no cereal surplus will have to depend upon other types of fodder or upon imports. The possibility of importing feed grains will remain but on account of the relatively high costs of transporting grain products, livestock industries dependent upon imports will tend to be uncompetitive.

References

Commonwealth Secretariat (1974) *Grain Bulletin*

FAO (1971) Agricultural Commodity Projections 1970–80, CCP 71/20, Rome.

Home Grown Cereals Authority (1974) Private communication.

USDA (1973) American Agriculture: Its Capacity to Produce, Economic Research Service Report.

World oilcake and meal supplies

P. J. R. Breslin
Tropical Products Institute, London

Summary

Oilcake is a valuable feed concentrate for livestock and a major raw material for the compound feed industry. At one time considered only as the by-product of seed crushing, oilcake has now taken on a more significant role. Soya beans for example, are crushed essentially to provide soya bean meal and the oil is regarded as the by-product.

World production of vegetable oilcakes in 1972 amounted to 56 mt, 50% higher than the average for 1961–65. Soya bean, two-thirds of which originates in the United States is the dominant oilcake and accounts for 56% of world production. The other principal vegetable oilcakes which figure in international trade are cotton seed, groundnut, sunflower seed, rapeseed, linseed, copra, sesame seed and palm kernel.

Developed countries account for 52% of world production of all oilcakes and meals (including fishmeal), developing countries for 27% and centrally planned countries for 20%. Production of vegetable oilcake in developed countries is partly dependent upon production of the raw materials in the developing countries. In 1972, 23% of total imports of oilseeds into developed countries originated from the developing countries.

World trade in vegetable oilcakes has shown considerable expansion. The volume of exports in 1972 was 88% higher than the average for the years 1961–65.

The growth and pattern of vegetable oilcake supplies and factors associated with this are examined.

Résumé

Les provisions mondiales de tourteaux d'huile et de pâture
Les tourteaux d'huile sont des concentrés alimentaires précieux pour le bétail et une matière première majeure pour l'industrie alimentaire combinée. Considéré jadis seulement comme un sous-produit de graines broyées, les tourteaux d'huile ont maintenant un rôle plus significatif. Les graines de soja par exemple, sont broyées essentiellement pour préparer des pâtures de graines de soja et l'huile est considéré comme un sous-produit.

La production mondiale de tourteaux d'huiles végétales s'élevait en 1972 à 56 mt, 50% plus grande que la moyenne pour 1961–65. Les graines de soja – deux tiers de la production provenant des Etat-Unis – procurent les tourteaux d'huile dominants et comptent pour 56% de la production mondiale. Les autres tourteaux principaux d'huile végétale qui figurent dans le commerce international sont les graines de coton, les cacahuètes, les graines de tournesol, les graines de colza, les graines de lin, le copra, les graines de sésame et les noyaux de palmier.

Les pays développés comptent pour 52% de la production mondiale de tourteaux d'huile et de pâtures (incluant la farine de poisson), les pays en développement pour 27% et les pays planifiés par une direction centrale pour 20%. La production de tourteaux d'huile végétale dans les pays développés est partiellement dépendante de la production de matières premières dans les pays en développement. En 1972, 23% des importations totales de semences huileuses dans les pays développés provenaient de pays en développement.

Le commerce mondial de tourteaux d'huiles végétales a montré une expansion considérable. Le volume des exportations en 1972 était 88% plus élevé que la moyenne pour les années 1961–65.

On examine la croissance et les types de provisions de tourteaux d'huile végétale et les facteurs associés avec ces provisions.

Resumen

Suministros mundiales de tortas de orujo y harina
La torta de orujo es un concentrado de pienso valioso para la ganadería y una materia prima importante para la industria de piensos compuestos. Considerado en

un tiempo solamente como el producto secundario del prensado de la semilla, la torta de orujo está desempeñando un papel más importante. La soja, por ejemplo, se prensa esencialmente para suministrar la harina de soja y el aceite es considerado como un producto secundario.

La producción mundial de tortas de orujo vegetal en 1972 ascendió a 56 millones de toneladas, el 50% más alta que el promedio durante 1961—65. La soja, de la que se producen dos tercios en los Estados Unidos, es la semilla dominante en la producción de tortas de orujo, y contribuye con el 56% a la producción mundial. Las otras tortas de orujo vegetales principales que figuran en el comercio internacional son las de la semilla del algodón, del cacahuete, de la semilla del girasol, de la colza, de la linaza, de la copra, la de la semilla del sésamo, y la de la semilla de la palma.

Los países desarrollados aportan el 52% a la producción mundial de todas las tortas de orujo y harinas incluyendo la harina de pescado), los países en vías de desarrollo aportan el 27% y los países planeados centralmente contribuyen con el 20%. La producción de tortas de orujo vegetales en los países desarrolados depende, en parte, de la producción de materias primas en los países en vías de desarrollo. En 1972, el 23% del total de las importaciones de semillas oleaginosas en los países desarrollados procedía de los países en vías de desarrollo.

El comercio mundial de tortas de orujo vegetales ha mostrado una expansión considerable. El volumen de las exportaciones en 1972 fué el 88% más alto que el promedio durante los años 1961—65.

Se examinan el aumento y la pauta de los suministros de tortas de orujo vegetales y los factores asociados con ellos.

Oilcake, the product which remains after extracting the oil from oil-bearing materials, is a valuable feed concentrate for livestock and a major raw material for the compound feed industry. This paper is concerned with the principal oilcakes which figure in international trade, of which there are nine viz soya bean, cotton seed, groundnut, sunflower seed, rapeseed, linseed, copra, sesame seed and palm kernel. Together they account for about 95% of world trade in oilcakes. However, as fishmeal is interchangeable with oilcakes as a source of protein for animal feed supplies of this must also be taken into account. For simplicity, the term 'oilcake' is used throughout to cover both cake and meal and refers to vegetable oilcake.

At one time, oilseeds were crushed primarily for their oil and the oilcake was regarded as a by-product. Now, however, with the increased demand for animal feedstuffs, oilcake has assumed a more significant role. Soya beans, for example, are now crushed primarily to provide the meal and the oil is regarded as the by-product.

World production of oilcake increased from an average of nearly 38 mt/a in the period 1961—65 to 48 mt/a in 1966—70 and in 1972 was estimated to have reached 56 mt, 50% higher than the average for 1961—65. Although nine different oilcakes make up this total, soya bean alone accounts for about 55% of total production. The relative importance of each oilcake in the pattern of world production is shown in Table 1.

Soya bean dominates world production of oilcake. Production of soya bean in 1972, at over 31 mt, was 76% above the average production of nearly 18 mt in the period 1961—65 and nearly four times greater than the tonnage of cotton seed cake produced (FAO, 1973). Although production of the other oilcakes is much more modest when compared with that of soya bean, nevertheless, with the exception of linseed, they have all recorded an increase over the past decade. Rapeseed cake, in fact, shows a percentage increase of 83%

TABLE 1. *World production of oilcake*

	% world production		
	1961—65 average	*1966—70 average*	*1972*
Soya bean	47	54	56
Cotton seed	19	16	15
Groundnut	10	9	8
Sunflower seed	7	7	6
Rapeseed	6	6	7
Linseed	5	4	3
Copra	3	2	3
Sesame seed	2	1	1
Palm kernel	1	1	1
	100	100	100

Source: FAO, 1973.

in 1972 over the average for 1961—65, although in absolute terms this represents an increase in tonnage of from only 2.3 mt to 4.2 mt.

Since oilcake is valued for its protein, it is relevant to consider what world production represents in terms of protein equivalent. World production of vegetable oilcake in the period 1961—65 was equivalent to an average production of nearly 16 mt of protein; by 1972 this figure had increased to nearly 24 mt, of which soya bean cake accounted for 59%.

Developed countries in 1972 accounted for 52% of world production of oilcake (including fishmeal), compared with 27% for developing countries and 20% for centrally planned countries. The share of developing countries shows a slight decline from 29% for the period 1961—65. In terms of tonnage of oilcake produced, however, the production of developing countries has increased by about 40% in the period under review, compared with 64% for developed countries and 24% for centrally planned countries.

It is the practice in determining exports of vegetable oilcakes to convert exports of individual oilseeds into cake equivalent and add this figure to exports of the oilcake itself. World average percentage rates for conversion of oilseeds to cake equivalent are: soya beans 79; cotton seed 69; groundnuts (shelled) 56; sunflower seed 37; rapeseed 57; sesame seed 52; copra 35; palm kernels 52; linseed 64 (UNCTAD/GATT, 1972).

World exports of vegetable oilcakes (including the cake equivalent of oilseeds exported) increased from an average of 14 mt in the period 1961–65 to 26 mt in 1972, an increase of some 88%. Soya bean increased its share of total exports from 46% to 66% in the same period. The volume of soya bean involved, from over 6 mt to 17 mt, shows a dramatic increase of 167%. Rapeseed records an even more dramatic increase of 225%, although this raised exports from only 0.4 mt to a modest 1.3 mt in the same period. The other oilcakes recorded very much smaller changes in volume over the period (FAO, 1973).

The developed countries increased their share of the world export trade in oilcake (including fishmeal) from 51% to 63% during the period reviewed, representing an increase in the volume exported of 130%. In the case of developing countries, on the other hand, although they increased their volume of exports by 42%, their share of the world export trade declined from 45% to 35%. The volume of exports from centrally planned countries declined by 29% and their modest share of the export trade was reduced even further, from 4% to 2%.

Although in the context of world supplies generally it is necessary to consider exports of oilcake on the basis of cake equivalent, it is also of interest to consider world exports of oilcake as such, excluding cake equivalent supplies. World exports on this basis show that in the period 1966–70 developed and developing countries had, on average, almost equal shares of the trade, 47% and 48% respectively. The actual volume of exports was 4.5 mt and 4.6 mt respectively. By 1972, the figures were 6.1 mt (50%) and 5.9 mt (49%), the balance having moved marginally in favour of developing countries (FAO, 1969–72).

World supplies of oilcake depend upon a number of interrelated factors viz world production of oilseeds, world demand for fats and oils, and the demand for livestock feeds as a function of the demand for livestock products. Oilcake utilisation is directly related to the growth of the compound feed industry which

has expanded in developed countries as livestock producers have become aware of the benefits to be derived from feeding a balanced ration to their stock. Consumption of meat is responsive to changes in income. As incomes have risen in the developed countries, so too has the demand for livestock products, meat in particular, and this has stimulated the utilisation of oilcake as a feed. The FAO has projected that world demand for meat will increase from 98 mt in 1970 to 133 mt in 1980 (FAO, 1971).

Effective demand for oilcake has been increasing much faster than for fats and oils. Consumption of fats and oils is largely determined by levels of income. In high income countries, consumption of fats and oils is approaching saturation level and there is little increase in total consumption consequent upon increased income, although variations in the pattern of consumption among fats and oils occur. The result of this has been a stronger trend in prices for oilcakes than for fats and oils, which in turn has favoured those oilcakes with a high cake and low oil content, particularly soya beans (USDA, 1971).

In Western Europe, for example, the output of the oilseed crushing industry 20 years ago consisted of 59% oilcake. The present pattern, with the increased emphasis on soya beans, results in an output of 73% oilcake and only 27% oil.

The increasing demand for oilcake as a protein supplement in animal feeds (the FAO has projected a demand for 71.4 mt in 1980) will require an increase in the production of oilseeds (see Table 2) to satisfy this demand at a price which is profitable to both the oilseed producer and the livestock producer. Although soya beans, mainly produced in the USA, are the major raw material for oilcake production, the production of oilcake in developed countries is partly dependent on the production of the raw material in the developing countries. In 1972, 23% of the oilseeds imported by developed countries originated from developing countries. World trade in oilseeds in 1972 amounted to some 20 mt compared with 14 mt in 1967 (FAO, 1969–72). The increase is mainly attributable to increased exports of soya beans. The developed countries are dependent on the developing countries for supplies of groundnuts, copra, palm kernels, cotton seed and sesame seed. Europe is the major market, absorbing about 12 mt, 60% of total world exports of oilseeds in 1972. Japan is the other major market, importing over 4 mt of oilseeds in 1972,

TABLE 2. *Oilseeds: world production (million tonnes)*

	1963	1964	1965	1966	1967	1968	1969	1970	1971	1972
Soya beans	31.6	32.3	36.4	39.0	40.6	43.8	45.0	46.5	48.5	53.0
Cotton seed	20.6	20.6	21.2	19.9	19.3	21.1	21.1	21.8	22.7	24.1
Groundnuts (in shell)	15.4	16.2	15.7	16.4	17.4	15.7	16.7	18.2	18.2	16.9
Sunflower seed	6.5	8.3	8.0	9.1	10.0	9.9	9.8	9.9	9.7	9.5
Rapeseed	3.9	4.0	5.2	4.8	5.4	5.7	5.0	6.7	8.1	6.9
Copra	3.3	3.3	3.3	3.5	3.2	3.3	3.3	3.5	3.9	4.4
Linseed	3.5	3.3	3.7	3.2	2.4	3.0	3.4	4.1	2.9	2.7
Sesame seed	1.6	1.7	1.6	1.6	1.7	1.6	1.8	2.2	2.0	1.9
Palm kernels	1.0	1.1	1.1	1.1	0.9	0.9	1.0	1.3	1.4	1.3

Source: FAO 1968–72.

80% of which consisted of soya beans. Manufacture of compound feeds in Japan has increased from under 3 mt in 1960 to around 15 mt at the present time. (In Japan, soya beans have a direct food use for human consumption in addition to their use in compound animal feeds.)

In considering what is likely to be the position in the future for supplies of oilcake it is necessary to consider the role of developing countries as both suppliers and consumers. As incomes rise in developing countries, so too will the effective demand for fats and oils. Current consumption per caput of fats and oils in developing countries is about 6 kg compared with an average of 25 kg in developed countries. This expected increase could be met in two ways: either by the importation of fats and oils or by domestic crushing of oilseeds. In developing countries where a seed crushing industry already exists, increased domestic consumption of output would simply reduce export availability. In those countries where no crushing industry exists at present, increased domestic consumption of fats and oils could act as a stimulus to the development of a crushing industry. Moreover, the establishment of crushing industries in the developing countries represents a means of increasing the value added to agricultural produce via processing and may also increase employment opportunities. The move towards establishing crushing industries in developing countries is already underway, and an extension of this is to be expected in the future.

A further factor which relates to this is the effort to develop the livestock sectors in developing countries, both to meet their own internal needs for meat and to supply increasing demand in export markets. As livestock development proceeds, it is not unreasonable to suppose that internal consumption of oilcake as a feed for livestock will increase. In areas of oilseed production with a crushing industry established this may well occur, though much will depend on the relative returns from consuming oilcake internally compared with the price which it will fetch on the world market. In areas with little or no domestic availability of oilcake, it might well be that other feed sources would be available for use which would be more economical and so restrict the need for imports of oilcake. It would not necessarily follow, therefore, that the development of the livestock sector in developing countries would result in expanding internal markets for oilcake.

The recent sharp increases in the price of feedstuffs, including oilcakes, on the world market, serves to underline the possibilities which exist for substitution of oilcakes by other materials which could provide a protein supplement for animal feeds. Price in relation to protein purchased is bound to be the factor which ultimately determines the material to be used. Non-traditional protein sources, such as single cell protein, may provide competition for oilcakes in the future. Urea is another material which might in the long run challenge oilcake as an animal feed. Clearly, it is not possible to assess with any confidence what the future impact of these non-traditional protein sources might be. But it is a well known fact among natural products generally that any continuous situation of shortages of supply, coupled with high prices, acts as a direct stimulus to research for alternative materials. And once these alternatives have been successfully and economically developed, and have made inroads into the market for the traditional product, it is a process which is generally non-reversible. Although the possibility of such substitution having a significant effect on the oilcake situation may be regarded as for the very long term, the pace of development may occur more quickly than expected and it is important to acknowledge the existence of such potential substitutes.

The projected supply of oilcakes, according to the FAO (1971) should increase up to 1980 by 3.4% per year, much slower than the 5% annual growth rate throughout the 1960's. Total production would then have increased by almost two-thirds from 45.5 mt in 1965 to 75 mt in 1980. With demand projected at over 71 mt this would anticipate a slight surplus.

References

FAO (1968–72). Production Yearbook Vol **22–26**.

FAO (1969–72). Trade Yearbook Vol **23–26**.

FAO (1971). Agricultural Commodity Projections 1970–80. CCP 71/20.

FAO (1973). Commodity Review and Outlook (1972–73). CCP 73/15.

UNCTAD/GATT (1972). The Major Import Markets for Oilcake. International Trade Centre, Geneva.

USDA (1971). World Supply and Demand for Oilseed and Oilseeds products.

Discussion

Mr Hone: Because of the oil crisis it looks as though the world can no longer be assured that all the inputs of herbicides, pesticides, diesel oil and fertilisers needed for cereal production will now be available. Perhaps Mr Low could give us the projections for world cereal production in 1974/75 and tell us what impact the oil crisis is likely to have on world production. Perhaps he could also make some projections on future prices for cereals.

Mr Low: It is difficult to make projections for the coming year let alone ten years forward. The most recent projections from the USDA indicate increases in production in the USA in 1974/75 leading to a better end of year stock position, but this will not produce a supply/demand balance such as existed during the 1960's. It is very difficult to quantify the effect of oil shortages, but it should be recognised that it is the price of oil and its pattern of marketing rather than the quantity available which has changed. It would be unwise to suggest that present levels of fertiliser production could not be maintained in future.

Dr Lengelle: I would like to add to the comments in the paper by Mr Breslin on the use of urea as animal feed in developed countries. A survey we recently carried out indicated that in 1972/73 between 850 and 900,000 t of urea was used for this purpose in the USA and about 100,000 t in other developed countries. This million tons of urea represents a protein equivalent of that supplied by 6 mt of soya meal.

Professor Oyenuga: It is noted from Mr Breslin's paper that increases in world production of oilseed cakes have been most marked for soya bean meal and rapeseed meal, which are largely produced in the developed countries, while production of other oilcakes from oilseeds produced in developing countries has only marginally increased. Is this difference due to the price structure of the market, and are there any prospects of more favourable prices for oilcakes from oilseeds grown in developing countries so as to stimulate greater production?

Mr Breslin: Most soya beans are produced in the USA where it has proved easier from the technical standpoint to increase production than in developing countries. Soya beans also have a lower oil content than other oilseeds which results in their yielding a greater proportion of cake or meal. It is not possible to predict the future prices of oilcakes under present circumstances.

It has to be borne in mind that when one commodity such as soya bean dominates a market it will inevitably become the price leader as far as consumers are concerned. For this reason it is perhaps inevitable that the prices of other oilseeds will always be determined by that of soya beans.

Dr Bhagwan: I refer to the statement in Mr Breslin's paper to the effect that consumption of fats and oils in high income countries is approaching saturation level and that as incomes have risen so too has the demand for livestock products. It is also mentioned that rising incomes in the developing countries will increase demand for fats and oils. In view of the prevalence of protein malnutrition in developing countries, is it not desirable that demand for protein-rich foods should also rise with increasing incomes as in the developed countries?

Mr Breslin: I would not disagree at all on this. I was contrasting the fact that in for example Europe demand of oils and fats has now reached saturation point and that demand can now only increase with the natural increase in the population, whereas in the developing countries there is a long way to go before the same situation is reached and oilseeds are regarded more as a source of protein than oil as in the case in Europe today. As incomes rise in developing countries demand for protein will increase as well as that for fats and oils. However, a distribution should be made between a demand in purely economic terms and a need in nutritional and humanitarian terms.

Dr Findlen: Farmers respond to prices and in 1973 the area devoted to soya bean production in the USA

54 increased in response to high prices. In 1974 it is likely that maize production in the USA will be increased at the expense of soya beans as maize is now a more profitable crop to grow. I would like Mr Breslin's views on the expansion of soya bean production in other countries such as Brazil and those with centrally planned economics.

Mr Breslin: Brazil is certainly increasing her production of soya beans very rapidly indeed, and there is enormous potential here. It is difficult to comment on the prospects for increased oilseed production in centrally planned countries as little is known about their policy decisions. However, in view of plans by for example the USSR to increase animal production there would appear to be a considerable advantage in increased oilseed production so that they have their own supplies of cakes and meals to satisfy their increased requirements for animal feed.

Dr Abou-Raya: In Egypt we have found that the best way to grow soya beans is to intercrop them with maize. I would like to ask Mr Breslin if he knows of any other countries where this system is used.

Mr Breslin: I am afraid that this is an agricultural question which I am not qualified to answer.

Produccion y utilizacion mundial de harina de pescado y factores que afectan su calidad y mercado

A. D. Bellido
Pesca, Peru

Summary

World production and utilisation of fishmeal and factors affecting quality and marketability.

World production of fishmeal reached a peak of 5,323,000 t during 1970 but declined to 3,956,000 t in 1972. The most important exporters are Peru, Norway, South Africa, Chile, Angola and Iceland, while the USA, Japan, Federal Republic of Germany, Great Britain and the USSR are the most important consumers.

The technology of the production of fishmeal is described, and its uses outlined. Fishmeal is included in poultry and pig rations mainly at minimum levels of around 2.5%. Its amino acid composition makes it particularly useful in obtaining rapid growth of animals. Inclusion rates of 8% for poultry and 7% for pigs are considered safe for the avoidance of taint in the meat.

The main factors in the evaluation of fishmeal quality are protein, lipid, mineral and vitamin content. Excessive heating during storage reduces protein quality, while the presence of unsaturated fish oils is undesirable. A high salt content may also be detrimental in animal feeding. Official standards and analytical methods of various international organisations such as the European Economic Community (EEC), Latin American Associates of Free Trade (ALALC) and the International Association of Fish Meal Manufacturers (IAFMM) are available to enable standards of quality to be maintained and contract requirements to be fulfilled.

Résumé

La production mondiale et l'utilisation de la farine de poisson et les facteurs affectant sa qualité et sa commercialisation

La production mondiale de farine de poisson a atteint un sommet de 5,323,000 t pendant l'année 1970 mais déclina à 3,956,000 en 1972. Les plus importants exportateurs sont le Pérou, la Norvège, l'Afrique du Sud, le Chili, l'Angola et l'Islande, tandis que les EUA, le Japon, la République Fédérale Allemande, la Grande Bretagne et l'URSS sont les plus importants consommateurs.

On décrit la technologie de la production de la farine de poisson et on esquisse ses utilisations. La farine de poisson est incluse dans les rations pour la volaille et les cochons, surtout à des niveaux minimum d'environ 2.5%. Sa teneur en aminoacides, la rend particulièrement utile pour obtenir une croissance rapide des animaux. Des proportions d'inclusion de 8% pour la volaille et de 7% pour les cochons sont considérées sûres pour éviter que la viande se gâte.

Les facteurs principaux dans l'évaluation des qualités de la farine de poisson sont le contenu en protéines, lipides, minéraux et vitamines. Le chauffage excessif pendant le stockage réduit la qualité des protéines, tandis que la présence des huiles de poisson non-saturées est indésirable. Un taux élevé de sel pourrait aussi être nuisible dans l'alimentation des animaux. Des standards officiels et des méthodes analytiques de différentes organisations internationales comme la Communauté Economique Européenne (EEC), Les Associés Latino-Américains du Commerce Libre (ALALC) et l'Association Internationale des Fabricants de Farine de Poisson (IAFMM) sont disponibles pour permettre de maintenir des standards de qualité et de satisfaire les exigences des contrats.

Resumen

Producción mundial y utilización de la harina de pescado y factores que afectan la calidad y comercialización

La producción mundial de harina de pescado alcanzó un máximo de 5.323.000 t durante 1970 pero bajó a 3.956.000 en 1972. Los exportadores más importantes son Perú, Noruega, Africa del Sur, Chile, Angola e Islandia, mientras que los Estados Unidos, el Japón, la República Federal Alemana, la Gran Bretaña y la URSS son los consumidores más importantes.

Se describe la tecnología de la producción de harina de pescado y se reseñan sus usos. Se incluye la harina de pescado en las raciones de las aves de corral y cerdos principalmente, a niveles mínimos de alrededor de 2,5%.

Su composición de aminoácidos la hace particularmente útil para obtener el crecimiento rápido de los animales. Se considera que no hay riesgo de que comunique ni olor ni sabor a la carne si se incluye en proporciones del 8% para las aves y del 7% para los cerdos.

Los principales factores en la evaluación de la calidad de la harina de pescado son el contenido de proteínas, lípidos, minerales y vitaminas. El calor excesivo durante el almacenamiento reduce la calidad de las proteínas mientras que no es deseable la presencia de aceites de pescado no saturados. Igualmente, un alto contenido de sal puede ir en detrimento de la alimentación del animal. Se dispone de los niveles oficiales y de los métodos analíticos de varias organizaciones internacionales tales como la Comunidad Económica Europea (CEE), la Asociación Latinoamericana de Libre Comercio (ALALC) y la Asociación Internacional de Fabricantes de Harina de Pescado (IAFMM) para facilitar los niveles de calidad que deben mantenerse y las exigencias de contrato que deben cumplirse.

En los últimos años se ha incrementado el conocimiento en nutrición animal en forma espectacular, particularmente en aves y cerdos, al mismo tiempo que el uso de computadoras en la formación de las dietas ha facilitado, sea por programación lineal o por otros medios, la fabricación de compuestos nutritivos para estos animales que, conteniendo todos los nutrientes que requieren para un rápido y buen desarrollo, al mismo tiempo económicamente representen la menor inversión posible.

Por estas razones los fabricantes de alimentos compuestos incluyen en sus productos aquellos que proporcionan al menor precio del mercado, todos los nutrientes requeridos para los pollos, cerdos, etc. Es decir, hay un criterio de selección por valor nutritivo y precio.

En estas circunstancias la harina de pescado es indudablemente un producto casi completo en lo que a nutrientes se refiere y su precio en condiciones normales atractivo, por lo que siempre se le trata de incluir en las dietas.

Por otra parte la disponibilidad de la harina de pescado se incrementa de año en año, desde hace una década, por lo que su uso en este tipo de alimentos se fué incrementando paulatinamente.

El cuadro N° 1 nos muestra la producción de los principales países que elaboran harina de pescado y en el que podemos advertir que en sólo una década la producción mundial ha llegado a ser de más de 5 mt y si bien los dos últimos años ha descendido, ésto se debe a que el principal país productor, el Perú, ha soportado cambios climáticos extraños e infrecuentes que redujeron la captura, pero que dadas las circunstancias actuales se espera una recuperación en el futuro.

Por otra parte, el consumo de harina de pescado se ha distribuído ampliamente por todo el mundo, pero los países con más alto nivel de desarrollo económico han sido los principales usuarios de este producto, tal como se puede ver en el cuadro N° 2.

CUADRO N° 2. *Consumo aparente de harina de pescado (mt)*

Países	1972	1971	1970	1969	1968
Estados Unidos	646	563	504	583	1,014
Japón	581	651	726	670	627
Alemania Occ.	502	545	562	607	594
Reino Unido	443	383	438	540	576
Unión Soviética	395	396	357	321	298
Polonia	217	166	155	149	133
España	167	152	151	181	152
Sudáfrica	137	131	143	123	102
Alemania Oriental	135	217	101	71	100
Otros países	1,260	1,196	1,239	1,252	1,170
Total mundial	4,483	4,400	4,376	4,497	4,766

Fabricacion de la harina de pescado

Por reducción de contenido de agua y aceite del pescado o de sus residuos y por cocción, prensado y deshidratación, lo que se continúa con una adecuada molienda, se obtiene un concentrado proteíco que conocemos como harina de pescado. Esta operación es efectuada por los

CUADRO N° 1. *Producciones mundiales de harina de pescado*

Países	1972	1971	1970	1969	1968	1967	1966	1965	1964	1963
Perú	897	1,935	2,253	1,611	1,922	1,816	1,470	1,282	1,552	1,144
Japón	380	417	405	370	312	226	226	360	353	282
Unión Soviética	410	406	369	348	326	300	238	120	144	140
Noruega	376	384	252	310	404	491	422	309	186	132
Estados Unidos	249	256	233	221	205	183	179	225	287	219
Sudáfrica	245	273	308	411	476	345	257	272	257	253
Dinamarca	257	248	245	247	244	149	108	120	124	87
Chile	116	263	197	181	232	130	194	70	144	90
Angola	131	52	64	99	47	44	54	47	59	32
Otros países	895	944	897	843	787	574	618	540	501	402
Total mundial	3,956	5,178	5,323	4,641	4,955	4,258	3,766	3,345	3,607	2,781

países principales productores y exportadores, en equipo moderno y altamente automatizado que garantiza una adecuada calidad y garantía sanitaria en su producto.

El pescado y los residuos provenientes de plantas de fileteo, congeladoras o conserveras, se depositan en grandes pozos desde donde son transportados a los cocinadores donde, por medio de vapor indirecto, se cocinan a fin de coagular la proteína dándole más firmeza, romper la célula grasa y facilitar así la posterior operación de prensado que se efectúa en prensas de tornillo de eje cónico de donde resulta una fuerte reducción del contenido de agua y aceite.

De allí la parte sólida se conduce a secadores rotativos, sea a fuego directo donde los gases calientes están directamente en contacto con el material, o en secadores a vapor indirecto en que por medio de vapor circulando por tubos o chaquetas, seca igualmente la harina al nivel deseado.

La parte líquida obtenida en el prensado se centrifuga y se recupera el aceite, que es usado tante en alimentación animal como humana y en la industria; el agua que se separa en esta operación, contiene sólidos solubles valiosos, que se recuperan en las llamadas plantas de agua de cola, donde se concentran a niveles adecuados, previamente se separan los sólidos en suspensión y ambos se reintegran a la torta de prensa que va al secador, obteniéndose así la harina entera.

El alto valor nutritivo de la harina de pescado es debido a su alto nivel proteíco y contenido adicional de grasa con un mínimo contenido de residuos indigeribles y porque su proteína es una fuente excepcional de aminoácidos esenciales, principalmente: Lisina, Metionina, Cistina y Triptófano, tal como puede verse en el cuadro N° 3.

La harina de pescado se usa para suplementar las dietas basadas en cereales y tortas oleaginosas y así tenemos que, para lechones de siete días de edad de 25 kg de peso, se usa de 8 a 10% de harina de pescado, para marranos de hasta 50 kg, de 8 a 12% y para cerdos en la última etapa se evitará problemas de olor y sabor en la carne incluyendo de 2 a 5%.

Para pollos parrilleros en dietas iniciales se ha sugerido como mínimo 2.5% de inclusión, llegándose a 15%, como en el caso del Perú, sin problemas en la alimentación y respuesta en el crecimiento; para dietas finales se limita

la inclusión a 2.5%, pero hay muchos países que llegan al 8 y 10%. La IAFMM recomienda no más de 8% para evitar problemas del 'Fishy Taint'. En ponedoras la cantidad que usualmente se incorpora es del orden del 10%, con variaciones menores y que en todo caso no influyen en el sabor final de los huevos.

Conviene recalcar que en todos los casos los niveles señalados son los mejores, tanto en rendimiento para carne o huevos como de conversión alimenticia, que es óptima, y el gran número de estudios disponibles que avalan esta información es notable, coincidiendo todos en señalar la bondad de la harina de pescado como elemento nutritivo en las dietas.

¿Qué factores afectan la calidad y por lo tanto la comercialización de la harina de pescado?. Podemos empezar por sus características físicas, donde la apariencia, color, textura y olor característicos nos darán una primera información de la calidad de la harina, si está muy grasienta, húmeda, quemada, mal molida, si ha sido mezclada con otras sustancias detectables a simple vista o ha sufrido infestación por parásitos ú hongos.

Luego se tratará de evaluar, mediante pruebas químicas, el valor de la harina, siendo las utilizadas comercialmente, la determinación del contenido de proteína o más propiamente N x 6.25 y que fluctúa alrededor del 65%, dependiendo de la clase de materia prima de la cual proviene. La grasa es otro de los componentes, admitiéndose hasta 10% en harinas corrientes y hasta 12% en las tratadas con antioxidantes. Harinas con mayor contenido de grasa que el mencionado, pueden causar problemas tanto en el transporte, almacenamiento, como en los animales que la consumen. La determinación se hace por extracción con éter etílico y aunque existen otros métodos propuestos tentativamente por diferentes organismos internacionales, la extracción con éter sigue siendo la preferida. Los minerales se determinan en conjunto y el fósforo y calcio son los principales aportes, pero existen otros minerales que en trazas son valiosos constituyentes de la harina. La arena residuo insoluble debe ser muy bajo y en la práctica lo es, teniendo valores límite alrededor del 2% que nunca se alcanzan con los métodos de fabricación moderna.

Los usuarios deseosos de encontrar otros criterios más exactos han venido utilizando otros análisis que brevemente mencionaremos.

CUADRO N° 3. *Contenido de aminoacidos esenciales criticos en la harina de pescado*

	Lisina	Metionina	Cistina	Triptófano
	(% harina)			
Harina de pescado de anchoa[2]	5.23	1.90	0.61	0.73
Harina de pescado de arenque[3]	5.69	2.10	0.71	0.85
Harina de pescado de sábalo[3]	4.68	1.75	0.56	0.66
Harina de pescado de sardina[3]	5.19	1.77	0.62	0.67
Harina de pescado blanco[3]	4.49	1.69	0.60	0.61
Harina de soja	3.20	0.72	0.73	0.64
Harina de semilla de algodón	1.70	0.52	0.64	0.52
Harina de sésamo	1.20	1.26	0.58	0.60
Harina de girasol	1.70	1.50	0.70	0.50
Harina de maíz	0.24	0.18	0.16	0.07
Harina de trigo	0.30	0.15	0.27	0.12
Harina de cebada	0.42	0.17	0.17	0.16

58 Dado que la proteína y los aminoácidos que la constituyen son la parte más valiosa y está sujeta a deterioración por el calor, sea durante el proceso de fabricación y/o almacenaje, principalmente en algunos aminoácidos más sensibles a dicha deterioración, se idearon métodos para establecer la calidad de la harina, tales como la digestibilidad por pepsina, que si bien dá una idea aproximada del grado de deterioración sufrida, no correlacionan los resultados con los tests biológicos, por lo menos entre harinas provenientes de distintas especies, por lo que esta prueba no ofrece mucha seguridad.

La determinación de la Lisina disponible ha tenido por algún tiempo muchos entusiastas seguidores, pero en primer lugar, lo laborioso del método que no puede ser rutinario, en segundo término las limitaciones que tiene al determinar el aminoácido libre y los N terminales, y por último, las dudas razonables que hoy se tienen sobre la correlación con los tests biológicos, ha hecho que hoy esté siendo revisado este método en forma exhaustiva, esperándose resultados en corto plazo.

El uso de antioxidantes ha reducido o prácticamente eliminado los problemas de oxidación de la grasa, traducidos en la combustión espontánea de la harina, la deterioración de la proteína que se manifiesta por cambios de color, agregación del producto y reducción de los aminoácidos disponibles para el animal.

En el aspecto sanitario, sea por infección con bacterias del tipo salmonella o por infestación por insectos, las medidas de seguridad que hoy en día se toman, garantizan que el producto se fabrique sin contaminarse de las fábricas, y durante su manipulación y transporte, los riesgos de contaminación son mínimos.

Resumiendo: La harina de pescado es un excelente concentrado protéico que se usa para la alimentación animal, principalmente aves y cerdos, como fuente de los principales nutrientes necesarios para ellos. Su calidad puede determinarse por medio de tests químicos, físicos y bacteriológicos, evitándos problemas en la comercialización y garantizándose un adecuado uso en la alimentación de los animales.

References
Cifras sacadas de Hubbell (1972).

Instituto WARF, Inc. Madison, Wisconsin, Análisis de Aminoácidos (1972). Laboratorios Agro-Scientíficos, Inc. Análisis, 1972. Hawthorne, California. Analista de Aminoácidos 1972.

Miller (1970) *Contenido Disponible de Aminoácidos en la Harina de Pescado*, Informe de Pesca de FAO, Nº 92.

The potential of single cell protein for animal feed

A. Tolan and J. F. Hearne
Ministry of Agriculture, Fisheries and Food, London

Summary

A variety of protein products or bioproteins have been produced from single celled organisms such as yeast and bacteria and from simple multicellular organisms such as microfungi and algae. Carbohydrate substrates such as sugars and starch and hydrocarbon substrates which include paraffins, methanol and methane have been used. Some of the developments have reached the stage of commercial production on a modest scale and larger scale production especially that from hydrocarbon substrates is aimed at the compound animal feed market.

The potential use of such products for animal feed will depend primarily on how they complete both nutritionally and economically with conventional feeds, the high protein concentrates such as fishmeal and soya being particularly in mind. The nutritional effectiveness of single cell products compared with these latter compounds is discussed briefly and an economic comparison taking into account recent and possible future trends is made.

Résumé

Le potentiel des protéines unicellulaires dans les pâtures pour les animaux

Une variété de produits protéiniques ou de bioprotéines ont été produites de microorganismes unicellulaires comme la levure et les bactéries et de microorganismes multicellulaires simples, comme les microchampignons et les algues. Les substrats de hydrate de carbone comme les sucres et l'amidon et les substrats hydrocarbonnés qui incluent la paraffine, le méthanol et le méthane ont été utilisés. Certains développements ont atteint le stade de production commerciale sur une échelle modeste et des provisions sur une large échelle, surtout celui de substrats hydrocarbonnés vise le marché des pâtures combinées pour les animaux.

L'usage possible de tels produits pour les animaux dépendra en principe comment ils rivalisent au point de vue nutritif et économique avec les fourrages conventionnels, en tenant compte surtout de concentrés riches en protéines comme la farine de poisson et le soja. On discute brièvement l'efficacité nutritive des produits unicellulaires comparée avec ces derniers composés et on fait une comparaison économique en tenant compte de tendances récentes et celles possibles dans l'avenir.

Resumen

El potencial de proteína unicelular para alimentacíon de animales

Se ha producido una variedad de productos de proteínas o bioproteínas de organismos unicelulares, tales como las levaduras y las bacterias, y de organismos multicelulares sencillos, tales como microhongos y algas. Se han usado substratos de hidratos de carbono tales como los azúcares y el almidón, y substratos de hidrocarburos que incluyen las parafinas, el metanol y el metano. Algunos de estos desarrollos han alcanzado la etapa de la producción comercial en una escala modesta y una producción más grande, especialmente de substratos de hidrocarburos, se dirige al mercado de piensos animales compuestos.

El uso potencial de tales productos para la alimentación animal dependerá, en primer lugar, de la manera en que compitan, tanto en valor nutritivo como económicamente, con los piensos convencionales, teniendo en cuenta, sobre todo, los concentrados de proteínas ricos tales como la harina de pescado y la soja. Se discute brevemente la efectividad nutritiva de productos unicelulares comparada con estos últimos compuestos, y se hace una comparación económica teniendo en cuenta las recientes y posibles tendencias futuras.

Single-cell protein (SCP) has been defined as 'a generic term for crude or refined sources of protein whose origin is unicellular or simple multicellular organisms' (Tannenbaum, 1970). Organisms which have been used to produce such material are yeast, bacteria and microfungi. The substrates used include carbohydrates such as sugar and starch and hydrocarbons such as paraffins and methanol. The object of this paper is to consider the potential for animal feed of those materials which have reached such a stage as to allow large-scale commercial development. On this basis consideration is limited to single-cell protein material produced from hydrocarbon substrates. Plans have been announced for the construction of 10,000 t/a plants for these materials.

In the context of animal feeds some other SCP developments should be mentioned initially which may also have a part to play in the future, these are:
(1) On-farm fermentation (Reade, 1972);
(2) Fermentation of carbohydrate material from the agriculture or food processing industries (Morris *et al*, 1973);
(3) Fermentation of waste materials from industrial and other sources (Snyder, 1970).

A protein material has been produced for non-ruminant feeding by growing filamentous microfungi on barley under farm conditions. Mixed with cereal it was readily eaten by pigs. This development is at an early stage and it would be premature to attempt to assess its potential but with improvements in technology and costs it would be worthy of serious consideration. Carbohydrates have been used to produce protein by fermentation for a number of years, eg molasses is used for this purpose in Cuba and Taiwan. However, in conjunction with nitrogen sources, materials such as molasses are being used on an increasing scale for direct feeding to farm animals and for other purposes. Other carbohydrate-rich materials cannot be used in such ways and are discarded as waste. One example is vegetable waste effluent from the food processing industry. SCP has been produced in Sweden from this material using the so-called Symba processes (Jarl, 1969). Another process is that of Tate and Lyle who have developed a method of producing microfungal protein from waste starchy material produced during the processing of carob beans in Mediterranean countries (Imrie & Vlitos, 1973). It is considered that such processes could find their value probably more in a local context.

Yeasts for use in animal feeds have been produced from waste materials for a numbers of years. For example wood waste in Russia and spent sulphite liquor from the pulp and paper industry in Rumania have been used. Figures of 250,000 t of SCP in Russia and 12,000 t in Rumania have been quoted (Wolnak, 1972). Recently plans were announced to construct a plant in Finland to produce SCP from spent sulphite liquors using a microfungus (de Castro, 1974). However, because of economic and other factors, the potential for such processes in other parts of the world would not appear to be great. The impetus for development of SCP from these and other waste materials may largely depend on environmental considerations and profits may be secondary.

In this paper, the potential of SCP from hydrocarbon substrates for animal feed is considered mainly from a general economic viewpoint and other factors which may affect its potential such as nutritional effectiveness, feed regulations etc. are only touched on briefly. It is assumed also that these materials are completely acceptable from a safety viewpoint. Finally it is probably only the manufacturers of these materials who have to take the economic risks that can provide the most realistic assessment of the potential of SCP.

Nutrient composition of single-cell protein

The proximate composition of various SCP and conventional feed materials is given in Table 1. It will be seen that yeast SCP contains roughly the same quantity of crude protein (N x 6.25) as anchovy meal, while the crude protein content of bacterial SCP is higher at 80%. Energy values of SCP are similar to those of equivalent conventional foods, while there appear to be no problems with regard to digestibility.

Table 2 shows the amino acid composition of SCP taken from figures published by the manufacturers. SCP appears to be adequate in all essential amino acids with the exception of perhaps the sulphur amino acids. Thus fishmeal contains 4.3 g/16 gN of methionine plus cystine compared with 2.9, 3.1, 3.9 and 3.5 for yeast protein, bacterial protein (from methanol), bacterial protein (from methane) and microfungal protein respectively. However, this would not appear to present any great problems

TABLE 1. *Proximate composition of some SCP materials (%)*

	BP Yeast n-alkane	BP Yeast gas oil	ICI protein	Fishmeal (anchovy)	Soya bean meal	Dried skim milk powder
Moisture	5	5	3	9	10	2
Crude protein	60–63	65–67	80	65	44	33
Lipids	7.6	1.4	9.5	3.8	0.5	0.5
Calcium	0.01	0.30	0.1	4.5	0.25	1.25
Phosphorus	1.50	1.60	3.2	2.80	0.60	1.00
ME (poultry) k cal/kg	3050	2550	3020	2530	2240	2500

TABLE 2. *Amino acid composition of some SCP materials*

	BP yeast n-alkane	BP yeast gas-oil	ICI bacterial	g/16gN Shell bacterial (on methane)	Tate & Lyle micro-fungal	Fishmeal anchovy	Soya bean meal	Dried skim milk powder
Isoleucine	5.3	5.3	4.4	4.3	4.3	4.6	5.7	6.4
Leucine	7.4	7.8	6.8	8.1	5.5	7.6	7.6	–
Phenylalanine + tyrosine	7.9	8.8	7.6	8.4	6.9	7.9	8.4	–
Threonine	4.9	5.4	4.6	4.6	5.1	3.8	3.9	5.3
Tryptophan	1.4	1.3	1.0	1.6	2.1	1.2	1.6	1.4
Valine	5.8	5.8	5.3	6.5	5.3	5.5	5.4	7.2
Arginine	5.1	5.0	4.5	6.2	–	5.2	7.7	3.3
Histidine	2.1	2.1	2.0	2.2	–	2.3	2.5	2.5
Lysine	7.5	7.8	6.3	5.7	5.8	6.7	6.6	7.0
Methionine	1.8	1.6	2.5	2.7	2.5	3.4	1.5	1.3
Cystine	1.1	0.9	0.6	1.2	1.0	0.9	1.5	3.0

since balancing up the amino acids in feed ingredients by suitable blending or addition of synthetic amino acids is an established practice in feed manufacture. The nutritional testing data available on SCP indicates its use in compound feed formulation in much the same way as conventional feeds.

Supplies of conventional feed materials

The potential of SCP must in large measure depend on the supply position and prices of conventional and perhaps other new sources of high protein feeds. This subject has been covered by other papers given at the present Conference (see Breslin, Bellido and Roberts), but it would be useful to consider here at least the question of price movements of the major conventional high protein feeds. Figure 1 shows how the prices of fishmeal, soya meal, groundnut cake and rapeseed meal have moved over recent years in the United Kingdom.

The yearly average unit values shown have been deflated to the 1963 base using the general Cost of Living Index. In 1963, the unit values per tonne for fishmeal, soya bean meal, groundnut cake and rapeseed meal were 51.5, 39.8, 35.6 and 25.2 respectively. Up to 1972, the value of fishmeal tended to fluctuate and was generally in the region of 50–65. Oilseeds appear to show a downward trend but this would seem to be marginal since on a price per unit protein basis soya bean meal was generally similar to fishmeal over this period. In 1973, the real value of fishmeal and soya bean meal almost doubled and at the present time, these are still at this level, with the cash price of fish-meal at about £250 /t and soya £95 /t.

The potential role of single-cell protein

Although there is no reason to suppose that SCP would not be suitable for ruminant feeding, it is not antici-pated that there will be any significant use of SCP in

Figure 1

Real unit value of certain imported feeding stuffs (deflated to 1963 baseline)

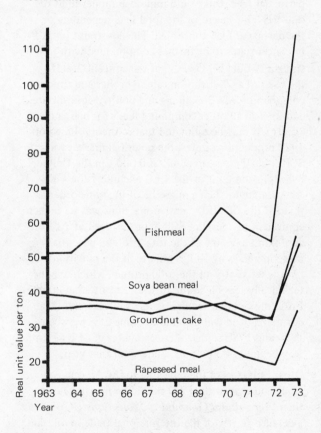

Source: Statistics Division II M.A.F.F.

this area since on a cost basis it is unlikely that SCP could compete with low quality concentrates and forages. Use in pre-ruminant calf and lamb rations has not been excluded but the best prospects lie in feeds for pigs and poultry. In intensive livestock operations they will therefore compete with soya bean meal and fishmeal.

Three factors which may be of importance in deter-mining the ultimate potential of SCP given that is acceptable from the point of view of safety are:

(1) Technical limitation on inclusion rates in animal feeds;

(2) Costs compared with conventional feed proteins;

62 (3) In the long term by availability of conventional high protein sources.

On the first point inclusion rates of the order of up to 10% of the rations of broilers, up to 20% for laying stock 10–12% in pig grower and finisher rations and 10% in veal calf diets have been quoted (Shacklady, 1973, ICI, 1973). In terms of total protein in the ration a 10% inclusion rate in broiler diets could amount to about 36% of the total protein. Evidence indicates that 50% may be feasible and in the long term with technical improvement even more (McLennan *et al*, 1973) with a knowledge of the amounts of compound feeds produced it is then possible to estimate the amount of SCP which could theoretically replace soya and fishmeal in animal feeds. In 1972 compound feed production for pigs and poultry in the enlarged EEC amounted to about 19.1 and 16.5 mt respectively, whilst in Japan the respective figures were 4.7 and 9.2 mt (CAFMNA, 1973). If an assumption on the conservative side is made that the level of SCP is constrained to 25% of the total protein content of any particular feed, this would impose an upper limit of 0.5–7.5% by weight of the total feed depending on the particular feed concerned. The constraint on SCP in concentrates to be further compounded with straights would be 25–30% of the concentrate. On the basis of a 5% inclusion rate in broiler and turkey compound feeds, 30% in pig and poultry concentrates and 2½% in all other compound feeds for pigs and poultry, it may be estimated that in theory, based on 1972 production figures, the potential market for SCP could amount to about 300,000 t in the United Kingdom and 2 mt in the EEC excluding the UK. An estimate for Japan might be of the same order as that for the UK. If the nutritional limit was set higher equivalent to a level of SCP in broiler feed of 10% or approximately 36% of the total protein, potential usage increases by half. Similarly, if the nutritional limit is set at 50% of the total protein, which may be technically feasible in the future, the potential doubles. FAO forecast that requirements of oilcake and meal for animal feed will increase by some 25% between 1970 and 1980 so that these estimates of SCP potential may be increased proportionately for that year.

Other estimates of the potential for SCP have been made and that of the FAO should be mentioned. In their *'Agricultural Commodity Projections 1970–1980'* a detailed review of the International trade in oilcakes and meals is given and some production demand balances for 1980 made (FAO, 1971).

It is estimated that fishmeal, soya and groundnuts will probably account for over 70% of the world protein production from oilcakes and meals in 1980 of which soya will comprise about 50% and fishmeal and groundnuts about 10% each. Western Europe and Japan will continue to be the major importing areas for oilseeds and fishmeals and in 1980 their total imports of these materials could exceed 20 mt. Because of possible supply difficulties with fishmeal and foreign exchange costs, it is considered that SCP may have a role to play. FAO estimate that it would be technically feasible for about 2.8 mt of SCP material to be used in pig and poultry feeds in the enlarged EEC and

Figure 2
World production of fishmeal and solubles

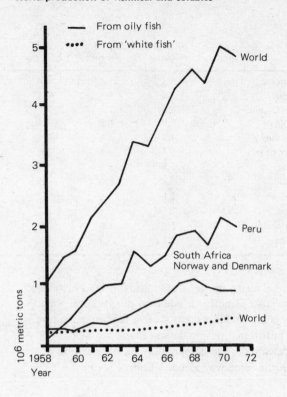

Source: FAO Yearbooks of fishery statistics, volumes 17, 23, 33 (1963), 1966, 1971)

Figure 3
World production of oilseeds

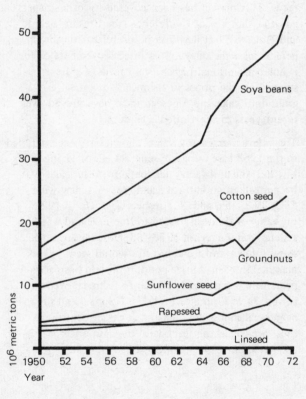

Source: FAO Production Yearbooks, volumes 23, 26 (1969, 1972)

Japan together, but the level likely to be attained by 1980 is put at 1.4 mt, the protein equivalent of 2 mt of oilcake containing on average 45% protein.

Another consideration is the supply and price situation with respect to alternative protein sources. The advantages of a stable supply of high protein feed material in times of rapid price fluctuations of conventional materials would seem obvious but such situations are unpredictable and may be infrequent in nature. Therefore real comparisons should only be made under normal supply and price conditions. In general terms, it might be anticipated that if only relatively small quantities of SCP were available, the selling price would be comparable with fishmeals. If more were available, the price might be intermediate between fishmeal and soya, and much of the fishmeal in rations could be replaced. The availability of large quantities might bring the price nearer to that of soya.

Turning to the potential in other parts of the world. There would seem to be little incentive for production in the USA with its large capability of producing conventional proteins. Conditions in Eastern Europe may be appropriate for a certain amount of development. The major part of the protein in average East European diets is of vegetable origin but it is likely that in the future the trend will be towards consumption of more animal protein which could result in the expansion of the livestock industry and a rising demand for high protein animal feeds. In other parts of the world also, demand for SCP production would seem to depend largely on the extent to which demand for animal protein in the diet increases and intensive livestock production methods are introduced, coupled with the ability to feed these livestock from indigenous or imported conventional sources. Such areas might include the Middle East and certain parts of Africa and South America. At the present time it is difficult to quantify this with any degree of reliability.

Looking into the more distant future, it seems likely that SCP plants commissioned in the mid or late seventies will have a working life into the mid or late nineties. The availability of protein supplies well into the eighties should therefore be considered. According to FAO it seems unlikely that world production of fishmeal can increase by more than 0.6% per annum over the period to 1980. Also it has been forecast that by 1985, world production of soya beans will amount to 81 mt and US production of 61 mt which would be approaching the maximum capacity of that country. Present world production is 59 mt of which 43 mt were produced in the United States. The increase in world demand for oilcake has been estimated by FAO to be about 2.5–3.0% /a up to 1980 which is equivalent to an additional 2 mt of oilcake per year. If the demand persisted at that rate up to 1985 or thereabouts, then the extra oilcakes required in that year above the present amount would be 20 mt. If this were wholly soya produced in the United States, the demand would be approaching the US maximum capacity for soya bean production.

Perhaps in the light of recent events in the Middle East, the most important factor governing the potential of SCP is that of feedstock costs. As far as crude oil prices are concerned at the end of 1970 the price was approximately £5/t and at the beginning of 1974

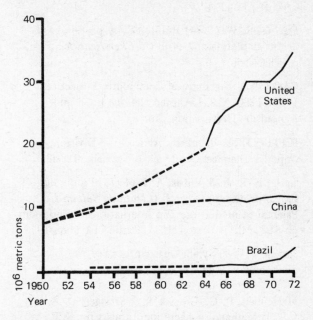

Figure 4
World production of soya beans

Source: FAO Production Yearbooks, volumes 23, 26 (1969, 1972)

about £36 /t (Anon, 1974). The price of other fuels such as natural gas has followed or will follow the trend in crude oil prices. It is not possible however to interpret these meaningfully in terms of SCP potential because of the many factors involved, and perhaps the only people who are in a position to do this are the manufacturers themselves.

Finally in the long term other developments may emerge which may help to increase world protein supplies. For example the extension of soya and other high protein crops into areas in which they have not traditionally been cultivated, and the development of animal feeding regimes which make more efficient use of protein sources. With improved technology, the better exploitation of the protein contained in some temperate climate crops may be possible. Such developments could presumably affect the long-term potential of SCP and indeed perhaps of conventional high protein sources themselves.

References

Anon (1974) Prices in transition. *The Petroleum Economist, XLT* (2) pp 42–44.

Bellido, A. (1974) World production and utilisation of fishmeal and factors affecting quality and marketability. *Proc. Conf. Animal Feeds of Tropical and Sub-Tropical Origin.* Tropical Products Institute, London.

Breslin, P. J. R. (1974) World oilcake and meal supplies. *Proc. Conf. Animal Feeds of Tropical and Sub-Tropical Origin,* Tropical Products Institute, London.

de Castro, G. (1974) Finns find feed in paper mill waste, *Farmers Weekly* April, 12, p. 77.

64   **CAFMA** (1973) Compound Animal Feedingstuffs Industry Information Service Bulletin 1973, Edition, CAFMA, London.

Fischer, R. W. (1974) Future of soy protein foods in the marketplace. *J. Am. Oil Chemists Soc., 51,* p 178A–180A.

FAO (1971) Agricultural Commodity Projections, 1970–1980. Vol. 1. United Nations Food and Agriculture Organisation, Rome.

ICI (1973) Handbook by Agricultural Division, Imperial Chemical Company, Billingham, Teeside.

Imrie, F. K. E. & Vlitos, A. J. (1973) Production of fungal protein from carob (*Ceratonia Siliqua L*). Paper presented at the 2nd International Symposium on SCP. M.I.T. Boston, USA. 29th–31st May, 1973.

Jarl, K. (1969) Symba yeasts process. *Food Technology, 23* (8), p.23.

Maclennan, D. G., Gow, J. S. & Stringer, D. A. (1973) Methanol – bacterium process for SCP, *Process Engineering,* June, p.22–24.

Morris, G. G., Imrie, F. K. E. & Phillips, K. C. (1973) The production of animal feedingstuffs by the submerged culture of fungi on agricultural wastes. Paper presented at the IVth International Conference on Global Impact of Applied Microbiology, Sao Paulo, Brazil, 23–28 July, 1973.

Reade, A. E. (1972) The use of barley as a substrate for the production of microbial protein of nutritive value to non-ruminant animals. Ph.D Thesis, University of Aberdeen, Scotland.

Roberts, R. (1974) Use of oilseed cakes and meals as animal feed. *Proc. Conf. Animal Feeds of Tropical and Sub-Tropical Origin,* Tropical Products Institute, London.

Shacklady, C. A. (1973) Responses of livestock and poultry to SCP. Paper presented at a Symposium on Single Cell Protein. Rome 7–9th Nov. 1973.

Snyder, H. E. (1970) Microbial sources of protein. In *Advances in Food Research,* Vol. 18, p. 85–140.

Tannenbaum, S. R. (1970) A Book Review, PAG Bulletin FAO/WHO/UNICEF Protein Advisory Group, 9 p 34.

Wolnak, B (1972) Protein sources of the future – single cell. Activities Report Vol. 24, (2) pp. 104–124.

Discussion

Mr Graves: From Dr Tolan's paper it would appear that on a reasonable projection the combined potential for single cell protein production by the countries of the EEC and Japan is in order of 2m t/a by 1980–85. This contrasts with the present availability of around 100, mt of protein rich materials which is likely to increase by 30% at least by 1985. It would appear, therefore, that single cell protein is unlikely to be the answer to the world scarcity of proteins, and that action should be directed to developing other sources of supply.

Dr Tolan: The figures I quoted for estimated potential production of single cell protein are purely our own impressions. I would not like our figures to be taken as the only ones, as other estimates have been made.

Chairman: I do not think that it matters if the figures are not completely accurate. The point that Mr Graves was making was the small part single cell protein was likely to be able to play in satisfying world demand for protein.

Prof Abou-Raya: I would be grateful if Ing. Bellido could give me some information on the economics of extracting the oil from fishmeal. The removal of oil is obviously good in many ways from the nutritional stand point but it raises production costs and the price of the final product will obviously be higher than that of unextracted fishmeal.

Ing Bellido: It depends very much on the market in which the fishmeal is to be sold. For some markets energy is important as well as protein content and fishmeal is sold on a protein plus energy basis. Other more sophisticated markets require fishmeal without oil. Peru produces both extracted and unextracted fishmeal, the former being some US $ 3–4 / t more expensive than the latter.

Chairman: I think it has been established that fishmeal can be had with or without oil but it will cost a little more without.

Chief Ashamu: My question refers to figures for the production of soya beans mentioned by Dr Tolan. As one coming from a developing country I am surprised that so little has been done to promote production of soya beans in suitable areas of the developing world. What is being done by organisations such as the TPI to encourage producers in the developing countries.

Chairman: There is some difficulty here as it is obviously not possible for Dr Tolan to comment on what the TPI is doing. Perhaps Dr Tolan would reply as best he can or comment on the question and a representative of TPI could then come in to answer the question.

Dr Tolan: I understand that there is talk of establishing a World Soyabean Research Institute, and a possible joint research and development programme between China and the USA. Apart from this I have no real expertise or knowledge in this area.

Mr Morgan-Rees: The Tropical Products Institute is concerned with post-harvest matters rather than those of production, but obviously maintains an interest in production problems. A considerable amount of activity is going on around the world to try to expand production of soya beans into regions where it is not currently grown. One of the best examples of this is Brazil where annual production has increased from virtually nil to more than 6 mt in only eight years. There are obviously other areas of the tropics and sub-tropics into which soya bean production could be expanded.

Mr Hancock: I am secretary of the Inter-Government Group on Seeds, Oils and Fats, which is very active in trying to help the wider growing of soya beans. There is a centre for soya beans research at the University of Illinois which is actively collecting genetic material from as wide a range of climatic conditions as possible. The University of Illinois is cooperating with the University of Puerto Rico in developing more suitable varieties for tropical climates. Soya bean production is now being increased in developing countries with the assistance of bilateral aid from the USA and support from many other countries and agencies. FAO has now strengthened its technical services with regard to the production of annual oilseeds and any enquiries in this area from developing countries will be treated with the utmost urgency.

Mr Delaitre: In Mauritius we are interested in producing beef using intensive feeding systems based on sugar cane or molasses. In these systems supplementation of the ration with a protein concentrate of low solubility such as a low quality heated fishmeal is necessary. I would like to ask Ing Bellido whether this type of fishmeal is produced to any extent and Dr Tolan whether single cell protein of low solubility is being investigated for ruminant nutrition.

Ing Bellido: We are successfully using a mixture of molasses and fishmeal. I am aware of work in progress in many parts of the world in this field. It has been found that there are problems in feeding fishmeal to ruminants especially with regard to fat content. If you require further information I suggest that you contact the International Association of Fish Meal Manufacturers which has an advisory centre in London.

Dr Tolan: I do not have sufficient details of the properties of single cell protein products which are available to answer this question. However, I would think that their high cost might make them too expensive for feeding to ruminants.

Dr Lengelle: Single cell protein is too expensive for ruminant feeding but it could be used in milk replacers for pre-ruminants. We made a projection of possible future demand for single-cell protein within the EEC using a computer, and arrived at a maximum figure of 4.5 mt/a by 1977/78. Of course this is not a realistic figure but it does show the FAO projection of 1.4 mt/a by 1980 to be reasonable. Another computer projection we made was that around 700,000 t/a of single cell protein would be needed to restrict imports of protein concentrates by the EEC at 1970 levels. On this basis it is felt that there are good prospects for several thousand tons of single cell protein being used each year within the EEC by 1980.

Recently, however, new factors have emerged which will effect the possible use of single cell protein. The high price of protein has caused imports by Western Europe and Japan to drop by as much as 10%. On the other hand exports of soya beans from the USA to developing countries have increased, keeping total exports at about the same level. The likelihood of additional supplies of synthetic lysine soon becoming available should also be mentioned.

Mr Knights: There is technically no reason why single cell protein should not be used for ruminants, but this is unlikely on economic grounds. The recent rise in oil prices have greatly increased production costs and it will be necessary to use single cell protein in high quality rations for pigs and poultry and preruminant calves. Single cell protein is not soluble in water.

Third Session

New sources of energy for animals in tropical countries

**Tuesday 2nd April
Morning**

Chairman
Dr C. Devendra
Malaysian Agricultural Research and
Development Institute (MARDI)
Selangor, Malaysia

Sugar cane as the basis for intensive animal production in the tropics

T. R. Preston

Technical Adviser and Director of Research, Livestock Nutrition Project,
Comision Nacional de la Industria Azucarera, Humboldt, Mexico

Summary

The advantage of sugar cane over other tropical crops is its high yield potential and efficient enzyme system for converting solar energy into carbohydrate. As animal feed, the by-products from normal sugar production are final molasses and bagasse. Sugar cane can also be used directly after a separating process to remove the rind; in this case, dry matter production (as derinded whole sugar cane) is equivalent to some 21 t/ha. Sugar cane and molasses are most conveniently given to ruminants, since their lack of protein is a severe constraint to their use for pigs and poultry. Ruminants, however, use the sugar in molasses and derinded cane to derive almost all their protein needs by microbial synthesis from simple nitrogen compounds like ammonia and urea.

The constraints to the use of molasses as a basis for intensive beef and milk production stem from the nature of the end products of digestion: namely, high butyrate/low propionate production in the rumen, limitation of amino-acid supply due to dependence on microbial synthesis from simple N compounds, and no onward passage of soluble carbohydrates to the duodenum due to complete fermentation of these in the rumen. As a result, availability of glucose and/or its precursors, becomes a major constraint to productivity, since a shortage of glucose (a) reduces voluntary intake and efficiency of feed utilization and (b) makes the animal susceptible to metabolic disorders. A further factor relating to molasses usage stems from its liquid nature, and therefore a need for an adequate forage intake to ensure efficient rumen function.

Derinded sugar cane, after supplementation with urea, is a more balanced feed since it contains the fibrous pith and the tops from the cane. There appear to be no metabolic problems associated with feeding this material, nevertheless, addition of starch in the form of cereal grain gives a non-additive response in terms of voluntary intake, live weight gain and feed efficiency. This implies that glucose availability at the metabolic level may also be a factor with this feed, although not to the same extent as with molasses.

In its raw state, as a by-product of normal sugar manufacture, bagasse is relatively valueless as animal feed since its digestibility is only some 28%. However, by short term (5 to 15 minutes) treatment with wet steam at 200°C, digestibility is increased to some 55%. Steam treated bagasse, supplemented with urea, has potential as a maintenance feed for beef cows and for supplying the roughage component in a combined feed with molasses for more intensive feeding systems.

Résumé

La canne à sucre comme base de production intensive d'animaux dans les régions tropicales

L'avantage de la canne à sucre sur les autres récoltes tropicales est son grand potentiel de rendement et son système efficace d'enzymes pour convertir l'énergie solaire en hydrate de carbone. Comme pâture pour les animaux, les sous-produits de la production normale de sucre sont les mélasses finales et la bagasse. La canne à sucre aussi peut être utilisée directement, après un procédé de séparation pour écarter l'écorce; dans ce cas, la production de la matière sèche (comme canne à sucre totale écorcée) est équivalente à environ 21 t/ha. La canne à sucre et les mélasses sont données plus convenablement aux ruminants, puisque le manque de protéines est une contrainte sévère pour leur usage pour les cochons et la volaille. Cependant, les ruminants utilise le sucre en mélasses et canne écorcée, pour recueillir presque tous les besions en protéines, par synthèse microbienne en partant des composés azotés simples comme l'ammonium et l'urée.

Les limitations pour l'usage des mélasses comme base intensive pour la production du bétail et du lait sont issues de la nature du produit fini de la digestion: nommément la production des grandes quantités de butyrate et des quantités réduites de propionate dans le rumen, la limitation de la provision d'amino-acides, due à la dépendance de la synthèse microbienne des simples composés de N et le manque de passage en avant des hydrates de carbone solubles dans le duodénum à cause de la fermentation de ces éléments dans le rumen. Comme résultat, la disponibilité de la glucose et/ou de ses précurseurs, devient une grande gêne pour la productivité, puisque un manque de glucose (a) réduit la prise volontaire et l'efficacité de l'utilisation des pâtures et (b), rend l'animal susceptible aux désordres métaboliques. Un autre facteur qui se rapporte à l'usage de mélasses, résulte de sa nature liquide et par conséquent le besoin d'une prise adéquate de fourrage pour assurer la fonction efficace du rumen.

La canne à sucre écorcée, aprés l'addition d'urée, est une pâture plus équilibrée puisqu'elle contient la substance fibreuse et les bouts des cannes. Il paraît qu'il n'y a pas de problèmes métaboliques associées avec l'ingestion de ce matériel; néanmoins l'addition d'amidon sous forme de graines de céréales donne une réaction non-additive, en terme de prise volontaire, de gain de poids utile et d'efficacité alimentaire. Ceci signifie que la disponibilité de glucose au niveau métabolique pourrait être aussi un élément dans cette nourriture, quoique pas dans la même mesure que les mélasses. Dans son état brut, comme un sous-produit de la fabrication du sucre normal, la bagasse est rélativement sans valeur comme pâture pour les animaux, puisque sa digestibilité est seulement de 28%. Toutefois, par traitement à court terme (5 à 15 minutes) par passage à vapeur à 200°C, la digestibilité est augmentée à environ 55%. La bagasse passée à la vapeur, completée avec de l'urée, a un potentiel comme pâture d'entretien pour le bétail — boeufs, vaches, — et pour fournir l'aliment brut dans une pâture combinée avec les mélasses, pour des systèmes d'alimentation plus intensifs.

Resumen

La caña de azúcar como base para la producción animal intensiva en los trópicos

La ventaja de la caña de azúcar sobre otros cultivos tropicales es su alto potencial de rendimiento y eficaz sistema de enzimas para convertir las energía solar en hidratos de carbono. Como alimento animal, los productos secundarios de la producción de azúcar normal son las melazas y bagazos finales. La caña de azúcar puede usarse también directamente después de un proceso de separación para quitar la corteza; en este caso, la producción de materia seca (como caña de azúcar entera descortezada) es equivalente a unas 21 t/Ha. Es preferible dar la caña de azúcar y las melazas a los rumiantes, porque su falta de proteína es un inconveniente grave en su uso para cerdos y aves de corral. Los rumiantes, sin embargo, usan el azúcar de las melazas y de la caña descortezada para obtener casi todas sus necesidades de proteínas por síntesis microbiana de compuestos de nitrógeno sencillos como el amoniaco y la urea.

Las limitaciones en el uso de las malazas como base para la producción intensiva de carne de vaca y de leche tienen su origen en la naturaleza de los productos finales de la digestión: es decir, producción alta de butirato/baja de propionato en el herbario; limitación del suministro de aminoácido debido a la dependencia en la síntesis microbiana de compuestos de N sencillos; y sin paso de hidratos de carbono solubles hacia el duodeno debido a la fermentación completa de éstos en le herbario. Como resultado, la disponibilidad de glucosa y/o de sus precursores se convierte en una limitación a la productividad puesto que una escasez de glucosa (a) reduce la toma voluntaria y la eficacia en la utilización del alimento y (b) hace al animal susceptible a desórdenes metabólics. Otro factor relacionado con el uso de las melazas tiene origen en su naturaleza líquida, y por consiguiente, hay necesidad de una toma de forraje adecuada para asegurar la eficaz función del herbario.

La caña de azúcar descortezada, después de suplementada con urea, es un pienso más equilibrado, puesto que contiene la médula fibrosa y las partes superiores de la caña. No parecen existir problemas metabólicos asociados a la alimentación con este material; sin embargo, la adición de almidón en la forma de granos de cereales da una respuesta no aditiva en cuanto a la toma voluntaria, aumento de peso en vivo, y eficacia en la alimentación. Esto significa que la disponibilidad de glucosa en el nivel metabólico puede ser también un factor con este alimento, aunque no en la misma magnitud que con las melazas.

En su estado de materia prima como un producto secundario de la fabricación del azúcar normal, el bagazo tiene un valor relativamente insignificante como pienso para animales, puesto que su digestibilidad es únicamente alrededor del 28%. Sin embargo, mediante un tratamiento corto, (de 5 a 15 minutos) con vapor húmedo a 200°C, se aumenta la digestibilidad a alrededor del 55%. El bagazo tratado con vapor, suplementado con urea, tiene un potencial como alimento de sostenimiento para vacas de carne y para suministrar los componentes brutos en un pienso combinado con melazas para sistemas de alimentación más intensivos.

Introduction

In almost all developed countries, the intensification of beef cattle production has been by the feeding of increasing quantities of cereal grain. There are agronomic reasons for this procedure since cereal production has, possibly to a greater extent than almost all other crops, been subjected to increasing mechanisation and the application of genetic and management techniques to raise yields. Furthermore, from the feedlot operators' point of view, grains are easily transported and stored, besides being readily adaptable to mechanised feeding practices.

What distinguishes the developing countries, and particularly those in tropical- and- sub-tropical regions, is the scarcity of feed grains. Apart from rice (which is grown entirely for human consumption), there has been little or no development of modern, intensive grain production. This situation has led most tropical advisers to try to develop livestock enterprises on the basis of pasture and forages, with the inevitable result that animal production in these regions has been characterised by low levels of productivity and efficiency.

The excessive preoccupation with the *problems* of the tropics, particularly the difficulty of producing conventional animal feed supplies, has been unfortunate. For in fact the tropics are potentially richer than any temperate region; the important difference is that their wealth lies not so much in pasture and

certainly not in grain, but rather in crops such as sugar cane, cassava and even bananas. These are typically food crops, and have been grown for centuries, both as a staple of the local human diet and for export. Sugar cane in particular has enormous potential for capturing solar energy. It is known that, along with certain other tropical grasses, sugar cane possesses an additional enzyme system not found in temperate type grasses or cereals, which provides it with this faculty for efficient transfer of solar energy into carbohydrate (Hatch and Slack, 1966). To the traditional sugar technologist it may appear heretical to suggest the growing of sugar cane for animals rather than humans. But there are convincing reasons to justify such a change of emphasis.

There are two approaches to the use of sugar cane as the basis of intensive cattle fattening in the tropics. One, is to utilise the by-products, which arise in the normal course of sugar manufacture, principally final (blackstrap) molasses and bagasse (Table 1); the other is to use sugar cane directly, after removal of the indigestible rind (Table 2). The technology for this latter process was developed by Miller and Tilby in Canada (Dion, 1973). They invented a machine which splits the sugar cane stalk, scoops out and grinds the pith containing the sugars and discards the two strips of rind. The coarsely ground pith is a creamy white palatable feed with the consistency of wet sawdust, which is readily eaten by cattle and is an excellent energy source for intensive beef production (Table 3).

The great advantage of using sugar cane and its by-products for animal feeding in the tropics lie in the high yield potential. Moreover, sugar cane technology is widely understood in most tropical countries, where the crop has been grown on a large scale for at least the last 200 years. Sugar cane can be planted, cultivated, and harvested by hand and is relatively unsophisticated in its requirement for management practices. It thus lends itself to the conditions that one encounters in most developing countries, where there is always a shortage of technology and skilled labour. Finally, derinded whole sugar cane, as well as the by-products of sugar production, being composed almost entirely of sugars and structural carbohydrates, are ideal substrates for the utilisation of non-protein nitrogen through the medium of the microorganisms in the digestive tract of ruminants, thus leading to the direct synthesis of animal protein from chemical nitrogen. Such a system can exist symbiotically with

TABLE 1. *By-products for animal feeding which arise during normal sugar production*

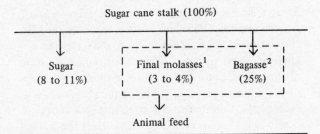

Sugar cane stalk (100%)

Sugar (8 to 11%) — Final molasses[1] (3 to 4%) — Bagasse[2] (25%) → Animal feed

[1] Contains 20% moisture
[2] Contains 50% moisture

TABLE 2. *Derinding of sugar cane for animal feeding*

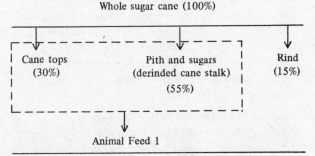

Whole sugar cane (100%)

Cane tops (30%) — Pith and sugars (derinded cane stalk) (55%) — Rind (15%) → Animal Feed 1

1   Both components combined can supply total diet needs for readily fermentable carbohydrate and roughage.
Source: Pigden, 1972

TABLE 3. *Average values (and range) for composition of derinded cane stalk alone and with the tops included*

	Derinded cane stalk	Derinded cane stalk plus tops
Dry matter (DM) %	30 (27 to 31)	32 (30 to 33)
Composition of DM (%)		
Crude protein (N x 6.25)	1.9 (1 to 2.5)	3.0 (1.8 to 4.2)
Cellulose	18 (15 to 20)	24 (21 to 38)
Total sugars	50 (40 to 63)	43 (40 to 48)
Digestibility of DM (%)	70 (68 to 73)	69 (68 to 71)

Source: Pigden, 1972

TABLE 4. *Yields of total digestible nutrients (TDN) from carbohydrate crops in selected tropical countries*

	Maize grain	Sorghum grain	Cassava tubers	Derinded whole sugar cane	Final molasses	Dry bagasse
	Total digestible nutrients (t/ha)				(t/ha)	
Peru	1.28	1.35	2.07	21.8	3.62	12.8
Ethiopia	0.88	0.56		21.6	3.59	12.7
Uganda	0.88	0.88	0.66	13.8	2.29	8.13
Taiwan	1.82	1.28	2.88	11.0	1.83	6.48
Ecuador	.40			10.4	1.73	6.13
Jamaica	.95		0.40	10.4	1.73	6.13
Mexico	.95	2.00	0.40	9.4	1.56	5.54
India	.80	0.40	2.35	7.2	1.20	4.24
Kenya	3.44	0.64	1.17	7.0	1.16	4.12

Source: Pigden 1972; FAO 1969

the needs of the human population, since the two do not compete for raw materials as is the case with pig and poultry production in the tropics, for these species must consume cereal grain and vegetable and animal proteins themselves the staple of the human diet in these regions, and almost always in short supply.

The yields per hectare in several tropical countries of the by-products from sugar production and of derinded sugar cane combined with cane top, are set out in Table 4; for comparison, yields of certain cereal grains and of cassava tubers are also included.

It is immediately obvious the enormous potential that sugar cane has as a basis for intensive beef production in the humid tropics. No other crop can approach its potential for yielding up to 20 t/ha of digestible nutrients; moreover, it has a high nutritive value mid-way between maize silage and maize grain, which makes it eminently suitable as a basis for intensive feeding.

Comparable per hectare yields can be obtained from many grasses, but this is always at the expense of feed value, the final product being of only moderate digestibility which will barely support maintenance when used as the sole energy component in the ration.

Even after extraction of sugar, the yield per hectare of only one of the by-products namely, final molasses, is greater than present average yields of cereal grains in the tropics.

Final molasses

Final molasses has been fed to beef cattle for many years, mainly as an additive to increase palatability or to improve pelleting characteristics in conventional dry mixed rations. It has also been used as a vehicle for various types of proprietary liquid feeds used as supplements for range cattle; in these cases the other components have been mainly urea and phosphoric acid (or other soluble sources of phosphorus) and occasionally other minerals and vitamins. These mixes were not designed for fattening cattle, as it was generally considered that the intake of molasses should be restricted to relatively low levels for fear of digestive disturbances and laxative effects.

The mixing of liquid molasses in dry feeds at the farm level is difficult, without special machinery, and several attempts have been made to produce a 'dry' molasses feed by combining it with a highly absorbent inert base such as bagasse pith. In such 'dry' mixes the molasses level can be as high as 70 to 75%. These formulae have been used commercially in South Africa (Cleasby, 1953) and to some extent in the USA, (Brown, 1962). The idea has not found widespread application, partly due to the poor handling qualities of the mixes, but more so to the cost and difficulties of the mixing operation, which requires special machinery. Considering the circumstances of the developing world/humid tropics, which is the origin of the greater part of the cane molasses that is produced, the question of specialized machinery and advanced technology presents particularly serious constraints. Moreover, another characteristic of the humid tropics is the ready availability of poor quality roughage, thus it makes little economic sense to expend energy and money in mixing and transporting poor quality roughage, as a component of a molasses based feed. Finally, in the sugar factories where it is produced, molasses is always handled in liquid form, and there would appear to be advantages in keeping it liquid in order to utilise these same factories at points of storage and distribution.

This was the reasoning behind the decision, in initiating the development programme in Cuba on the use of molasses, to accept from the start that the molasses should be mixed, transported and fed as a liquid.

The composition of final molasses is given in Tables 5 and 6. It appears to be reasonably homogeneous in composition from country to country, except with respect to its potassium content, which probably reflects fertiliser practices and soil composition in the original cane lands. With respect to its suitability as the major component in an intensive fattening ration, attention should be directed to four factors:

(1) It has no 'roughage' characteristics, in contrast to other high carbohydrate feeds such as cereal grains.

(2) It contains very little nitrogenous material (less than 5% of N X 6.25), in the dry matter, and of this only one third is considered to be in the

TABLE 5. *Composition data for final molasses (%)*

	Mauritius[1]	Rhodesia[2]	USA[3]	Cuba[4]
Dry matter	80.4	80.0	74.5	76.9
Sucrose	33.6	26.9	52.2	35.0
Reducing sugars	13.5	17.0		17.0
N X 6.25	5.06	4.0	4.30	3.40
Minerals	9 to 10	7.8	8.10	5.54
Potassium	3.42	3.54	2.38	2.00
Calcium	1.11	0.66	0.89	0.71
Phosphorus	0.10	0.07	0.08	0.06
Magnesium	0.60	0.34	0.35	0.45
Sodium		0.13	0.17	

[1] MSIRI (1961)
[2] Fincham (1966)
[3] NRC (1956)
[4] Institute of Animal Science, Havana, Cuba, unpublished data.

TABLE 6. *Mineral composition of molasses from four different sugar factories in Mauritius (Sansoucy; 1973)*

Sample	Ash %	Calcium %	Phosphorus %	Potassium %	Sodium %	Magnesium %	Zinc mg/kg	Iron mg/kg	Manganese mg/kg	Copper mg/kg
1	11.1	1.16	0.08	3.20	0.060	0.45	13	180	59	11
2	9.9	0.95	0.05	2.80	0.039	0.45	14	142	36	22
3	10.7	0.93	0.06	3.10	0.054	0.45	10	106	17	6
4	11.3	0.85	0.06	3.45	0.036	0.53	12	120	21	5
Average	10.9	0.94	0.06	3.175	0.046	0.47	12.25	137	33.2	11.0

form of aminoacids, and furthermore these appear to be in highly soluble form (undisclosed source, 1970). At best then, the existing nitrogenous material in molasses cannot be considered as other than a source of N for microbial growth.

(3) It is a good source of all the major and minor mineral elements, with the exception of phosphorus, in which it is highly deficient in relation to animal requirements even for fattening, and sodium, the need for which is enhanced due to the presence of so much potassium; in certain circumstances there may also be a need for additional managanese, copper, cobalt, zinc and selenium, one or all of which have been detected in low concentrations in molasses arising from specific regions.

(4) The form of the readily available carbohyrate in molasses is entirely as highly soluble sugars, mainly sucrose and the reducing sugars glucose and fructose, which has important consequences in relation to the pattern of rumen fermentation associated with high levels of molasses feeding.

The successful utilisation of high levels of molasses in the fattening of cattle requires an understanding of the above four factors, particularly the need for, and the effect of specific supplements, namely roughage, protein, minerals and non-sugar carbohydrates. In discussing these various factors, it is assumed that, except where stated otherwise (eg in the section on urea/protein), the molasses in the experiments referred to contains 2 to 3% of urea.

Roughage

Although molasses is generally considered to be highly palatable to cattle, in fact, this is not true over the whole range of feed intake. The data in Table 7 refer to experiments in which Zebu bulls had free access to molasses/urea and either ground sorghum grain or elephant grass forage. When the alternative food was grain, the bulls ate only 11% of the total diet ME in the form of molasses; when freshly cut forage was the alternative feed, then more molasses was consumed (58% of diet ME), however, animal performance was considerably inferior. Restricting the intake of forage to a level of 1.5% of

TABLE 7. *Performance of Zebu bulls given ad libitum molasses with 3% urea and either ground sorghum grain or freshly cut elephant grass*

	Molasses-urea and	
	grain	forage
Number of bulls	80	246
Initial weight (kg)	194	218
Final weight (kg)	368	398
Daily gain (kg)	0.97	0.57
Daily feed-intake (kg)		
Molasses	1.0	5.3
Grain	6.1	
Forage		56
% of total ME consumed in form of molasses	11	58
Feed conversion (Mcal ME/kg gain)	18.8	43

TABLE 8. *Performance of Zebu bulls given ad libitum molasses with 3% urea and either ad libitum or restricted maize forage (1.5% of live weight) and concentrate supplement*

	Forage ad libitum	Forage restricted
Number of bulls	24	23
Initial weight (kg)	220	210
Final weight (kg)	385	390
Daily gain (kg)	0.63	0.75
Feed intake (kg molasses)	3.36	5.6
Feed intake (kg forage)	30.6	4.3
Consumption of ME in form of molasses (%)	33	72
Feed conversion (Mcal/ kg gain)	32.8	20.6
Killing out (%)	51.7	52.3

animal live weight per day (Table 8) brought about a marked increase in the consumption of molasses as a proportion of the diet (27% of total ME); individual animal performance was also considerably improved as compared with *ad libitum* forage. A rate of 1.5% of live weight implies an average forage intake of approximately 15% of total diet DM. Figure 1 shows that total DM intake is at a maximum for forage concentrations between 20 and 28% of dietary DM; however, above or below this range, intake falls rapidly. At increasing forage concentrations, the reduction in dry matter intake is understandable as a dilution effect, due to decreasing digestibility. But at the low forage levels, some metabolic and/or physiological mechanism is presumably involved. The results of experiments carried out by Peron and Preston (1971) and by Losada and Preston (1972) suggest that the principal effect is physiological, probably due to the stimulatory effect of 'roughage' on rumen motility and hence on rumen turnover time and finally on voluntary intake. Peron and

Figure 1

Effect of forage content of a molasses-based diet on voluntary intake of DM by milking cows $Y = 1.767 + 1.27X - 0.027X^2$

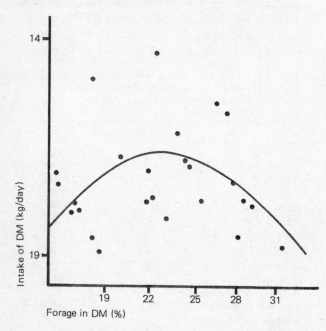

Source: Clark (1971)

TABLE 9. *Effect of artificial roughage, or none, on voluntary intake of molasses and total diet by young bulls (basal diet was ad libitum molasses, protein supplement and the respective forage sources)*

	No forage	Plastic roughage	Fresh forage	Dried ground forage
Peron and Preston (1971)				
Molasses (kg DM/day)	2.32	3.17	4.92	
Total diet DM (kg day)	2.68	3.53	5.64	
Losada and Preston (1972)				
Molasses (kg DM/day)	1.53		1.91	1.32
Total diet, DM (kg day)	1.81		2.38	2.04

Preston (1971) supplemented molasses/urea with fresh forage (1.5% of live weight), plastic simulated roughage, or no roughage; intake of molasses was highest with fresh forage, lowest with no forage and intermediate with the plastic (Table 9). Losada and Preston (1972) gave fresh forage, ground dehydrated forage, or none, and reported declining molasses intakes in the same order. There was also increased incidence of molasses toxicity with declining roughage characteristics of the diet, even though absolute molasses intakes on these diets were lower.

As would be expected, in the light of the above findings, the absolute amounts of forage required to maximise intake (and avoid toxicity) varies according to the nature of the roughage. Sansoucy *et al* (1973) gave fresh forage (elephant grass) at 3% of live weight, fresh bagasse *ad libitum* or the latter plus 1% live weight as fresh forage (Table 10). The bagasse/forage ration supported equal growth rate and slightly higher molasses intake as the 3% forage diet, although the former contributed only 15% roughage to the diet DM compared with 27% for the latter. On bagasse alone, total roughage was only 7% of dietary DM, yet molasses intakes as percent of live weight, were comparable. Redferne (1972) also found that molasses intakes were higher with the same roughage DM intake in the form of wheat straw, as compared with freshly cut immature forage sorghum. Similar findings were reported by Martin and Preston (1972);

TABLE 10. *Effect of type of roughage on feed intake and performance of Zebu bulls fed molasses/urea based diets in Mauritius*

	Forage 3% of live weight	Forage 1% of live weight Bagasse ad. lib.	Bagasse ad. lib.
Molasses intake			
kg/day	6.21	6.48	5.84
% liveweight	2.23	2.33	2.30
% DM	68	78.6	86.2
Roughage intake (kg DM/day)			
Bagasse		.31	.38
Forage	1.41	.50	–
Total	1.41	.81	.38
Total (as % of diet DM)	26.6	15.2	6.9
Average daily gain (kg)	.53	.52	.37

Source: Sansoucy *et al.* (1973)

0.95 kg/day of DM as bagasse supported higher total DM intakes than 1.2 kg/day of DM in the form of rice husks (7.32 vs 6.08 kg/day).

In all these reports the low quality dry roughages were fed *ad libitum* and the cattle voluntarily restricted themselves to much lower intakes of roughage DM than when fresh forage was given, *ad libitum.* This has some relevance to construction of feedlots, since reduced trough feeding space is required (and therefore construction costs are less) when feeding is *ad libitum.*

The above discussion relates entirely to conditions of drylot feeding where forage is transported to the cattle. Similar results are obtainable, in terms of animal performance and molasses intakes, when cattle are allowed to harvest their own roughage, the restriction being applied by limiting the number of hours spent grazing, usually 3 to 4 hours daily preferably in two periods, morning and evening of 1 to 2 hours each. Results for this system are compared with those for the conventional drylot in Tables 15 and 16.

In conclusion, it would appear that for molasses fattening, the roughage input is critical, as a means of maximising animal performance. Wherever possible it is preferable to use a type of forage that supplies maximum 'roughage' characteristics, ie contains long fibres that will induce adequate stimulation of the rumen wall, and yet is sufficiently palatable to be eaten in the minimum required amount — probably about some 10% of total DM, in the case of a roughage of optimum physical consistency such as wheat or rice straw. The use of restricted grazing as a roughage source, in conjunction with simple fenced enclosures for confining the animals and giving the molasses, is particularly attractive as a low investment system which is particularly suited to dry season fattening; accumulation of mud and difficulties in transporting molasses tankers, restrict the use of this system in the wet rainy season.

Urea/protein

In designing the nitrogen component of a molasses-based ration, it is important to appreciate that the constraints which apply to this aspect of the diet formulation are quite different from those which apply to a grain or forage-based diet.

All of the soluble sugars present in molasses are fermented in the rumen and none reaches the abomasum and duodenum (Geerken and Sutherland, 1969; Kowalczyk *et al*, 1969; Ramirez and Kowalczyk, 1971). First then, one must satisfy the nitrogen needs of the microorganisms so that these grow as fast, and as efficiently as possible so as to maximise rate of production of microbial cells which will be the major supplier of protein to the animal. The work of Hume *et al.* (1970) shows the microbial growth is maximised when ammonia nitrogen is approximately 2.5% of the carbohydrate fermentable

in the rumen in terms of 80° Brix molasses; this is equivalent to a level of about 1.5%. Assuming the nitrogen in the molasses is about 0.5% of which half is available to microorganisms, then the need for supplementary nitrogen is 1.25% which can be supplied most economically and conveniently by fertiliser grade urea (45%N) added at the level of 2.5% of the molasses (80° Brix).

Theoretical considerations leads us to expect that microbial growth in the rumen is not a sufficiently efficient process to cover the entire protein needs of the fast growing ruminant, since the latter is physiologically unable to consume the required amounts of fermentable carbohydrate. With an average molasses intake of some 2.5% of live weight, the dietary carbohydrate that is fermentable in the rumen will be sufficient to supply some 60% of total protein requirements as microbial protein, according to the theoretical conversion rates for this process in a molasses-based diet, estimated in vivo by Ramirez and Kowalczyk (1971). Supplementary dietary protein would therefore need to supplied to the extent of 40% of total needs; moreover, this should be as 'protected' or insoluble protein to avoid it being degraded in passage through the rumen. Confirmation of this hypothesis is provided by the data given in Figure 2. The experiment was carried out with a typical molasses fattening ration except that the composition of the nitrogen fraction, over and above that present in the forage and molasses, was varied between 100% urea N and 100% fish meal N, the latter being considered as a naturally insoluble protein due to the heat treatment received in its manufacture. The response to the fish meal protein was curvilinear with the biological optimum at some 40% of fish meal N in the diet. In view of the much greater cost of fish meal compared to urea nitrogen, the economic optimum is closer to 20% fish meal N, ie the equivalent of some 4% of fish meal in the dietary DM. A similar trial was carried out with

Figure 2

Effect of replacing urea N with fishmeal N on growth rate and feed conversion in Holstein x Brahman bulls fattened on a molasses-based diet

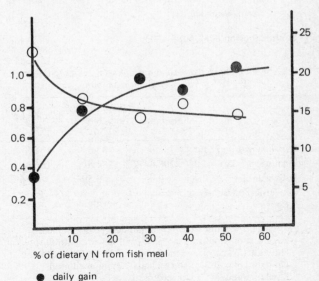

% of dietary N from fish meal

● daily gain
○ feed conversion
Source: Preston & Martin (1972)

yeast protein (*Torulopsis candida*) by Preston and Muñoz (1971). The nature of the response curve (Figure 3) was broadly similar to that with fish meal, the only difference being that a greater total amount of protein was needed to reach maximum animal performance, a finding possibly related to the lower level of sulphur aminoacids in this particular protein source.

The importance of the insolubility of the supplementary protein is emphasised by results (Table 11) from an experiment where the supplementary protein for a molasses/urea diet was supplied by either solvent-extracted rapeseed meal, fish meal, or a mixture of the two (Preston and Molina, 1972).

Animal performance on the rapeseed ration was no better than that expected for urea alone and less than half that recorded on the fish meal diet. The rapeseed meal was found to be 80% 'soluble' in the rumen fluid, and therefore likely to be degraded rapidly by rumen organisms. Subsequent trials have shown that 'expeller' rapeseed meal, which is less soluble due to the heating received in the extraction process, is much more suitable as a protein supplement for molasses/urea diets (Donefer pers. comm. 1972).

The importance of the solubility of the protein source seems to be a function of its level in the diet. Thus, Redferne (pers. comm. 1972) obtained good results with a mixture of maize germ and cottonseed meal in a molasses-based fattening diet, but in this case no urea was given and all the supplementary nitrogen came from protein. Similarly, Preston (1973) found that gains of 0.9 kg daily could be obtained from a diet of 60% molasses, 20% whole cottonseed; again, at this level of supplementary protein (no urea was given), a relatively soluble protein source was acceptable.

Other data, relevant to the general hypothesis regarding requirements for supplementary 'protected' protein, refer to programming or phase feeding (Table 12); supplying the protein at higher levels during the first month, supported better animal performance and more effecient use of both the molasses and the protein. In some respects such findings could be interpreted as indicating a period of adaptation to urea utilisation, however we prefer to agree with Burroughs *et al* (1970) that the implied adaptation is more likely to be related to the energetic components of the diet, for which there is adequate documentation (Turner and Hodgetts, 1955; Marty and Sutherland, 1970), and that this in turn affects microbial growth, in view of the energy limiting nature of this process.

Feed intake always increases after the transition period, thus providing for greater microbial synthesis, and hence reduced requirement for supplementary protein. There is also the fact that protein requirement relative to energy decreases with time on feed, because of the changing composition of the tissue being laid down.

Non-sugar carbohydrate

It has been shown that the pattern of rumen fermentation on molasses-based diets is characterised by abnormally high levels of butyrate and low levels of pro-

Figure 3

Effect of yeast protein on performance of bulls fed a molasses/urea based ration

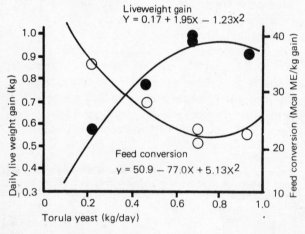

Liveweight gain
$Y = 0.17 + 1.95x - 1.23x^2$

Feed conversion
$y = 50.9 - 77.0x + 5.13x^2$

Source: Preston & Muñoz (1971)

TABLE 11. *Rapeseed meal as a supplement to molasses-urea diets for cross-bred Brahman bulls*

	Fishmeal[1]	Fishmeal and rapeseed (50:50)[2]	Rapeseed meal[3]
Initial weight (kg)	135.6	133.6	134.8
Final weight (kg)	208.4[a]	184.6[b]	166.0[c]
Daily gain (kg)	0.85[a]	0.59[b]	0.36[c]
Conversion (Mcal ME/kg gain)	11.4[c]	13.8[b]	20.8[a]

abc = Means without letter in common differ at P < 0.05.
[1] Basal diet of ad libitum molasses (with 2% urea), fresh forage (1.5% of LW daily), and minerals, plus 450 g fishmeal daily.
[2] Basal diet plus 200 g fishmeal and 392 g rapeseed meal.
[3] Basal diet plus 785 g rapeseed meal.

Source: Preston & Molina (1972)

TABLE 12. *Effects of programming the protein supplement supply for Holstein x Brahman bulls given ad libitum molasses-urea and restricted grazing (Morciego et. al., 1972)*

	Allowance of fish meal in successive months (g)	
	300:172:57:57	400:57:57:57
Number of bulls	700	700
Liveweight (kg)		
Initial	270	262
Final	345	330
Daily gain	0.581	0.641
Conversion (kg feed/ Kg gain)		
Molasses	9.64	8.80
Fishmeal	0.275	0.231
Mean daily intake of fishmeal (kg)	0.160	0.148

Source: Morciego et. al. (1972)

TABLE 13. *Molar composition of the VFA in rumen liquour from cattle given high levels of molasses compared with values from the literature for more conventional diets*

| Class and number of animals | Diet | Total VFA (ME) | acetic | Molar proprotions % | | |
				propionic	butyric	others
Dairy cows						
2	Pasture	131–148	65–67	18–19	11–12	3–4
–	High-grain	148	58	20–21	15–16	4–5
8	High-molasses	114	36	24	29	10
Fattening cattle						
2	Alfalfa hay	107	75	18–19	7–8	
5	High-grain	115	39	40	21	
8	High-molasses	143	31	19	51	9
Weaned calves						
4	High-grain	115	50	37	13	
8	High-molasses	96	28	20	37	14

Source: Marty & Preston (1970)

pionate (see Table 13). Such a situation could lead to low voluntary intakes and to decrease efficiency of energy utilisation for fattening. Addition of a source of starch to a molasses-based ration theoretically would be expected to increase propionate levels and, in fact, this hypothesis has been proved experimentally in dairy cows (see Table 14). While conclusive data

TABLE 14. *Effect on rumen fermentation of substituting maize with molasses in a low-forage diet given to dairy cows*

| | Maize: molasses (% of diet MD) | | | |
	63.8	42.25	20.45	0.61
VFA (molar %)				
Acetic	57.4	56.8	55.8	51.9
Propionic	29.3	23.9	19.9	18.0
Butyric	10.7	17.4	21.1	25.8
Valeric	0.6	1.2	2.6	3.7
Blood ketone bodies (mg/100 ml)	4.6	3.9	4.9	7.0

Source: Clark (1971)

are lacking for such effects on the rumen fermentation pattern in fattening bulls there is evidence that supplementary starch has a non-additive (ie stimulatory) effect on animal performance. A linear improvement in voluntary feed intake, live weight gain and feed conversion to added maize grain was reported by Preston *et al.* (1973) for Zebu bulls fattened on a molasses-based ration in Mauritius (see Figure 1). A curvilinear response to increments of maize bran (including the germ) in a molasses ration was observed by Redferne (pers. comm.) in Kenya (Figure 5).

Since, in the early phase of supplementation with cereal grain, the response appears to be in excess of the substitution value of the ME of the starch source (there is also an increase in voluntary intake), it will usually be economic to include some cereal grain (or grain by-product) at low levels in a molasses-based ration despite the higher price of the energy in cereal grain or by-product, relative to that in molasses. There is also the added advantage that inclusion of some grain product appears to give protection against molasses toxicity since this syndrome also seems to

be caused by low levels of propionate production (Losada *et al.,* 1973).

Figure 4

Effect of supplementary maize grain on performance of Zebu bulls fed a diet based on molasses/urea

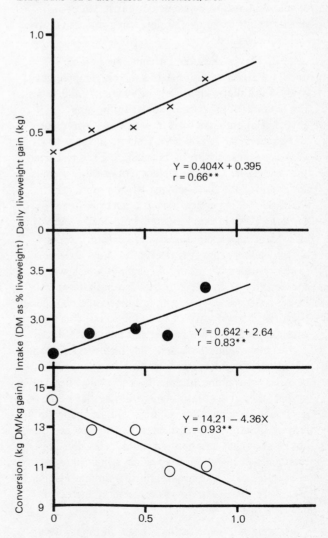

$Y = 0.404X + 0.395$
$r = 0.66**$

$Y = 0.642 + 2.64$
$r = 0.83**$

$Y = 14.21 - 4.36X$
$r = 0.93**$

Source: Preston *et al* (1973)

Figure 5

Effect of adding increasing amounts of maize germ meal on performance of Zebu steers fed a molasses-based diet

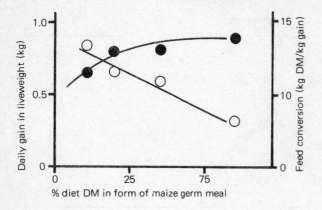

Source: Redferne (1972)

Input-output data for molasses

There can be no one formula, or combination of ingredients, applicable to all the varied situations where molasses is produced. For each particular region, input-output studies must be carried out in order to determine the most economic combination of inputs. The information given in the preceding part of this section aims to provide a guide line as to the nature of the responses to be expected, but it can be no substitute for experimental data using the actual raw materials available.

In 1970/71, the molasses fattening programme in Cuba was based on fish meal as the only supplement to the basal diet of molasses/urea minerals and forage. The data in Tables 15 and 16 refer to the input-output relationships obtained with these ingredients in a commercial 10,000 head feedlot, and in a series of dry-season fattening units employing the restricted grazing system. For the feedlot, comparative figures are given for the previous year when a more conventional feeding system was in operation, based on *ad libitum* forage supplemented with smaller quantities of concentrates and molasses. The information is self-explanatory, but it is relevant to draw attention to the considerably greater throughput which was one consequence of changing to the high molasses-low forage programme. This simply reflects the considerable logistics problem, involved in trying to cut and transport large amounts of green forage; and how much easier it is to move liquid molasses. The high mortality and emergency slaughter figures in the first year of the molasses programme also merit some explanation. In part, they reflect management difficulties which are bound to arise when making a radical change in feeding systems. Nevertheless, it should also be pointed out that despite the greater losses, the overall economics of high molasses feeding remained attractive, as in terms of all the major-inputs, ie animals, labour, machinery, feed-output was considerably increased, inclusive of the losses.

There were no important differences in input-output relationships between the restricted grazing and the

TABLE 15. *Input-output data (January to June inclusive) for fattening bulls on molasses-based (1970 and 1971) compared with forage-based (1969) diets in a 10,000 head capacity feed-lot*

	Forage-based, 1969	Molasses based 1970	1971
Total daily liveweight gain (kg) in the feed-lot	3,724	8,295	13,797
Daily liveweight gain (kg) per:			
Bull	0.43	0.88	0.89
Worker	14.3	51.8	82.2
Tractor	85.6	420	282
Conversion (kg feed/ kg gain)			
Forage	34.7	11.9	10.3
Molasses	3.10	10	9.62
Urea	0.23	0.32	0.31
Concentrates	3.84		
Fishmeal		0.41	0.41
Minerals		0.13	0.10
DM	15.4	10.8	9.82
Mortality (%)	0.1	1	0.21
Emergency slaughter (%)	0.4	3.04	1.31

Source: Muñoz *et al.* (1970); Muñoz (1971)

TABLE 16. *Input-output data for fattening bulls given ad libitum molasses-urea and restricted grazing and fishmeal supplementation (3,500 bulls in 11 units)*

	Best unit	Mean of all units	Worst unit
Daily liveweight gain (kg)	1.04	0.83	0.74
Conversion (kg feed/kg gain)			
Molasses	5.9	9.1	14.7
Fish meal	0.32	0.45	0.54
Urea	0.19	0.29	0.47
Mortality (%)	0.0	0.38	1.33
Emergency slaughter (%)	0.0	0.44	1.33

Mean initial and final weights were 313 and 403 kg; breeds were Brahman and Holstein x Brahman.

Source: Morciego *et al.* (1970)

feedlot system; in fact, health problems were considerably reduced on the former, probably reflecting a better control over the forage input, since the animal under conditions of free grazing can select more effectively this portion of its diet.

Derinded sugar cane

In contrast to the high molasses fattening system, there are only limited animal data available on the use of derinded sugar cane. The development work has been carried out exclusively in Barbados, and mainly with Holstein steers, by-products from the local dairy industry. The scope of the research facilities available did not allow comprehensive evaluation of the different inputs associated with this feeding method; nevertheless, certain dietary variables have been studied and in general, the results obtained indicate that in most respects similar constraints apply to derinded sugar cane, as to final molasses, when it is used as the major component in a fattening programme.

Forage

Although derinded cane obviously has better 'roughage' characteristics than liquid molasses, animal performance seems to be improved by incorporating additional forage in the form of the cane tops (Table 17). As with molasses, supplementary forage seems to exert its major effect in increasing voluntary intake, and, as a result, rate of live weight gain; but with slight deterioration in efficiency of feed conversion. The ratio of tops to derinded cane was fixed at that normally found in whole sugar cane; it is not known if slightly narrower ratios, ie a lower proportion of tops, might not be more beneficial.

TABLE 17. *Derinded sugar cane as the basis for intensive feeding of Holstein steers*[1]

| | Derinded cane stalk | | Improvement due to cane tops% |
	Alone	Plus tops	
No. of calves	12	13	
Liveweight (kg)			
Initial	105	102	
Final	257	271	
Daily gain	0.59	0.66	12
Feed intake (kg/day)			
Derinded cane stalk	12.8	10.9	
Cane tops		4.3	
Derinded whole cane	12.8	15.2	19
Protein supplement	1.1	1.1	
Total DM	4.72	5.63	19
Conversion (kg DM/kg gain)	8.0	8.5	−6

[1] It was considered that in this trial the live weight gains were lower than would normally be expected, due to use of immature cane and effect of flooding on cane quality.

Source: James, (1973)

Cereal grain

Two carbohydrate supplements, maize grain and molasses, were added to rations based on derinded whole sugar cane (Table 18). Both supplements, which were supplied at the rate of 1% of live weight daily, brought about consistent increases in voluntary intake and in live weight gain. Gain was increased over 25% by maize, and there was also an important improvement in feed conversion (11% on average). Molasses increased gain by a lesser amount (about 9%) and caused a deterioration in feed conversion. Data are not available on the rumen fermentation on derinded sugar cane, but it is probable that it will show the same constraints as on molasses ie propionate production will be limiting, since the beneficial effects of added maize grain are of the same magnitude on both feeds.

The level of maize supplementation in these trials was high (32% of the dietary DM), and in most cases would not be economic. It is probable that smaller amounts would be more appropriate, since it can be expected that the response curve to maize supplementation will be curvilinear, with the effect being observed in the early phase of substitution.

TABLE 18. *Effect of added (1% of LW) maize or final molasses on performance of Holstein steers fattened on a basal diet of derinded sugar cane*

| | | Improvement over control % | |
	Control[1]	Molasses	Maize
Initial weight (kg)			
Trial 1	308		
Trial 2	322		
Trial 3	380		
Final weight (kg)			
Trial 1	459		
Trial 2	407		
Trial 3	426		
Mean daily gain (kg)			
Trial 1	0.99	9	27
Trial 2	0.95	13	24
Trial 3	1.02	3	32
DM intake (% liveweight)			
Trial 1	2.40	29	17
Trial 2	2.47	14	11
Trial 3	2.47	7	4
Feed conversion (kg DM/kg gain)			
Trial 1	9.1	−16	8
Trial 2	10.1	0.0	11
Trial 3	9.9	−15	15

[1] Control diet was (% DM basis): derinded cane stalk: 52; cane tops: 28; protein/mineral/vitamin supplement: 20.

Source: James, (1973)

Protein

In all the trials reported with derinded cane, urea was added to supply from 50 to 60% of the total N requirements, however, the absolute quantities of protein nitrogen in the diet were higher than in the Cuba work with molasses. This might account for the apparent absence of response to fishmeal in comparison with the more soluble rapeseed meal (Table 19). Further trials are needed with more critical levels of protein, before definitive conclusions can be made on this aspect.

TABLE 19. *Effect of the type of protein supplement on the performance of Holstein steers fed a basal diet of derinded sugar cane*[1]

	Fishmeal	Fishmeal/ rapeseed	Rapeseed
Liveweight (kg)			
Initial	146	148	147
Final	303	308	309
Daily gain	.90	.91	.91
Feed conversion (Kg DM intake/Kg gain)	8.3	8.3	8.7

[1] The diets were (DM basis, %): 60 derinded cane plus tops, 12 protein supplement, and 27 maize grain or molasses, for the fishmeal diet; and 54 derinded cane plus tops, 19 protein supplement, and 27 maize or molasses for rapeseed meal; the mixed protein diet was intermediate in proportions of the two protein sources.

Source: James (1973)

TABLE 20. *Effect of ensiling derinded whole sugar cane on performance of Holstein steers (James, 1973)*

| | | | Derinded whole sugar cane | | | |
| | No. supp. | Molasses | Maize | No supp. | Molasses | Maize |
		Fresh			Ensiled	
No. of steers	8	8	8	8	7	8
No. of days	95	95	95	42	95	95
Liveweight (kg)						
Initial	322	310	385	290	325	325
Final	407	405	440	304	408	415
Daily gain	0.89	1.00	1.09	0.31	0.87	0.94
Feed intake (kg DM/day)						
Derinded whole cane	7.45	5.86	5.31	4.45	4.45	3.22
Protein supp.	1.54	1.54	1.59	1.45	1.54	1.54
Energy supp.	–	2.81	3.09	–	2.86	3.04
Conversion (kg DM/kg gain)	10.1	10.1	9.1	18.8	10.1	8.2

Source: James (1973)

Ensiling

One disadvantage of using derinded sugar cane as animal feed is that, because of its high moisture and sugar content, it ferments quickly, necessitating that it be produced and fed daily. A similar problem exists with the feeding of whole crop maize, and which has been solved by ensiling the material; in this form maize stores easily and is well accepted by cattle. Unfortunately, the ensiling process appears to be much less suitable for derinded cane. The data in Table 20 show that there is a serious fall off in animal performance, specifically gain and feed conversion, when fresh derinded cane is replaced by the ensiled material; the effect can be related in greater part to depression in voluntary intake of the ensiled material and to reduced efficiency of its utilisation. Both these effects can be corrected by addition of either molasses or maize. Rate of gain on ensiled derinded cane plus energy supplement was comparable with that on fresh unsupplemented material. Nevertheless both gain and intake remained below the levels reached when these energy supplements were added to fresh derinded cane.

Input-output data for a feedlot using derived sugar cane

Available data are summarised in Table 21. Caution must be used in interpreting this information, since only limited numbers of animals were involved; and the same group did not continue throughout the full feeding period. The overall level of performance is similar to that recorded on molasses-based rations. It should be remembered that the breed used was Canadian Holstein; limited observations on Zebu steers (James, 1973) indicated that average gains on similar rations were some 18% poorer than for the Holstein.

TABLE 21. *Input-output data for fattening Holstein steers given diets based on derinded whole sugar cane*[1]

Days in feedlot	446		
Liveweight (kg)			
Initial	102		
Final	459		
Daily gain	0.80		
Feed intake (kg)		As fed	Dry matter
Derinded cane stalk		5620	1713
Cane tops		2300	854
Protein supplement		676	598
Total			3165
Conversion rate (kg DM/kg gain)			8.87

[1] The information relates to 13 animals over the first part of the growth curve, to 271 kg; and only 6 animals from 308 to 459 kg. Performance from 271 to 308 kg was assumed to be at the average rate for the range 308 to 459 kg.

Bagasse

Early attempts to use bagasse fractions for animal feeding were as a carrier for molasses, and little attention was given to the possibility of the bagasse itself being a source of nutrients. Work on this aspect is of recent origin, and was stimulated by the findings of Bender *et al.* (1970) that sawdust and chippings from hardwood wastes could be processed by steam under pressure, to the point of having an energetic feeding value equivalent to a moderate quality hay. Bender (pers. comm. 1971) postulated that bagasse, because it possessed an arrangement of structural carbohydrate similar to that in hardwood, should respond equally to steam treatment.

The first trials, carried out in Mauritius, used an old steam chest erected in a sugar factory. Fresh bagasse was put in the chest and then treated with superheated steam at 10.5kg/cm². The temperature could not be measured in these early trials but it was likely to have been about 180 to 200°C. Steaming was carried out for 1.5 hours after which the bagasse was removed and compressed in a baling machine to reduce the moisture content from some 70% (as it emerged from the digester) to 50%. Evaluation was initially by the nylon bag in vivo technique, which measures dry matter loss of the sample when immersed in the rumen of a fistulated bull for a period of 72 hours (Rodriguez, 1968). The results given in Table 22 (Figure 6) show almost a threefold increase in digestibility, compared to untreated bagasse.

TABLE 22. *Steam treated bagasse as a roughage for growing heifers*

	Dry matter content of bagasse (%)		
	51	30	61
No. of animals	5	5	5
No. of days	54	28	31
Mean liveweight (kg)	145	133	138
Voluntary intake (kg/day)			
Bagasse (dry basis)	3.27	2.58	4.39
Molasses	.51	.46	.61
Urea	.13	.12	.15
Bagasse (% liveweight)	2.26	1.94	3.33

Subsequent experimental work was in two directions. One aimed to determine the optimum physical parameters, governing the steaming process which would result in a processed end product with maximum feeding value. Results of these investigations are described by Wong You Cheong *et al.* (1974).

The second series of trials was with growing and pregnant cattle, using as the basic energy source the treated bagasse resulting from processing in the steam chest as described earlier. It was realized that the feeding value of the bagasse treated by this procedure

Figure 6

Rumen digestibility of bagasse steamed at different temperatures and times (band within dotted lines indicates range of values found for dehydrated *Setaria* grass cut at preflowering stage)

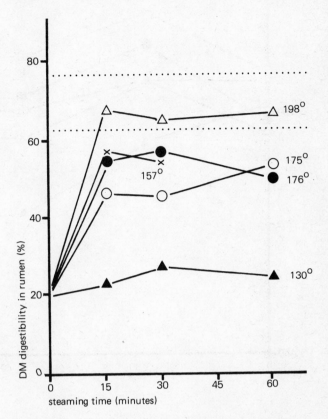

might not be at the maximum level attainable; nevertheless, it was considered that preliminary results from feeding such material would be a valuable guide to the planning of more comprehensive feeding trials later.

Figure 7

Liveweights of two groups of five calves fed maintenance rations based on steam processed bagasse with different contents of dry matter (DM)

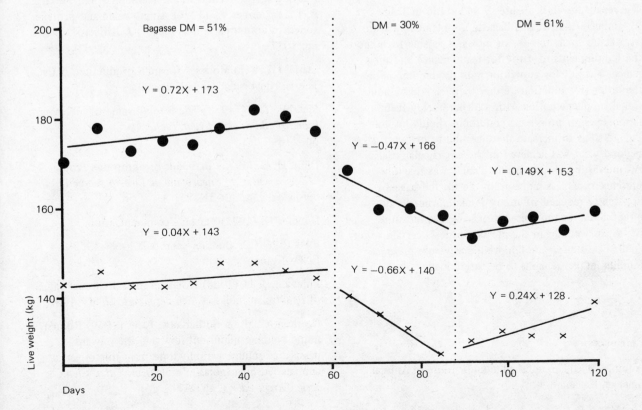

Figure 8

Liveweights of five pregnant Zebu cows fed a maintenance ration of steam processed bagasse and urea as the basal diet (arrows indicate date of calving)

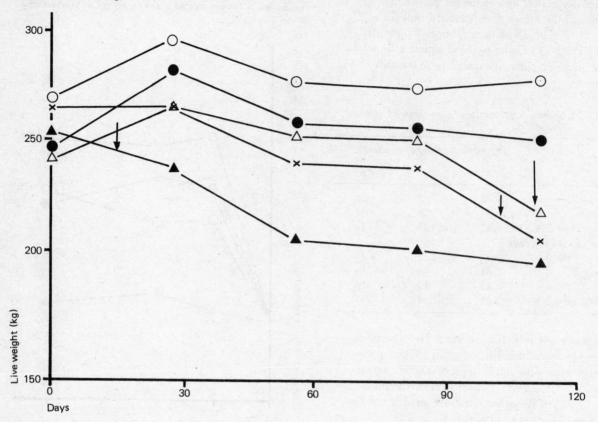

The diet used in both trials had the following composition: Treated bagasse (dry basis) 85%, molasses 12% and urea 3%. It was fed *ad libitum* and the animals also had access to a complete mineral mixture. Vitamin A was also given. Trial 1 was with two groups each of five eight-month-old Creole heifers; while in trial 2, a group of five pregnant Zebu cows was used. The feeding period was for 113 days in trial 1 and for 140 days in trial 2.

The results, given in Figures 7 and 8 and in Table 22, indicate that treated bagasse, supplemented with urea and a small amount of molasses, is able to support the requirements of cattle for maintenance and pregnancy, but not for growth or milk production. For the latter two functions, either there should be additional supplementation with a higher energy feed, ie molasses, or processing conditions should be improved so as to increase the digestibility of the bagasse beyond what was achieved in these particular trials. An interesting observation in Trial 1 was that the presence of excess moisture in the treated bagasse, apparently reduced voluntary intake. Whether this effect is due to the presence of excessive moisture per se, or whether in the pressing process certain water-soluble substances are removed, and whose presence inhibits intake, remains to be ascertained.

References

Anon., (1970) Unpublished data, Institute of Animal Science, Havana.

Bender, F., Heaney, D. P., & Bowden, A. (1970) Potential of steamed wood as a feed for ruminants. *Forest products J.* 20, 35 col. 26.

Bender, F. (1971) Personal communication.

Brown, P. B. (1962) Sugar cane bagasse — blackstrap molasses rations for beef cattle *Proc. 11th Cong. I.S.S.C.T. Mauritius,* pp. 1216—1224.

Burroughs, W., Ternus, G. S., Trenkle, A. H., Vetter, R. L., & Cooper, C. (1970) Amino acids and proteins added to corn-urea rations (abstr.) *J. Anim. Sci.* 31 pp. 1037

Clark, J. (1971) Molasses for milk production. N.Cs. Thesis, Universidad de la Habana.

Cleasby, T. G. (1953) The feeding value of molasses *Proc. of the S. Afr. Sugar Tech. Ass.,* pp. 113—117

Dion, H. G. (1973) Barbados breakthrough *Proc. CIDA Seminar on Sugar cane as Livestock Feed, Barbados,* Jan. 30—31.

Donefer Q. (1972) Personal communication.

FAO (1969) Production yearbook 1968—1969 FAO—Rome.

Fincham, J. E. (1966) Notes on the nutritional value of Rhodesian molasses. *Rhodesia agric. J.* 63, 105.

Geerken, C. M., & Sutherland, T. M. (1969) Rumen liquid volume, liquid outflow, and the onward passage of soluble carbohydrate from this organ in animals fed high molasses diets. *Rev. cuba. Cienc. agric. (Engl. ed.)* 3, pp. 217

Hatch, M. D., & Slack, C. R. (1966) Photosynthesis by sugar cane leaves. *Biochem J.* **101**: 10.

Hume, I. D., Moir, J. J., & Somers, M. (1970) Synthesis of microbial protein the rumen: I. Influence of the level of nitrogen intake. *Aust. J. agric. Res.* **21**, pp. 283.

James, L. A. (1973) Comfith in rations for livestock. *Proc. CIDA Seminar on Sugar Cane as Livestock Feed, Barbados,* Jan. 30–31

Kowalczyk, J., Ramírez, A., & Geerken, C. M. (1969) Studies on a composition and flow of duodenal contents in cattle fed diets high in molasses and urea. *Rev. cuba. Cienc. agric. (Engl. ed.)* **3**, pp. 221.

Losada, H., Dixon, F., & Preston, T. R. (1971) Thiamine and molasses toxicity. 1. Effect with roughage-free diets. *Rev. cuba. Cienc. agric. (Engl. ed.)* **5**, 369.

Losada, H., & Preston, T. R. (1972) Effect of forage on some rumen parameters in calves fed molasses-based diet. *Rev. cuba. Cienc. agric. (Engl. ed.)* **6**

Martin, J. L., & Preston, T. R (1972) Different forage sources for bulls fed a molasses-based diet. *Rev. cuba. Cienc. agric. (Engl. ed.)* **6**

Marty, R. J., & Preston, T. R. (1970) Molar proportions of the short chain volatile fatty acids (VFA) produced in the rumen of cattle given high-molasses diet. *Rev. cuba. Cienc. agric. (Engl. ed.)* **4**, 183.

Marty, R. J., & Sutherland, T. M. (1970) Changes in sucrose and lactic acid metabolism in the rumen of cattle during adaptation to a high-molasses diet. *Rev. cuba. Cienc. agric. (Engl. ed.)* **4**, 45

Morciego, S., Muñoz, F., & Preston, T. R. (1970) Commercial fattening of bulls with molasses/urea and restricted grazing. *Rev. cuba. Cienc. agric. (Engl. ed.)* **4**: 97

Morciego, S., Muñoz, F., Martin, J. L., & Preston, T. R. (1972) A note on the effect of different levels of fishmeal and urea in a molasses-based diet for fattening bulls under commercial conditions. *Rev. cuba. Cienc. agric. (Engl. ed.)* **6**

MSIRI (1961) 'By-products of the Sugar Industry in Mauritius'. *Mauritius Sugar Industry Research Institute – Technical Circular* No. 18 p. 147

Muñoz, F., Morciego, F., & Preston, T. R. (1970) Commercial fattening of bulls with molasses-urea, fishmeal and restricted forage under feedlot conditions, *Rev. cuba. Cienc. agric. (Engl. ed.)* **4**: 91

Muñoz, F. (1971) Unpublished data.

NRC (1956) National Research Council Publ. **No.** 449 (NRC, Washington).

Peron, N., & Preston, T. R. (1971) Effect of synthetic and natural roughage on tissue weights and contents of the intestinal tract in bulls fed liquid diets based on molasses/urea. *Rev. cuba. Cienc. agric. (Engl. ed).* **5**, 49

Pigden, W. J. (1972) Evaluation of Comfith as a commercial livestock feed in the Caribbean, *Proc. CIDA Seminar on Sugar Cane as Livestock Feed, Barbados,* Jan. 30–31.

Preston, T. R., & Muñoz, F. (1971) The effect of giving increasing quantities of torula yeast protein to bulls fattened on a molasses-based diet. *Rev. cuba. Cienc. agric.* **5**, pp.9

Preston, T. R., & Molina, A. (1972) Rapeseed meal in molasses/urea-based diets for fattening cattle. *Rev. cuba. Cienc. agric. (Engl. ed.)* **6**

Preston, T. R., & Martin, J. L. (1972) Different forage sources for bulls fed on a molasses-based diet. *Rev. cuba. Cienc. agric. (Engl. ed.)* **6**

Preston, T. R., Sancoucy, R., Nielsen, S. A., & Delaitre, C., (1973) Effect of supplementary maize grain on performance of Zebu bulls fed a molasses-urea based ration. In preparation.

Preston, T. R., (1973) Unpublished data.

Ramírez A., & Kowalczyk, J. (1971) Synthesis of microbial protein in bulls fed a protein-free diet based on molasses/urea. *Rev. cuba. Cienc. agric.* **5**, pp. 21

Redferne, D. (1972) Personal communication.

Rodríguez, H. (1968) The in vivo bag technique in digestibility studies. *Rev. cuba. Cienc. agric. (Engl. ed.)* **2**, 77.

Sansoucy, R., Nailsen, S. A., Delaitre, C., & Preston T. R. (1973) Bagasse as a source of roughage in molasses-based diets. In preparation.

Turner, A. W., & Hodgetts, V. E.(1955) Buffer systems in the rumen of the sheep. II Buffering properties in relation to composition. *Austr. Jn. agric. Res.* **6**, 125.

Wong You Cheong, Y., d'Espaiget, J. T., Deville, P. J., Sansoucy, R., & Preston, T. R. (1974) The effect of steam treatment on cane bagasse in relation to its digestibility and furfural production. Páper to be presented at the *XV ISSCT Congress, Durban, South Africa.*

Cassava as a total substitute for cereals in livestock and poultry rations

Z. Müller, K. C. Chou and K. C. Nah
F.A.O., Pig and Poultry Research and Training Institute, Singapore

Summary

The annual world production of cassava roots amounts to about 100 mt. The world shortage and high cost of feed-grain during 1973 gave an impressive stimulus to cassava exports from tropical countries to feed manufacturers in temperate zones.

The cassava plant has an unusually high photosynthetic potential for the conversion of solar energy into starch. About 3 to 4 tons of fresh cassava roots make up one ton of cassava meal; the roots, however, are perishable and deteriorate within a few days of harvesting if they are not dried or adequately stored. Dry cassava products are briefly described.

A nutritive profile of the cassava root, and leaf meal, is given with particular emphasis on basic nutrients, amino acids, minerals and vitamins. The low level of amylolytic activity of root meal, in comparison with maize and rice bran, is discussed; likewise the importance of sulphur-containing amino acids, other donors of methyl groups and lipotropic compounds in high cassava level diets.

The feasibility of cassava root meal as a total substitute for cereals in respective rations for pigs, poultry and cattle is discussed in detail.

Attention is given to factors (other than glucosides) which limit the use of cassava, namely, volume, texture and a deficiency in essential nutrients. Pelleting of cassava-based diets is recommended, as this not only ensures that the volume of ration is equated to grain-based diets, but also increases feed intake and digestibility producing the same (if not better) performance in animals as those on cereal-based diets.

Résumé

Le manioc comme substitut total pour les céréales dans les rations pour le bétail et la volaille

La production mondiale annuelle de racines de manioc s'élève à environ 100 millions de tonnes. Le manque mondial et les prix élevés des céréales pendant 1973 a stimulé d'une manière impressionante l'exportation du manioc des pays tropicaux vers les fabricants dans les zones tempérées.

Le manioc a un potentiel photosynthétique extraordinairement élevé pour la conversion de l'énergie solaire en amidon. Environ 3–4 tonnes de racines fraîches de manioc produisent une tonne de pâture de manioc; cependant les racines sont périssables et se détériorent dans quelques jours après la récolte si elle ne sont pas séchées ou stockées proprement. On décrit brièvement les produits secs de manioc. On donne un profil nutritif de la racine de manioc et de la pâture de feuilles avec un accent particulier sur les éléments nutritifs de base, les amino-acides, les minéraux et les vitamines. On discute le bas niveau d'activité amylolytique de la pâture de racines, en comparaison avec le son de mais et de riz; également on discute l'importance des amino-acides contenant du soufre, d'autres donneurs des groupes méthyliques et des composés lipotropes dans les diètes contenant un niveau élevé de manioc.

La praticabilité de la pâture de racines de manioc comme substitut total pour les céréales dans les rations respectives pour les cochons, la volaille et la bétail, est discutée en détail.

On fait attention aux facteurs (autres que les glucosides), qui limitent l'usage du manioc, nommément le volume, la texture et une déficience en nutriments essentiels. On recommande des diètes basées sur des boulettes de manioc, parceque ces-ci non seulement assurent que le volume est mis en parallèle avec les diètes basées sur les graines, mais augmentent aussi la prise de pâtures et la digestibilité, en produisant le même rendement (sinon meilleur) chez les animaux que les diètes basées sur les céréales.

Resumen

El cazabe como sustituto total de los cereales en las raciones para el ganado y las aves de corral

La producción mundial anual de raíces de cazabe asciende a alrededor de 100 millones de toneladas. La escasez mundial y el alto coste de granos para la alimentación durante 1973 estimuló de manera impresionante las exportaciones de cazabe de los países tropicales a los fabricantes de piensos en las zonas templadas.

La planta de cazabe tiene un alto potencial fotosintético poco corriente para la conversión de la energía solar en almidón. Alrededor de 3 a 4 t. de raiz de cazabe fresco producen una tonelada de harina de cazabe; sin embargo las raíces son perecederas y se deterioran a los pocos días de cosechadas si no se secan o se almacenen adecuadamente. Se describen brevemente los productos de cazabe seco.

Se incluye un esquema nutritivo de la raíz de cazabe, y de la harina de hojas, con atención particular a los elementos básicos nutritivos, aminoácidos, minerales y vitaminas. Se discute el bajo nivel de actividad amilolítica de la harina de raíz, en comparación con

el salvado de maíz y de arroz; de igual modo, la importancia de los aminoácidos que contienen azufre, otros donantes de grupos de metilo y compuestos lipotrópicos en dietas de nivel de cazabe alto.

Se discute con detalle la posibilidad de la harina de la raíz de cazabe como un sustituto total de los cereales en las raciones respectivas para cerdos, aves de corral y ganado. Se concede atención a factores (diferentes de los glucósidos) que limitan el uso del cazabe, es decir, el volumen, textura y una deficienca en elementos nutritivos esenciales. Se recomienda el granulado de las dietas basadas en cazabe, porque esto no solamente asegura que el volumen de la ración se asemeje al de las dietas basadas en granos, sino que también aumenta la toma de alimento y la digestibilidad, produciendo en los animales el mismo resultado (si no mejor) que en los que se alimentan con dietas basadas en cereales.

Introduction

Cassava or tapioca are widely used English names for dried-root products of the cassava plant *Manihot utilissima* Pohl (*Manihot esculenta* Crantz) of the family *Euphorbiaceae* (bitter types of cassava). *Manihot palmata aipi* Muell is synonymous with *Manihot dulcis, Jatropha dulcis, parodi, Manihot palmata,* etc., which are the sweet types of cassava.

Cassava is synomymous with Spanish yuca or mandioca, manioc or cassave in French and mandioca in Portuguese; ubi kettella or kaspe in Indonesia, mandioca or aipim in Brazil, manioca, rumu or yuca in Spanish speaking America, manioc in Madagascar and French speaking Africa, tapioca in India, tapioca and ubi kayu in Malaysia; cassava or cassade in English speaking regions of Africa, Thailand and Ceylon. Originally the name tapioca was derived from tipioca. Tupi is the Indian name for tuber meal and tipiocet for pellets (Grace, 1971).

Cassava is an all-season crop of the humid tropics, growing as a subsistence food for the poorest of people (Clarke and Haswell 1964; Wood, 1967; Coursey and Haynes, 1970). Cassava originated in South America where it has been growing throughout the ages, but the plant was not known in the Western Hemisphere before the discovery of America (Barrett, 1910; Normanha, 1970).

Although the plant has a high photosynthetic potential, an ability to tolerate drought and poor soils, and has a high resistance to weeds and pests, it is only in recent years that more systematic and sophisticated methods of cassava plantation and cultivation have been employed (Benny, 1969; Mahendranathan, 1971). Its toxicity due to the presence of cyanogenetic glucosides has, however, always been the main factor retarding its exploitation and real potential (Nestel, 1973).

Cassava requires a sandy soil and a warm tropical climate. It is also fairly resistant to dry periods but adaptable to high humidity. The optimum growing temperature is about 27°C. When the temperature drops to 15°C, growth stops and at 8 or 10°C the plant dies. The optimum rainfall is 700–1,000 mm (Choleva, 1968).

The plant is photosensitive, with a high sunshine requirement but not requiring too much rain. The cyanogenetic glucoside content increases with nitrogen fertilisation and drought but decreases with potassium or organic fertilisers. Shading young plants increases the glucoside content in the leaves but not in the root. The bark of tubers has a relatively higher level of glucoside than the inner part (Bruijn, 1973).

In Africa, cassava is sometimes planted together with maize, yams, citrus, coffee or cocoa trees (Jones, 1959; Jennings, 1970), while in Indonesia it is planted with yams, rubber, oil palms and on the bunds surrounding padis. In South and Central America, it is often planted together with legumes and melons (Rankine and Houng, 1971).

The cassava plant has a reputation for exhausting soils, the uptake of nutrients at 50 t yield of cassava tubers being approximately 120 kg P_2O_5, 450 kg K_2O and about 250 kg CaO. This explains the significance of fertilisation (Seemanthani, 1962; Chew, 1970; Williams, 1973). Small-scale farmers in South East Asia usually start to grow cassava on virgin forest clearings, and when productivity drops two to three years later, they desert the land.

Some of the new commercial varieties contain as much as 38% of starch and an average yield of 40 t/ha. (Chan, 1969). A number of improved varieties have recently been planted in South America, Africa and Asia, thanks to immense efforts by international organizations in the development of cassava production (IITA in Nigeria, CIAT in Columbia and the UN particularly the FAO).

To produce one tonne of dry product requires about three to four tonnes of fresh cassava roots (Aw-Yong,

1967). The fresh tubers are highly perishable and deteriorate within two or three days of harvest, if not dried or properly stored.

The following are the main dry commercial feeding products of the cassava plant:
1. Cassava chips: 4 or 5cm irregular slices of roots
2. Broken cassava roots: 12 to 15cm long with irregular thickness
3. Cassava roots in form of rectangular bars: 0.8 x 0.8 x 5.0cm (Roa and Cock, 1974)
4. Cassava cubes: about 1 x 1 x 1cm (Zahri and Tan, 1974)
5. Cassava pellets: in cylindrical form, about 2cm long and 0.5 to 0.8cm diameter
6. Cassava meal: fine powder from the manufacturing of chips and ground cassava roots
7. Cassava refuse (or waste): residual pulps separated from the starch
8. Cassava leaf meal: dried aerial part (or only leaves) of the cassava plant

The pelleting of tapioca products is becoming very popular because it decreases the volume by about 25%. This makes it easier to transport, handle, store, load and unload, and produces a uniform product which is less fragile during shipment overseas. (Lee, 1963, Pramanik, 1970; Grace, 1971).

It is estimated that the annual world production of cassava roots is about 100 m/t, and that Brazil (30%), Indonesia (20%) and Zaire (10%) are the largest producing countries. Although Brazil is the largest world producer, most of the crop is utilized locally and only a very small portion is exported (Vries et al., 1967). On the other hand, Thailand, which produces only about 3% of total world production, is the largest exporter. One of the biggest importers is the EEC (mainly Germany) consuming about 1 mt of tapioca meal as a feed for livestock (Bonney, 1971; Drysdale and Zahri, 1972). However, recent feed shortages have given a great stimulus to demand for tapioca as a substitute for grain in cattle feedlots and manufactured feed in general.

The nutritive profile

The principal chemical composition of the root product is given in Table 1. It varies considerably according to variety, age (see Figure 1) of the plant and also

Figure 1

Trend of the nutrient content in cassava tubers during 3 year growing period

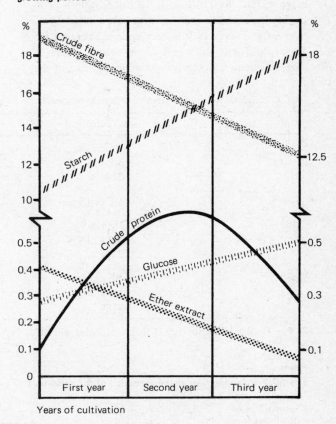

Years of cultivation

TABLE 1. *Chemical composition of cassava root meal in comparison with maize and composition of a mixture of cassava root meal with soya bean meal*

Constituent	Cassava root meal	Maize (yellow, grnd)	Mixture of 85% cassava and 15% soya bean meal*
Moisture	12.10	13.50	11.78
Crude protein (Nx6.25)	2.50	8.50	8.88
Ether extract	0.30	3.80	0.39
Crude fibre	3.50	2.00	3.64
Ash	1.80	1.10	2.40
NFE	79.80	71.10	72.91
Calcium	0.18	0.03	0.20
Phosphorus	0.09	0.27	0.18
Lysine	0.042	0.250	0.473
Methionine + cystine	0.019	0.260	0.226
Threonine	0.055	0.350	0.332
Tryptophan	0.011	0.050	0.099
Cost per £ (Singapore – Jan. 1974)	39.09	68.18	57.98

*Cost of soya bean meal Jan., 1974 was £165.00/t.

the processing technology (Johnson and Raymond, 1965; Mesa *et al.*, 1970). The caloric value of roots (see Table 2) and the digestibility of cassava starch are relatively high compared to cereals (Yoshida *et al.*, 1966; Yoshida, 1970, Maust *et al.*, 1972b; Tillon and Serres, 1973).

The protein, mineral and vitamin content of the root product is nutritionally insignificant. Amylolytic activity of tapioca meal is about one third of that found in maize and about half the level of rice bran (Müller *et al.*, 1972).

Cyanides are the most undesirable constituents of the cassava plant. The content in fresh tubers varies between 0.01 to 0.04% with the bitter varieties containing 0.02–0.03% and the sweet ones having less than 0.01%. Free hydrogen cyanide is liberated from the cyanogenetic glucosides by the action of the enzyme linamarase which is naturally present in the plant. However, the glucosides and linamarase only come into contact when the plant tissue is damaged during storage, processing and handling (Baker *et al.*, 1950; Serres and Tillon, 1970).

The importance of methionine as a moderator of toxic side effects of cassava products in poultry has been acknowledged by many authors, and recent studies by Maner and Gomes (1973) have also demonstrated the detoxication effect of methionine on rats and pigs.

The growing aerial parts of cassava may offer a new perspective for livestock and poultry development on a subsistence basis (Figure 2). As cassava leaves produce about 10 to 15 t of dry matter per hectare they could become, in future, an important source of protein in the form of protein extract for monogastrics while the fibrous residues could be used as forage for cattle.

TABLE 2. *Caloric values of dry matter of cassava and maize*

Animal	Item	Cassava root meal	Maize, yellow, grain
Pig	DE kcal/kg	4,000	4,055
	ME kcal/kg	3,800	3,810
	TDN %	91	92
Chicken	ME kcal/kg	3,650	3,660
Cattle	DE kcal/kg	3,970	4,010
	ME kcal/kg	3,250	3,290
	TDN %	90	91
Sheep	DE kcal/kg	3,750	4,320
	ME kcal/kg	3,070	3,540
	TDN %	85	98

Source: Draft Feeding Standard, Republic of Singapore, 1972.

Figure 2

The commercial potential of the utilisation of the aerial part of cassava (modification of the PRO—XAN process)

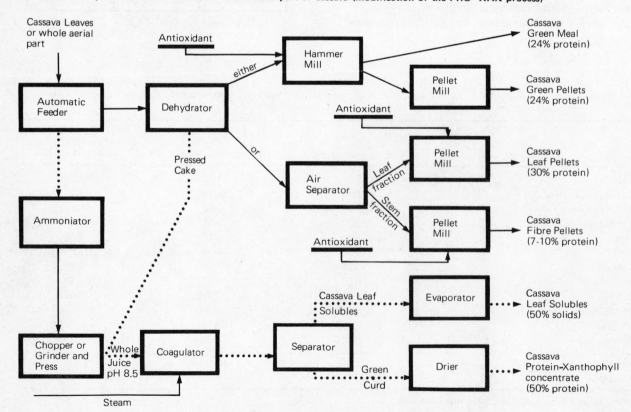

Although the protein value of cassava leaves is well established, the usual objection to their use as feed for animals has been the content of cyanide. In human nutrition this problem was elucidated by Rogers and Milner (1963). Recent studies (Ross and Enriquez, 1969; Annuar, 1969; Eggum, 1970; Mahendranathan, 1971; Hew and Hutagalung, 1972; Hutagalung, 1972) have shown that cassava leaves have also a good potential as an alternative source of protein for animals. The protein quality indicates similar amino acid patterns as in a grass or legume meal (see Table 3).

Cassava leaf protein is deficient in methionine, marginal in tryptophan and in isoleucine, but rich in lysine. The nutritional availability of the amino acids is somewhat variable and only 60% of the total methionine was found to be available (Eggum, 1970).

Cassava-based diets for pigs

The first serious scientific approach to substituting cereals by cassava in commercial pig rations began during the early years of World War II (Leite, 1939; Kok and Robeiro, 1942, 1943; Raymond et al., 1941. It was soon recognised by European farmers that cassava could be a substitute for grain especially in Germany (Sperling, 1954). This was acknowledged by Oyenuga and Opeke (1957) in Nigeria, Zarate in Philippines (1956) and many others. Later, nutritional studies in pigs were geared to the maximum substitution of cereals by raw or dried cassava root products (Noland et al., 1957; Oyenuga, 1955; Salmon-Legagneur and Fevrier, 1959; Mejia, 1960; Oyenuga, 1961; Van Vaerenbergh, 1961; Anon, 1962; Modebe, 1963; Nehring et al., 1963; Castillo et al., 1963; Hansen, 1964; Buitrago, 1964; Velloso et al., 1967; Jeffers and Haynes, 1967; Maner et al., 1967, 1969, 1970; Peixoto, 1968, 1969; Maust et al., 1969; Peraza-Castro, 1970 and others). It was generally recommended that cassava could replace cereals at 20–40%

level of dry matter calculated on the whole pig ration (Serres and Tillon, 1973).

However, in some experiments, even low levels of cassava noticeably depressed growth. Thus, for example, Velloso et al. (1967) observed that even 22% cassava root meal in pig diets had significantly affected the growth and feed efficiency. In experiments on rats, Jeffers and Haynes (1967) concluded that cassava and other root crops produced satisfactory growth and feed efficiency only at levels up to 20%. Maust et al. (1969) replaced maize by 36% tapioca meal. Pigs fed on maize had significantly better performance than a tapioca-based diet which resulted in a poor appetite and rapid decline of body weight gain. Pigs on cassava-based diets developed parakeratosis during the 4th week, but this disorder was eliminated by the addition of 100 mg $ZnCO_3$ per kg of diet (Maust et al., 1972a).

Maner et al. (1969, 1970) fed fresh chopped cassava with protein supplements offered ad libitum. These trials clearly demonstrated a satisfactory feed intake and a liveweight gain, but feed efficiency did not show the expected results in the majority of cases. This was inevitable because of the feeding system and because of a number of variable dietary factors involving palatability of protein supplements, dry matter intake, level of digestible energy, amino acid balance and probably because of a higher level of calcium in some diets, supplemented by bone and fish meal. However, Maner's team (1972, 1973) has conclusively proved on a large number of pigs that cassava may fully replace cereals without any negative effects when diets are properly balanced.

In a short term experiment (seven weeks) conducted in Venezuela (Chicco, et al. 1972), maize was gradually substituted by cassava meal up to 58.5%. Levels of tapioca over 40% reduced liveweight gain, but the feed efficiency at 40% and 58.5% levels of inclusion of tapioca was the same. There were no differences in digestibility of organic matter and nitrogen retention

TABLE 3. *Protein value of dehydrated aerial part of cassava and some tropical grasses compared to that of soya bean (dry basis)*

Constituent	Cassava (Manihot utilissima) leaves	Cassava (Manihot utilissima) leaves and stems	Napier grass (Pennisetum purpureum)	Gatton panic (Panicum maximum)	Soya bean meal (Solv. extr.)
Crude protein	27.0	20.3	12.6	11.9	45.7
Amino acids (g/16g nitrogen)					
Arginine	5.21	3.89	6.10	5.64	7.41
Cystine	1.18	0.98	0.51	Nil	1.52
Glycine	4.92	5.10	5.85	5.00	5.23
Histidine	2.47	2.32	2.54	2.82	2.39
Isoleucine	4.12	4.40	4.32	3.45	5.45
Leucine	10.09	8.75	8.64	7.55	6.97
Lysine	7.11	5.89	6.02	4.82	6.32
Methionine	1.45	1.83	1.86	1.36	1.52
Phenylalanine	3.87	4.37	5.42	5.82	4.79
Threonine	4.70	5.70	4.41	4.73	4.14
Tryptophan	1.09	1.24	–	–	1.30
Tyrosine	3.97	4.12	3.73	3.18	3.27
Valine	6.18	8.43	6.27	5.18	5.23

Source: Draft Feeding Standard, Republic of Singapore, 1972

between diets based on maize and cassava. The carcass evaluation and chemical analysis of carcass did not show any differences with the comparative treatments.

Hansen's experiments (1960) showed that pigs, having a good growth potential, perform better on tapioca-based diets (up to 30% cassava level) than pigs with lower growth potential. The latter did not tolerate any level of cassava meal, and increasing levels of cassava significantly affected their growth and feed efficiency.

In Malaysia, Hutagalung et al. (1973) carried out experiments on pigs, replacing maize with cassava from 15% up to 60% of the diet. All cassava-based diets depressed weight gain and feed efficiency compared to maize-based diets.

Aumaitre (1967, 1969, 1972) compared pig diets based on wheat, barley, maize and cassava. A cassava-based diet was clearly superior in weight gain. feed efficiency, and there was a drop in the frequency of diarrhoea.

Numerous experiments have also been conducted by the Pig and Poultry Research and Training Institute in Singapore during the period 1969 to 1974 (Müller and Chou, 1971; Chou and Müller, 1972). The replacement of cereals by cassava meal from 40% up to 75% of the total diet had no adverse effects on the pigs. Cassava-based rations presented in mash form, especially when introduced per se, were generally disliked by pigs (Henry, 1970). In contrast, pelleted diets were more readily accepted by the pigs than conventional maize-based diets. The results of nine experiments have conclusively proved that overall performance, health and carcass quality of pigs have not been influenced by any level of tapioca meal when diets were pelleted and carefully balanced in all limiting factors (see Table 4). Similar results with a 50% tapioca-based diet, fed either in mash or pelleted form were recently reported by Peixoto and Farias (1971).

Cassava-based diets for poultry

Experiments with cassava meal in poultry dates back to 1935 in the Philippines (Tabayoyong, 1935). A greater emphasis was given to the replacement of cereals in poultry diets during the last war (Temperton and Dudley, 1941). Diets having more than 10% cassava meal showed visibly poorer performance than cereal-based diets. Similarly in Germany (Klein and Von Barlowen, 1954; Sperling, 1954; Wegner, 1961; Vogt and Pennen, 1963; Vogt and Stute, 1964) and also in other countries (Squibb and Wyld, 1951; Maciel, 1958; Soares, 1965) a significant growth depression was recorded when the level of cassava meal was higher than 10–20% of the total diet. In some experiments, there was already a tendency to substitute cassava for cereals up to 50% but the performance of birds having more than 30% cassava meal in their diet favoured cereal-based diets (Olson et al., 1969a). In was suggested that glucosides still remained in the cassava, and that a phosphorylase inhibitor in the rind of the cassava tuber could also have been responsible for the growth depression (Vogt, 1966). Somewhat better results were obtained when cassava-based diets were fortified by methionine (Enriquez and Ross, 1972; Olson et al., 1969b).

The higher levels of cassava meal in poultry diets have only recently been introduced by Singapore and Malaysian workers (Müller and Chou, 1971; Chou and Müller, 1972; Hutagalung, 1972; Hew and Hutagalung, 1972 and Hutagalung et al., 1973). While pelleted diets containing cassava meal (up to 58% of broiler rations and up to 75% of rations for replacement birds) resulted in a similar performance as maize-based diets, poorer growth and feed efficiency of all cassava-based diets were recorded when they were fed mashes rather than pellets. The summarized results of the experiments carried out with broilers are shown in Table 5.

TABLE 4. Comparative results of experiments on pigs fed maize and cassava diets (as mash or pellets)

Main energy source of diet	Form of diet	Level of cassava in diet %	Relative specific weight of diet (maize diet = 100)	Live-weight gain head/ day (in g)	Feed conversion	No. of pigs in trials	Breed
Maize	mash	nil	100	440	4.2	95 (4 replications)	local cross breeds
Cassava	mash	38–40	81	425	4.3		
Maize	mash	nil	100	463	3.8	94 (4 replications)	imported breeds
Cassava	pellets	40	109	499*	3.5**		
Maize	mash	nil	100	699	3.1	16	imported cross breeds
Cassava	pellets	60–75	109	676	3.0		

*P < 0.05   **P < 0.025

TABLE 5. *Comparison of rations containing different levels of cassava with maize-based diets for broilers*

Levels of cassava meal in diet, (%)	Specific weight of diets (g/l)		10 weeks liveweight (kg)	Feed conversion	Mortality (%)
	Mash	Pellets			
0 (Maize diet)	620	670	2.04	2.61	9.2
20	560	670	2.05	2.59	3.0
40	500	680	2.03	2.61	3.0
58	510	680	2.04	2.53	5.0

TABLE 6. *Cassava as a total substitute for cereals in recycling animal waste to beef cattle*

Diet	Constituents of silage			Quality of silage			
	Cassava meal (%)	Animal waste (%)	Others (%)	pH	Lactic acid (%)	Butyric acid (%)	Apparent palatability
1	60	38 (PL)	2	4.6	2.5	1.7	very high
2	50	38 (PL) and 10 (CM)	2	4.4	2.4	1.4	very high
3	43	38 (PL) and 15 (CL)	4	4.6	1.9	1.9	very high
4	18	63 (PL) and 15 (CM)	4	4.8	2.3	1.7	very high
5	80	15 (PM)	5	6.6	0.6	1.6	low
6	45	40 (PM)	15	5.1	3.1	0.6	very high

Note: 1. PL = poultry litter; CM = cattle manure; CL = cattle litter; PM = poultry manure; Wood shavings were used as a litter material. All animal wastes were undried.

2. Other constituents were molasses, phosphoric acid, sodium chloride and other minerals. These constituents were added in variable quantities to balance the respective diets.

3. Acetic acid was not present or in negligible quantities with the exception of diet 6 where its level was 1.1%.

The possibility of replacing maize by cassava meal in diets for laying hens was also studied recently (Chou and Müller, 1972); Enriquez and Ross, 1967). It was concluded that the inclusion of cassava meal at levels of up to 50% of the diet for laying birds did not affect either the performance or the quality of eggs. A significant drop in yolk pigmentation could be easily overcome by adding synthetic xanthophylls.

The literature on the use of tapioca in poultry rations thus gives a very inconclusive picture due to the fact that cassava-based diets must meet not only the physiological aspect of nutrient balance but also the physical requirement of the diet. Therefore, volume as a factor controlling feed intake has greater importance than the nutritive make-up of the diet. This is of extreme importance particularly in the tropics where feed intake is already seriously affected by climatic interference.

Potential of cassava in cattle feeding

The current shortage of feed grains has given an important stimulus to the use of cassava root products in all-mash feedlots and in feed concentrates for dairy cattle. The fact that the release of energy from cassava starch, and nitrogen from urea, have a very close timing, is of great importance for the maximum utilisation of non-protein nitrogen (NPN) compounds in cattle diets and thus a significant saving in protein feedstuffs.

However, even in green-lots, a simple feed concentrate, for instance with the composition of 85% cassava root

meal, 6% molasses, 8% urea and 1% mineral supplement would match up well with tropical grasses and supply all deficient nutrients required for optimum performance and utilisation of the nutritive potential of green forage.

In dry lots, the combination of two feed ingredients, such as tapioca pellets and grass meal pellets, may totally substitute for cereals. Fresh cassava roots, when properly processed, may also serve as a basic calorie source for intensive cattle feeding. In tropical regions, cattle development is usually hampered by inadequate forage the quality of which is changing too rapidly to meet the physiological requirements and economic performance of ruminants (Kay *et al.*, 1972; Karue, 1973). Under such conditions, cassava, either in fresh form or preferably as a dry product, may play an important role.

The recycling technology applied by the Pig and Poultry Research and Training Institute (UNDP/SF programme) in Singapore, uses cassava meal and animal waste as the only feeding ingredients for beef cattle. Cassava, rich in digestible energy, poor in protein and fibre, matches up well with animal waste which is low in energy and high in other feed constituents like protein, fibre and ash. The results of these experiments are given in Table 6.

These silages are fed initially to calves of 55 kg liveweight as a substantial proportion of the diet and later (from 120 kg) as an exclusive diet up to slaughter weight without any traditional forage such as grass or hay. The apparent quality of the silage, feed intake and palatability are very high. The performance of cattle for the overall period (daily gain 0.75–1.2 kg;

feed conversion 1:7.5) under this new feeding system is extremely satisfactory despite continuous exposure to the tropical climate. These experiments in Singapore based on the recycling technology, offer an entirely fresh approach to the use of cassava as the most effective counterpart to animal and other organic wastes for intensive cattle production, the more so in view of the increasing cost of conventional feed materials and their competition with human consumption, the necessity for abating pollution, and the need to conserve energy resources.

Conclusions

The following conclusions can be drawn from the worldwide experiments conducted on substituting cassava for cereals in pig and poultry diets:

1. Cassava-based diets must be either pelleted or fed in wet or liquid form. While pelleted diets allow for a once-a-day feeding system, wet or liquid diets would require two, or preferably three, feeding intervals in order to obtain a reasonable performance. However, pelleted diets will always produce, under field conditions, more reliable results.

2. The pellets eliminate irritation of the respiratory organs and eye infections, and ensure an optimum feed intake. Thus, with poultry, the feeding time required for eating pellets is about 20—30% of that for mash.

3. Pelleted diets have a higher digestibility of starch and fibre than diets fed in mash form. The digestibility of fibre increases, probably because of the heat generated by both pressure and steam treatment which release the cellulose from the lignin-cellulose bond.

4. The pelleting of diets to some extent enables the use of cassava products of a lower quality because of the requirement for nutrients is regulated by a higher feed intake.

5. Heat generated by steam treatment and high pressure during pelleting destroy many of the growth inhibitors of microbial origin (moulds, fungi and other contamination during sun drying or storing), and probably also destroys part of the glucosides. This was supported by observations showing that, while pigs on mash diets suffered almost continuously from a mild type of diarrhoea, this was not the case when the same cassava-based diets were fed in pelleted form.

6. Cassava-based diets must be formulated with more care than cereal-based diets. Great attention must also be paid to the level of crude fibre and ash in cassava root meal as this product is often being adulterated by lime, chalk, etc. An excess of crude fibre and ash decreases the digestibility of the ration as a whole, and limits the choice of other feed ingredients which are high in these components. Therefore, cassava roots, having more than 4% fibre, are not suitable as a total substitute for cereals in high energy diets. Special care must be paid to the balance of limiting amino acids, essential fatty acids (linoleic acid specifically for laying hens), basic

minerals and microelements (such as zinc and iron) and vitamins.

7. Poultry diets based on cassava roots must be fortified by natural and/or synthetic xanthophylls (preferably a combination of both yellow xanthophylls from grass/leguminous meals and synthetic red xanthophylls).

8. Economic considerations are, however, of prime importance and the replacement of some cereals by cassava is only feasible when the price of the nutritionally equivalent mixture of cassava and protein feeds is lower than that of the respective cereal grain to be substituted.

References

Anon (1962) Tapiokamehl ein sehr gefragter Artikel. *Schweinezucht Schweinemast,* pp. 10—19.

Anuar, Q. (1969) Preliminary report on the use of tapioca or cassava and sago as feedstuffs for poultry. Paper presented at the *Malaysian Vet. Assoc. Meet.* Penang, Malaysia, Oct. 1969.

Aumaitre, A. (1967) Futterverwent von Tapioka und verschiedenen Getreidearten in Futterrationen für frühzeitig abgesetzte Ferkel. *Zeit. Tierphysiol. Tierernähr. Futtermittelk,* **23** pp. 41—43.

Aumaitre, A. (1969) Valeur alimentaire du manioc et de différentes cereales dans les régimes de sevrage precoce du porcelet: Utilisation digestive de l'aliment et effet sur la croissance des animaux. *Ann. Zootech.,* **18** (4) pp. 385—398.

Aumaitre, A. (1972) Le manioc convient-il aux porcs? *La Revue de L'elevage* **27** pp. 83—87.

Aw-Yong, K. K. & Mooi, S. W. (1967) Cultivation and production of tapioca in Perak. *Ann. Dep. Inf. Pap.* No. 7 Div. Agric., W. Malaysia.

Baker, F., Nasr, H., Morrice, F., & Bruce, J. (1950) Bacterial breakdown of structural starches and starch products in the digestive tract of ruminant and non-ruminant mammals. *J. Path. Bact.* **62** pp. 617—638.

Barrett, O. W. (1910) Promising root crops for the South. Yautias, Taros and Dasheens, *Bull. U.S. Bureau Pl. Ind.* No. **164** pp. 1—37.

Benny, J. M. (1969) Mechanisation of tapioca. *Malaysian Crop Diversification Con.* Reprint No. **15**.

Bonney, J. (1971) German firm offers to buy all the tapioca Malaysia can produce. *The Straits Times,* Singapore. July 6.

Bruijn, G. H. (1973) The cyanogenic character of cassava (*Manihot esculenta*), in *Chronic Cassava Toxicity,* Proc. of an interdisciplinary workshop, London, 29—30 Jan. pp. 43—48.

Buitrago, J. (1964) Utilization of yuca (Manihot) in diets for growing finishing pigs. Thesis, Columbia.

Castillo, L. S., Aglibut, F. B., Javier, T. A., Gerpacio, A. L., Garcia, G. V., Puyacan, R. B. & Ramin, B. B. (1963) Camote and cassava tuber silage as replacement for corn in swine growing-fattening rations. *Philippines Agric.* 47 pp. 460–474.

Chan, S. K. (1969) Tapioca (*Manihot utilissima*) investigations at the Federal Experiment Station, Serdang, Fed. Experiment, St. Serdang, Malaysia, 1969.

Chew, W. Y. (1970) Varieties and NPK fertilisers for tapioca (*Manihot utilissima Pohl*) on peat. *Malay, Agric. J.* 47 pp. 483.

Chicco, C. F., Garbati, S. T., Müller-Haye, B. & Vecchionacce, H. I. (1972) La harina de yuca en el engorde de cerdos. *Rev. Agron. Trop.,* 12 (6) pp. 599–603.

Choleva, E. (1968) *Rostlinná výroba tropů a subtronů (Okopaniny),* St. Pedag, Nakl., Praha, Czechoslovakia.

Chou, K. C. & Müller, Z. (1972) Complete substitution of maize by tapioca in broiler ration. *Proc. Australasian Poult. Sci. Conv.,* New Zealand, Auckland, pp. 149–160.

Clarke, C. & Haswell, M. R. (1964) The Economics of Subsistence Agriculture. London: MacMillan.

Coursey, D. G. & Haynes, P. H. (1970) Root crops and their potential in the tropics. *Wld. Crops* 22 (4) pp. 261–265.

Drysdale, J. & Zahri, A. (1972) Feasibility study of a projected tapioca pelletization industry for East Java. Singapore, Asia Research Pte. Ltd. pp. 1–49.

Eggum, E. O. (1970) The protein quality of cassava leaves. *Br. J. Nutr.* 24 (3) pp. 761–768.

Enriquez, F. Q. & Ross, E. (1972) Cassava root meal in grower and layer diets. *Poultry Sci.* 51 pp. 228–232.

Enriquez, F. Q. & Ross, E. (1967) The value of cassava root meal for chicks. *Poultry Sci.* 46 pp. 622–626.

Grace, M. (1971) Processing of Cassava. *Agric. Serv. Bull.* No. 8 Ags/ASB/71/2. pp. 1–124 Rome: FAO.

Hansen, V. (1964) Le manioc dans l'alimentation des porcs (danois). *Ugeskrift for Landmaend.* 109 (35) pp. 562–563.

Henry, Y. (1970) Effects nutritionnels de l'incorporation de cellulose purifice dans le regime du porcen croissance – finition, *Ann. Zootech.* 19 (2) pp. 117–141.

Hew, V. F. & Hutagalung, R. I. (1972) The utilization of tapioca root meal (*Manihot utilissima*) in swine feeding, *Mal. Agric. Res.* 1 pp. 124–130.

Hutagalung, R. I. (1972) Nutritive value of tapioca leaf meal, tapioca root meal, normal maize and *opaque-2* maize and pineapple bran for pigs and poultry. *17th Ann. Conf. Mal. Vet. Ass., Univ. Malaya,* Dec., 1972.

Hutagalung, R. I., Phuah, C. H. & Hew, V. F. (1973) The utilization of cassava (tapioca) in livestock feeding. Papers presented at the *Third Inter. Symp. on Trop. Root Crops,* Nigeria, IITA, Ibadan, 2–9 Dec., 1973.

Jeffers, H. F. & Haynes, P. H. (1967) A preliminary study of the nutritive value of some dehydrated tropical roots. *Proc. Int. Symp. on Trop. Root Crops,* Univ. W. Indies, St. Augustine, Trinidad. 2 (6) pp. 72–89.

Jennings, D. L. (1970) Cassava in Africa. *Field Crop Abstract* 23 (3) pp. 271–278.

Johnson, R. M., & Raymond, W. D. (1965) The chemical composition of some tropical food plants. IV. Manioc. *Trop. Sci.* 7 (3) pp. 109–115.

Jones, W. C. (1959) Manioc in Africa. California, USA Stamford Univ. Press.

Karue, C. N. (1973) Metabolism of dietary urea with cassava when used to supplement dry season roughage in Kenya. *Proc. III World Conf. Anim. Prod.* Melbourne, Australia, 1 5 (c) pp. 44–45.

Kay, M., Macdearmid, A. & R. Massie (1972) Intensive beef production 13. Replacements of concentrates with root crops. *Anim. Prod.* 15 (1) pp. 67–73.

Klein, F. W. & V. Barlowen, C. (1954) Tapiokamehl in Aufzuchtfutter *Arch. Geflügelk.* 18 pp. 415.

Kok, E. A. & G. A. Robeiro (1942) A Farelo de raspa de mandioca em comparacaeo com a quierera de milho na alimentacao dos suinos. *Bol. Ind. Animal.* São Paulo (N.S.) 5 (4) pp. 86–124.

Kok, E. A. & G. A. Robeiro (1943) Mandioca crua em comparacao com a quirera de milho na engoroa de porcos. *Bol. Ind. Animal,* São Paulo (N.S.) 6 (1–2) pp. 24–45.

Lee, S. Y. (1963) Thailand's tapioca, *Far Eastern Econ. Review,* pp. 230–235.

Leite, A. C. (1939) Contribuição para o estudo da mandioca e da araruta na alimentacão dos porcos de engorda. *Bol. Ind. Anim.,* São Paulo (N.S.) 2 (2) pp. 3–26.

Maciel, E. (1958) Contribuicão ao estudo da mandioca na alimentacão animal. *Bol. Din. da Prod. Anim.* DIPAN. 119/20 (11) pp. 23–40.

Mahendranathan, T. (1971) Potential of tapioca (Manihot utilissima pohl) as a livestock feed – a review. *Mal. Agric. J.* 48 (1) G426, pp. 77–89.

Maner, J. N., Buitrago, J. & Jimenez, I. (1967) Utilization of yuca in swine feeding. *Proc. Int. Symp. on Trop. Root Crops,* Trinidad, Univ. W. Indies, St. Augustine, 2 (6) pp. 62–71.

Maner, J. H., Buitrago, J. & Jimenez, I. (1969) Utilization of yuca in swine feeding. Paper presented at the *Int. Symp. on Trop. Root Crops.* Trinidad Univ. of the W. Indies.

94   **Maner, J. H., Buitrago, J. & Callo, J. T.** (1970) Protein sources for supplementation of fresh cassava (Manihot esculenta) rations for growing-finishing swine. *J. Anim. Sci.* **31**, (1), pp. 208 abs. 203.

Maner, J. H. (1972) La yuca en la alimentación de cerdos. *Seminar Report on Sistemas de Production de Porcinos en America Latina* Colombia, CIAT.

Maner, J. N. & Gomes, G. (1973) Implications of cyanide toxicity in animal feeding studies using high cassava rations, in *Chronic cassava toxicity.* Proc. of an interdisciplinary workshop, London, 29–30 Jan. pp. 133–120.

Maust, L. E., Warner, R. G., Pond, W. G., & McDowell, R. E. (1969) Rice bran-cassava meal as a carbohydrate feed for growing pigs. *J. Anim. Sci.* **29** (1) pp. 140 abs. 149.

Maust, L. E., Pond, W. G. & Scott, M. L. (1972a) Energy value of a cassava-rice bran diet with and without supplemental zinc for growing pigs. *J. Anim. Sci.* **35** pp. 953–957.

Maust, L. E., Scott, M. L. & Pond, W. G. (1972b) The metabolisable energy of rice bran, cassava flour, and blackeye cowpeas for growing chickens. *Poult. Sci.* **51** pp. 1397–1401.

Mejia, T. R. (1960) Valor comparative entre la yuca y el maiz en la alimentación de cerdos. *Rev. Fac. Nacional de Agronomia,* **55** (20) pp. 95–113.

Mesa, J., Maner, J. H., Opando, H., Portela, R. & Callo, J. T. (1970) Nutritive value of different tropical sources of energy. *J. Anim. Sci.* **31**, (1) pp. 208, abs. 206.

Modebe, A. N. A. (1963) Preliminary trial on the value of dried cassava (Manihot utilissima Pohl) for pig feeding. *J. W. Afr. Sci. Ass.* **7** pp. 127–133.

Müller, Z. & Chou, K. C. (1971) Different levels of tapioca meal in broiler rations. *UNDP/SF Project SIN 67/505,* Singapore Pig & Poultry Research & Training Institute, Nut (Pou) R-871 pp. 1–26.

Müller, Z., Chou, K. C., Nah, K. C. & Tan, T. K. (1972) Study of nutritive value of tapioca in economic rations for growing/finishing pigs in the tropics. UNDP/SF Project SIN 67/505, Singapore Pig & Poultry Research & Training Institute, (Pigs) R-672 pp. 1–35.

Nehring, K., Hoffmann, L. & Schiemann, R. (1963) Die energetische Verwertung der Futterstoffe. 3 Die energetische Verwertung der Kraftfutterstoffe durch Schweine. *Arch. Tierenähr.,* **13** pp. 147–161.

Nestel, B. (1973) Current utilisation and future potential for cassava in *Chronic Cassava Toxicity:* Proc. of an interdisciplinary workshop London, 29–30 Jan. pp. 11–26.

Noland, P. R., Vega, E. H. & Stanziola, L. C. (1957) Resultados de pruebas sobre alimentacien de puercos. Institute Nacional de Agriculture, Divis R. P. Folleto No. **31**.

Normanha, E. S. (1970) General aspects of cassava root production in Brazil. *2nd Int. Symp. on Root and Tuber Crops* **1** pp. 61. Univ. Hawaii.

Olson, D. W., Sunde, M. L. & Bird, H. R. (1969a) The metabolisable energy content and feeding value of mandioca meal in diet for chicks. *Poultry Sci.* **48** pp. 1445–1452.

Olson, D. W., Sunde, M. L. & Bird, H. R. (1969b) Amino acid supplementation of mandioca meal chick diets. *Poultry Sci.* **48** pp. 1949–1953.

Oyenuga, V. A. (1955) The composition and nutritive value of certain feeding stuffs in Nigeria. I. Roots, Tubers and Green Leaves. *Empire J. Expt. Agric.* **23** pp. 81–95.

Oyenuga, V. A. & Opeke, L. K. (1957) The value of cassava rations for pork and bacon production. *W.A.J. Biol. Chem.* **1**, (1) pp. 3–14.

Oyenuga, V. A. (1961) Nutritive value of cereal and cassava diets for growing and fattening pigs in Nigeria. *Brit. J. Nutr.* **15** (3) pp. 327–338.

Peixoto, R. R. (1968) Comparison of cassava meal and maize as feed for growing and fattening pigs, **19**, Nov. 1965. Escola Agronom. Eliseu Maciel, Pelotas, Brazil.

Peixoto, R. R. (1969) Substituicao do milho ao nível de 50% farinha de mandioca na alimentacão de suínos em crescimento e engorda, *Boletim Técnico,* **5**, ρ. 1–20.

Peixoto, P. R. & Farias, J. V. da Silva (1973) Estudo da influencia da prensagem (pellets) em racces com elevado teor de farinha da mandioca pra porcos em crescimento e terminacao. *X. Reuniao Anual da S.B.Z. I Congresso Brasileiro de Forrageiras, P.A.R.S.* (16-29/7/73) pp. 241–242.

Peraza-Castro, C. (1970) Etude de la digestibilite du manioc par le porc en croissance – finition. These doctorat veterinaire, Mexico, 1970

Pramanik, A. (1970) Prospects for tapioca cultivation and pelletization in Malaysia. Central Oil Palm Industries Berhad. pp. 240–246.

Rankine, L. B. & Houng, M. H. (1971) A preliminary view of cassava production in Jamaica. *Dept. Agric. Econ. Occasional Series* **6**, Trinidad Univ. West Indies.

Raymond, W. D., Jojo, W. & Nicodemus, Z. (1941) The nutritive value of some Tanganyika foods. II – Cassava. *Afr. Agric. J.* pp. 154–159.

Roa, G. & Cock, J. H. (1974) Natural drying of cassava. Unpublished Ph.D. thesis. Michigan Agricultural Engineering Dept. Michigan State University, East Lansing.

Rogers, D. J. & Milner, M. (1963) Amino acid profile of manioc leaf protein in relation to nutritive value. *Econ. Bot.* **17** pp. 211–216.

Ross, E. & Enriquez, F. Q. (1969) The nutritive value of cassava leaf meal. *Poultry Sci.* **48** pp. 846–853.

Salmon-Legagneur, E., & Fevrier, R. (1959) Les preferences alimentaires des porcelets. III. Appetence de quelques cereals. *Ann. Zootech.,* 8 pp. 139–146.

Seemanthani, K. B. (1962) Two new double-yielding — tapioca, *Indian Farm,* pp. 10–12,

Serres, H. & Tillon, J. P. (1973) Cassava for pig feeding. I. Possibilities and use limits. *Rev. Elec. Med. Vet. Pays Trop.* 26 (2) pp. 225–228.

Soares, P. R. (1965) Whole mandioca meal and wheat mill feed in the feeding of chicks. Master's thesis. Brazil Universidade Rural Do Estado De Minas Gerais, Vicosa.

Sperling, D. (1954) Tapiokamehl. *Futter u. Fütterung,* 44 pp. 343.

Squibb, R. L. & Wyld, M. K. (1951) Effect of yuca meal in baby chick rations. *Turrialba,* 1 pp. 298–299.

Tabayoyong, T. T. (1935) The value of cassava refuse meal in the ration for growing chicks. *Philippine Agric.* 24 pp. 509–518.

Temperton, H., & Dudley, F. J. (1941) Tapioca meal as food for laying hens. *Harper Adams Utility Poultry J.* 26 pp. 55–56.

Tillon, J. P. & Serres, H. (1973) Cassava for pig feeding. II Cassava digestibility in various forms. *Rev. Elev. Med. Vet. Pays Trop.,* 26 (2) pp. 229–233.

Van Vaerenbergh, R. (1961) La Coix lacryma-jobi en remplacement du mais jaune dans l'engraissement du porc. *Bull. agric. Congo,* 52 pp. 271–277.

Velloso, L., Rodrigues, A. J., Becker, M., Neto, L. P., Scott, W. N., Kalil, E. B., Melotti, L. & Da Roche, G. L. (1967) Substitucao parcial e total do milho pelo farelo de mandioca em racões se suinos em crescimento e engorda. *Bolm. Ind. Anim.* 23, (1965/66) pp. 129–137.

Vogt, H. & Pennen, W. (1963) Inclusion of tapioca and cassava meals in feed for fattening chickens. *Arch. Geflügelk* 27 pp. 431–460. Cited by *Nutr. Abst. & Revs.* 34 pp. 886.

Vogt, H., & Stute, K. (1964) Prüfung von Tapiokapellets im Geflügelmast-I-Alleinfutter. *Arch. Geflügelkde,* 78 pp. 342–358.

Vogt, H. (1966) The use of tapioca meal in poultry rations. *World Poultry Sci.* J. 22 pp. 113–125.

Vries, C. A. de., Ferwerda, J. D. & Flach, M. (1967) Choice of food crops in relation to actual and potential production in the tropics. *Neth. J. Agr. Sci.* 15 pp. 241.

Wegner, R. M. (1961) Zur Verbilligung von Kukenmastfuttermischungen. *Kraftfutter.* 44 pp. 84–88.

Williams, C. N. (1973) Production and productivity of tapioca. Ph.D. Thesis. Malaysia, University of Malaya.

Wood, T. (1967) Cassava, *Home Econ. Quart. Rev. Nutr. and Food Sci.* 6 pp. 16–18.

Yoshida, M., Hoshii, H., Kosaka, K. & Morimoto, H. (1966) Nutritive value of various energy sources for poultry feed. IV. Estimation of available energy of cassava meal. *Jap. Poultry Sci.* 3 pp. 29–34.

Yoshida, M. (1970) Bioassay procedure of energy sources for poultry feed and estimation of available energy of cassava meal. *Jap. Agr. Res. Quart (JARQ).* 5 (4) pp. 44–47.

Zahri, A. & Tan, S. A. (1974) Personal communication.

Zarate, J. J. (1956) The digestibility by swine of sweet potato vines and tubers, cassava roots and green papaya fruits. *Philippines Agric.* 40 pp. 78.

Discussion

Professor Abou-Raya: In Egypt we are using alkali treated sugar cane bagasse as animal feed. Treatment with calcium rather than sodium hydroxide is preferred as it is safer and can be employed by ordinary farmers. We also make a mixed feed from bagasse pith, molasses and urea. Three per cent of urea would appear to be nearly a toxic level. I would be grateful if Dr Preston could comment on possible urea toxicity and also give figures for nitrogen retention rather than metabolisable nitrogen.

Dr Preston: The utilisation of urea is always controlled by the amount of fermentable carbohydrate in the ration. Toxicity at high levels of urea is no problem if adequate fermentable carbohydrate is also made available in the ration. With regard to metabolisable nitrogen the terminology in my paper was that used by Dr Bowes, Iowa State University to describe protein which escaped degradation and contributed as a source of amino acids at the metabolic level. He used the expression 'metabolisable protein' for this. This term describes the amount of protein which escapes degradation in the rumen, and has nothing to do with retained nitrogen or percentage intake. I never use nitrogen retention as there are more problems associated with the use of nitrogen balance studies than there are advantages.

Dr Babatunde: I would like to ask Dr Preston if nitrogen levels in diets were equalised when maize was substituted for molasses in the experiment described in his paper (see Table 14). I would also like to mention that blood meal is used successfully as an ingredient of pig and poultry diets in Nigeria. However, the quality of blood meal depends on conditions of processing.

Dr Preston: Additional maize protein could be of importance especially as maize protein is not very soluble in the rumen. In this particular experiment the ration contained no urea, the total nitrogen requirements being covered by cotton seed cake. Protein was not limiting in the diet so I do not think the small amount of maize protein affected things. If it had been a urea/molasses ration the maize protein might well have had an effect. It may be that protein fed in high urea rations may be just as important as a source of glucose as amino acids to ruminants.

My remarks about blood meal were intended to indicate perhaps that some at least of this material should be fed to cattle rather than pigs or poultry.

Mr Woo: In Singapore we are being forced to use cassava for pig starter rations due to the current high prices of cereals. I would like Dr Müller to comment on the use of cassava in pig rations. I would also like Dr Preston's comments on the use of 'dried molasses' in pig and poultry rations.

Dr Müller: I do not think there is any problem in using cassava as an ingredient of pre-start and starter rations for pigs. It has been shown by French workers that cassava meal can be included at up to 40% in these diets with good results.

Dr Preston: I can see little positive value in drying molasses, and I consider that this is mainly for the purpose of merchandising.

The use of waste bananas for swine feed

H. Clavijo

Instituto Nacional de Investigacions Agropecurias (INIAP), Ecuador

and

J. Maner

Centro Internacional de Agricultura Tropical (CIAT), Colombia

Summary

Reject or waste bananas (*Musa sapientum L.*) which contain an average of 20% DM, 1.0% crude protein, 1.0% fibre, 0.17% ether extract, 0.08% calcium and 0.28% phosphorus have been used to provide the major source of dietary energy for swine during the entire life-cycle. Properly supplemented with protein, vitamins and minerals, fresh ripe bananas can be used during all phases of the life-cycle of the pig except in lactation during which time the sow, because of limited gastrointestinal capacity, will not consume adequate quantities of fresh bananas to meet her energy needs. Because of bitter taste and poor palatability which significantly limit daily intake, fresh green bananas should not be fed if maximum voluntary consumption is required. Since ripe bananas have poor drying characteristics, banana meal was prepared from green bananas. Meal prepared in this manner can be used to supply up to 75% of the pigs' diet. During the growing-finishing period each increase in banana meal substitution for maize is associated with a small linear depression in growth rate and feed conversion efficiency. This depression in performance results from the reduced daily consumption of metabolisable energy which is only 3,200 kcal/kg DM for bananas as compared to 3,800 kcal/kg for maize. Performance equal to that obtained with control diets based on cereal grains has been obtained during gestation and lactation when banana meal replaced up to 50% of the diet.

Résumé

Les dechets de bananes comme pâture pour les cochons

Les bananes rejetées ou les déchets (*Musa sapientum L.*) qui contiennent une moyenne de 20% matière sèche, 1.0% protéine brute, 1.0% fibre, 0.17% extrait éthéré, 0.08% calcium et 0.28% phosphore ont été utilisées pour fournir la source majeure d'énergie diététique pour les cochons, durant le cycle entier de vie. Convenablement complétées avec protéines, vitamines et minéraux, les bananes fraîches, mûres. peuvent être utilisées pendant toutes les phases du cycle de la vie du cochon, exceptant pendant la lactation, quand la truie, à cause de sa capacité gastro-intestinale limitée ne peut pas consumer des quantités adéquates de bananes fraîches, pour satisfaire les besoins d'énergie. A cause du goût amer et de la saveur désagréable, qui limitent d'une manière significative la ration quotidienne, les bananes vertes, fraîches, ne doivent pas être données quand on a besoin d'une consommation volontaire maxima. Puisque les bananes fraîches ont des caractéristiques mauvaises déssicatives, les pâtures de bananes ont été préparées avec des bananes vertes. Les pâtures préparées de cette manière peuvent être utilisées pour fournir jusqu'à 75% de la diète du cochon. Pendant la période finale de croissance, chaque augmentation de la ration de pâture de bananes substituant le mais, est associée avec une petite dépression liniaire du rythme de croissance et efficacité dans la conversion de la pâture. Cette dépression du rendement résulte de la consommation quotidienne réduite d'énergie métabolisable qui est seulement de 3,200 kcal/kg MS pour les bananes, comparée avec 3,800 kcal/kg pour le mais. Un rendement égal à celui obtenu avec des diètes-témoins, basées sur des graines de céréales a été obtenu pendant la gestation et la lactation, quand les pâtures de bananes ont remplacé jusqu'à 50% de la diète.

Resumen

El uso de plátanos desechados en la alimentacion de los cerdos

Los plátanos rechazados o desechados (*Musa sapientum L.*) que contienen un promedio del 20% de materia seca ((DM) = dry matter)), el 1% de proteína en bruto, el 1% fibra, el 0,17% de extracto de éter, el 0,08% de calcio y el 0,28% de fósforo han sido usados para suministrar la fuente principal de energía dietética para cerdos durante el ciclo de vida completo. Se pueden utilizar plátanos frescos maduros suplementados adecuadamente con proteínas, vitaminas y minerales durante todas las fases del ciclo de vida del cerdo excepto en la lactancia durante cuyo tiempo la cerda, debido a una capacidad gastrointestinal limitada, no

100 consumirá cantidades adecuadas de plátanos frescos para satisfacer sus necesidades de energía. Debido al sabor amargo y el mal gusto que limitan de manera significativa la toma diaria, no se debe alimentar con plátanos frescos verdes si se requiere un consumo voluntario máximo. Puesto que los plátanos maduros tienen características de secado pobres, se preparó la harina de plátano con plátanos verdes. La harina preparada de esta manera puede usarse para suministrar hasta el 75% de la dieta de los cerdos. Durante el período final del crecimiento cada aumento en la sustitución de maíz por harina de plátano se asocia con una pequeña depresión lineal en el tipo de crecimiento y en la eficacia de la conversión del pienso. Esta baja en el resultado se origina a causa del consumo diario reducido de energía metabolizable que es únicamente de 3.200 kcal/kg de materia seca (DM) para los plátanos en comparación con 3.800 kcal/kg para el maíz. Se ha conseguido rendimiento igual al obtenido con dietas de control basadas en granos de cereales durante la gestación y lactancia cuando se reemplazó hasta el 50% de la dieta con harina de plátano.

Introduction

The continually rising demand for food with which to feed an ever-increasing human population makes it imperative that every available feed resource be utilised in the most efficient manner to overcome and prevent hunger. Within the tropical environments there are many feedstuffs that could contribute greatly to increased protein production if properly utilised in animal agriculture. Reject bananas are exemplary of this unrealised potential.

The banana belongs to the genus *Musa*, comprising 32 or more distinct species and at least 100 subspecies. The standard varieties of bananas handled in the trade belong to *M. sapientum* L. (Gros Michel) and *M. cavendishii* (Cavendish). The plantain (*M. paradisiaca* L.) or starchy, cooking banana is used locally but seldom enters into international trade.

Although bananas and plantains are grown largely for export and domestic consumption by the human population, large quantities of this fruit are available for livestock feed. At the packing plants in areas of commercial banana production, the bananas that are too small or too large, slightly bruised, have spots or off colour, or are not in an optimal stage of maturity for shipping are rejected for export. The 'reject' or waste bananas along with smaller quantities of farm produced bananas and plantains constitute a good source of carbohydrates for livestock.

It is estimated that a world total of more than 28 mt of bananas is produced annually. Of this production, more than 65% is produced in Latin America, 25% in the Far East and 7% in Africa. The exact quantity of the total production available for livestock is not known, but depending upon seasons, supply, demand and peculiarities of the market, the total amount available for uses different from human consumption may represent as much as 50% of total production within a country or region. This large percentage of unmarketable fruit probably represents the extreme under very poor marketing conditions and a figure of 25 to 30% would probably better represent the true quantity. Data from Costa Rica (Anon, 1963) indicate that only 68% of all fruit produced is marketed. The remainder could be used for home consumption and for livestock feed. Even if only 20% of the world production were available for livestock feed, this would represent more than 5.5 mt of fresh bananas or approximately 1.1 to 1.3 mt of air-dried material.

The fresh whole banana with the peel contains approximately 80% water, 20% dry matter, 1.0% protein, 1.0% fibre, 0.2% fat, 1.0% ash and 16.8% nitrogen-free-extract. The crude fibre of whole bananas consists of 60% lignin, 25% cellulose and 15% hemicellulose. The ripe pulp contains 0.50% lignin, 0.21% cellulose and 0.12% hemicellulose (von Loesecke, (1950)). Although the more commonly grown commercial varieties of bananas, Gros Michel and Cavendish, conform to this generalisation as to chemical composition, locally grown bananas and plantains vary in their chemical make-up (Bressani *et al* (1961)).

Fresh bananas

The banana can be utilised fresh or as a dried meal. The degree of ripeness of fresh bananas greatly affects the results obtained when this fruit serves as the major source of energy in growing/finishing swine rations. Ecuadorian studies (Hérnandez & Maner, pers. comm.) have clearly demonstrated that the pig will daily consume large quantities of bananas if they are allowed to ripen before they are fed (Table 1). If they are fed green in combination with a 30% protein supplement, the pig will voluntarily consume only about 50% (8.85 vs 4.25 kg per day) as much of green bananas as ripe bananas. These animals partially compensate for this reduced banana consumption by increasing their intake of protein supplement. The net result of an overconsumption of protein supplement and a low consumption of green bananas was to depress daily consumption of air-dried feed which reduced the growth rate.

Cooking of the green bananas significantly improved both banana consumption and pig performance; however, even this process failed to improve the performance of the pigs to a level obtained when ripe bananas were fed.

Viteri *et al*. (pers. comm.) demonstrated that when growing pigs were fed equal daily quantities of either ripe or green bananas along with a controlled quantity of protein supplement, growth rate and efficiency of

TABLE 1. *Performance of growing/finishing pigs fed ripe, green or cooked green bananas[a]*

Treatments Parameters[b]	1 Control Maize + supplement	2 Ripe	3 Green	4 Cooked-green
		30% Protein Supplement + Bananas		
Av. daily gain, (kg)	0.68[a]	0.56[b]	0.46[c]	0.50[c]
Av. daily feed, (kg)				
Bananas, (kg)	–	8.85[a]	4.25[c]	6.20[b]
Supplement, (kg[c])	–	0.71[a]	1.04[c]	0.88[b]
Total dry feed, (kg)	2.31[a]	2.48[a]	1.89[b]	2.13[a]
Feed/gain	3.41[a]	4.44[b]	4.16[b]	4.26[b]

Source: Hérnandez & Maner, (pers. comm. 1965)

[a]Eighteen pigs per treatment in two replications of nine pigs per group. Av. initial weight 28.5 kg. Each group removed from experiment when obtained average weight of 92.0 kg.

[b]Means in the same line with different superscripts are significantly different (P<0.05).

[c]30% protein supplement composed of fishmeal, cottonseed meal, maize, vitamin, minerals and antibiotics.

feed conversion were almost identical for both groups (Table 2). Complementary studies (Clavijo and Maner, Viteri and Maner, pers. comm.) have shown that the level of consumption of ripe and green bananas is associated basically with differences in palatability of the two forms of presentation. These data were confirmed by Clavijo and Maner (1973) who showed that the digestion coefficients of green and ripe bananas are not different.

TABLE 2. *Performance of growing pigs fed equal quantities of either green or ripe bananas*

Parameters[a]	Treatment	
	Ripe bananas	Green bananas
Initial weight, (kg)	31.9	32.1
Final weight, (kg)	61.1	60.3
Av. daily gain, (kg)	.463	.449
Av. daily cons. (kg)	1.05	1.05
Av. daily banana, (kg)	2.96	2.72
Feed/gain[a]	3.55	3.55

Source: Viteri, Oliva & Maner, (per. comm., 1971).

[a]Parameters not statistically different (P> 0.05).

The difference in palatability of ripe and green bananas is immediately obvious to the consumer. The green banana is dry and bitter to the taste while the ripe banana is more moist and sweeter. Palatability depends on the chemical composition of the fruit. Many chemical changes occur within the banana during the ripening process which greatly affect palatability.

A feature of the unripened banana is its strongly astringent taste. This taste is the result, in part, of the presence of tannins in the fruit. It has been suggested that the total amount of tannins in bananas remains practically constant during ripening and that the loss or reduction of astringence is associated with a change in the state or chemical form of the tannins (Von Loesecke, 1950). It is suggested that tannins exist in the banana in two forms, (a) 'free' or active tannins which impart a strong bitter taste to the fruit and (b) 'bound tannins' or 'vegetable tannates' which are insoluble and supposedly inert and which have little or no effect on palatability. During the ripening

process, the level of free tannin decreases because the tannins are slowly bound in an insoluble form. The level of free or active tannin is much higher in the peel than in the pulp but both are significantly reduced by the time the fruit has matured to an 'eating-ripe' state (Table 3) (Von Loesecke, 1950). Since the quantity is greater, the reduction in active tannin of the peel is much greater than in the pulp.

TABLE 3. *Changes in amount of 'active' tannin in the pulp and peel of bananas during the ripening process expressed as units per 100g of tissue*

Days	Fruit condition	Pulp	Peel
0	Green	7.36	40.5
1	Green	8.01	34.0
2	Green	7.57	28.3
3	Green	4.30	25.4
4	Green	5.02	25.9
5	Coloring	4.30	16.5
6	Coloring	3.87	18.1
7	Coloring	1.95	11.2
8	Eating-ripe	2.84	4.6
9	Eating-ripe	1.99	4.7
10	Over-ripe	2.00	4.5
11	Over-ripe	1.32	3.5

Source: Von Loesecke (1950).

Although the water content of the pulp is increased during the 10 to 11 days required for the banana to ripen to the eating stage, the most conspicuous change in the maturation of the banana is the conversion of starch to sugar (Table 4) (Stratton and Von Loesecke, 1930). During ripening there is a decline in starch content and a corresponding increase in sugar content. There is a gradual shift in the carbohydrate fraction from almost all starch to almost all sugar. In the banana, 10 to 11 days are required for this shift. Because of respiration needs there is a small but definite decrease in total carbohydrates in all varieties during ripening.

The majority of the sugars are present as sucrose, glucose and fructose, although maltose has been identified to be present in trace amounts. The predominant sugar in bananas is sucrose, some of which

TABLE 4. *Changes in the starch, sugar and total carbohydrate content of banana and plantain pulp during the ripening process expressed as percentage of fresh pulp*

	Number of ripening days							
Variety	0	3	5	7	9	11	14	17
Gros Michel								
Starch	20.65	12.85	6.00	2.93	1.73	1.21	–	–
Total sugars	0.86	7.66	13.76	16.85	16.87	17.91	–	–
Total carbohydrates	21.51	20.49	19.72	19.78	18.60	19.62	–	–
Plantain								
Starch	32.20	31.68	30.90	30.48	28.52	20.17	11.69	6.12
Total sugars	0.82	0.85	1.02	0.92	3.84	9.78	18.89	21.10
Total carbohydrates	33.02	32.53	31.92	31.40	32.36	29.95	30.58	27.22

Source: Stratton and Von Loesecke, (1930).

TABLE 5. *Average distribution of sugars in the banana at various stages of ripeness of Gros Michel bananas*

Stage of Ripeness		% of fresh pulp			% of total sugars		
Days	Colour	Glucose	Fructose	Sucrose	Glucose	Fructose	Sucrose
0	Green	2.24	1.45	6.65	19.24	12.46	68.30
3	Very slightly green	3.09	2.50	10.61	19.07	15.43	65.49
6	Green tip	3.99	2.75	12.00	21.29	14.67	64.03
9	Slightly speckled	4.21	3.24	12.08	21.56	16.59	61.85

Source: Adapted from Von Loesecke (1930).

TABLE 6. *Performance of growing/finishing pigs fed ripe bananas and either a 30 or 40% protein supplement free choice*[a]

		Treatments		
	Control	Banana	+	supplement
Parameters[b]	1	2		3
Protein in concentrate, %	16	30		40
Av. daily gain, kg[c]	0.87[a]	0.77[b]		0.66[c]
Av. daily fresh bananas, kg	–	8.29		8.85
Av. daily bananas, kg, DM[d]	–	1.84		1.97
Av. daily supplement, kg	–	0.82[a]		0.62[b]
Total daily air-dry feed, kg	2.64	2.66		2.59
Feed/gain	3.04[a]	3.47[a]		3.92[b]
Feed in mixture consumed, %	16.0	12.4		13.0

[a]Calles *et al.* (10)
[b]Means in the same line with different superscripts are significantly different (P<0.05)
[c]Seventy-two pigs, 4 replications of 6 pigs per treatment. av. initial weight 23.2 kg, av. final weight 90.1 kg.
[d]Bananas expressed on 10% dry matter basis.

is hydrolysed to glucose and fructose (Table 5) (Von Loesecke, 1950). Glucose accounts for approximately 58% of the total reducing sugars present and fructose for about 42%.

The carbohydrate values of the banana peel are much smaller in magnitude than those of the pulp, but their changes from starch to sugars follow a similar pattern to that demonstrated for the pulp.

Ripe bananas which are very palatable and readily consumed by the pig are fed whole with the peel. When offered fresh, ripe, whole bananas, the pig will first consume the banana pulp leaving much of the peel. If the quantity offered is in excess of its daily capacity, it will tend to consume more pulp and less peel. However, if the total daily quantity offered is controlled, it will consume both pulp and peel.

The low protein content and high moisture present in the banana require that a supplemental source of both protein and energy as well as vitamins and minerals be supplied. Several studies have been conducted to determine the voluntary consumption patterns and performance of growing finishing pigs and lactating sows fed ripe bananas and protein supplement containing varying levels of crude protein.

One such study (Calles *et al.*, (1970)) indicated that the average daily gain of growing finishing pigs fed whole ripe bananas to appetite was significantly improved (770 vs 600 g) when a 30% protein supplement was supplied instead of a 40% protein supplement (Table 6). The improvement in gain was assumed to be the effect of the increased daily consumption of metabolisable energy.

Although the level of protein supplement consumed daily does not vary greatly during the entire period of growing and finishing, as the pig increases in size and weight there is a marked increase in daily voluntary consumption of ripe bananas. This change in the daily consumption pattern is especially evident during the first two to three weeks of the feeding period and may be associated not only with the needs to adapt to taste changes but also to the need to develop a greater stomach capacity with which to handle the high moisture feed.

When allowed a voluntary choice of the quantity of both ripe bananas and either a 30 or 40% protein supplement, the pig will consume a diet during the growing finishing period that contains 12.4 to 13.0% crude protein. From the results of a number of studies it is evident that an average of 8.0 to 8.8 kg per day of ripe bananas is all that the growing finishing pig will consume. The young pig (25–30 kg) will consume as little as 5 to 6 kg of ripe bananas per day, and it appears that the finishing pig approaching market weight will not consume more than 10 to 11 kg daily. Because of this inadequate level of banana consumption, when a 40% protein supplement is fed energy becomes limiting. The increased consumption (820 vs 620 g) of a 30% protein supplement that supplies more carbo-hydrate and less protein calories partially corrects the energy deficiency.

Later studies from the same station (Clavijo, 1972) also showed that no further improvement in pig per-formance or efficiency of feed utilisation was obtained when a 20% protein supplement was used to replace the 30% protein supplement when ripe bananas were fed (Table 7).

Although pig performance was similar when either the 20 or 30% protein supplement was fed with bananas, the level of supplement consumption was increased and the daily consumption was reduced when the 20% supplement was used.

Under practical feeding conditions the recommended level of protein in the supplement will depend upon price relationships of bananas, proteins and grains or grain substitutes. Locally available energy sources such as maize, milo, rice bran, sugar and molasses can be efficiently used as energy sources for supplement preparation and dilution.

Fresh bananas can be used efficiently for sows during gestation. During this period the daily feed is con-trolled to meet the needs of the sow and maximum feed consumption is not required; therefore, either green or ripe bananas can be utilized. Clavijo et al. (1971) fed gestating sows maintained on pasture a ration of ripe bananas and protein supplement and compared the performance of these sows to similar sows fed a basal 16% protein diet based on maize, wheat, wheat bran, alfalfa meal and fish meal. The control sows were fed 1.5 kg of feed per day from the day of breeding until the 76th day of gestation. During the remainder of the gestation period and until the sows were removed to farrowing crates (76th to 110th day) daily feed was increased to 2.0 kg per day.

The sows receiving bananas were fed 4.5 kg of ripe bananas and 600 g of 40% protein supplement per day from the day of breeding until the 76th day of gesta-tion. The ration was then altered to supply 6.0 kg of bananas and 800 g of supplement from day 76 until the 110th day of gestation. The supplement employed was composed of 55.44% fishmeal, 20% maize, 12.34% alfalfa meal, 2.82% bonemeal and 9.4% vitamins, minerals and salt premix.

The reproductive performance of both groups of sows was not different. The number, weight and vigour of the pigs were similar for both treatment groups (Table 8). Sows from both treatment groups entered the farrowing crates in good condition. The banana fed sows gained an average of 11 kg more weight during gestation than the sows fed the control diet. After farrowing no differences in pig livability or perform-ance were observed.

Contrary to the limited daily feed intake required for good performance by the gestating sow, the lactating sow must consume 5 to 6 kg of good quality air-dried feed if she is to meet the nutrient demands for maintenance and milk secretion. If an equivalent quantity of dry matter were supplied by fresh bananas

TABLE 8. Performance of gestating sows fed a diet based on ripe bananas and a 40% protein supplement[a]

| | Treatments | |
Parameters[b]	Control 16%	Bananas + Supplement
Av. daily concentrate, kg	1.66	0.67
Av. daily bananas, kg	–	5.00
Av. daily air-dried feed, kg[c]	1.66	1.67
Av. daily crude protein, kg	.266[b]	.318[a]
Av. no of pigs per litter	8.9	8.4
Av. pig weight at birth, kg	1.22	1.26
Av. daily sow gain, (1–100 day), kg	26.08[b]	37.04[a]

Source: Clavijo et al. (1971).
[b]Means in the same line with different superscripts are significantly different (P<0.05).
[c]Calculated on basis of approximately 10% moisture.

TABLE 7. Comparison of 20 and 30% protein supplements for growing/finishing pigs fed supplement and ripe bananas free-choice[a]

| | | Treatments | |
| | | Bananas + | supplements |
Parameters[b]	16% protein	20% protein	30% protein
Av. daily gain, kg	.70[a]	.61[b]	.64[b]
Av. daily consumption of fresh bananas, kg	–	5.97[b]	7.37[a]
Av. daily consumption of supplement, kg	2.46[a]	1.91[b]	1.15[c]
Feed/gain[b]	3.52[a]	4.42[b]	4.27[b]

Source: [a]Clavijo (1972)
[b]Means in the same row without a common superscript are significantly different from each other (P<0.05).

and supplement, a sow would be required to consume at least 20 kg of bananas and 2.0 kg of a 40% supplement per day. This quantity appears to be greater than the sow's physical capacity.

Clavijo and Maner (1971) fed ripe bananas and a 40% protein supplement to lactating sows in the tropical banana zone of Ecuador and reported smaller and lighter litters when the sow's rations were based on ripe bananas (Table 9). The control sows voluntarily consumed an average of 3.66 kg per day of a 16% grain-supplement diet. The test sows were fed a mixture of bananas and supplement. In order to provide a 16% diet, the proportion of supplement to ripe bananas was 1.0 kg of supplement to 11 kg of bananas. The sows were allowed voluntary consumption of the mixture and consumed an average of 1.02 kg of supplement and 11.22 kg of bananas per day. Because of the inability of the sow to consume adequate fresh bananas to meet her energy needs and also because of the extremely laxative effect observed when the sow consumes a ration containing 14 to 15 kg of ripe bananas, fresh bananas are not generally recommended as the major energy source for the lactating sow.

Fresh plantains are similar to bananas in appearance, but their chemical composition is somewhat different. They have a higher dry matter content that is principally represented by a 10 to 12% increase in the level of carbohydrates in the plaintain. On the basis of their chemical analysis, similar or better results than those obtained with bananas might be expected; however, results from studies in Ecuador (Clavijo, 1972) where plaintains were compared to bananas (Table 10)

seem to indicate that the performance of growing finishing pigs fed plaintains was inferior to that of similar groups of pigs fed bananas.

The reason for the slower rate of growth, lower level of plantain consumption and depressed feed efficiency is not readily apparent, but may have been associated with the chemical stage of ripeness. As is shown in Table 4, compared to bananas six to eight days longer are required for the plantain to reach an ideal stage of ripeness. After 10–12 ripening days, only about one-third of the plantain starch has been converted to sugars whereas almost 90% of the banana carbohydrate at that time is in the form of sugar. If a similar pattern exists for the contents of tannins, then the low performance as well as reduced voluntary consumption of plantains might be explained.

Dried bananas

It is difficult to dry ripe bananas and plantains; however, the green fruit dries readily in the sun or in drying ovens. Once dried, the banana slices are ground to a form green banana meal. Banana meal prepared from whole, unpeeled green bananas should contain approximately 12% moisture, 4.3% protein, 2.8 fat, 3.0% fibre, 4.3% ash and 73.6% nitrogen-free-extract. Plantain meal from whole, unpeeled green plantains should contain approximately 10.0% moisture, 4.3% protein, 1.0% fat, 6.2% fibre, 4.5% ash and 74.0% nitrogen-free-extract.

Celleri et al. (1971) used dried green banana meal as a substitute for grain to supply 0, 25, 50 or 75% of the total diet for growing finishing pigs. The protein of the 16% protein diets was supplied by a combination of fishmeal and cottonseed meal. As shown in Table 11, each increase in level of dried banana meal in the diet caused a small but significant linear reduction in average daily gain, and a linear increase in daily feed intake and in feed required per unit of gain.

Similar studies (Oliva, 1970 and Oliva et al., 1971) employed, 0, 12, 24, 36 and 48% banana meal and confirmed the previous findings. Although daily feed intake was increased, a linear depression in both growth and efficiency of feed conversion was observed.

The reason for this linear depression in pig performance when increasing levels of banana meal replaced grain

TABLE 9. *Performance of lactating sows fed either a complete concentrate or ripe bananas plus a protein supplement*[a]

Parameters[b]	Treatments	
	Control	Banana + Supplement
Av. pigs per litter, no.	8.5	8.7
Av. weight at birth, kg	1.31	1.24
Av. pigs weaned, no.	6.3[a]	5.9[b]
Mortality, %	26.4	30.3
Av. daily concentrate, kg	3.66	1.02
Av. daily cons. banana, kg	–	11.22
Av. daily protein cons., kg	.586	.520
Weight loss of sows, kg	9.5[a]	11.3[b]

[a]Clavijo and Maner (1971).
[b]Means in same line with different superscript are significantly different (P<0.05).

TABLE 10. *Comparison of ripe bananas and plantains as the principle source of energy in diets for growing finishing pigs*[a]

Criteria[b] [c]	1 Control diet	2 30% protein supplement banana	3 30% protein supplement plantain
Av. daily gain, kg	.48[a]	.46[a]	.43[b]
Av. banana cons. per day, kg	–	3.8[a]	2.4[b]
Av. cons. concentrate per day, kg	1.89[a]	1.26[b]	1.25[b]
Feed/gain	3.63[a]	4.39[b]	4.69[c]

Source: Clavijo (1972)
[b]Ten pigs per treatment in two replications of five pigs per group. Average initial weight, 11.0 kg.
[c]Means in the same line with different superscript are significantly different (P<0.05).

TABLE 11. *Performance of growing/finishing pigs fed diets containing varying levels of green banana meal*

Treatments	1	2	3	4	5
Celleri et al.[a]					
Level of green banana meal, %	0	25	50	75	–
Av. days to slaughter, No.[c]	119	121	124	128	–
Av. daily gain, kg[c]	.67	.65	.63	.61	–
Av. daily feed, kg	2.45	2.54	2.54	2.55	–
Feed/gain[c]	3.66	3.88	4.04	4.19	–
Oliva et al.[b]					
Level of banana meal, %	0	12	24	36	48
Days on trial, No.[c]	126	126	128	131	143
Av. daily gain, kg[c]	.62	.60	.61	.59	.54
Av. daily feed, kg	2.62	2.59	2.78	2.82	
Feed/Gain[c]	4.24	4.35	4.36	4.48	5.23

Sources:[a]Adapted from Celleri *et al.* (1971) who utilised dehydrated, commercial banana meal pellets.
[b]Adapted from Oliva *et al.* (pers. comm.) who utilised sun dried, green banana slices for preparing meal.
[c]Significant linear response (P<0.05).

TABLE 12. *Coefficients of digestibility and digestible and metabolisable energy values of fresh and dried ripe and green bananas*[a]

	Fresh		Meal	
Digestibility	ripe	green	ripe	green
Dry matter, %	84.25	76.93	−50.52	83.63
Protein, %	−42.65	−102.00	−126.61	3.38
Crude fibre, %	78.01	56.98	39.40	78.35
Ether extract, %	32.40	−24.87	24.50	22.09
Nitrogen-free extract, %	92.43	92.74	68.60	92.51
Total digestible nutrients, %	81.51	83.13	57.39	80.94
Digestible energy, kcal/kg DM	3114	3119	1703	3207
Metabolisable energy, kcal/kg DM	2967	3141	1520	3173

Source: Clavijo and Maner (1971).

was elucidated in studies by Clavijo and Maner (1973). The results of these studies (Table 12) clearly demonstrated that although the fresh ripe and green bananas were equal in both digestible and metabolisable energy as was green banana meal, the level of metabolisable energy of all these types of presentation is inferior to that of maize (3,200 vs 3,800 kcal/kg of dry matter).

These studies further showed that ripe bananas will not dry at 60°C but because of the sugar content must be dried at a higher temperature. When treated in this manner, dry matter, nutrient and energy digestibility is severely reduced.

Two experiments with gestating sows (Oliva *et al.*, pers. comm.) provide evidence to show that banana meal can supply at least 40% of the total diet without affecting reproductive performance of the sow (Table 13). Sow weight gains, post-partum weight losses,

and pig numbers and weight were not different when fed either the 16% protein control diet based on grain or a similar test diet in which 40% green banana meal substituted for an equal quantity of grain.

Dried green banana meal can also be used to supply at least 50% of the diet of lactating sows. Studies by Ecuadorian workers (Hernandez and Maner, and Celleri *et al.*, pers. comm.) indicate that the performance of both sows and their offspring are not different whether fed a control 16% protein, lactation diet based on maize, wheat and fishmeal or a test diet in which 50% of the maize was replaced by a similar level of banana meal (Table 14). Both groups of sows consumed a similar level of total diet. In doing so, the

TABLE 13. *Reproductive performance of gestating sows fed diets containing green banana meal with peel*

Diets	1	2
Criteria[b]	Control	4% banana meal
Sow weight gain (1–110 days), kg	41.3	40.0
Post-partum weight loss, kg	26.3	27.6
Av. pigs born per litter, no.	8.9	9.0
Av. pigs weight at birth, kg	1.4	1.4

Source: Clavijo, (1972).
[b]Means not different (P>0.05).

TABLE 14. *Performance of lactating sows fed diets containing green banana meal with peel*

Treatments	1	2
Criteria[b]	Control diet	50–53% banana meal
Number of sows	24	24
Av. number of pigs per litter, no.	9.2	9.5
Av. birth weight, kg	1.1	1.1
Av. number of pigs weaned, no.	7.1	7.3
Av. weight at weaning, kg	11.5	11.4
Av. sow gain, kg	2.1[a]	7.6[b]
Av. daily feed intake, kg	5.8	5.8

Source: Clavijo (1972).
[b]Means in the same line with different superscripts are significantly different (P<0.05).

banana meal fed sows consumed less digestible energy and, as a consequence, lost an average of 7.6 kg body weight each during the 56-day lactation while the control sows gained an average of 2.1 kg.

These studies provide information to indicate that fresh bananas can adequately provide the major source of energy for pigs during the growing/finishing and gestation periods of the life-cycle if properly supplemented. However, because of the high moisture content which prevents adequate energy consumption, it is not recommended as the sole source of energy for lactating sows.

The poor palatability of fresh green bananas, which significantly limits daily consumption, prevents the efficient use of green fruit in swine rations. Banana meal is prepared from the green fruit. However, because of its lower content of metabolisable energy as compared to grains for which it is substituted, it supports a lower level of pig performance.

References

Anonymous (1963) Censos Agropecuarios de Costa Rica.

Bressani, R., Aguirre, A., Arroyave, R., & Jarquin, R. (1961) La composición química de diversas clases de banano y el uso de harinas de banano en la alimentación de pollos. *Turrialba* 11 (4) pp 127.

Calles, A., Clavijo, H., Hervas, E., & Maner, J. H. (1970) Ripe bananas (*Musa*) as energy source for growing finishing pigs. *J. Animal Sci.* 31: 197 (Abstr.)

Celleri, H., Oliva, F. & Maner, J. H. (1971) Harina de banano verde en raciones de cerdos en crecimiento y acabado. *ALPA Mem.* 6: 148 (Abstr).

Celleri, H. *et al.* Unpublished data.

Clavijo, H., Maner, J. H., & Calles, A. (1971) Banano maduro en dietas para ceraos en gestación. *ALPA Mem.* 6: 146 (Abstr.)

Clavijo, H., & Maner, J. H. (1971) Banano maduro endietas para cerdos en lactancia. *ALPA Mem.* 6: 147 (Abstr).

Clavijo, H. (1972) Utilización de banano y platano en alimentación de cerdos. *Seminario sobre Sistemas de Producción de Porcinos en América Latina.* Cali, Colombia, Sept. 18–21, 1972.

Clavijo, H., & Maner, J. H. (1973) Factores que afectan la digestibilidad y el valor energético del banano para cerdos. *IV Reunion de ALPA.* Guadalajara, Mexico, June 26, 1973.

Clavijo, H. & Maner, J. H. Unpublished data.

Hernández, F. & Maner, J. H. (1965). Unpublished data.

Hernández, F. & Maner, J. H. Unpublished data.

Oliva, F. (1970) Evaluación de la harina de banano verde con cascara, en crecimiento y acabado de cerdos en confinamiento. Thesis. Quito Faculted de Ingeniería, Agronomía y Medicina Veterinaria, Universidad Central, Quito, Ecuador.

Oliva, F., Viteri, J., Calles, A., & Maner, J. H., (1971) La harina de banano verde con cascara como reemplazo del maíz para cerdos en confinamiento durante el periodo de crecimiento y engorde. Guía de Alimentación, INLAP, Quito, Ecuador.

Oliva, F. *et al.* Unpublished data.

Stratton, F. C., & Von Loesecke, H. W. (1930) A chemical study of different varieties of bananas during ripening. United Fruit Co., Research Dept. Bul. 32.

Viteri, J., Oliva, F. & Maner, J. H. (1971). Unpublished data.

Viteri, J. & Maner, J. H. Unpublished data.

Viteri, J. *et al.* Unpublished data.

Von Loesecke, H. W. (1950) Bananas. Chemistry, Physiology, Technology.

The use of coffee processing waste as animal feed

R. Bressani, M. T. Cabezas, R. Jarquín and Beatríz Murillo
Institute of Nutrition of Central America and Panama (INCAP), Guatemala

Summary

Processing of coffee cherries yields two by-products viz coffee pulp (CP) and coffee hulls (CH). About 1 m/t of dry CP and 500,000 t of dry CH are available in Latin America. Dry CP contains about 10% of crude protein, 21% of crude fibre and 45% of nitrogen-free extract. CP contains about 77% of mositure when fresh. CH are much poorer in nutritive value containing less than 3% of crude protein, 70% of crude fibre and 19% of nitrogen-free extract on a dry basis. The essential amino acid content/g N of CP is similar to that of soya protein, lysine being high while total sulphur amino acids are relatively low. Dry CP contains 1.2% of caffeine and 3.4% of tannins while potassium content is also high.

Trials with young and adult ruminants fed levels of dry CP above 20% of the total ration showed decreased weight gain, feed efficiency, nitrogen retention, and dry matter and protein digestibility. Caffeine has been shown to be responsible for this when its level in the diet surpasses 0.24%, but tannins fed in amounts equal to those present in the ration when 30% of dry pulp was included were without effect. Animals seem to gradually adapt to CP consuming increasing amounts with time. Although sun dehydration and ensiling decrease caffeine and tannin content animal performance does not correspondingly improve. High free consumption has not yet been possible.

Feeding of CP to pigs resulted in good performance when included in proportions of up to 24% of the total ration, results being similar to those for ruminants. Despite its negative effects CP is already being used as animal feed in Central America. Its wider use in this way would represent a significant gain to the coffee industry.

Résumé

L'usage des déchets du traitement du café comme pâture pour les animaux

Le traitement des drupes de café donne deux sous-produits, nommément la pulpe du café (PC) et les cosses du café (CC). Environ 1 million de tonnes de PC sèche et 500,000 de CC sèches sont disponibles dans l'Amérique Latine. La PC sèche contient environ 10% de protéines brutes, 21% de fibre brute et 45% d'extrait exempt d'azote. La PC, quand fraîche, contient environ 77% d'humidité. Les CC sont beaucoup plus pauvres en valeur nutritive, contenant moins de 3% de protéines brutes, 70% de fibre brute et 19% d'extrait exempt d'azote, sur une base sèche. Le contenu en aminoacides essentiels/g N de la PC est similaire à celui des protéines de soja, la lysine se trouvant en grande quantité, tandis que le total d'aminoacides sulfurés est rélativement bas. La PC sèche contient 1.2% de caféine et 3–4% de tanins, tandis que le contenu en potassium est aussi très élevé. Des essais avec des ruminants jeunes et adultes, recevant des proportions de PC sèche dépassant 20% de la ration totale, ont montré une diminution de poids gagné, d'efficacité des pâtures, de rétention d'azote et de digestibilité de la matière sèche et des protéines. On a trouvé que la caféine était responsable pour ces choses, quand son niveau dans la diète dépassait 0.24%, mais les tanins en quantités égales à celles présentes dans la ration, quand 30% de pulpe sèche était incluse, étaient sans effet. Les animaux avec le temps paraissait s'adapter graduellement à des quantités augmentées de PC consumées. Quoique la déshydratation par le soleil et la mise en silo décroît le contenu en caféine et tanin, le rendement des animaux ne s'améliore pas d'une manière correspondante. Une grande consommation libre n'a pas été encore possible.

Le nourrissage des cochons avec la PC résulta dans un bon rendement quand incluse dans des proportions jusqu'à 24% de la ration totale, les résultats étant similaires à ceux des ruminants. En dépit de ses effets négatifs, la PC est déjà utilisée comme pâture dans l'Amérique Centrale. Son usage étendu de cette manière pourrait représenter un gain significatif à l'industrie du café.

El uso de los desechos de la elboración del café como piensos para animales

El tratamiento a que se somete a las cerezas del café deja dos productos secundarios, a saber, la pulpa del café (CP) y las cáscaras del café (CH). En América Latina se dispone de alrededor de 1 millon de toneladas de CP seca y de 500.000 toneladas de CH secas. La CP seca contiene alrededor del 10% de proteína en bruto, el 21% de fibra en bruto y el 45% de extracto libre de nitrógeno. La CP contiene alrededor del 77% de humedad cuando está fresca. Las CH son mucho más pobres en valor nutritivo y contienen menos del 3% de proteína en bruto, el 70% de fibra en bruto y el 19% de extracto libre de nitrógeno, sobre una base seca. El contenido/gN de aminoácido esencial de CP es similar al de la proteína de la soja, siendo alta la lisina al tiempo que el total de los sulfuroaminoácidos es relativamente bajo. La CP seca contiene un 1,2% de cafeína y 3–4% de taninos al tiempo que es también alto el contenido de potasio.

Pruebas con rumiantes jóvenes y adultos alimentados con niveles de CP seca por encima del 20% de la ración total, mostraron disminución en la ganancia de peso, en la eficacia de la alimentación, en la retención del nitrógeno y en la digestibilidad de la materia seca y de las proteínas. Se ha comprobado que la cafeína origina esto cuando su nivel en la dieta excede del 0,24% pero no produjeron efecto los taninos incluídos en el pienso en cantidades iguales a las presentes en la ración cuando se incluía el 30% de pulpa seca. Parece que con el tiempo los animales se van adaptando gradualmente a cantidades crecientes de consumo. Aunque la deshidratación por el sol y el ensilaje disminuyen el contenido de cafeína y de tanino, el desarrollo de animal no mejora en correspondencia. El alto consumo libre no ha sido todavía posible.

El suministro de CP a los cerdos produjo buenos resultados cuando se incluía en proporciones de hasta el 24% de la ración total, siendo similares los resultados a los de los rumiantes. A pesar de sus efectos negativos, la CP está siendo usada ya en América Central como alimento de los animales. Su uso más amplio de esta manera representaría un beneficio significativo para la industria.del café.

Introduction

Research on the utilisation of coffee pulp as an animal feed goes back some 30 years, at least, in Central America, where Choussy (1944) reported that cows consumed rations containing coffee pulp well, without adverse effect on milk production. On the basis of these observations, Choussy recommended the use of coffee pulp as a component of diets for milking cows.

Other reports have been published since then, on the use of coffee pulp as a feed for growing fattening cattle (Work *et al.*, 1946; Osegneda Jiménez *et al.*, 1970; Echavarría; 1947; Madden, 1948; Squibb, 1950; and Bolanas, 1953) and its nutritive value has been evaluated for sheep and goats (Rogerson, 1955; Van Severen & Carbonell, 1949). These studies suggest that coffee pulp has a good potential as animal feed but until very recently, its use has been very limited.

Even though the chemical composition of coffee pulp is well documented, its use is not very extensive for various reasons. One is the problem of the handling and direct use of the material as it is obtained from coffee processing plants. Furthermore, the relative abundance of other more stable feeds which are more easily obtained probably contributed in the past to the lack of interest in making use of this resource. A second reason is that there has not in the past been a continuous research effort on the problems coffee pulp feeding presents.

The need to eliminate the public health problems present methods of coffee pulp disposal are causing, feed availability pressures, and the need to increase efficiency in meat and milk production, has stimulated interest in utilising coffee processing by-products, particularly coffee pulp. The present paper describes this potential animal feed in terms of its chemical composition, nutritional value and processing needed to convert it into a component of animal feeds.

Description of the by-products of coffee cherries

(a) General

The two main coffee cherry by-products are of course well known in coffee producing countries, but for the benefit of those not familar with coffee production and processing it would be useful to briefly describe the processes by which these are obtained.

Coffee cherries are harvested in Latin America from late August to May, depending on the locality of the coffee plantations. They are picked from the plants when the colour of the cherry is dark red. After harvesting, the coffee cherries are dumped into a water tank to be washed with running water, which transports the cherries to the pulping machines. These remove the pulp from the beans which remain covered by mucilage and the hulls. The beans are then allowed to ferment for about 48 to 72 hours or are treated chemically to remove the mucilage; they are washed and partially sun-dried before being subjected to final drying in a revolving drum with hot air. The final step is to thresh the beans to remove the hulls.

The pulp is moved by water to fall on to a trailer, and is then thrown into rivers or used as organic fertiliser in the coffee plantation.

TABLE 1. *Distribution of coffee pulp, bran and beans in coffee cherries*

	Fresh weight (g)	Weight (%)	Moisture (%)	Dry weight (g)	%
Coffee cherry	1,000	100.0	65.5	345	–
↓ pulper					
Coffee pulp	432	43.2	77.0	99	28.7
+					
Coffee beans					
+ mucilage					
+ coffee hulls	568	56.8	56.0	250	72.2
fermentation					
and washing					
Mucilage	–	–	–	17	4.9
+					
Coffee beans					
+ coffee hulls	450	–	50.0	225	–
↓ dehulled					
Coffee hull	61	6.1	32.0	41	11.9
+					
Coffee beans	389	38.9	51.0	191	55.4

Source: Bressani *et al.* (1972).

The material balance as obtained in the laboratory is presented in Table 1. The process is described on the left of the table. The table includes the moisture content of the whole cherry and of parts. From 1,000 g of cherries, 432 g of coffee pulp are obtained which, on a dry weight basis, represents about 29% of the weight of the whole cherry. Further processing of the beans yields 61 g of coffee hulls or 41 g dry weight equivalent to about 12% of the cherry (Bressani *et al.*, 1972).

On the basis of the entire Latin American production, and using the figures in Table 1, it has been estimated that there are approximately 1.5 mt of dry coffee pulp available and 0.5 mt of hulls. Obviously, these quantities have some economic significance and should serve a useful purpose in the area.

(b) Chemical composition of coffee pulp

The gross chemical composition of coffee pulp has been described often enough and representative values from our laboratories (Bressani *et al.*, 1972) are shown in Table 2. Three groups of analyses are shown corresponding to fresh pulp, dehydrated pulp and a sample of the pulp as found two to three days after obtaining it from the cherry.

Attention is called to its high water content. For purposes of utilisation, this moisture level constitutes a problem in its handling, transportation, and direct use as an animal feed. When coffee pulp is dried, it leaves a material with about 10% crude protein, 21% crude fibre, 8% ash and 44% of nitrogen-free extract. These values change according to the variety of coffee, location and agricultural practices.

Results on the fractionation of the cellular wall and structural polysaccharides of coffee pulp are shown in Table 3. Cellular content of 63% indicates that the material has relatively high levels of nutrients, while the levels of lignocellulose, hemicellulose, cellulose and lignin show the product to be superior to various

types of feeds. Of the protein, 3% is found in a lignified form, and is probably not readily available (Murillo *et al.*, 1974).

TABLE 2. *Chemical composition of coffee hulls (%)*

	Fresh	Dehydrated	Naturally fermented and dehydrated
Moisture	76.7	12.6	7.9
Dry matter	23.3	87.4	92.1
Ether extract	0.48	2.5	2.6
Crude fibre	3.4	21.0	20.8
Crude protein			
N x 6.25	2.1	11.2	10.7
Ash	1.5	8.3	8.8
Nitrogen-free			
Extract	15.8	44.4	49.2

	Dry weight %
Total pectic substances	6.5
Reducing sugars	12.4
Non-reducing sugars	2.0

Source: Bressani *et al.* (1972).

TABLE 3. *Cellular wall constituents and structural polysaccharides in coffee pulp (g %)*

Cellular content	63.2
Neutral detergent fibre	36.8
Acid detergent fibre	34.5
Hemicellulose	2.3
Cellulose	17.7
Lignin	17.5
Lignified protein	3.0
Crude protein	10.1
Insoluble ash	0.4

Source: Murillo *et al.* (1974).

The mineral breakdown of the ash fraction is shown in Table 4. The Ca/P ratio is much toward calcium; however, the availability of either one is not known. Attention should be drawn to the high potassium concentration which could very well have nutritional implications. Levels of minor elements are quite low (Bressani *et al.*, 1972).

TABLE 4. *Ash and mineral content of coffee pulp*

Component	Content
Ash (g %)	8.3
Ca (mg %)	554
P (mg %)	116
Fe (mg %)	15
Na (mg %)	100
K (mg %)	1,765
Mg	traces
Zn (ppm)	4
Cu (ppm)	5
Mn (ppm)	6.25
B (ppm)	26

Source: Bressani *et al.* (1972).

Finally, Table 5 summarises the average amino acid composition of the protein of two samples (Bressani *et al.*, 1972), together with those of other products for comparison purposes.

TABLE 5. *Amino acid content of coffee pulp protein (g/16g N)*

Amino acid	Coffee pulp	Maize	Soyabean meal	Cottonseed meal
Lysine	6.8	1.7	6.3	4.3
Histidine	3.9	2.8	2.4	2.6
Arginine	4.9	3.1	7.2	11.2
Threonine	4.6	3.3	3.9	3.5
Cystine	1.0	1.0	1.8	1.6
Methionine	1.3	1.6	1.3	1.4
Valine	7.4	5.0	5.2	4.9
Isoleucine	4.2	4.3	5.4	3.8
Leucine	7.7	16.7	7.7	5.9
Tyrosine	3.6	5.0	3.2	2.7
Phenylalanine	4.9	5.7	4.9	5.2
Hydroxyproline	0.5	–	–	–
Aspartic acid	8.7	–	–	–
Serine	6.3	–	–	–
Glutamic acid	10.8	–	–	–
Proline	6.1	–	–	–
Glycine	6.7	–	–	–
Alanine	5.4	–	–	–

Source: Bressani *et al.* (1972).

It is of interest to note the relatively high level of lysine present in coffee pulp protein which is as high as that found in soyabean meal protein. Coffee pulp protein is like that of soya deficient in sulphur-containing amino acids. At present, the biological availability of these amino acids is not known. This is an aspect which requires some investigation because of the relatively high tannin content of coffee pulp and the known fact that tannins may combine with protein to make its component amino acids biologically unavailable to the animal.

(c) Chemical composition of coffee hulls

This anatomical fraction of coffee cherries has received little attention as far as uses are concerned. A small part is utilised as fuel and the rest is discarded into coffee plantation fields where it slowly decomposes. Its decomposition is very slow because it is covered by a waxy film (of unknown chemical composition) which protects the cellulosic base of the hulls from breakdown. Table 6 summarises the chemical composition of coffee hulls compared to that of corn cobs and cottonseed hulls (Bressani *et al.*, 1972). Protein content is similar among the three by-products, while crude fibre is significantly higher in coffee hulls. The nitrogen-free extract content of coffee hulls is very low and its value as a feed is questionable. Table 7 summarises some results on the fractionation of the carbohydrates of coffee hulls. The lignin concentration is very high, as well as pentose and hexose content suggesting that with some chemical treatment the energy value of this potential feed could be increased (Murillo *et al.*, 1974).

TABLE 6. *Chemical composition of coffee hulls*

Component	Coffee hulls	Corn cobs	Cottonseed hulls
Moisture (%)	7.6	8.1	10.4
Dry matter (%)	92.8	91.9	89.6
Crude fat (%)	0.6	0.9	1.1
Nitrogen (%)	0.39	0.39	0.58
Crude fibre (%)	70.0	38.9	45.7
Ash (%)	0.5	1.6	2.5
Nitrogen-free extract (%)	18.9	48.1	36.7
Ca (mg)	150	765	160
P (mg)	28	274	80

Source: Bressani *et al.* (1972).

TABLE 7. *Fractionation of the carbohydrates of coffee hulls (%)*

Soluble carbohydrates, hexoses	0.45
Structural carbohydrates	
Pentoses	20.30
Hexoses	45.90
Lignin	24.40
Total	91.05
NFE + crude fibre	96.21

Source: Murillo *et al.* (1974).

Finally, Table 8 gives information on the cellular wall constituents and fractionation of the crude fibre of coffee hulls. The results show, as indicated before, that there might be some feed value in this by-product of coffee (Murillo *et al.*, 1974).

TABLE 8. *Cellular wall constituents and fractionation of the crude fibre of coffee hulls (%)*

Cellular contents	11.8
Neutral detergent fibre	88.2
Acid detergent fibre	67.5
Hemicellulose	20.7
Cellulose	44.5
Lignin	17.7
Insoluble ash	5.3

Source: Murillo *et al.* (1974).

Nutritional value of coffee pulp to ruminants

A material such as coffee pulp, the dry matter of which contains around 10% crude protein and less than 25% crude fibre, has considerable potential as a feed for ruminants.

The majority of studies on the nutritive value of coffee pulp have been carried out with the use either of sun-dehydrated pulp, or coffee pulp silage, with the objective to learn of its effects on growing and fattening cattle (Work *et al.*, 1946; Osegneda Jiménez *et al.*, 1970; Echavarria, 1947; Squibb, 1950; Bolañas, 1953; Jarquin *et al.*, 1973; Bara *et al.*, 1970). The results of these studies show that feed intake and average weight gain decreases in direct relation to the level of pulp in the diet. These effects are more marked when the level surpasses 20% as shown in Table 9. With Zebu-Criollo steers it has been reported (Fiores Recinos, 1973) that coffee pulp decreased dry matter intake even when fed in diets containing up to 24% protein. However, the effects of coffee pulp on weight gain were not as marked when included in diets with higher levels of dietary protein. Similar findings have been reported with rats as experimental animals (Bressani *et al.*, 1973).

TABLE 9. *Average weight gain and feed intake of 78 day old calves fed increasing levels of sun-dried coffee pulp in the diet**

	Coffee pulp, % in diet			
	0	10	20	30
Initial wt (kg)	90.5	89.6	89.2	90.5
Final wt (kg)	170.6	167.3	155.6	146.9
Av wt gain (kg/day)	0.95[a]	0.92[a]	0.79[b]	0.67[b]
Feed intake (kg/day)	5.9	5.9	5.3	4.5
Kg feed (kg weight gain)	6.2	6.4	6.7	6.7

* Duration of study: 84 days.
[a,b] Values with different letters are statistically significant ($P < 0.5$).

The results of studies carried out with lactating cows (Choussy, 1944; Work *et al.*, 1946) also showed that levels below 20% of dietary coffee pulp did not affect milk production. In view of these results, various cattle growers in Central America are already using small quantities of dehydrated pulp in the feeding of dairy cattle.

Coffee pulp silage could be an adequate solution to both the high moisture content problem and the seasonal production of pulp, since the resulting silage even after 4 months has very good organoleptic characteristics, particularly, if ensiled with molasses. However, after exposure to air, the coffee silage becomes dark in colour, with a concomitant decrease in palatability, which reduces the free intake by the animal. This dark coloration is probably due to polyphenols present in coffee pulp. This problem is being investigated since its solution could represent a significant advance in the utilisation of coffee pulp. A possibility under investigation for the purpose of improving the palatability is the production of a grass-coffee pulp silage.

Materials which control oxidation of polyphenol oxidase activity, are also under investigation. Results obtained so far show, however, that ensiling and dehydrating (Braham *et al.*, 1973) do not substantially modify the factors present in coffee pulp which adversely affect the animals, as indicated in Table 10. In this example, animal performance remained essentially the same for calves fed either type of coffee pulp preparation when compared to the control group.

Up to the present time, the factors responsible for the adverse effects are not known. Some investigators have attributed the adverse effects observed to caffeine (Choussy, 1944; Bressani *et al.*, 1973; Jaffé, 1952). Recent studies (Estrada, 1973) have revealed that coffee pulp contains relatively high levels of caffeine which is able to decrease feed intake and the growth of the animals. However, it was also found that possibly there are other factors which may act independently or in association with caffeine to produce such negative effects as those shown in Table 11. On the other hand, tannins did not have any effect on animal performance (Estrada, 1973).

Among these factors, one which could interfere with utilisation is the relatively low digestibility and feed efficiency of the nutrients present in coffee pulp. The available information is scarce and contradictory. The results of studies with goats (Van Severen & Carbonell, 1949) indicated digestibility coefficients of 33.44, 76.28, 76.39, 97.91 and 86.67% for crude protein, dry matter, nitrogen free extract, fat and crude fibre, respectively. On the other hand, with sheep (Rogerson, 1955) values from 7–13% were reported for the digestibility of crude protein, 48% for the NFE and 26% for crude fibre.

TABLE 10. *Calf performance when fed diets containing 30% sun-dried fresh and ensiled coffee pulp**

		Treatments	
	Control	30% dried coffee pulp	30% dried coffee pulp silage
Initial wt (kg)	143.8	144.4	144.5
Final wt (kg)	259.0	234.8	223.2
Av wt gain (kg/day)	1.37[a]	1.08[b]	0.94[b]
Feed intake (kg/day)	10.9	9.2	8.4
Feed efficiency (kg feed/kg wt gain)	7.9	8.5	9.0

* Duration of study: 84 days.
[a,b] Values with different letters are statistically significant.

TABLE 11. *Performance of 100 day old calves fed sun-dried coffee pulp or equivalent amounts of caffeine**

| | Control no coffee pulp | 30% dehydrated coffee pulp | Caffeine % of diet | |
			0.12**	0.24**
Initial wt (kg)	95.3	95.5	95.6	96.0
Final wt (kg)	215.0	195.1	215.1	191.2
Av. wt gain (kg/day)	1.21[a]	1.00[a]	1.21[a]	0.96[b]
Feed intake (kg/day)	8.22	7.35	8.10	6.78
Feed efficiency (kg feed/kg weight gain)	6.8	7.3	6.6	7.0

* Duration of study: 99 days.

** Amounts of caffeine equivalent to that in 30 and 60% coffee pulp (caffeine in coffee pulp: 0.4%).

[a,b] Values with different letters are statistically significant ($P < 0.05$).

More recent studies (Cabezas *et al.*, 1974) have shown that diets which contain 24% coffee pulp produce a significant decrease in nitrogen retention by young calves, resulting from an increased excretion of urine nitrogen in comparison with diets without coffee pulp as indicated in Table 12. The lower efficiency of nitrogen utilisation in coffee pulp diets was the result of increases in water intake and a large excretion of urine. It was considered that caffeine was responsible in view of its well known diuretic effect and that greater nitrogen loss in urine takes place upon high water intake (Bressani & Braham, 1964).

As with other feeds, ruminants tend to adapt to coffee pulp feeding which results in increased feed intake and greater weight gain after the animals have been fed for some weeks on coffee pulp diets (Cabezas *et al.*, 1974).

In view of the importance that adaptation to coffee pulp could have with respect to a more efficient use of its nutrients, studies have been carried out (Bolanos, 1953) to obtain more information on the effect of time and level of intake of coffee pulp on its utilisation by growing calves. In one study of 102 days duration, three adaptation periods lasting 34 days each were analysed. The results for weight gain are summarised in Figure 1. The overall results show that weight gain decreased when daily pulp intake increased, which confirm previous findings (Oseguda Jiménez *et al.*, 1970; Jarquin, 1973; and Bressani *et al.*, 1973). Figure 2 summarises feed intake results which are similar to those shown previously for weight gain. The behaviour of the animals in the three periods, indicated that in order to adapt them to consume higher intakes of pulp, it was necessary to initiate them on a very low intake. Under the conditions of this study the minimum level of intake was 2.8 kg per day when the diet contained 30% coffee pulp. From the results in the figures it will be noticed that lower amounts were not capable of adapting the animals to consume higher intakes.

The factors responsible for controlling the minimum amount of pulp necessary to adapt them to higher intakes are not known. Likewise, the changes occurring

TABLE 12. *Nitrogen balance of calves fed 0, 12 and 24% sun-dried coffee pulp in the diet*

| Coffee pulp | | | | Nitrogen balance | | |
% of diet	Intake	Faecel (mg/kg/day)	Urine	Absorption % of intake*	Retention % of intake	Urine volume (ml/day)
0	653	334	116	48.8	31.1[a]	2,652
12	650	346	125	46.8	27.5[a]	2,892
24	643	358	188	44.3	15.1[b]	3,832

* Apparent nitrogen digestibility.

[a,b] Values with different letter are statistically significant ($P < 0.05$).

Figure 1

Average weight gain of calves fed diets containing variable levels of coffee pulp

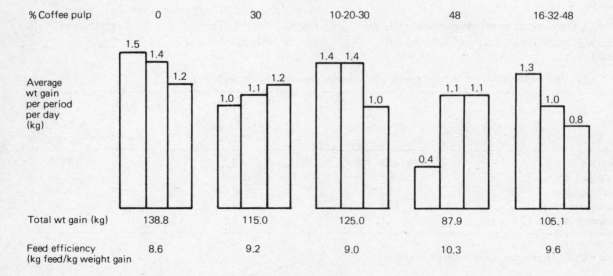

% Coffee pulp	0	30	10-20-30	48	16-32-48
Total wt gain (kg)	138.8	115.0	125.0	87.9	105.1
Feed efficiency (kg feed/kg weight gain	8.6	9.2	9.0	10.3	9.6

Figure 2
Feed intake of calves fed diets containing variable levels of coffee pulp

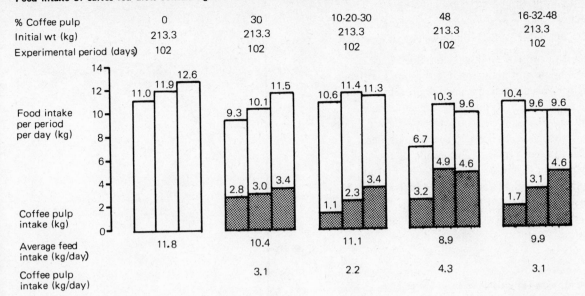

% Coffee pulp	0	30	10-20-30	48	16-32-48
Initial wt (kg)	213.3	213.3	213.3	213.3	213.3
Experimental period (days)	102	102	102	102	102

Average feed intake (kg/day): 11.8, 10.4, 11.1, 8.9, 9.9

Coffee pulp intake (kg/day): 3.1, 2.2, 4.3, 3.1

in the digestive and metabolic processes which contribute to such adaptation are unknown. This is an area where research is needed to make possible the increased and better use of coffee pulp in ruminant feeding.

Use of coffee hulls

As with coffee pulp, the use of coffee hulls in animal nutrition is not common. As shown previously, coffee hulls are characterised by a high level of crude fibre and in this respect they are similar to various other by-products used as fillers in animal feeds.

Cellulose as is well known can be utilised by ruminants as a source of energy, however, its utilisation is limited by the presence of lignin, silica and other compounds. Lignin content is around 24% and cellulose about 70% in coffee hulls. Fractionation of the cellular walls by the methods proposed by Van Soest (1967), showed that the cellular contents of coffee hulls amount to about 12%, while cellular wall components are as high as 67.5% with 17.7% lignin (Murillo *et al.*, 1974).

To increase the metabolic ulitisation of coffee hulls, it would be necessary to hydrolise the cellulose and similar compounds. Some preliminary results show that alkaline treatment increased soluble carbohydrates as shown in Table 13. Hydrolysis by 10% solutions of sodium or calcium hydroxide, reduced crude fibre from about 62.1% to about 34.5% (Murillo *et al.*, 1974).

TABLE 13. *Alkali treatment: effect on acid detergent fibre content of coffee hulls*

Treatment	Sodium hydroxide (%)			
	0	2.5	5.0	10.0
Sodium hydroxide	62.1	50.5	42.5	34.1
Calcium hydroxide	62.1	52.6	45.6	35.8

Source: Murillo *et al.* (1974).

Coffee hulls have been tested in ruminant diets and some results are summarised in Table 14. As shown, weight gain and feed efficiency were inversely related to level of coffee hulls in the diet. This type of performance was improved slightly by increasing energy levels in the diet with molasses (Jarquin *et al.*, 1974).

TABLE 14. *Calf performance when fed on diets containing coffee hulls*

	Treatment		
	Control	15% coffee hulls	30% coffee hulls
Initial wt (kg)	87.6	87.9	87.8
Final wt (kg)	183.9	177.9	168.4
Av. gain, (kg/day)	1.06	1.00	0.90
Feed intake, (kg/day)	6.15	6.20	5.76
Feed efficiency (kg feed/kg wt gain)	5.8	6.2	6.4

These results suggest, therefore, that coffee hulls may be used in ruminant diets and that hydrolysis could improve utilization. It can be effectively pelleted in spite of its physical form and chemical composition.

Coffee pulp in swine feeding

The use of coffee pulp for the purpose of feeding swine is not known. However its chemical composition suggests that it has good possibilities as a source of nutrients, although its major limitation is its high crude fibre content (as is well known, swine do not have the necessary microbiological and physiological conditions essential for the utilisation of high levels of fibre). The essential amino acid content of coffee pulp protein (Bressani *et al.*, 1972) which is similar to that of soyabean, suggests on the other hand, that it could be used to replace part of these protein sources, as well as the cereal grains in some diets with economic benefits.

114 Studies designed to learn of the possibilities of using coffee pulp (Rosales, 1973) in swine feeding were carried out with diets containing 18, 15 and 12% protein. These were fed to growing swine from 12–30, 34–60 and 65–90 kg body weight, respectively.

At each level of protein, the diets contained 0, 8.2, 16.4 and 24.6% of sun dehydrated and ground coffee pulp, which replaced equal amounts of protein from a basic soya-maize blend. Crude fibre was equalised between diets. The same animals were used throughout the study, although at each stage, the animals were randomised into new groups. The results of the study are summarised in Figure 3. It can be seen that at each growth stage, weight gain and feed efficiency were inversely related to coffee pulp level in the diet. The high coffee pulp level showed a significantly lower performance.

Figure 3

Performance of swine fed variable levels of coffee pulp

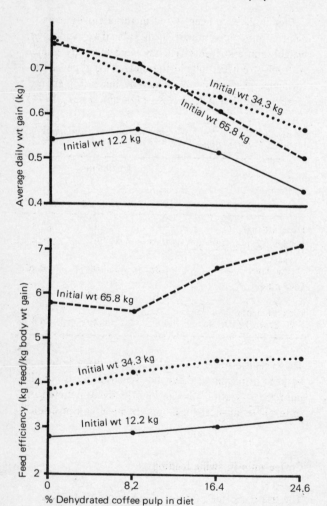

In this same study it was found that plasma-free fatty acids increased as coffee pulp in the diet increased. Other blood values remained essentially the same with respect to the control group.

Results of metabolic studies with swine are shown in Figure 4. As with other animal species used, nitrogen absorption and nitrogen retention decreased with respect to coffee pulp levels in the diet. Even though it was attempted to keep feed intake equal between groups, those consuming diets containing 24.6% pulp

Figure 4

Nitrogen balance of swine fed variable levels of coffee pulp

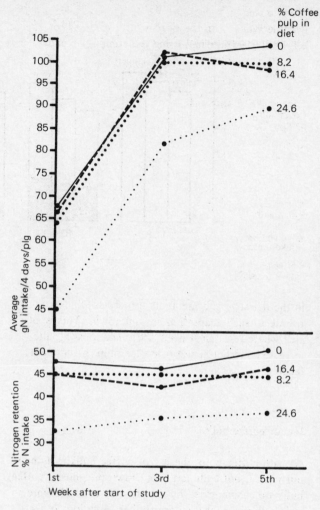

showed 14% less feed intake than other groups. This decrease is of course responsible, in part, for the lower nitrogen retention values obtained with the group receiving diets with higher levels of coffee pulp. Dry matter digestibility was also inversely related to coffee pulp in the diet. Na and K balances showed Na excretion to decrease and K to increase with respect to coffee pulp intake. In spite of these results, levels as high as 16% of coffee pulp in the diet could be used with good results.

Processing of coffee pulp

As indicated earlier, the main problem in the utilisation of coffee pulp is its high water content, which interferes in its transportation, handling, preservation and direct use as an animal feed. Some attention has therefore been given to overcome this problem and various approaches have been considered. The efficacy of any one of such approaches depends on the site and mode of use of the by-product. At the present time, two methods have been studied. One is by ensiling and the second is by dehydration, either by solar or by fuel energy.

(a) Ensiling

This method of coffee pulp preservation and utilisation is being studied extensively at INCAP, and results will become available in the near future. Ensiled coffee pulp has been prepared using the common practices for maize or other forages. The steps followed are shown in Figure 5. The coffee pulp containing

Figure 5
Steps in the preparation of coffee pulp silage

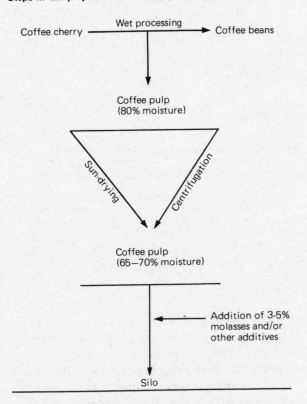

around 80% water when obtained from the coffee cherry processing plant is spread on a clean surface to dehydrate to a water content of about 65%. Under ideal conditions of sunlight and air, with occasional mixing, 35% dry matter content is achieved in about 8–10 hours. It is then packed into a trench silo with additions of 3–5% of sugar cane molasses. When the material is well packed into the silo, each cubic metre will contain 1,073 kg of coffee pulp (Gonzales, 1973). As the ensiled pulp is removed from the silo, it retains its original dark red coloration, but, becomes dark soon after. The gross chemical composition is

practically the same as for fresh coffee pulp but caffeine and total tannin content decreases significantly. In spite of these changes the nutritional value does not change as indicated by the results shown in Table 15 for coffee pulp ensiled for 4, and 14 months, fed to young ruminants as 30% of the total diet. Ideally, the ensiled coffee pulp should be offered without further processing, the animals intake being good when fresh but decreasing after a few days. To stimulate intake a mixed silage prepared from grass and coffee pulp is being tested. The material looks highly acceptable, but no feeding trials have so far been carried out. Results from other laboratories have, however, not been encouraging. This is an area where more research is needed, and studies are under way on the effects of various other additives besides sugar cane molasses, such as sodium metabisulphite, salt, urea or mixtures of these.

(b) Dehydration

Two methods of dehydration have been developed but tests have been carried out only on coffee pulp dried in the sun.

For sun dehydration, the material is spread on a clean surface to a thickness of about 5–8 cm. It is mixed about three to four times daily and in about 24 to 32 hours, moisture content is reduced to about 12%. It looks red and as dehydration proceeds it becomes almost black. This is the type of material used in our studies. It can be preserved at a moisture content of 12% indefinitely and results indicate that storage time decreases the antiphysiological effects at least with respect to mortality rates as tested in rats, shown in Table 16. In this particular study caffeine concentration decreased to about 0.45% which could very well be responsible for the lower mortality seen (Bressani et al., 1973).

Sun dehydration represents the cheapest way to dehydrate pulp. However there are various disadvantages, such as the large area needed to spread the pulp, the time involved, and the colour changes which are suspected to decrease intake by the animal.

A second approach to dehydration is by means of heated air or by contact with heated surfaces (Molina et al., 1974). Results of our studies with heated air

TABLE 15. *Performance of calves fed on coffee pulp silage*

Age of coffee pulp silage	Av. wt gain (kg/day)	Feed intake (kg/day)	Av. wt gain (kg/day)	Feed intake (kg/day)
	4 months		14 months	
Control	1.21*	8.2	1.19**	9.1
30% sun-dried coffee pulp	1.00*	7.3	0.98**	7.7
30% dehydrated coffee pulp silage	1.06*	7.1	1.08**	7.5

Source: Estrada (1973).
* Average initial weight: 95.4 kg.
** Average initial weight: 130.9 kg.

TABLE 16. *Effect of storage on caffeine content of coffee pulp and consequent weight gain and mortality in rats fed increasing levels of coffee pulp*

Storage time (months)	Caffeine (%)		% coffee pulp in diet					
			0	10	20	30	40	50
5	0.90	Final wt (g)	158	146	133	104	88	0
		Mortality (%)	0	0	0	0	60	100
9	0.80	Final wt (g)	151	166	163	144	118	0
		Mortality (%)	0	0	0	0	16.7	100
15	0.70	Final wt (g)	116	122	124	114	109	99
		Mortality (%)	0	0	0	0	0	25
17	0.46	Final wt (g)	123	113	128	116	97	67
		Mortality (%)	0	0	0	0	12.5	12.5

Source: Bressani *et al.* (1973).

indicate that given appropriate conditions coffee pulp can be satisfactorily dehydrated in 60 min (see Figure 6).

A modification of this method shown in Figure 6 involves grinding of the wet coffee pulp, followed by a continuous centrifugation step, which is optional, to reduce moisture, and then dehydration in a drum dryer. The material resulting from this type of processing has the best appearance of all products with some highly desirable characteristics. Typical figures for the composition of coffee pulp processed by different techniques are shown in Table 17. Sun dehydration and ensiling resulted in coffee pulp with lower concentration of caffeine and tannins when compared to the drum dried pulp.

Up to the present time no comparative studies have been carried out to determine which process gives the product with the least negative effects and higher nutritional value.

TABLE 17. *Partial chemical composition of coffee pulp processed by different techniques (%)*

	Sun-dehydrated	Ensilaged and dried	Drum-dried
Moisture	11.5	10.1	23.2
Dry matter	88.5	89.9	76.8
Protein (N x 6.25)	14.4	11.2	8.8
Caffeine	0.53	0.42	0.75
Tannins	2.94	1.47	4.06

A final way to process coffee pulp which is at present under initial study involves the inoculation of fresh coffee pulp with yeast or bacteria. At the moment, laboratory studies are being carried out with *Pencillium crustosum* which has been reported to utilise caffeine (Schwimmer & Kurtzman, 1972). Similarly, caffeine removal by successive water washings has resulted in a material practically free of this substance. The final outcome of such process awaits evaluation, both nutritional and economical.

Finally, some consideration is already being given to the economics of the use of coffee pulp. Although a more detailed analysis needs to be made the results of swine feeding in which 16% of the maize could be replaced by dehydrated coffee pulp indicate that there are potential savings in feed costs to be made. In this connection the cost of maize is assessed to be about five times that of dehydrated coffee pulp, and this further emphasises the need to further investigate the possibilities of using coffee pulp in animal feeding.

Figure 6
Steps in the dehydration of coffee pulp

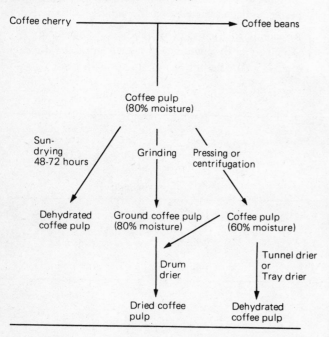

References

Bara, M., Espinosa, F. M., & Guevara, M. S. (1970) Determinación del nivel adecuado de pulpa de café en la ración de novillos. *Boletín informativo, Instituto Salvadoreño de Investigaciones del Café* **92** pp. 1–8.

Bolaños, J. R. (1953) La pulpa de café como alimento para ganado. *Café de El Salvador* **23** pp. 217–218.

Braham, J. E., Jarquín, R., González, J. M. & Bressani, R. (1973) Pulpa pergamino de café. III. Utilización de la pulpa de café en forma de ensilaje. *Arch. Latinoamer. Nutr.* 23 pp. 379–388.

Bressani, R. & Braham, J. E. (1964) Effect of water intake on nitrogen metabolism in dogs. *J. Nutr.* 82 pp. 469–474.

Bressani, R., Estrada, E. & Jarquín, R. (1972) Pulpa y pergamino de café. I. Composición quimica y contenido de aminoacidos de la proteina de la pulpa. *Turrialba* 22 pp. 299–304.

Bressani, R., Estrada, E., Elías, L. G., Jarquín, R., & Urrutia de Valle, L. (1973) Pulpa y pergamino de café. IV. El efecto de la pulpa de café deshidratada en la dieta de ratas y pollos. *Turrialba* 23, 403–409.

Cabezas, M. T., González, J. M. & Bressani, R. (1974) Pulpa y pergamino de café. V. Absorción y retención de nitrógeno en terneros alimentados con raciones elaboradas con pulpa de café. *Turrialba.*

Cabezas, M. T., Murillo, B., Jarquín, R., González, J. M., Estrada, E., & Bressani, R. (1974) Pulpa y pergamino de café. VI. Adaptacion, del ganado bovino a la pulpa de café. *Turrialba*

Choussy, F. (1944) La pulpa de café como alimento del ganado. *Anales del Instituto Tecnologico de El Salvador* 1 pp. 265–280.

Echavarría, G. (1947) La pulpa del café como alimento para ganado. *Revista Cafetera de Colombia* 8 pp. 3310–3313.

Estrada, E. (1973) Cafeína y taninos como factores limitantes en el uso de la pulpa de café en la alimentación de terneros. Thesis (Magister scientificae). Guatemala: Universidad de San Carlos de Guatemala/ INCAP.

Flores Recinos, F. (1973) Respuesta bio-económica de novillos en engorde alimentados con diferentes niveles de pulpa de cafe ensilada y proteína. Thesis (Magister scientificae). Turnalba: Instituto Interamericano de Ciencias Agrícolas, Centro Tropical de Enseñanza e Investigación, Departamento de Ganadería Tropical.

González, J. M. (1973) Boletín informativo de la División de Ciencias Agrícolas y de Alimentos del INCAP, VIII. Preparación de ensilaje de pulpa de café. *AGA* 16 pp. 16–19.

Jaffé, W. & Ortiz, D. S. (1952) Notas sobre el valor alimenticio de la pulpa de café. *Agro* 23 pp. 31–37.

Jarquin, R., Gonzalez, J. M., Braham, J. E. & Bressani, R. (1973) Pulpa y pergamino de café. II. Utilización de la pulpa de café en la alimentación de rumiantes. *Turrialba* 23 pp. 41–47.

Jarquín, R., Murillo, B., González, J. M. & Bressani, R. (1974) Pulpa y pergamino de café. VII. Utilización de pergamino de café en alimentación de rumiantes. *Turrialba.*

Madden, D. E. (1948) The value of coffee pulp silage as feed for cattle. Thesis (Magister scientificae). Turrialba: Instituto Interamericano de Ciencias Agricolas.

Molina, M. R., de la Fuente, G. & Bressani, R. (1974) Pulpa y pergamino de café. IX. Estudios sobre la deshidratación y procesamiento de la pulpa de café. (In preparation).

Murillo, B., Estrada, E. & Bressani, R. (1974). Pulpa y pergamino de café. VIII. Composición de carbo-hidratos estructurales y contenido de cafeína y taninos de pulpa y pergamino de café diferentes tratamientos. INCAP, Guatemala.

Osegueda Jiménez, F. L., Quiteño, R. A., Martínez, R. A. & Rodriguez, M. (1970) Uso de la pulpa de café seca en el engorde de novellos en confinamiento. *Agricultura de El Salvador* 10 pp 3–9.

Rogerson, A. (1955) Nutritive value of coffee hulls. *E. African Agron. J.* 20 pp. 254–255.

Rosales, F. (1973) Uso de la pulpa de café deshidratada en la alimentación de cerdos. Thesis (Magister scientificae). Guatemalai Universidad de San Carlos de Guatemala/INCAP.

Schwimner, S., & Kurtzman, R. H., Jr. (1972) Fungal decaffeination of roast coffee infusions. *J. Food Sci.* 37 pp. 921–924.

Squibb, R. L. (1950) Present status of dried coffee pulp and coffee pulp silage as an animal feedstuff. *Bulletin Instituto Agropecuario Nacional, Guatemala, C.A.*

Van Severen, M. L. & Carbonell, R. (1949) Estudios sobre digestibilidad de la pulpa de café y de la hoja de banano. *Cafe de El Salvador* 19 pp. 1619–1624.

Van Soest, P. J. (1967) Development of a compre-hensive system of feed analyses and its application to forage. *J. anim. Sci.* 26 pp. 119–128.

Work, S. H., Van Severen, M. L. & Escalón, L. (1946) Informe preliminar del valor de la pulpa seca del cafe, como sustituto del maiz en la ración de vacas lecheras. *Café de El Salvador* 16 pp. 773–780.

Discussion

Dr Osuji: I note that in the paper by Clavijo and Maner it was reported that liveweight gains by gestating sows fed on bananas were poor. I would like some comments on this, and could the cost of feed be related to the experimental gains achieved.

Dr Gomez: It is not good to allow gestating sows to gain too much weight as this would cause it to produce pigs that would be fat later on. For this reason I do not consider that performance during gestation was bad. One problem is, however, that we do not know what effect prolonged feeding over several gestation/lactation periods has on the sow. With regard to economics, in Ecuador fresh bananas cost about US $0.02/kg which means a dry matter cost of between US $0.08 and 0.10/kg, while maize costs about US $0.16/kg. It will be seen, therefore, that there is a considerable financial advantage in substituting bananas for maize in pig rations. Obviously the price of the product must also be taken into account as well. We have estimated that the banana-based rather than maize based rations will result in a saving of at least 15% on feed costs in Ecuador.

Dr Weselowski: I would like to ask if all the experiments reported in the paper by Clavijo and Maner were carried out on the unpeeled bananas? Also are the tannins present in the pulp or the peel, and is any banana meal currently being produced in South America on a commercial basis? The poor performance of the sow during lactation was probably due to her lack of physical capacity to consume the necessary amount of bananas to support her lactation. I would also like to ask Dr Bressani if he has considered removing or inactivating caffeine in coffee pulp.

Dr Gomez: Tannin content is normally a lot higher pulp in the peel than the pulp of the banana. However, by the 10th or 12th day of the ripening period there are only very low levels of free tannins in both pulp and peel. Dr Clavijo has carried out many experiments using peeled and unpeeled bananas, and there seems to be very little difference in nutritive value. Dr Clavijo is currently working on the problem of the lactating sow for which it will be necessary to only supply part of the energy requirement as bananas. I understand that there are one of two plants producing banana meal in Ecuador.

Dr Bressani: We are very much concerned with the effect of caffeine on animals fed with coffee pulp, and its removal is in our programme. We have found that washing with water will remove much of the caffeine, while we are also carrying out experiments to remove it by fermentation with a fungus, which can utilise caffeine as a nitrogen source. However, it should be mentioned that caffeine and chlorogenic acids are also responsible for antiphysiological effects and in fact we found better correlation between the content of these and toxicity rather than caffeine.

Dr Were: In developing countries the pulp of bananas is used for edible purposes and only the peel is a by-product. Has any consideration been to feeding the peel alone to pigs and is there any information on the feed value of banana leaves?

Dr Gomez: Dr Clavijo has mainly used export reject bananas (peeled or unpeeled) in his experiments. I agree that it would be interesting to find out what happens with the peel alone but no work has been done on this. I do not know of any work on the nutritive value of banana leaves.

Dr Topps: Has Dr Bressani any evidence to suggest that either coffee pulp or caffeine itself has an adverse effect on the rumen micro-flora when coffee pulp is fed to ruminants.

Dr Bressani: We are currently studying this but have no information to present at this stage.

120   *Mr Djoukam:* Do you have any information on the storage properties of dried coffee pulp, with particular reference to possible attack by pests?

Dr Bressani: Coffee pulp can be readily stored as silage, and we have now kept some for 18 months. Dehydrated coffee pulp stores well and is not attacked by insects or subject to moisture uptake.

Fourth Session

Fats and oils and miscellaneous materials for use as animal feed

**Tuesday 2nd April
Afternoon**

Chairman
Mr A. Sunkwa-Mills
Ministry of Agriculture
Accra, Ghana

The use of fats and oils as a source of energy in mixed feeds

H. L. Fuller

Department of Poultry Science, University of Georgia, USA

Summary

Since *energy* takes up most of the room in the feed bag the use of fat as a concentrated source of energy makes it possible to increase the density of mixed feeds thus reducing the total weight and volume of feed that must be transported, mixed, stored and handled.

When nutrient-energy ratios are held constant, feed efficiency is usually improved more than proportionally by the use of fats, depending upon the energy values employed, for several reasons: (a) it permits the animal to meet its energy needs with less energy expended in eating, (b) metabolisable energy values underestimate the net or usable energy value of fats because of their lower heat increment (HI), and (c) the associative dynamic effect of fats lowers the HI of the entire diet resulting in a further improvement in energetic efficiency of the diet, far greater than would be anticipated from using the sum of individual energetic values.

In experiments at the University of Georgia it was found that energy and nutrient intake could be increased during heat stress by replacing carbohydrate calories with fat calories and further by lowering total protein level of the diet. Both of these manipulations serve to lower HI, indicating that HI may serve to restrict feed intake to a greater extent than total dietary energy especially during heat stress. Growth rate was greatly improved by these dietary manipulations further suggesting that energy may be a limiting factor in the reduced growth rate associated with hot weather even while birds are forced to dissipate heat as heat increment.

Résumé

L'usage des graisses et des huiles comme source d'energie dans les pâtures mixtes

Puisque 'l'énergie' occupe le plus grand espace dans le sac de pâture, l'usage de la graisse comme une source concentrée d'énergie rend possible d'augmenter la densité des pâtures mixtes, en réduisant ainsi le poids total et le volume de nourriture qui doit être transportée, melangée, mise en dépôt et manipulée. Quand les proportions de nourriture-énergie sont maintenues constantes, l'efficacité des pâtures d'habitude est améliorée plus que proportionnellement par l'utilisation des graisses, dépendant des valeurs d'énergie utilisées et ceci pour plusieurs raisons: (a) permet à l'animal de satisfaire ses besoins d'énergie avec moins d'énergie dépensée en mangeant, (b) les valeurs d'énergie métabolisable sousestiment l'énergie nette ou utilisable des graisses à cause de leur quantité différentielle de chaleur plus réduite (heat increment (HI)) et (c) l'effet dynamique associatif des graisses diminue le HI de l'entière diète, résultant dans une amélioration supplémentaire de l'efficacité énergétique de la diète, beaucoup plus grande qu'il serait anticipé en utilisant la somme des valeurs énergétiques individuelles.

Dans des expériences faites à l'Université de Georgia, on a trouvé que l'énergie et les prises nutritives pouvaient être augmentées pendant les perturbations dues à la chaleur, en replaçant les calories de hydrate de carbone avec des calories de graisses et puis par la diminution du niveau des protéines totales de la diète. Ces deux manipulations servent à abaisser le HI, indiquant que le HI pourrait servir à restreindre la prise de nourriture dans une plus large mesure que l'énergie diététique totale, surtout pendant le stress dû à la chaleur. Le rythme de croissance était grandement amélioré par ces manipulations diététiques, suggérant de plus que l'énergie pourrait être un facteur limitant du rythme réduit de croissance associé avec le temps chaud, tandis que même les oiseaux sont forcés de dissiper de la chaleur comme quantité différentielle de chaleur.

Resumen

El uso de grasas y aceites como fuente de energía en los peinsos compuestos

Puesto que la energía ocupa el lugar principal en el pienso el uso de la grasa como fuente concentrada

124   de energía hace posible aumentar la densidad de los piensos compuestos, reduciendo de esta manera el peso y el volumen totales del pienso que debe ser transportado, mezclado, almacenado y manejado.

Cuando se mantienen constantes las proporciones de alimentos nutritivos—energía, la eficacia de la alimentación se mejora generalmente más que proporcionalmente mediante el uso de grasas, dependiendo de los valores de energía empleados, por varias razones: (a) permite al animal satisfacer sus necesidades de energía con menos gasto de energía al comer, (b) los valores de energía metabolizable subestiman el valor de la energía neta o utilizable de las grasas a causa de su aumento de calor más bajo (HI), y (c) el efecto dinámico de la asociación de las grasas reduce el HI de la dieta completa, produciendo una mejora más en eficacia energética de la dieta, mucho más grande que la que podría

esperarse del uso de la suma de valores energéticos individuales.

En los experimentos verificados en la Universidad de Georgia, se averiguó que la energía y la ración nutritiva podrían aumentarse durante incomodidad a causa del calor sustituyendo las calorías de hidratos de carbono por calorías de grasa y además bajando el nivel total de proteínas de la dieta. Ambas operaciones sirven para restringir la ración de pienso aún más que la energía dietética total, sobre todo durante las ocasiones de incomodidad por el calor. Se mejoró mucho el índice de crecimiento mediante estas operaciones dietéticas, indicando además que la energía puede ser un factor limitador en el índice de crecimiento reducido asociado con el tiempo cálido aún cuando los pájaros se ven obligados a disipar calor a medida del incremento del calor.

Fats serve a three-fold purpose in mixed feeds: (a) they provide the essential fatty acids; (b) they are a rich source of energy; and (c) they improve the physical characteristics, handling qualities and palatability of feeds. This presentation will be devoted primarily to the use of fats and oils as sources of energy, evaluating their energetic efficiency and their particular place in feeds under conditions of heat stress.

In the highly industrialised animal agriculture characterised by the broiler and feedlot industries, the performance of feeds can be evaluated realistically. Feed costs are considered only as a part of the ultimate cost of production of meat, milk or eggs. It then becomes economically advantageous to formulate feeds to specified energy/nutrient ratios, ie by specifying nutrient levels on a 'per megacalorie' basis rather than the familiar percent or 'per kg' basis. If the appropriate energy/nutrient ratios are specified, the density of the diet can be altered by simply changing the weight restriction in the formula. This makes it possible to use ingredients of greater nutrient and energy density without having to make up the difference in weight with a filler. In the simplified example, Table 1, 110 kg of maize have been replaced with 40 kg of fat and 20 kg of soyabean meal.

TABLE 1. *Replacement of one nutrient (maize) by one with a higher energy density (animal fat)*

Feed	1	2
	(kg)	(kg)
Maize	600	490
Soyabean meal	340	360
Mineral and vitamin mix	50	50
Animal fat	10	10
Total weight	1000	950
Metabolisable energy (mega calories)	3000	3000
Protein (kg)	230	230
Minerals and vitamins (kg)	50	50

By permitting total weight to float downward the same amount of energy and nutrients find their way into a smaller weight of feed. Since it is calories and nutrients that produce meat and eggs, the 950 kg package of feed (No. 2) will do the same job as the original 1,000 kg (No. 1) and this will be reflected in a saving in feed conversion of at least 5%. For a 1 m broiler operation, which is not uncommon in the broiler growing areas of the USA, a saving of 5% in feed conversion would amount to approximately 200 t of feed per growing cycle or 1,000 t of feed per year. Such a saving would be much more worthwhile in those areas of the world where a large part of the feed ingredient supply is imported, where a major part of the feed cost is represented by transportation, handling, storage and mixing operations in which weight and volume are important.

In this example, as in most feed formulation for poultry, metabolisable energy (ME) is used as the measure of calorific value since there is no other suitable set of values for describing all ingredients. Unfortunately, it does not fairly evaluate fats and oils. Forbes & Swift (1944), in studies with rats, demonstrated that fats have the lowest heat increment or 'specific dynamic effect' of the three major classes of nutrients. Carbohydrates were intermediate and proteins the highest. Furthermore, when the different classes of nutrients were fed in all combinations, the resulting heat increment was always less than the values calculated from the specific dynamic effect of the individual feedstuffs. Fats exerted the greatest sparing effect on the expense of energy utilisation of carbohydrates and proteins. In later work, Forbes *et al.* (1946) studied the effect of dietary fat levels on the heat increment of complete diets containing 2.5, 10 and 30% fat. The heat increment decreased in the order of increasing fat content. Since net energy is defined as metabolisable energy minus heat increment, then fat, with the lowest heat increment of three classes of nutrients, should have a much higher net energy value in relation to its metabolisable energy than do the other nutrients. This phenomenon has actually been demonstrated on a number of occasions.

Carew & Hill (1964) demonstrated that replacement of dietary carbohydrate with corn oil increases the metabolic efficiency of energy utilisation by chicks. Based upon this and other reports in the literature they suggested that the beneficial effect of corn oil on efficiency of energy metabolism is mediated through the heat increment component rather than the basal component of heat production. At that time, however, it was not clear whether this was a general property of well-utilised fats or due to an unidentified substance present in corn oil.

Jensen et al. (1970) described what they termed an 'extra calorific effect' of fat when animal fat was used in place of carbohydrate calories in diets for growing turkeys. The fat was substituted into the turkey rations isocalorifically based upon metabolisable energy (ME) values. Addition of fat-improved feed efficiency more than would be expected from the calculated ME values of the fat-supplemented diets. When the authors assumed that the same number of ME calories would be required per unit of gain, the added fat was calculated to have a ME value of 10,165 kcal/kg in contrast to the 7,700 kcal used to formulate the ration. Obviously, the higher of the two values exceeds gross energy values for fat and, therefore, would be unrealistic. However, such a value could be explained on the basis of the associative dynamic effect of the added fat in the various diets wherein the added fat has a sparing effect, resulting in greater efficiency of utilisation of the carbohydrate and protein calories of the diet as well as those provided by the fat itself. They stated that a much closer correlation between productive energy values and relative gain was observed.

In a similar experiment, Potter (1967) compared diets containing 2 or 8% added fat for growing turkeys. The high fat diet resulted in a 9.8% decrease in feed consumption, 2.5% increase in body weight gain and a 14% improvement in feed efficiency. This was equivalent to an improved feed efficiency of 2.3% for each 1% increase in added fat, whereas a 1.5% increase in efficiency would be expected based upon the extra metabolisable energy added in the form of fat. Thus the acceptance or rejection of fat as a feed ingredient based solely upon its relative metabolisable energy is not justified, even when fat only is considered as a source of energy. Allowance should be made for the specific and associative dynamic effects, which are difficult to quantify for all conditions and diets.

Synergistic effect of fats

Even when various fats are compared with one another, published ME values should be viewed with some question. Differences in ME value of fats are caused primarily by differences in their absorbability. Factors influencing absorbability have been studied in great detail by workers at Cornell University and reported by Young (1964). They include:

(a) Chain length of the fatty acids
(b) degree of unsaturation
(c) degree of esterification
(d) position of the fatty acid on the glyceride moiety
(e) ratio of saturated to unsaturated fatty acid.

All of these factors interact with one another so that differences attributable to any of the factors alone are difficult to measure in practical rations.

Renner & Hill (1961) showed that palmitic (16:0) and stearic (18:0) acids were utilised poorly when fed as free fatty acids in the absence of other fats; however, when they were fed as intact triglycerides, utilisation was greatly improved. Young (1964) demonstrated the beneficial effect of oleic and linoleic acids (18:1 and 18:2, respectively) in the diet upon the absorbability of 16:0 and 18:0. When the ratio of unsaturated to saturated fatty acids was greater than one, absorbability of these long chain saturated fatty acids was greatly improved. Since they are seldom found at high levels in the free state in feed grade fats, their low absorbability may be more academic than practical. Even when 16:0 and 18:0 appear at relatively high levels in the intact fats, as in tallow, the unsaturated to saturated ratio is approximately 1:1. When the tallow is added to a practical ration the ratio is further widened to 2:1 or 3:1 in the total dietary fat depending upon the amount of fat added, because of the residual fat contributed by the grain portion of the diet.

Utilisation of fat also varies with the nature of the diets in which it is used. Kalmbach & Potter (1959) reported ME values for corn oil and tallow of 9.04 and 7.28 kcal/g, respectively, when fed at the expense of a portion of the basal diet and only 8.60 and 6.77 kcal, respectively, when added to the diet at the expense of 'Cerelose'. Sibbald et al. (1961, 1962) demonstrated a synergistic effect between tallow and soya bean oil. Tested individually the soya bean oil and tallow had ME values of 8.46 and 6.94 kcal/g, respectively. When fed together in a 1:1 ratio, the mixture was found to provide 8.4 kcal ME/g. If the soya bean oil remained unchanged, the tallow would have to contain 8.36 kcal/g under these conditions. This is why tables of ME values are not very meaningful.

Very little experimental work has been done to compare the net or productive energy of various types of fat relative to each other. Such an investigation was carried out by the Lohman and Company Experimental Centre at Cuxhaven (Volker & Amich-Gali, 1967). They compared a control ration with no added fat and three others with either tallow, beef grease or soya bean oil. On the basis of the energetic efficiency of body weight gains they found no difference in what they termed 'net energy for production' between diets containing vegetable oil or tallow but the diet containing beef grease was considerably higher. The beef grease described by the authors appears to be similar to what is classified as *feed grade animal fat* by the rendering industry.

In an extensive study at the Belgian Research Station at Merelbeka, DeGroote *et al.* (1971) found that the metabolic efficiency of ME utilisation for maintenance plus growth was higher for all of the fats tested against glucose. The relative net availabilities were respectively 103.1% for lard, 102.9% for degummed soya bean oil, 107.3% for fancy tallow, 108.1% for prime tallow and 102.9% for brown grease. The efficiencies were significantly higher for the two tallows relative to glucose. Conversion rates could not be found to be significantly different among the different fats. Recognising the influence of observed variability in carcass fat deposition as a percentage of digested test fat intake on the availability of the ME of the different fats, they concluded that no important differences exist between the commonly used vegetable and animal fats, soya bean oil, lard, fancy and prime tallow and brown grease, concerning the utilisation of their ME for maintenance plus growth by chicks.

They also found that tissue energy retention per unit ME intake was significantly higher for the high fat diets (37.1–40%), compared to the low fat diets (33.2–34.7%). A significantly higher tissue energy gain on the prime tallow diet (40%) relative to the lard and soya bean oil diets (37.1 and 37.7% respectively) was obtained, suggesting that tallow might be more efficiently utilised for growth than lard and soya bean oil.

If metabolisable energy values continue to be used in feed formulation, and it looks as if there is no good alternative at present, then a set of correction factors is needed for the various fats which will evaluate their actual energetic contribution and enable the nutritionist to take full advantage of their potential value in his feeds.

Response of chickens to heat increment of the diet under heat stress.

Throughout much of the temperate zone, including the USA and Europe, climatic conditions are highly variable from summer to winter. The event of hot summer weather is usually accompanied by a decline in feed consumption especially noticeable in broiler growing operations in the southern areas of the USA. We are inclined to take this for granted since it is generally conceded that animals require less dietary energy as environmental temperature increases. This seems logical and is reinforced by casual observation of the eating habits of animals from summer to winter. The distressing part is the accompanying decline in growth rate which should not happen if the birds were simply adjusting energy intake to meet thermostatic heat requirements. The problem lies in removal of basal heat production which becomes a burden at temperatures above the zone of thermal neutrality. This requires increased energy expenditure from such activity as increased respiration (panting) and increased peripheral circulation, creating still more heat and greater discomfort in a vicious cycle. In extreme cases this process continues until death from hyperthermia occurs, but in less severe situations the animals simply reduce feed consumption to avoid the added discomfort of the heat increment associated with eating.

Working with swine which are quite sensitive to heat and cold, Andrews (1957) observed that feed intake is reduced when the temperature of the environment increases above the critical zone because the animals need to lower the burden of heat dissipation. Brobeck (1960) noted that food intake of rats decreases with increasing ambient temperature, to a point where they will not eat if the temperature is so warm as to induce hyperthermia. Consolazio & Johnson (1971) showed that energy cost of humans increased in an extremely hot environment. In a study conducted in the desert, they found that energy expenditure was significantly higher in either the hot sun or hot shade compared to the cool indoor environment of 26°C. The National Research Council, on the basis of that work, suggests that energy requirements in humans should be increased by at least 0.5% for each degree increase in environmental temperature between 30 and 40°C.

If this is true in avian species, then the reduction in feed intake that invariably occurs during prolonged periods of high temperature may simply create a deficiency of usable energy, thus contributing to the reduced growth rate.

In experiments at the University of Georgia (Fuller & Mora 1973), an attempt was made to increase voluntary feed consumption of young chickens subjected to heat stress by manipulating the diet in such a manner as to reduce its heat increment.

Young male chickens 4 or 5 weeks of age were grown at two temperature regimes (hot and cool) for either 2 or 3 weeks. The diets varied in density, in proportion of total calories provided by fat, and in protein level. The results are shown in Table 2.

Increasing the energy and nutrient density of the diet by replacing carbohydrate calories with fat calories reduced feed intake almost proportionally but increased body weight gains at both temperatures.

Feed efficiency and efficiency of utilisation of energy and protein were also improved. The improvement was greater under high temperatures. Reducing protein level in this diet below that of the basal diet while maintaining the increased energy density and proportion of ME calories furnished by fat resulted in an increase in energy intake of 3% in the cool, and 6% in the hot, environment. Despite the greatly lowered protein intake weight gains were improved by 8 and 9% in the cool and hot rooms, respectively.

TABLE 2. *Relative performance of chicks at two environmental temperatures*

	A	B	C	E	F
Kcal ME/kg	3091	3434	3427	3089	3091
Protein (%)	23.6	26.3	22.0	23.7	20.1
ME from fat (%)	12.2	33.2	33.3	30.0	11.5
Rel. density[1]	100	111	111	100	100
Rel. intake:[2]					
Feed (cool)	100	90	93	91	98
(hot)	100	91	95	96	100
ME (cool)	100	100	103	91	98
(hot)	100	100	87	91	83
Protein (cool)	100	100	87	91	83
(hot)	100	101	89	96	85
Rel. wt. Gains:					
(cool)	100	102	108	100	101
(hot)	100	106	109	102	92

Source: Fuller & Mora (1973).

[1] Density of ME and all nutrients relative to diet A.

[2] All values expressed as percent of performance of diet A which is taken as 100.

Although the increase in energy calories coming from fat as the main energy source improved growth in the high temperature, such improvement was of sufficient magnitude to overcome only a small part of the adverse effect of high temperature on growth (Table 3).

TABLE 3. *Feed intake and weight gains of chicks in hot environment (as percent of that in cool)*

	A	B	C	E	F
Kcal ME/kg	3091	3434	3427	3089	3091
Protein (%)	23.6	26.3	22.0	23.7	20.1
ME from fat (%)	12.2	33.2	33.3	30.0	11.5
Hot as percent of cool:					
Feed intake	79	80	81	83	–
Weight gains	74	78	75	75	–

Source: Fuller & Mora (1973).

The increased energy intake resulted in improved growth rate partially overcoming the effect of heat stress on growth.

It was concluded that the net energy intake of chickens can be increased during high temperature stress by manipulating the diet so as to reduce its heat increment. This can be accomplished by replacing carbohydrate calories with fat calories and by reducing total protein level while maintaining appropriate levels of critical amino acid in relation to energy level.

This further indicates that the chicken may have difficulty in consuming sufficient net energy for optimal growth during heat stress when fed diets containing high heat increment.

In a subsequent experiment in our laboratory, an attempt was made to estimate the heat increment of the experimental diets described earlier (Fuller & Mora 1973). A modification of the body balance technique first described by Swift *et al.* (1934) was used. Triplicate pens of three chicks each (4½ weeks of age) were assigned to each of five dietary treatments within each of three temperature treatments. Ten similar chicks were selected and sacrificed for initial carcass analysis. The chicks were fed the experimental diets (to which chromic oxide was added) for 2 weeks during which time total excreta were collected.

Excreta and feed were analysed for gross energy, nitrogen, moisture and chromic oxide. All birds were sacrificed at the end of 2 weeks and the entire carcasses analyzed, along with those of the birds in the initial sample for moisture, nitrogen, ash, total lipids and gross energy.

Heat increment cannot easily be separated from heat generated by normal activity, referred to as 'activity increment' (AI); therefore, one value was calculated for HI + AI by the following formulae:

$$HP = GEi - (GEexc + GEg)$$
$$HI + AI = HP - HBM$$

Where:
HP = heat production
HI = heat increment
HBM = heat of basal metabolism
AI = activity increment (energy expended in activity)
GEi = gross energy consumed
GEg = gross energy gained
GEexc = gross energy of excreta

In Table 4 is shown the partition of energy consumed. HBM was calculated by the formula $Q = 72W^{75}$ where Q = HBM in kcal/day and W = weight of the birds in kg. Deducting this value from total heat production provided an estimate of heat increment plus the activity increment.

TABLE 4. *Heat production of chicks (hot and cool combined)*

Diets:	A	B	C	E	F
			(kcal/chick/day)		
GEi	395	427	395	379	377
GEexc	90	109	110	105	82
GEg	68	88	74	64	75
HP	237	230	213	210	221
HBM	87	86	86	82	82
HI + AI	150	144	127	128	139
HI + AI/GE cons.	0.38	0.34	0.32	0.34	0.37

Values for the cool variable environment were eliminated because of aberrant results on two of the diets. Values for the remaining two temperature treatments were combined, since differences between them were not significantly different.

It is reasonable to assume that the activity increment would be similar for all groups so that the value 'HI + AI' should be a fair estimate of the *relative* heat increments of the various diets. This lends support to the conclusions drawn earlier that replacing carbohydrate calories with fat calories

128 and reducing crude protein levels resulted in greater intakes of *net* energy first by reducing the energy lost as HI and further by improving feed intake. If the chickens were actually suffering from an energy deficiency, as surmised, then the increase in net energy intake would account for the improvement in growth rate. It is also suggested that the heat increment of the diet may exert a greater influence on feed intake than the requirement for energy at uncomfortably high environmental temperatures.

References

Andrews, F. N., (1957) The effects of climate environment on livestock production. *Proc. AFMA Nutr. Council,* Dec. 1957, pp. 7–12.

Brobeck, J. R., (1960) Hormones and metabolism. Food and temperature. *Recent Prog. Horm. Research* 16, 439–459.

Carew, L. B., Jr. & Hill, F. W. (1964) Effect of corn oil on metabolic efficiency of energy utilization by chicks. *J. Nutr.* 83, pp 293–299.

Consolazio, C. G. & Johnson, H. L. (1971) Measurement of energy cost in humans. *Fed. Proc.* 30, pp 1444–1453.

De Groote, G., Reyntens N. & Amich-Gali, 'J. (1971) Fat studies. 2. The metabolic efficiency of energy utilization of glucose, soya bean oil and different animal fats by growing chicks. *Poultry Sci.* 50, pp 808–819.

Forbes, E. B. & Swift, R. W. (1944) Associative dynamic effects of protein, carbohydrate and fat. *J. Nutr.* 27, pp 453–468.

Forbes, E. B. Swift, R. W. Elliott, R. F. & James, W. H. (1946) Relation of fat to economy of food utilization. II. By the mature albino rat. *J. Nutr.* 31, pp 203–212.

Fuller, H. L. & Mora, G. (1973) Effect of diet composition on feed intake and growth of chicks under heat stress. *Feedstuffs* 45 (no. 45 of 29 Oct 73), p.30.

Jensen, L. S., Schumaier, G. W. & Latshaw, J. D. (1970) "Extra caloric" effect of dietary fat for developing turkeys as influenced by calorie-protein ratio. *Poultry Sci.* 49, pp. 1697–1704.

Kalmbach, M. P. Potter, L. M. (1959) Studies in evaluating energy content of feeds for the chick. 3. The comparative values of corn oil and tallow. *Poultry Sci.* 38, pp. 1217.

Potter, L. M. (1967) Research in turkey nutrition at VPI during 1966. *Proceedings, 20th Virginia Feed Convention and Nutrition Conference, Feb. 21–22.*

Renner, R. & Hill, F. W. (1961) Factors affecting the absorbability of saturated fatty acids in the chick. *J. Nutr.* 74, pp 254–258.

Sibbald, I. R., Slinger S. & Ashton, G. C. (1961) Factors affecting the metabolizable energy content of poultry feeds. 2. Variability in the ME values attributed to samples of tallow and undegummed soya bean oil. *Poultry Sci.* 40, pp 303–308.

Sibbald, I. R., Slinger S. J. & Ashton, G. C. (1962) The utilisation of a number of fats, fatty materials and mixtures thereof evaluated in terms of metabolizable energy, chick weight gains and gain: feed ratios. *Poultry Sci.* 41, pp. 46–61.

Swift, R. W., Kahlenberg, O. J., Voris L. & Forbes, E. B. (1934) The utilisation of energy producing nutrient and protein as affected by individual nutrient deficiencies. I. The effect of cystine deficiency. *J. Nutr.* 8, pp. 197.

Völker, L. & Amich Gali, J. (1967) Comparative values for metabolizable energy and net energy for production resulting from the addition of tallow and other fats to broiler rations. Rome: National Renderers' Association (Auspices: Foreign Agr. Sec., USDA). *Publ.* E–110, 6 pp.

Young, R. J. (1964) Quality of fats with regard to absorption. *Proc. Ga. Nutr. Conf.,* Atlanta, Feb. 1964, pp. 66–75.

The use of low quality forage in ruminant diets

V. J. Clarke and H. Swan
University of Nottingham, England

Summary

Studies on the improved utilisation of low quality roughages and by-products is of increased interest in developing regions, where intensification of ruminant production is desirable and in developed regions where a diminished grain surplus may mean less dependence on high concentrate feed systems.

Methods of processing and chemical treatment to improve the utilisation of low quality roughage are discussed together with the practical importance of changes in ruminant digestive function which occur with roughage inclusion in diets. The potential for use of non-protein nitrogen is considered together with by-product use for rangeland and more intensive supplementary feeding.

Some aspects of the changes in carcass composition and killing out percentage are discussed with their relevance to increase in edible meat output.

Résumé

L'usage du fourrage de qualité inférieure dans les diètes des ruminants

Des études sur l'utilisation améliorée des fourrages de qualité inférieure et des sous-produits sont d'un intérêt accru dans les régions en développement où l'intensification de la production des ruminants est désirable et dans les régions développées où un surplus réduit de graines pourrait signifier une moindre dépendance de systèmes de pâtures hautement concentrées.

On discute les méthodes de traitement et les procédés chimiques pour améliorer l'usage des fourrages de qualité inférieure, de concert avec l'importance pratique des changements dans la fonction digestive des ruminants qui se produisent avec l'inclusion des fourrages dans les diètes. Le potentiel pour l'usage de l'azote non-protéinique est considéré, ensemble avec l'usage des sous-produits pour les pâturages et pour des pâtures supplémentaires plus intensives.

On discute quelques aspects des changements dans la composition des carcasses et le pourcentage de l'abattage en rélation avec l'augmentation du rendement de viande comestible.

Resumen

El uso de forraje de baja calidad en las dietas de los rumiantes

Han adquirido interés creciente los estudios sobre el perfeccionamiento de la utilización de forrajes difíciles de digerir y de productos secundarios de baja calidad en las regiones en vías de desarrollo, en las que la intensificación de la producción de rumiantes es deseable y en regiones desarrolladas donde un excedente de granos reducido puede significar menos dependencia en los sistemas de piensos concentrados altos.

Se discuten los métodos de preparación y tratamiento químico para mejorar la utilización de forrajes de baja calidad difíciles de digerir junto con la importancia práctica de cambios en las funciones digestivas de los rumiantes que ocurren con la inclusión de dichos forrajes en las dietas. Se considera el potencial para el uso de nitrógeno no proteínico junto con el uso de productos secundarios para pastizales y más alimentación suplementaria intensiva.

Se discuten algunos aspectos de los cambios de composición en las carcasas y en el porcentaje de matanza con su relación al aumento de la producción de carne comestible.

130　Domestic ruminants need not compete with man for nutrients. The unique digestive and metabolic apparatus they possess enables them to utilise roughage and by-product feeds which we cannot use. Some 64% of the world's agricultural land (Reid, 1970) is considered to be non-arable and suitable for grazing, so we might expect the ruminants' ability to digest cellulose and hemicellulose to be an advantageous feature both for their survival, and their value to man as a food source in the coming years.

An estimate of cattle population in temperate and tropical areas (Table 1: FAO, 1972) shows that at least half of the world's cattle population is located in the tropics. In terms of productivity of meat per head per year, however, the most significant gains in output by cattle have been achieved in developed regions, (Table 2: FAO, 1970). This has largely been due to increased concentrate inputs from grain surpluses with consequent reductions in the energy costs of maintenance.

TABLE 1. *Estimated world cattle population*

Temperate		Millions
N. America		151
S. America		61
Europe		123
USSR		99
Asia		40
New Zealand		9
	Total	483
Tropical		
Central America and Caribbean		35
S. America		142
Africa		158
Asia		250
Oceania		25
	Total	610
	World Total	1,093

Source: FAO 1970/71

TABLE 2. *Production of beef cattle per head of cattle population (1966/67)*

	Kg/head/year
North America	87
Europe inc. USSR	59
Oceania	47
Latin America	27
Africa	14
Asia inc. China	11

Source: FAO 1970

In developing regions cereal grains and protein crops are required for internal consumption or for foreign exchange. An understanding of the use of low grade roughages and by-products would be of considerable value in regions wishing to intensify their husbandry systems where surplus grains are less plentiful or non-existent for ruminant feeding.

High voluntary food intake is the key to intensification for, without this, rapid liveweight gains cannot be achieved. Voluntary food intake of poor quality long roughage is usually low. Processing of dry roughage such as chopping, grinding or pelleting, which reduces the particle size of the feed, is well known to give increased feed intake when compared with the same material fed in the long form (Minson, 1963; Moore, 1964).

Montgomery and Baumgardt (1965) hypothesised that ruminants will adjust voluntary food intake in relation to physiological demand for energy if fill or rumen load does not limit their consumption. The intake of roughages is suggested to be related to the amount of digesta in the reticulo rumen, which is a function of the rate of digestion of food particles and their rate of passage out of the rumen (Balch and Campling, 1962). It is therefore to be expected that processing roughages to increase rate of ruminal digestion or rate of passage out of the rumen could lead to enhanced voluntary food intake.

Several experiments have been conducted to investigate the performance of beef cattle offered diets based on processed barley straw at variable concentrate to roughage ratios, (Figure 1). Dry matter intake has been plotted against liveweight gain in kilograms per day.

Figure 1

Effect of dry matter intake on liveweight gain in beef cattle

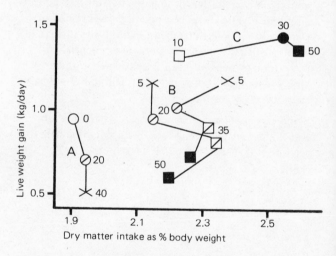

Although the experiments incorporated similar roughage to concentrate ratios, striking differences in animal performance were reported.

Increments of roughage from 0 to 20 and then 40% did not result in increased DM intake in compensation for energy dilution of the ration (Figure 1 Plot A). In fact intake remained almost constant when expressed as a percentage of liveweight. The barley straw included in this experiment (Forbes et al. 1969) was chopped in 25–75 mm lengths and fed in a loose mixture with the concentrate portion of the ration.

In the experiment of Kay (1972) and co-workers (Plots B) chopped barley straw was added to a pelleted concentrate fraction of the ration and this showed a more varied response, though generally intake did not increase uniformly with increase in the chopped straw fraction.

The third plot (C) shows the results of Swan and Lamming (1970.) Complete pelleted rations were fed, in which the added roughage was milled through a 3.17 mm screen before being pelleted in the ration. In this case increments in roughage level resulted in increased intake with no fall off in animal performance at the 30% level. Thereafter increments in roughage levels stimulated greater voluntary food intakes but these were insufficient to maintain high liveweight gain.

It is clear from the fuller analysis of the results of such experiments (Table 3), that a range of methods was used to chop the straw or the straw was milled through different screen sizes. There is a wide variation in the particle size of roughage added to the rations actually fed. (The situation is complicated when such rations are pelleted, causing a further reduction in particle size).

In experiments where high metabolisable energy intakes have been achieved, added roughage has been finely ground and mill screen sizes have been at or near 3.17 mm.

In a performance trial at Nottingham (Lockwood *et al* 1973), Friesian steers were fed diets containing 30% ground barley straw which was milled through various screen sizes from 3.17 mm up to 12.7 mm. They showed that an optimum screen size of 6.35 mm gave performance superior to either the finer or coarser grinding, and higher dry matter digestibility.

It appears that particle size can influence the different digestibility and food intake of diets based on low quality forage.

Chemical treatment

Another approach to the improvement of utilisation of poor quality forages has been by chemical treatment. This has been shown to increase the amount of digestible carbohydrate (Ferguson, 1942).

Sodium hydroxide treatments to increase the digestibility of straw have been known for many years (Ferguson, 1942; Beckman, 1921), and there has been considerable experimentation in India (Singh & Jackson, 1971; Chandra & Jackson, 1971) and Canada (Wilson & Ridgen, 1964; Mowat & Cladalade, 1970).

Earlier treatment methods have been considered uneconomic in the past and have often suffered from the drawback of loss of the fraction made soluble by soaking.

Danish workers (Thomsen *et al.*, 1973a & 1973b) are developing a method in which 4% (w/w) NaOH is added to ground straw which is then pelleted retaining all components in the pellet. The heat and pressure in the pellet die is adequate for a high temperature, short-time chemical reaction which renders additional components of the straw digestible.

Table 4 shows analyses of the chemical composition of sodium hydroxide treated barley straw using the

TABLE 3. *Roughage processing and ration preparation*

		Size (mm)	
Lamming *et al.* (1966)	Milled	3.17	Pellets
Swan & Lamming (1967)	,,	2.17	Pellets
Pickard *et al.* (1969)	,,	1.58, 4.75, 7.92	Pellets
Kay (1972)	,,	3.17	Straw loose, pellet concentrate
Swan & Lamming (1970)	,,	3.17	Pellets
Lockwood *et al.* (1973)	,,	3.17, 6.34, 9.51, 12.7	Loose, 20% molasses
Forbes *et al.* (1969)	,,	25.4	Loose, rolled barley
Raven *et al.* (1969)	,,	25.4	Loose
Forbes *et al.* (1969)	Chopped	25-75	Loose, meal concentrate
Kay (1972)	,,	75	Loose

TABLE 4a. *Weende analysis of barley straw treated with NaOH*

NaOH%	Crude Protein	Oil	% Crude Fibre	Nitrogen-Free Extract	Ash
0	4.28	1.41	44.18	44.82	5.32
2	3.39	1.78	42.43	46.02	6.38
3	3.96	1.63	40.78	45.88	7.75
4	3.88	1.39	40.94	44.42	9.37
5.9	3.33	1.19	40.33	42.97	12.19

TABLE 4b. *Van Soest's analysis of barley straw treated with NaOH*

NaOH%	Neutral detergent fibre	% Hemicellulose	Cellulose	Lignin
0	84.71	33.05	43.63	8.03
2	80.02	28.68	43.23	8.11
3	75.37	25.34	41.77	8.26
4	73.01	22.75	41.82	8.44
5.9	66.78	16.78	41.33	8.08

conventional Weende (Horwitz, 1970) system of analysis and Van Soest's (Goering & Van Soest, 1970) method for fibre determination. It will be seen that the conventional Weende analysis gives no real indication of any change in composition due to treatment, except ash, which is to be expected from the addition of sodium hydroxide. The more detailed breakdown of fibre analysis shows, however, a considerable reduction of the neutral detergent fibre (NDF) fraction and also a reduction of the hemicellulose fraction with treatment and illustrates that these are the fractions rendered soluble.

The same Danish workers have shown that both in the *in vitro* and *in vivo* digestibilities of barley straw are increased by NaOH treatment (Figure 2). The residence time in the rumen is associated with the amount of digestion which occurs, and the Danish *in vitro* results (Figure 3) indicate the importance of the time factor. It may well be that differential rumen residence time accounts for the somewhat better results achieved by workers in India (Singh & Jackson, 1971) when chaffed, treated straw proved

better than ground, treated straw. Larger particles will tend to stay in the rumen longer and be subjected to more extensive breakdown (Ferguson 1942).

Digestion and utilisation

It is important to consider the mechanisms by which roughage influences ruminant gut function. Apart from influences of roughage inclusion on appetite there are more specific effects on digestive function. Digestive disorders which have occurred in intensively fed cattle on concentrate diets include bloat and acidosis (Preston *et al.*, 1963). The occurrence of 'blue' or abscessed livers is also known to increase on all concentrate diets (Haskins *et al.*, 1969).

The practical farmer has often contended that ruminants require some roughage. The prophylactic effects and stimulus of feed intake due to low levels of roughage inclusion, often regardless of roughage quality, should not be underestimated. In the tropical context the prophylactic effects can be of paramount importance, particularly in intensive feeding of mature tropical animals.

Investigations of the stable pH conditions in hourly fed sheep illustrate the effect of roughage addition to an otherwise concentrate barley diet (Figure 4). At conditions of low pH near 5 only a small variation of food intake is required to induce a lactic ferment and the development of clinical acidosis.

Interrupted feeding in a beef feedlot is a good practical example of this phenomenon, where animals may suffer from clinical acidosis when feed is restored to them.

Low rumen pH is known to be associated with diets high in starch and the condition sometimes results in feed refusal and decreased food intake (McCullough, 1969).

Figure 2
Digestibility of treated straw

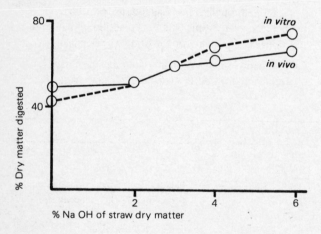

Figure 3
Rate of *in vitro* digestion for treated and untreated straw

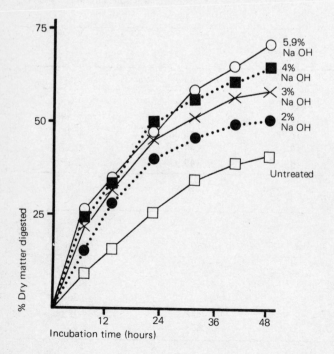

Figure 4
Effect of roughage addition on stable pH conditions in hourly fed sheep

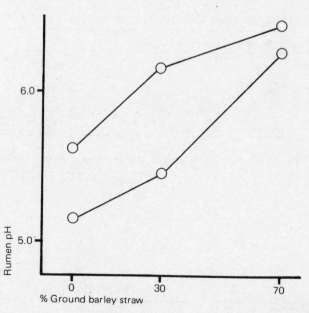

The importance of the prophylactic effects in tropical feed systems has been demonstrated. In a molasses/urea feed system 0.3 kg of dry matter per 100 kg liveweight was given daily as green forage (Preston & Willis, 1970). With feed based on dried cassava chips, dried waste sweet potato tubers and groundnut cake (Clarke, pers. comm.) fed to N'dama cattle over a weight range from 175–275 kg live-weight, it was found necessary to feed 10 kg per head per day of fresh Napier fodder (*Pennisetum purpureum* Schumach) to avoid scouring and loss of appetite. Supplementary phosphate in the form of dicalcium phosphate added to the ration and free access to salt blocks was also required.

Rumen fermentation

Changes in rumen fermentation are known to occur when roughage is included in ruminant diets (Blaxter, 1967). Figure 5 shows the dramatic change in relative proportions of the three major volatile fatty acids that occur in barley based diets containing ground

Figure 5
Molar percentages of acetic, propionic and butyric acids in rumen liquor of sheep fed barley diets with various proportions of ground straw

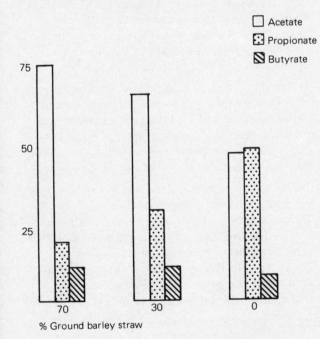

% Ground barley straw

straw when fed to sheep. The ratio of acetic to propionic acid particularly illustrates the change from 1:1 concentrate to 3.25 :1 on a high roughage diet.

Several workers including Hovell and Greenhalgh (1972), and Kay (1972) at the Rowett Research Institute have suggested that a better utilisation of metabolisable energy is possible when the end products of fermentation alter slightly in favour of acetate through a change in efficiency with which these are utilised.

Digestive tract function

Studies with fistulated sheep cannulated in the rumen, and with duodenal re-entrant cannulae show that ground low quality roughage in the diet reduces the total DM digestion which takes place in the rumen. This means that for a given DM intake the higher the roughage inclusion the greater will be the passage of dry matter to the hind gut. From numerous measures of DM content of duodenal fluid from sheep on diets ranging from all concentrate to 70% straw the DM content of the fluid was within a range from 3–7% with a mean value of 5% regardless of dietary type.

The percentage of apparent DM digestion that takes place in the rumen of sheep on a range of diets including ground straw is shown in Table 5.

Duodenal flow appears to be more closely related to the indigestibility of the food in the rumen rather than to its digestibility, in view of the 5% DM suspension observed.

As water intake and output (faeces and urine) did not differ greatly with different experimental diets, then an increased flow at the duodenum would reflect to some extent increases in the rate of cyclical movement of fluid within the animal via the digestive tract (Figure 6). The major fluid return pathways to the rumen for fluid absorbed from the hindgut are twofold; either by passage across the rumen wall or via the salivary glands and their secretions swallowed back to the rumen. There is considerable evidence to suggest that the salivary route is of major importance in the return of fluid from the hindgut to the rumen.

Increase in volume of salivary flow can be stimulated by either greater volume of food and hence a longer period eating the material or increased rumination

TABLE 5. *Apparent dry matter digestion in the rumen of sheep on a range of diets containing ground straw*

Diet	DM Intake g/24 hr	Apparent rumen digestibility %	Proportion total apparent digestibility
70% straw	1,240	25	47
30% straw	1,207	40	56
Concentrate	1,212	66	77
Diet	Ruminally indigestible	Duodenal flow 1/24 hr	Rumen pH
70% straw	75	20.86	6.3
30% straw	60	15.49	5.5
Concentrate	34	9.03	5.2

Figure 6
Fluid flow in the ruminant

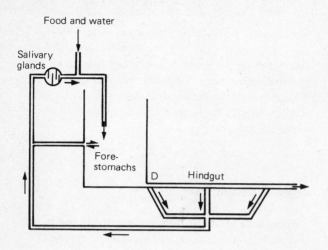

time (Minson, 1963; Moore, 1964). The secretions of the parotid salivary glands in particular, provide most of the total volume of salivary secretion in ruminants (Kay, 1966) and as far back as 1893 Eckhard (1893) demonstrated that the parotid gland continued to secrete even when all nerve attachments were severed. Eckhard suggested that another source of stimulus to the gland apart from nervous control was probable.

It is clear from the data published by Kay (1966) that the effects of salivary secretion on rumen pH may not alone account for all the pH variation seen with different diets. It is possible that saliva does not simply neutralise volatile fatty acids produced within the rumen, at least with calculations based on measured salivary flow data.

To summarise it seems that higher inclusions of ground roughage lead to more salivary secretion and higher flow rates at the duodenum. Water alone will not increase rumen volume or duodenal flow in the long term.

Egan (1965) observed the relationship of these factors with intake in sheep on chaffed oaten hay diets when he noted that lowest mean DM intake was associated with the longest retention time for stained particles and the lowest number of rumination chews, whereas the highest intakes were associated with the highest number of rumination chews and the shortest mean retention time for stained particles.

The satisfactory results of Randel (1970) devising cattle feeding systems in Puerto Rico using the mixtures of sugar cane bagasse and concentrates may well be due to the effect of the complete ration on gut function.

Use of non-protein nitrogen

Animals grazing on tropical rangeland in the dry season are often obliged to eat forage which is characteristically high in fibre and low in protein. Indeed the crude protein levels of some natural forage can be as low as 2% (Table 6; Rains, 1963).

The existence of an interaction between voluntary food intake and crude protein level is well known (Egan, 1965; Dougall, 1960). Where crude protein levels of ingested food fall below 6% there is good evidence to suggest that animals would benefit from increased nitrogen intakes to assist microbial synthesis, and breakdown of cellulose in the rumen and maintain appetite.

Faecal nitrogen forms a major route of nitrogen loss in ruminants. Blaxter (1964) has estimated that the egestion of faecal nitrogen on nitrogen free diets is about 0.45 g/100 g DM ingested. Practical values of faecal nitrogen loss on normal ruminant diets show a range dependent on dietary type. With concentrate diets low in fibre the obligatory faecal loss may be as low as it is in monogastric species but it increased with increasing fibre content of the diet (Blaxter, 1964).

Regardless of the form of faecal nitrogen loss, whether undigested food residues or nitrogen 'trapped' in micro-organisms from rumen (Blaxter, 1964) or faecal (Mason & White, 1971) fermentation, and whether the original nitrogen source was protein food or urea (Blaxter, 1964; Campling et al., 1962) the obligatory loss of nitrogen in faeces is a major determinant of nitrogen requirements.

The results of the classic experiment by Campling et al. (43), where oat straw intakes were measured in cows supplied with continuous intraruminal urea infusions, illustrate that voluntary food intake may be increased by supplementary urea through increased rumen fermentation and breakdown of the roughage.

Just as high carbohydrate rations may lead to digestive tract dysfunction in feedlot animals, so intakes of poor quality forage low in nitrogen may lead to digestive tract dysfunction in tropical animals through the development of rumen inpaction. Weston's results (Weston, 1966) show the striking reduction and even cessation of eating which can occur when indigestible materials are introduced into the rumen.

Apart from the introduction of legumes into the pasture (Haggar et al., 1971) which is outside the scope of this paper, practical methods by which non-protein nitrogen may be applied include liquid licks

TABLE 6. *Analysis of dry season fodder*

	Dry matter	Crude Protein	% Oil	Crude Protein	Ash
Andropogon gayanus hay	87.5	4.75	1.55	30.45	5.22
Groundnut hay	89.8	8.25	1.26	29.25	4.07
Foggage (*Pennisetum pedicellatum*)		1.91	0.63	33.76	10.22

TABLE 7. *Analysis of some natural waste and by-products (%)*

	Dry matter	Crude Protein	Oil	Crude Fibre	Nitrogen-Free Extract	Ash
Cottonseed with 'fuzz'	92.83	19.53	17.53	23.50	28.77	3.50
Groundnut dust	96.43	13.94	14.28	19.62	39.84	9.75
Rice and sorghum bran	89.18	5.81	2.29	21.34	45.22	14.52

with urea — molasses (Haggar *et al.*, 1971) where excessive intake is controlled through salt addition to the mix; or feed blocks which also contain minerals and may be advantageous in circumstances where mineral supplementation, particularly phosphate, is required (White *et al*, 1959; Davidson & Purchase, 1961). Blocks are likely to be safer to use where urea toxicity is to be avoided.

Low grade roughages supplemented with urea do not always enable stock to maintain liveweight in the dry season (Coombe *et al.*, 1971) and some additional energy source is usually recommended for improved utilisation of urea (Balch, 1967).

Natural waste and by-products are sometimes available apart from commercial oilseeds (Table 7); for example cottonseed from small gins where no commercial disposal operates. This may be fed whole or mixed with a low grade energy source, such as rice or sorghum bran. Groundnut residues, particularly residual haulms and groundnut dust are sometimes available. Groundnut dust consists of inorganic materials, husks of groundnut kernels and a variable fraction of broken kernel, approximately 1% separated in the decorticating process.

Such feeds are useful if fed on dry range and experience with N'dama cattle shows that with *ad libitum* water and 24 hour access to range they help to maintain condition of stock (Clarke, pers. comm; Maloiy & Taylor, 1971).

Other natural nitrogenous feed sources include pods and browse from leguminous trees (Rains, 1963; Oyenuga, 1968). Non-toxic species are usually known to the stockmen and are sometimes lopped for dry season forage. Three notable West African examples include Locust Bean (*Parkia* spp), *Acacia albida,* and *Prosopsis africana.* Such species should continue to be protected from further depravation by man in his attempts to clear land for cultivation.

Carcass composition

The addition of low quality roughage may affect growth rate but does not lead to inferior carcass composition. DM intake is frequently greater in animals receiving a mixture of concentrate and roughage than with concentrate alone though live-weight gain may decrease above 10% (Kay, 1972) or sometimes as high as 30% (Swan & Lamming, 1970). inclusion.

Under United Kingdom conditions the use of rations which include ground straw up to 50% DM intake had no effect on carcass composition at equal esti-mated carcass weights (Swan & Clarke, 1974). Killing out percentages, however, decreased as dietary roughage levels increased (Table 8). The proportion of gut fill is greater as the roughage level is increased and the effect is more marked for chopped than ground straw (Kay, 1972).

TABLE 8a. *Effect of roughage on killing out percentage*

	% ground barley straw in diet		
	10	30	50
Carcass weight (kg)	262	257	245
Killing out %	57.2	54.7	53.7

TABLE 8b. *Effect of roughage levels on carcass chemical composition*

	% ground barley straw in diet		
	10	30	50
Chemical components as percentage of dry matter in the dissected side, %			
Ash	4.59	4.67	4.79
Fat	28.5	27.0	27.6
Protein	15.9	16.1	16.0
Water	51	52	52

The considerable influence of gut fill in determination of killing out percentage remains important in tropical animals as evidenced by slaughter data for a well fed West African N'dama bull (Clarke, 1970). Killing out percentage on an unstarved basis was 52% whereas the same index calculated on an empty gut basis gave a value of 63%. A normal range of killing out percentage of 52.6—54.1% is reported for N'dama cattle reared on improved pastures in Nigeria (O'Korie *et al*, 1965).

It is clear that killing out percentage as usually expressed does not give a reliable index of carcass yield or quality. Killing out percentage calculated on an empty gut basis may give a more reliable index of carcass yield but not carcass quality.

In conclusion nutritive value of low grade roughages may be enhanced by mechanical or chemical processing and the use of non-protein nitrogen with an additional energy source may also be advantageous in certain circumstances.

136 References

Balch, C. C. & Campling, R. C. (1962) Regulation of voluntary food intake in ruminants, *Nutr. Abstr. Rev.,* **32**, 669–686.

Balch, C. C. (1967) Problems in predicting the value of non-protein nitrogen as a substitute for protein in rations for farm animals. *Wld. Rev. Anim. Prod.* **3**. pp. 84–91.

Beckman, E. (1921) Conversion of grain straw and lupins into feeds of high nutrient value, *Festschr, Kaiser Wilhelm Ges. Forderung Wiss. zehnjährigen Jubiläum,* pp. 18–26, (*Chem. Abstr.,* **16**, pp 765).

Blaxter, K. L. (1964) The Nutritional Significance of the endogenous secretion of nitrogen in ruminants, in Munro, H. N. (Ed) The Role of the gastrointestinal tract in protein metabolism, Oxford, Blackwell Scientific Publications, pp. 143–153.

Blaxter, K. L., (1967). The Energy Metabolism of Ruminants, London, Hutchinson & Co. Ltd., 2nd impression.

Campling, R. C., Freer, M. & Balch, C. C. (1962) Factor affecting the voluntary intake of food by cows 3. The effect of urea on the voluntary intake of oat straw, *Br. J. Nutr.,* **16** pp. 115–124.

Chandra, S. & Jackson, M. E. (1971). A study of various chemical treatments to remove lignin from coarse roughages and increase their digestibility. *J. agric. Sci., Camb.,* **77** pp. 11–17.

Clarke, V. J. Unpublished data.

Clarke, V. J. (1970) Thirteenth Annual Report of the specialist services, Yundum Agricultural Station for the year 1968–69, Dept. of Agriculture, The Republic of the Gambia, p. 56.

Coombe, J. B., Christian, K. R. & Holgate, M. D. (1971) The effect of urea on the utilisation of ground, pelleted roughage by penned sheep. *J. agric. Sci., Camb.,* **77**, pp 159–174. III. Mineral supplements.

Davidson, R. L. & Purchase, H. S. (1961) A preliminary comparison between two methods of urea supplementation of winter feed for beef cows. *Jl. S. Afr. vet. med. Ass.,* **32** pp. 85–89 cited by Whyte, R. O. (1967) Milk Production in Developing Countries, London, Faber & Faber, p. 94.

Dougall, H. W. (1960) Average nutritive values of Kenya feeding stuffs for ruminants *E. afr. Agric. For. J.,* **26** pp. 119–28, cited by Whyte, R. D. 1967 *Milk Production in Developing Countries* London, Faber & Faber, p. 93.

Eckhard, C. (1893) *Zbl. Physiol.* **7**, pp 365, cited by Kay, R. N. B., (1966) (below).

Egan, A. R. (1965) Nutritional status and intake regulation in sheep. III The relationship between improvement of nitrogen status and increase in voluntary intake of low protein roughages by sheep. *Aust. J. Agric. Res.,* **16**, pp 463–472.

FAO (1970) Production yearbook. **24**.

FAO (1972) Production yearbook. **26**, pp 186–188.

Ferguson, W. S. (1942) The digestibility of wheat straw and wheat straw pulp *Biochem. J.,* **36**, pp 786–789.

Forbes, T. J., Irwin, J. H. D., & Raven, A. M. (1969) The use of coarsely chopped barley straw in high concentrate diets for beef cattle *J. Agric. Sci., Camb.,* **73**, pp. 347–354.

Forbes, T. J., Raven, A. M. & Irwin, J. H. D. (1969) The use of coarsely milled barley straw in finishing diets for young beef cattle. *J. agric. Sci., Camb.,* **73**, pp 365–372.

Freer, M., Campling, R. C. & Balch, C. C. (1962) Factors affecting the voluntary intake of food by cows. 4. The behaviour and reticular motility of cows receiving diets of hay, oat straw and oat straw with urea. *Br. J. Nutr.,* **16**, pp. 279–295.

Goering, H. K. & Van Soest, P. J. (1970) Forage fibre analyses, U.S.D.A., A.R.S., Agr. Handbook No. 379.

Haggar, R. J., De Leeuw, P. N. & Agishi, E. (1971) The production and management of *Stylosanthes gracilis* at Shika, Nigeria II. In savanna grassland. *J. agric. Sci., Camb.,* **77** pp. 437–444.

Haskins, B. R., Wise, M. B., Craig, H. B., Blumer, T. N. & Barrick, E. R. (1969) Effects of adding low levels of roughage or roughage substitutes to high energy rations for fattening steers. *J. Anim. Sci;* **29**, pp 345–353.

Horwitz, W. (Ed.) (1970) Official methods of Analysis of the Association of Official Analytical Chemists, Eleventh Edition, Association of Official Analytical Chemists, Washington, D.C.

Hovell, F. D. & De B., Greenhalgh, J. F. D. (1972) Utilisation of salts of acetic, propionic and butyric acids by growing lambs *Proc. Nutr. Soc.,* **31**, 68A.

Kay, R. N. B. (1966) The influence of saliva on digestion in ruminants, *Wld. Rev. Nutr. Diet.,* **6**, pp 292–325.

Kay, M. (1972) Processed roughage in diets containing cereal for ruminants in Cereal Processing & Digestion, US Feed Grains Council, London, pp 39–52.

Lamming, G. E., Swan, H. & Clarke, R. T. (1966) Studies on the nutrition of ruminants, I. Substitution of maize by milled barley straw in a beef fattening diet and its effect on performance and carcase quality, *Anim. Prod.,* **8**, pp. 303–311

Lockwood, J., Swan, H. & Wilton, B. (1973) Diets for beef cattle based on barley grain and ground barley straw, *Proc. Br. Soc. Anim. Prod.,* New Series 2, pp 65.

Maloiy, G. M. O. & Taylor, C. R. (1971). Water requirements of African goats and haired sheep *J. agric. Sci., Camb.* **77** pp. 203–208.

Mason, V. C. & White, F. (1971) The digestion of bacterial mucopeptide constituents in the sheep 1. The metabolism of 2, 6 – diaminopimelic acid, *J. agric. Sci., Camb.,* **77**, pp 91–98.

McCullough, T. A. (1969) A study of factors affecting the voluntary intake of food by cattle *Anim. Prod.*, 11, pp 145–153.

Minson, D. J. (1963) The effect of pelleting and watering on the feeding value of roughage – a review, *J. Br. Grassld Soc.*, 18, pp 39–44.

Montgomery, M. J. & Baumgardt, B. R. (1965). Regulation of food intake in ruminants, 1. Pelleted rations varying in energy concentration, *J. Dairy Sci.*, 48, pp 569–574.

Moore, L. A. (1964) Symposium on forage utilisation: Nutritive value of forage as affected by physical form, Part 1. General principles involved with ruminants and effect of feeding pelleted or wafered forage to dairy cattle. *J. Anim, Sci.*, 23, pp 230–238.

Mowat, D. N. & Cladalade, R. G. (1970) Effect of level of NaOH treatment on digestibility and voluntary intake of straw. *Can. Soc. Anim. Prod. Proc.*, 35.

O'Korie, I. I., Hill, O. H. & MacIlroy, R. J. (1965) The productivity and nutritive value of tropical grass/legume pastures rotationally grazed by N'dama cattle at Ibadan, Nigeria, *J. Agric. Sci., Camb.*, 64 pp. 235–245.

Oyenuga, V. A. (1968) Nigeria's Foods and Feeding stuffs, Ibadan University Press.

Pickard, D. W., Swan, H. & Lamming, G. E. (1969) Studies on the nutrition of ruminants. 4. The use of ground straw or different practicable sizes for cattle from twelve weeks of age. *Anim. Prod.* 11, pp 543–550.

Preston, T. R., Aitken, J. N., Whitelaw, F. G., MacDearmid, A., Philip, Euphemia B. & MacLeod, M. A., (1963). Intensive beef production. 3. Performance of Friesian steers given low fibre diets. *Anim. Prod.* 5 pp 245–249.

Preston, T. R. & Willis, M. B. (1970), Intensive Beef Production, Oxford, Pergammon Press, pp. 324–328.

Rains, A. B., (1963). Grassland Research in Northern Nigeria 1952–62, *Samaru Miscellaneous Paper* No. 1. Institute for Agricultural Research, Ahmadu Bello University, Nigeria.

Randel, P. F. (1970) Dairy beef production from mixtures of sugar cane bagasse and concentrates *J. Agric. Univ. P. Rico. LIV* (2) pp. 237–246.

Raven, A. M., Forbes, T. J. & Irwin, J. H. D. (1969) The utilisation by beef cattle of concentrate diets containing different levels of milled barley straw and of protein *J. agric. Sci., Camb.*, 73 pp 355–363.

Reid, J. T. (1970) The future role of ruminants in animal production, in Physiology of Digestion and Metabolism in the Ruminant, ed. Phillipson, A. T., Oriel Press, Newcastle-upon-Tyne, pp 1–22.

Singh, M. & Jackson, M. G. (1971) The effect of different levels of NaOH spray treatment of wheat straw on consumption and digestibility by cattle, *J. agric. Sci., Camb.*, 77. pp 5–10.

Swan, H. & Lamming, C. E. (1967) Studies on the nutrition of ruminants, II. The effect of level of crude fibre in maize-based rations on the carcass composition of Friesian steers, *Anim Prod.*, 96, pp 203–208.

Swan, H. & Lamming, G. E. (1970) Five studies on the nutrition of ruminants. 5. The effect of diets containing up to 70 per cent ground barley straw on the liveweight gain and carcase composition of yearling Friesian cattle, *Anim. Prod.*, 12, pp 63–70.

Swan, H. & Clarke, V. J. (1974) The use of processed straw in rations for ruminants, *Eighth Nutrition Conference for Feed Manufacturers.* January, 1974.

Thomsen, K. V., Rexen, F. & Kristensen, V. F. (1973) Forsøg med natriumhydroxidbehandling af Halm, 1. Bedhandlingens inflydelse pa halmens fordøjelighed, *Ugeskr, f. Agronomer og Hortonomer*, 25 pp 436–440.

Thomsen, K. V., Rexen, F. & Kristensen, V. F. (1973) 2. Den bemiske sammensaetning af behandlet halm og metoder til maling af oplunkningsgraden *Ugeskr, f. Agronomer og Hortonomer* 26–27 pp 467–470.

Weston, R. H. (1966) Factors limiting the intake of feed by sheep, I. The significance of palability, the capacity of the alimentary tract to handle digesta, and the supply of glucogenic substrate. *Aust. J. agric. Res.* 17 pp. 939–954.

Whyte, R. O., Moir, T. R. G. & Cooper, J. P. (1959) Grasses in agriculture. F.A.O., p. 160.

Williamson, E. & Payne, A. J. A. (1965) *An Introduction to Animal Husbandry in the Tropics*, 2nd ed., London, Longman Green & Co. Ltd., pp. 208, 210.

Wilson, R. K. & Ridgen, W. J. (1964) Effect of NaOH treatment on the utilisation of wheat straw and poplar wood by rumen micro-organisms *Can. J. Anim. Sci.*, 44, pp 122–123.

Discussion

Dr Babatunde: I would like to ask Professor Fuller what precisely he meant; by hot versus cool environment. Is this in fact temperate versus tropical countries? It is difficult to raise fat levels in diets in tropical countries because of the problem of rancidity. What levels of anti-oxidant are needed to prevent this rancidity in diets. We have experienced great difficulty in producing acceptable poultry rations containing more than 5% of fat or more than 2.5% for pig rations in experiments at Ibadan.

Professor Fuller: The cool environment was around 25°C with the diurnal variation occurring under natural conditions. The hot environment was between 33 and 37°C with a diurnal variation. We use adoxiquin as an antioxidant at levels of 0.0125%. The rendering industry in the USA uses butylated hydroxyanisole or butylated hydroxytoluene at levels of around 0.24% but I cannot be sure about this without referring to my notes. Ethoxyquin at levels of 0.0125 to 0.05% is also suitable as an antioxidant. With regard to fat levels, the experiments described in my paper used 10 or 11% with very good effects. We have fed levels of up to 30% of the diet but there were deleterious effects at this level. When feeding high levels of fat in diets it is essential to ensure that adequate levels of nutrients other than energy are present and that the fat is properly stabilised.

Dr Babatunde: We did in fact take all the precautions mentioned by Professor Fuller with regard to other nutrients and the addition of antioxidants, but despite these poor results were still obtained.

Professor Abou-Raya: I would like to mention that we have tried treating roughages with both sodium hydroxide and calcium hydroxide. The effect of the treatment on nutritive value varied with the type of roughage and negative results were recorded with some due to losses of constituents. We have found that some good quality roughages such as dry sweet potato vines and groundnut haulms also benefited greatly from this type of treatment. I would be grateful if Mr. Clarke could inform me as to the method used for determining fineness of grinding of the straw used in the experiments he described, and also the method of milling used together with some information as to its efficiency and economics.

Mr Clarke: In the experiments carried out at Nottingham the straw was very finely ground through a screen size of almost 1 mm using a large hammer mill. This type of processing will of course never be used in commercial practice, and was purely for experimental purposes to see what effect this roughage had on digestive tract function. When normal roughage is put through a chopping process there will be a range of particle sizes. I think that perhaps you are referring to a comment in my paper on fineness of grinding and the effect of this on some beef experiments at Nottingham where an optimum size for fineness of grinding was found. The question of optimum size is rather difficult to explain as the objective is to obtain as much fermentation as possible in the rumen. The good results obtained by Indian workers with sodium hydroxide treated material was because the material was not in fact finely ground thus increasing the period of retention in the rumen. I would like to pass the question on the cost of milling straw to the co-author of my paper Dr Swan who is more competent to answer this than I am.

Dr Swan: It is possible in Britain to work out if it is worthwhile to feed rations diluted with treated straw as opposed to all concentrate rations. I think the metabolic data tends to show that rations including some roughage usually lead to higher food intake than rations without roughage. Dr. Preston's work with high molasses diets showed the beneficial effect of roughage.

Mr MacMichael: I think it is unfair to compare weight gains of cattle in North America with those obtained in Asia. It takes a long time to increase genetic potential by cross-breeding programmes based on exotic cattle imported into tropical countries. Livestock development in humid tropical countries is initially a question of breeding the right stock, and there is generally no shortage of feed once this is done. It is in theory possible to maintain greater numbers of cattle per unit area in the humid tropics than in the temperate developed countries.

Dr Osuji: I would like to ask Professor Fuller to what extent does he consider that the heat increment effect of fats observed for poultry also occurs in ruminants. Also I would like his comments on the use of high fat diets in the tropics to take advantage of the reported lower heat increment associated with feeding energy as fat rather than carbohydrate *vis à vis* the depression of feed intake caused by heat stress.

Professor Fuller: With regard to your second question on the use of high fat diets in the tropics, this subject was dealt with in a series of seminars held in South-East Asia during 1973. Mixed reactions were obtained and in some areas attempts were being made to remove fat from diets rather than including it. I consider that the conditions used in our experiments were much like those in the tropics, and we certainly had our best response to the lowered heat increment by increasing the energy supplied as fat. With regard to ruminants I would expect similar results but perhaps at lower levels, but I cannot comment more than this. Perhaps Mr Clarke could answer this question better.

Mr Clarke: Fat entering the rumen is subjected to a high degree of hydrogenation and nutritional availability is very low. The presence of fat in the rumen can also depress digestibility of roughage and in dairy cows insufficient acetate may be produced to sustain milk fat levels. I do not think we should try to draw direct parallels between poultry and ruminants with regard to the effect of fat in diets because of this massive modifying effect of the rumen which tends to reduce in many circumstances the real energy of fat.

The value of meat and bone meal as the major protein concentrate in the diet of weaners and fattening pigs in the tropics

G. M. Babatunde, B. L. Fetuga and V. A. Oyenuga
University of Ibadan, Nigeria

Summary

Two trials were carried out to investigate the suitability of locally produced meat and bonemeal as an ingredient of rations for pigs in Nigeria. Fifty weaners (average initial liveweight 14.91 kg) and 50 growing-fattening pigs (average initial liveweight 55.2 kg) were used for the trials.

It was found that high levels of meat and bonemeal in the diets of weaners significantly depressed growth rates, feed consumption and feed efficiency, and that supplementation with zinc or copper did not improve matters. Similar trends were observed with the growing-fattening pigs.

There was no evidence that high levels of meat and bonemeal in diets adversely affected carcass quality, and in fact the contrary was indicated. It was concluded that meat and bonemeal is not a satisfactory protein concentrate for use alone in the diets of weaners or growing-fattening pigs.

Résumé

La valeur des pâtures de viande et d'os comme concentrés protéiniques majeurs dans les diètes de cochons sevres et des cochons en voie d'engraissement, dans les régions tropicales

Deux expériences ont été entreprises pour investiguer la convenance des pâtures de viande et d'os produites localement comme un ingrédient des rations pour les cochons en Nigeria. Cinquante cochons sevrés (poids utile initial moyen 14.91 kg) et 50 cochons en voie d'engraissement (poids utile initial moyen 55.2 kg) ont été utilisés pour les expériences.

On a trouvé que des quantités élevées de pâture de viande et d'os dans les diètes des cochons sevrés, baissaient le rythme de croissance, la consommation et l'efficacité des pâtures, et que les suppléments de zinc ou cuivre n'amélioraient aucunement la situation. Des tendances similaires ont été observées avec les cochons en voie d'engraissement.

Il n'y a pas d'évidence que des quantités élevées de pâture de viande et d'os dans les diètes affectent défavorablement la qualité de la carcasse et en réalité, le contraire a pu être démontré. On est arrivé à la conclusion que la pâture de viande et d'os n'est pas un concentré protéinique satisfaisant pour être utilisé seul dans la diète des cochons sevrés ou en voie d'engraissement.

Resumen

El valor de la harina de carne y de huesos como el principal concentrado de proteínas en las dietas de los cerdos destetados y en engorde en los trópicos

Se llevaron a cabo dos pruebas para investigar la adecuación de la harina de carne y de huesos producida localmente como ingrediente de las raciones de los cerdos en Nigeria. Se usaron para las pruebas cincuenta cerdos destetados (promedio de peso en vivo inicial 14, 91 kg) y 50 cerdos de engorde en proceso de crecimiento (promedio de peso en vivo inicial 55, 2 kg).

Se halló que niveles altos de harina de carne y de heusos en las dietas de los cerdos destetados reducían los índices de crecimiento de manera significativa, el consumo de alimento y la eficacia del alimento, y que el suplemento con cinc o cobre no mejoraba las cosas. Se observaron tendencias semejantes en los cerdos de engorde.

No hay evidencia de que altos niveles de harina de carne y de huesos afectasen de manera adversa la calidad de la carcasa, y, en realidad, se demostró lo contrario. Se llegó a la conclusión de que la harina de carne y de huesos no es un concentrado de proteína satisfactorio usado sólo en las dietas de los cerdos destetados y de los que están en proceso de crecimiento y engorde.

Introduction

With the present trend of rising prices of animal feed-stuffs all over the world, greater attention is inevitably being paid to the use of cheaper sources of protein, especially in the developing countries that can ill afford the luxury of feeding fishmeal or milk powder to pigs and poultry. There are quite a few of these cheap sources of protein, but one of the most promising is meat and bone meal (MBM), not only because it is reasonably high in protein, but also because it is an animal protein concentrate. The adequacies of meat and bonemeal in the diets of pigs and poultry have been investigated by Batterham & Holder (1969a, 1969b), Gartner & Burton (1965), Todd & Daniels (1965), Luce et al. (1964) and Batterham & Manson (1970). Henson, et al (1954) had raised doubts about the adequacy of meat and bonemeal in sustaining good growth when fed as the only source of protein supplement in the rations of growing-finishing pigs because it is deficient in tryptophan, and its protein also has a low biological value due to a high level of connective tissue. It also has a high calcium content due to the presence of a large proportion of bone which can induce zinc deficiency in diets containing MBM at high levels.

MBM is, therefore, a product that can easily exhibit a wide degree of variability in its quality, depending on such factors as the amounts of connective tissue and bone included in the product during processing. Consequently the MBM samples produced and used in the temperate countries will not necessarily give the same responses to the types produced in the tropical countries. The investigations reported here were, therefore, carried out to look into the possibility of using the locally-produced MBM samples as the major protein concentrate in the practical diets of weaners and growing-fattening pigs in Nigeria.

Experimental procedures

The breed of the pigs used in both trials was the Yorkshire, and both sexes were represented, although the males had earlier been castrated at least three weeks before they were used in the trials. The first trial involved 50 weaners of average initial liveweight of about 14.91 kg. They were randomly divided into five equal treatment groups on the bases of sex and liveweight, so that each group had equal initial live-weight, and the sex ratios were the same, 6 gilts to 4 barrows. They were then allocated to pens equipped with individual feeding units and were fed diets containing approximately 22% protein compounded as shown in Table 1. Diet 1 was the standard weaners' diet used as the control and contained no MBM, while diet 2 was the same as the control diet, except that MBM replaced the groundnut meal in diet 1 so as to supply the same level of protein. Diets 3, 4, and 5 all contained 33% MBM as the only protein concentrate, diets 4 and 5 also being supplemented with 200 ppm Zn and 80 ppm of Cu respectively.

In trial 2, the same procedures of alloting the animals to treatments as in trial 1 were adopted and in the final grouping, the average initial liveweight was about 55.2 kg, and the sexes were the same. The diets used were formulated as shown in Table 1. Diet 1 was again the control standard growers' diet containing no MBM, while diets 2 and 3 contained 18.5% and 24.5% MBM respectively supplying approximately 50% and 60% of the total dietary protein. Diets 4 and 5 contained 15.5 and 21.0% of groundnut meal respectively, supplying 50% and 60% of the total dietary protein.

All the pigs were individually fed weighed quantities of feed four times daily, at 07.00 hr, 11.00 hr, 15.00 hr and 18.00 hr, each feeding period lasting one hour. After each feeding, they were driven out into the play pens where they drank water ad libitum. Records were kept of all feeds consumed and liveweight gained, the pigs being weighed once every week before the morning feed was given. Trial 1 was stopped after 84 days while trial 2 were carried up till the pre-determined slaughter weight of about 91.0±1.0 kg was reached. At that time, the pigs were taken off feed and fasted for 18 hours before being slaughtered and subjected to conventional grading. In this the carcass length was taken from the first rib to the tip of the aitch bone, the average back-fat thickness being the average of the three measurements taken opposite the first rib, last rib, and the lumbar vertebra. The loin eye area was the planimeter tracing of the vertical section of the longissimus dorsi muscle sectioned across the tenth rib, and the loin, ham and shoulders were all sectioned, trimmed and weighed. All these carcass measurements were made on the chilled carcass after the warm carcass had been chilled in the cold room maintained at about 5°C.

All the data except those of the average initial and final liveweights were subjected to an analysis of variance (Steel & Tovie, 1960) followed by the LSD test where significant F values were indicated.

Results

A summary of the performance characteristics of the pigs on both trials appears in Table 2.

Average daily gain

In trial 1, highly significant differences in performance ($P < 0.01$) were obtained. The pigs on high levels of MBM (treatments 3, 4 and 5) gained at a highly significantly slower rate ($P < 0.01$) than those pigs on the control diet or diet 2. Pigs on diet 2 also grew at a highly significantly slower rate than those on the control diet, while the pigs on diet 5 with copper supplementation grew at a significantly slower rate than those on diet 4 with zinc supplementation ($P < 0.05$).

TABLE 1. *Composition of rations used in trials 1 and 2, (%)*

Ingredients	Trial 1 Treatments					Trial 2 Treatments				
	1	2	3	4	5	1	2	3	4	5
Yellow maize	64.0	59.5	61.0	61.0	61.0	70.5	72.5	66.5	73.5	68.00
Rice bran	3.0	3.0	3.0	3.0	3.0	7.5	7.5	7.5	7.5	7.5
Meat and bonemeal	–	26.5	33.0	33.0	33.0	–	18.5	24.5	–	–
Groundnut meal	20.0	–	–	–	–	15.0	–	–	15.5	21.0
Fishmeal	3.0	3.0	–	–	–	1.5	–	–	–	–
Blood meal 8	4.5	4.5	–	–	–	2.0	–	–	–	–
Molasses	1.5	1.5	1.5	1.5	1.5	1.0	1.0	1.0	1.0	1.0
Oyster shell	0.5	–	–	–	–	0.5	–	–	–	–
Dicalcium phosphate	1.5	–	–	–	–	1.5	–	–	–	–
Vit. min. premix*	++	++	++	++	++	++	++	++	++	++
Salt	0.5	0.5	0.5	0.5	0.5	0.5	0.5	0.5	0.5	0.5
Supplemental zinc (ppm)	–	–	–	200.0	–	–	–	–	–	–
Supplemental copper (ppm)	–	–	–	–	80.0	–	–	–	–	–
TOTAL	100.0	100.0	100.0	100,0	100.0	100.0	100.0	100.0	100.0	100.0
Crude protein content by analysis (% dry weight)	20.50	22.45	22.50	22.41	22.34	15.65	14.80	16.80	14.40	16.73
Calcium (Ca) content by analysis (% dry weight)	0.75	3.70	3.82	3.888	3.81	0.85	2.67	3.35	0.96	1.34
Phosphorus (P) content by analysis (% dry weight)	0.63	1.68	1.92	1.96	1.90	0.65	1.26	1.79	0.68	0.70

*A Pfizer product supplying the following per kg diet: Vit. A, 9823 IU; Vit. D_3, 1965 IU; Vit. B_{12}, 10 ug/t; Riboflavin, 41 mg, Niacin, 246 mg; Pantothenic acid, 98 mg; Copper, 244 mg; Manganese, 341 mg; Zinc, 100 mg; Fe, 100 mg; Iodine, 20 mg; Folic acid, 10 mg; and Oxytetracycline hydrochloride, 20 g/t.

TABLE 2. *Performance characteristics of pigs feed locally-produced meat and bone meal diets in Nigeria*

Characteristics	Mean figures for each treatment					SE of means
	1	2	3	4	5	
Trial 1: Weaners						
Average initial liveweight (kg)	15.06	14.80	14.94	14.94	14.80	–
Average final liveweight (kg)	62.94	44.20	27.54	31.74	24.04	–
Average daily gain (kg)	0.57	0.35	0.15	0.20	0.11	0.016*
Average daily feed consumed (kg)	1.56	1.28	0.85	1.00	0.71	0.074*
Average feed consumed/kg of weight gained	2.76	3.67	5.92	5.14	6.48	0.20*
Trial 2: Fatteners						
Average initial liveweight (kg)	54.80	54.80	55.00	56.80	55.20	
Average final liveweight (kg)	93.04	90.07	85.75	90.43	92.39	
Average daily gain (kg)	0.65	0.45	0.41	0.55	0.60	0.03*
Average daily feed consumed (kg)	2.30	2.01	1.95	2.26	2.36	0.11*
Average feed consumed/kg of weight gained	3.54	4.46	4.76	4.11	3.93	0.31*

*Highly significant treatment differences among the means $P < 0.01$. See text for details.

TABLE 3. *Carcass quality measurements of pigs fed meat and bonemeal diets in Nigeria*

	Mean figures for each treatment					SE of means
	1	2	3	4	5	
Average warm carcass weight (kg)	70.12	68.72	65.33	68.91	70.05	–
Average backfat thickness (cm)	3.71	3.39	3.05	3.62	3.42	0.23 (NS)
Average carcass length (cm)	78.93	78.54	76.23	78.31	78.86	1.23 (NS)
Loin as per cent of chilled carcass	14.20	14.80	15.12	14.57	14.61	0.49 (NS)
Ham as per cent of chilled carcass	24.81	25.33	25.94	24.90	24.82	0.80 (NS)
Shoulder as per cent of chilled carcass	20.70	21.50	21.70	20.81	20.83	0.64 (NS)
Average yield of lean cuts (%)	59.75	61.63	62.76	60.28	60.26	1.40 (NS)
Average loin eye area (cm^2)	23.20	21.81	20.62	22.58	23.60	1.03*
Average trimmed fat in chilled carcass (%)	16.68	15.00	13.40	15.84	16.80	1.26 (NS)

NS = No significant treatment differences among the means.
* = Highly significant differences among the means: ($P < 0.01$).

In trial 2, the differences in performance were also highly significant ($P < 0.01$). The best gainers were the fatteners on the control diet while the poorest gainers were those on the 24.5% MBM (diet 3) which grew at a significantly slower rate than those pigs on diets 1, 4 and 5 ($P < 0.01$) Pigs on the 18.5% MBM (diet 2) also grew at a significantly slower rate than those on the control diet ($P < 0.01$) and diet 5 ($P < 0.05$).

Average daily feed consumed

Highly significant differences in feed consumption were also obtained ($P < 0.01$). The best figure for average daily feed consumed was obtained for diet 1, which was conspicuously better than the figures for diets 3, 4 and 5 ($P < 0.01$) and 2 ($P < 0.05$). The next best diet consumed was diet 2, the average daily feed consumption figure of which was highly significantly better than those for diets 5 ($P < 0.01$) and 3 and 4 ($P < 0.05$).

In trial 2, feed consumption for the treatments were also highly significantly different from one another. The lowest consumption figure of 1.95 kg (diet 3) was highly significantly lower than those for pigs on diets 1, 4 and 5 ($P < 0.01$), while the 2.01 kg recorded for diet 2 was also significantly lower than that for diet 5 ($P < 0.05$).

Average efficiency of feed utilisation

There were highly significant differences in the efficiency of feed utilisation in both trials. In trial 1, the best converters of feed were pigs on diet 1, closely followed by the pigs on diet 2, and the feed conversion ratios for both diets 1 and 2 were highly significantly better than those for pigs on diets 3, 4 and 5 ($P < 0.01$). Similarly, pigs on diet 4 had a highly significantly better feed conversion ratio than those on diet 5 ($P < 0.01$) while that for pigs on diet 1 was significantly better than that for pigs on diet 2 ($P < 0.05$).

In trial 2, the best feed conversion ratio was also from pigs on the control diet, and the poorest was from those pigs on 24.5% MBM. The poorest feed conversion ratio of 4.76 obtained for pigs on diet 3 was highly significantly poorer than those obtained for pigs on diets 1, 4, 5 ($P < 0.01$), while the feed conversion ratio of 4.46 obtained for diet 2 was significantly poorer than that for diet 2 ($P < 0.05$).

Carcass measurements

Table 3 summarises the results obtained from the examination of the carcasses of the pigs. There were no significant differences between the carcasses from the different treatments except that the average loin eye areas showed highly significant differences (P<0.01). The smallest loin eye area measurement of 20.62 cm² obtained for the pigs on diet 3 was highly significantly smaller than the averages for pigs on diets 1 and 5 (P<0.01), and diet 4 (P<0.05). All items of lean cuts, the loin, the ham, and the shoulder cuts individually or jointly expressed as percentages of chilled carcass favoured the higher MBM diet, followed closely by the 18.5% MBM diet. On the other hand, all the fat measurements including the backfat thickness and the per cent trimmed fat were higher for the control and groundnut meal diets than for the MBM diets, although the differences were not significant.

Discussion

The performance characteristics of the weaners and the growing fattening pigs on these various MBM diets were definitely all inferior to the pigs on either the control diets or the groundnut meal diets. These results are in good agreement with the earlier findings of Bloss *et al.* (1953) who found that the inclusion of MBM into wearners' diets at levels of up to 20% was highly unsatisfactory for the growth of weaner pigs unless they were supplemented with DL-tryptophan, and with the findings of Hensen *et al.* (1954), Batterham & Holder (1969a, 1969b), Sathe *et al.* (1964) among others. These high levels of MBM not only depressed growth rate significantly, but also significantly depressed appetite and reduced the efficiency of feed utilisation (as reported by Gartner & Burton, 1965, Batterham & Holder, 1969a, 1969b, and Peo & Hudman, 1965). The degree of depression depended on the level of inclusion of the MBM in the diets as was also reported by the workers cited above.

A few plausible explanations could be advanced for these depressions by high levels of MBM. There appeared to be a greater growth depression with the present diets containing MBM than those reported elsewhere, perhaps because the quality of the MBM used in the present trials was below those used in the temperate countries, Todd & Daniels (1965) have shown for example great differences in the qualities of different samples of MBM in Queensland. Besides, the present MBM samples had higher calcium levels (about 12%) than the levels commonly reported in the literature due presumably to the higher levels of bone inclusion in the finished product as suggested by Gartner & Burton (1965) and Sathe *et al.* (1964). With such high calcium levels in the diets, appetite would be reduced as our investigations showed, and hence feed consumption was significantly depressed with subsequent growth depression. High calcium levels would also accentuate the need for high levels of zinc, and with this in mind, zinc was added to one of the high MBM diets for fatteners but while it did show slight improvement, it did not significantly change the situation. Addition of copper even worsened the situation, even though the level of copper supplementation was far below the level

suggested by the National Research Council (1968) at which it could be toxic. This finding contradicts that of Batterham (1970) who reported a slight, though not significant improvement when he supplemented his MBM diet with 250 ppm Cu.

The other plausible explanation for these poor responses was that of amino acid deficiencies, notably tryptophan and methionine, and sometimes lysine. Bloss *et al.* (1953), Henson *et al.* (1954), Luce *et al.* (1964) among others, have conclusively shown that diets containing MBM given better responses when supplemented with these limiting amino acids, particularly tryptophan and methionine.

The carcass data were also in accordance with the findings of Batterham and Holder (1969b), although treatment differences were not significant. The same trend towards carcass improvement with increase in levels of MBM was evident. This, however, could have been due to the depressed feed intake which led to depressed rate of gain, and if the pigs could not satisfy their caloric requirements per day, then they would not be able to lay down so much fat as that laid down by the pigs on the control and groundnut meal diets that were well consumed.

In conclusion, meat and bonemeal is a highly unsatisfactory source of protein to be used alone in the diets of weaners and growing-fattening pigs. It appears to be better tolerated by the fattening pigs than the weaners, but its use as the major source of protein especially when included at levels higher than 15% is not recommended. It can be combined with other protein supplements for better results but if it has to be used at high levels, it may require adequate fortification with the limiting amino acids and zinc.

References

Batterham E. S. (1970) A nutritional evaluation of diets containing meal meat and bonemeal for growing pigs. 3. The effects of amino acid, copper and vitamin B supplements. *Austr. J. Expt. Agric. Anim. Husb.* 10, (1) 27–31.

Batterham, E. S. & Holder, J. M. (1969a) A nutritional evaluation of diets containing meat meal for growing pigs. 1. The effect of calcium level in wheat-animal protein diets. *Austr. J. Expt. Agric. Anim. Husb.* 9. (36), 43–46.

Batterham, E. S. & Holder, J. M. (1969b) A nutritional evaluation of diets containing meat meal for growing pigs. 2. The effects of level of meat meal or meat and bonemeal in wheat based diets. *Austr. J. Expt. Agric. Anim. Husb.* 9, (39), 408–412.

Batterham, E. S. & Manson, M. B. (1970) A nutritional evaluation of diets containing meat meal for growing pigs. 7. The value of meat meal as a protein supplement to oats, barley, sorghum and wheat based diets. *Austr. J. Expt. Agric. Anim. Husb.* 10, (46), 539–543.

146 Bloss, R. E., Luecke, R. W., Hoefer, J. A., Thorpe, F. & McMillen W. N. (1953) Supplementation of a corn-meat and bone scrap ration for weaning pigs. *J. Anim Sci.* 12, (1), 102–106.

Gartner, R. J. W. & Burton, H. W. (1965) Evaluation of meat and bonemeals in rations for growing chickens. I. Effect of varying levels of blood and bone. *Queensland J. Agric. Anim. Sci.* 22, (1), 1–15.

Henson, J. W., Beeson, W. M. & Perry, J. W. (1954) Vitamin, amino acid and antibiotic supplementation of corn-meat by-product rations for swine. *J. Anim. Sci.* 13, (4), 885–898.

Luce, W. G., Pao, E. R. & Hudman, D. B. (1964) Effect of amino acid supplementation of rations containing meat and bone scraps on rate of gain, feed conversion and digestibility of certain ration components for growing-finishing swine. *J. Anim. Sci. J. Anim.* 23, (2), 521–527.

National Academy of Sciences – National Research Council (1968) Nutrient Requirements of Domestic Animals. No. 2, Nutrient Requirements of Swine. 6th Revised ed.

Peo, E. R. & Hudman, D. B. (1962) Effect of levels of meat and bone scraps on growth rate and feed efficiency of growing-finishing swine. *J. Anim. Sci.* 21, (4), 787–790.

Sathe, B. S., Cumming, R. B. & McClymont, G. L. (1964) Nutritional evaluation of meatmeals for poultry. I. Variation in quality & its association with chemical composition and ash and lipid factors. *Austr. J. Agric. Res.* 15, (1), 200–213.

Steel, R. G. D. & Torrie, J. H. (1960) Principles and procedures of statistics with special reference to biological sciences. New York, Toronto, London, McGraw-Hill Book Co. Inc.

Todd, A. C. E. & Daniels, L. J. (1965) Comparison between fishmeal and two Queensland meat and bone-meals in bacon production. *Austr. J. Expt. Agric. Anim. Husb.* 5, (19), 404–409.

Utilisation of pyrethrum waste product as a livestock feed

H. R. Were

Senior Animal Husbandry Research Officer, Ministry of Agriculture, Kenya

Summary

An account of work carried out in Kenya on the utilisation of pyrethrum marc (the waste remaining after the extraction of insecticides from pyrethrum flowers) is given. Experimental results to date indicate that pyrethrum marc can be used at levels not exceeding 50 percent of the total ration in rations for sheep and dairy cattle, and that it can also replace a substantial proportion of maize silage in cattle rations in feedlot operations. On the basis of these results pyrethrum marc is now being used as an ingredient of cattle rations in commercial feedlots in Kenya.

Résumé

L'usage des déchets de pyrèthre comme pâture pour le bétail

On fait un compte rendu des recherches entreprises en Kenya sur l'utilisation du marc de pyrèthre (le déchet restant après l'extraction de l'insecticide des fleurs de pyrèthre). Les résultats expérimentaux actuels indiquent que le marc de pyrèthre peut être utilisé dans des proportions qui ne dépassent 50% de la quantité totale des rations pour les moutons et le bétail de ferme et qu'il peut aussi remplacer une proportion substantielle de fourrage de mais dans les rations pour les bétail. Sur la base de ces résultats, le marc de pyrèthre est utilisé maintenant comme un ingrédient dans les rations pour le bétail, inclu dans les pâtures commerciales en Kenya.

Resumen

Utilización de los productos de desecho de pyrethrum como pienso para el ganado

Se hace una relación del trabajo llevado a cabo en Kenya sobre la utilización de pyrethrum marc (los restos que quedan después de la extracción de insecticidas de las flores de pyrethrum). Los resultados experimentales hasta la fecha indican que puede usarse el pyrethrum marc a niveles que no excedan del 50% de la ración total en piensos para corderos y vacas de leche, y que puede también reemplazar una proporción sustancial de ensilaje de maíz en raciones para el ganado en operaciones de lotes de pienso. Sobre la base de estos resultados, el pyrethrum marc se está usando ahora como un ingrediente en las raciones de ganado en los lotes de pienso comercial en Kenya.

Kenya like, most other developing countries, suffers greatly from a constant shortage of livestock feeds, especially those supplying protein and energy. This situation tends to be highly magnified due to two main reasons. Firstly, a very high competition exists between stock and human population for the same source of food particularly energy feeds such as maize, sorghum grains, cassava and wheat on one hand, and the protein feeds such as soya bean, and fish, on the other hand. Whereas these sources of feeds form the basic constituents of most livestock rations particularly for the non-ruminants, they also form the major source of human food. The choice therefore has to be made between feeding human population for bare survival or feeding pigs, poultry or dairy cows.

Secondly, the existence of uncertain weather conditions in large areas of East and Central Africa with frequent periods of drought has become a major cause of constant food deficit and the situation has often caused concern and at present the situation is serious. The main worry in most of the countries mentioned is to keep the population alive. Already, several thousands of animals (both domesticated and wild) are dying

TABLE 1. *Chemical analysis of pyrethrum marc (%)*

	Fresh material	Air-dry	Material
Moisture	21.75*	0*	0**
Ash	5.55	7.10	12.7
Digestible crude protein	–	–	8.1
Fat	0.40	0.50	–
Total digestible nutrients (TDN)	–	–	66
Nitrogen-free extract (NFE)	43.60	55.75	63
Crude fibre	18.50	23.60	22.9

Sources: * Ayre-Smith, (1956)
        ** Dougall (1960)

every week and the decimation will continue unless the weather changes its present pattern. Signs of kwashiakor and marasmus are a common feature in the children, and general malnutrition a concern in pregnant mothers, old women and men.

The above factors explain the need for livestock producers to be looking into ways and means of utilising, for livestock production purposes, those feeds which have little direct competition for human food. In recent years therefore, research workers in Kenya have been looking into the value and use of certain farm by-products such as molasses (UNDP/FAO, 1973), maize and barley stover (Rogerson, 1956a), coffee waste (Rogerson, 1955), pineapple residues (Rogerson, 1956b), and pyrethrum by-products (Ayre-Smith, 1956), for feeding ruminant animals and, a small extent non-ruminants such as pigs.

The present paper gives some account of the advances made in the use of pyrethrum waste (generally known as pyrethrum marc) as a livestock feed in Kenya.

Pyrethrum marc is the waste product obtained from dried pyrethrum flowers (*Chrysanthemum cinerariae-folium*) after they have been ground and the pyrethrins extracted with petroleum ether. To purify the pyrethrum marc, it is further steam treated to remove any residual solvent and also to destroy any residues of pyrethrins remaining after extraction as these are poisonous to stock.

Currently Kenya produces large amounts of pyrethrum marc annually, and the production is expected to increase in the future due to the anticipated expansion in the production of pyrethrum flowers to meet the world demand for pyrethrin insecticides.

In recent years pyrethrum marc has been of little economic value to Kenya's livestock feeding. It was not until 1956 when small quantities of the material were fed to cattle that it was noted animals accepted the material quite readily. There were also no observed off-flavours in milk. Typical values for the major constituents of pyrethrum marc are given in Table 1 (Ayre-Smith, 1956; Dougall, 1960) and the mineral status in Table 2 (Ayre-Smith, 1956).

These results of mineral status compare very well if not better in some cases than those found by Chamberlain (1955) in his study of the major and trace element composition of some East African feeding stuffs. Of notable importance are the phosphorus and copper levels which are high and the slightly low calcium level.

TABLE 2. *Mineral status of pyrethrum marc, (dry matter basis*)*

Calcium as CaO (%)	0.739
Magnesium as MgO (%)	0.421
Sodium as Na_2O (%)	0.600
Potassium as K_2O (%)	3.110
Copper as Cu (ppm)	14.500
Phosphorus as P_2O_5 (%)	0.559

Source: * Ayre-Smith (1956)

It will be noted from Table 3 that the percentages of crude protein and digestible crude protein in pyrethrum marc are only comparable to that of sunflower heads with seed. The nutritive ratio compares very well with that found in many green fodders in Kenya, while the energy level is also highly satisfactory as a stock feed.

TABLE 3. *Comparison of composition of pyrethrum marc with other farm by-products*

Component	Crude protein	Digestible crude protein	Nutritive ratio	Total digestible nutrients	Gross digestible energy	Crude fibre
Coffee hullings (waste)	2.0	0.2	138.0	28	28	35.9
Cottonseed husks	3.2	6.6	59.0	41	41	46.1
Maize cobs without grain	1.3	0.1	–	20	–	36.3
Maize on the cob meal	5.5	2.2	26.3	58	59	28.7
Pyrethrum marc	12.7	8.1	66.8	63	66	22.9
Sisal waste	5.3	2.1	29.5	64	65	34.7
Sorghum stalks with head and some grain	7.8	4.1	15.9	69	71	16.1
Sunflower heads with seed	14.0	9.2	5.6	61	65	23.9
Sunflower head without seed	7.4	3.7	17.0	67	68	20.7
Wheat straw	5.3	1.8	25.7	48	49	41.9

Source: Dougall (1960)

Because of the excellent feeding value of pyrethrum marc compared with other farm by-products great attention has now been focussed on it mainly as a source of roughage but also as a source of some protein. Ayre-Smith (1956) reported good performance in the growth rate of sheep fed 33.3% of pyrethrum marc over those fed 66.6% pyrethrum marc in rations also containing lucerne meal, oat meal and barley meal.

In another trial where Naivasha star grass (*Cynodon plectostachyum*) was compared with pyrethrum marc as a feed for milking cows at inclusion rates of 30 and 50% in the ration, those cows fed rations containing 30% pyrethrum marc had better milk yields than the cows on either star grass or on rations containing 50% of pyrethrum marc.

In both the sheep and dairy trials, feeding pyrethrum marc at levels of more than 50% of the total ration depressed growth rates and milk yields. Conclusions drawn from the results of the above trials indicated that pyrethrum marc could be usefully employed as part of balanced rations for ruminants.

To follow up this work, a series of trials was set up at the Beef Research Station, Nakuru, Kenya, to determine the value of wheat straw, pyrethrum marc and sunflower husks as substitutes for part of the silage component of rations fed to improved Boran steers in feedlot operations at the station. The composition of control rations used in Trials 1 and 2 are shown in Table 4. These rations are generally known as Beef Research Station (BRS) Ration 4 and 5 as there are other standard rations ranging from Rations 1 to 6 for feedlot fattening operations. In Trial 1, 196 improved Boran steers were used in the feedlots with the following treatments:

Lot 1 Control: Normal ration 4 based on maize silage as source of roughage.

Lot 2 Wheat straw: Straw and maize silage each contributed 50% of the roughage on a dry matter basis

Lot 3 Pyrethrum marc: Pyrethrum marc and maize silage each contributed 50% of the roughage on a dry matter basis.

Lot 4 Sunflower hulls: Sunflower hulls and maize silage each contributed 50% of the roughage on a dry matter basis.

Equal numbers of the cattle were selected at random for slaughter on two different dates to facilitate the work on carcass analysis, the results of which are given in Table 5.

TABLE 4. *Calculated ration components for the control rations 4 and 5 used in Beet Research Station trials 1 and 2*

Ingredient	%	ME	NE_m	NE_g	CP	DP
Ration 4						
Maize	36.66	1.21	0.82	0.53	3.63	2.71
Forage	50.17	1.27	0.81	0.48	33.61	21.16
Mineral urea molasses (MUM)	10.52	0.32	0.20	0.12	2.21	1.50
Cottonseed cake (CSC)	2.65	0.07	0.05	0.03	1.16	0.94
Total	100.00	2.87	1.88	1.16	10.61	7.31
Ration 5						
Maize	52.83	1.17	1.18	0.77	5.23	3.91
Forage	33.42	0.85	0.54	0.32	2.41	1.44
MUM	11.10	0.34	0.21	0.13	2.33	1.59
CSC	2.65	0.07	0.05	0.03	1.16	0.94
Total	100.00	3.00	1.98	1.25	11.13	7.88

TABLE 5. *Performance of 196 improved Boran Steers fed maize silage, wheat straw, pyrethrum marc and sunflower hulls and Ration 4 as control*

Treatment	Maize silage	Wheat straw	Pyrethrum marc	Sunflower hulls
Average no. of days on feed	105.5	105.5	105.5	105.5
No. of animals on feed for 2 periods	48	50	47	45
Average initial weight (kg)	268.5	268.5	269.5	270.
Average final weight (kg)	401	385	403	402.5
Average gain (g)	132.5	116.5	133.5	132.5
Average daily gain (g)	1255.0	1105.0	1262.5	1260.0
Cold carcass weight (kg)	204	190.5	208.5	200
Dressing (%)	50.85	49.4	51.85	49.7
Average adj. daily gain (g)	1187	980	1315	1150
Average grade ratings	4.46	4.13	4.68	4.45
Carcass Assessment				
Fat (%)	22.65	21.4	24.45	22.5
Fat (kg)	46.	41.0	51.0	45.0
Lean (%)	60.6	61.35	59.5	61.5
Lean (kg)	123.5	116.5	124	121
Bone (%)	16.75	17.25	16.05	16.8
Bone (kg)	34.	33	33	34

TABLE 6. *Performance of 192 large crossbred steers fed maize silage, wheat straw, pyrethrum marc and sunflower hulls and Ration 5 as control*

Treatment	Maize	Wheat straw	Pyrethrum marc	Sunflower hulls
No. of days on feed	119.5	119.5	119.5	119.5
No. of animals on feed for two periods	45	47	46	46
Average initial weight (kg)	218.5	292	289	289
Average final weight (kg)	426	422	425	422.5
Average total gain (g)	137.5	130.5	136	133.5
Average daily gain (g)	1150	1092.5	1140	1070
Cold Carcass weight (kg)	220	211.5	224.5	213.5
Dressing (%)	51.6	50.5	52.8	50.6
Average adjusted daily gain weight (g)	1180	1020	1250	1060
Average grade	4.48	4.28	4.47	4.09
Carcass Assessment				
Fat (%)	21.2	19.2	22.9	20.25
Fat (kg)	46.5	42	51.5	43.5
Lean (%)	61.4	62.2	60.45	61.95
Lean (kg)	135.5	131.5	135.5	132.0
Bone (%)	17.35	18	16.65	17.8
Bone (kg)	38	38	37.5	38

In trial No. 2. 192 large cross-bred steers were assigned to four lots as in Trial 1 above but fed Ration 5 of the BRS. Results of the trial are shown in Table 6.

In both trials, the animals fed on pyrethrum marc performed as well as those fed the normal ration containing maize silage as the whole source of roughage and at the same time produced carcasses with a higher fat percentage and of a generally better grade than the control group. Sunflower hulls and wheat straw did not substitute as well as pyrethrum marc for silage. The animals fed straw performed worst and only gave a fairly satisfactory performance, grading an average of 4.27 compared to 4.58 for pyrethrum marc and 4.47 for maize silage. The average adjusted daily gain of the two breeds and four feeding periods was 1,183g for control and 1,283g for the group on pyrethrum marc but for straw fed animals dressing averaged 49.7% as compared to 51.2% for silage and 52.5% for pyrethrum marc fed groups.

This was reflected in the amount of lean meat obtained. The performance of animals fed sunflower hulls was very similar to that of the animals fed straw.

The practical implications of these results are now being used on a very wide scale in the feedlot operations in Kenya. At the moment, there are over 10 commercial feedlots with stock ranging from 2,000 head at any one time to 10,000 head per feedlot. Because drought conditions have caused a shortage of silage, most of the feedlots are now substituting pyrethrum marc for a very substantial part of the silage in the ration. In Ration 4, pyrethrum marc constitutes 18% of the total ration with a dry matter of 89%, while silage constitutes 65% of the ration with a dry matter of 20%. In Ration 5, silage forms 38% of the total ration and dry matter of 26% and pyrethrum marc forms 13.3% by composition and a dry matter of 89%. These compositions with pyrethrum marc have proved quite satisfactory for fattening animals in feedlots without loss in grade or quality.

References

Ayre-Smith, R. A. (1956) Pyrethrum waste as a stock feed. *Field and Farm*, May.

Chamberlain, G. T. (1955) The major and trace element composition of East African Feeding stuffs *E. Afr. Agric. J.*, 21, 103.

Dougall, H. W. (1960) Average nutritive values of Kenya feeding stuffs for ruminants. *E. Afric. Agric. J.*, 26, 119–128.

Rogerson, A. (1955) Nutritive value of coffee hulls. *E. Afr. Agric. J.*, 20, 254–255.

Rogerson, A. (1956a) Feeding value of local barley, maize and oat straws. *E. Afr. Agric. J.*, 21, 159–160.

Rogerson, A. (1956b) Nutritive value of pineapple residues (dried). *E. Afr. Agric. J.*, 21, 163.

UNDP/FAO (1973) Beef industry development project, Ken/72/008. Working paper on intensive beef production; feeding trials using maize based rations. Nov. 1973.

Tuber meals as carbohydrate sources in broiler rations

A. L. Gerpacio, D. B. Roxas, N. M. Uichanco, N. P. Roxas, C. C. Custudio, C. Mercado, L. A. Gloria and L. S. Castillo

Department of Animal Science, University of the Philippines, Los Baños, College of Agriculture, Laguna

Summary

Six trials were carried out on the use of locally produced tubers as energy sources for broilers. The tubers used were sweet potatoes (*Impomea batatas*), cassava (*Manihot utillissima*), gabi (*Colocasia esculentum*), pongapong (*Amorphallus campanulatus*) and ubi (*Dioscorea alata*). It was shown that sweet potato and cassava meals can partially replace yellow maize in broiler rations. However, the two meals were inferior to maize in promoting weight gain. These findings are in accordance with those of earlier work carried out elsewhere in which it was observed that maize could only be replaced by sweet potato meal to the extent of 50–57%.

Birds fed ubi, pongapong and gabi meals performed poorly with significantly lower rates of gain and feed efficiencies. Gabi meal, however, was very palatable and relished by the birds.

Résumé

Les pâtures de tubercules comme source de hydrate de carbone pour les rations des poulets a rôtir

On a entrepris six expériences sur l'utilisation des tubercules produits localement comme sources d'énergie pour les poulets à rôtir. Les tubercules utilisés étaient des patates (*Impomea batatas*), manioc (*Manihot esculenta*), gabi (*Colocasia esculentum*), pongapong (*Amorphallus campanulatus*) et ubi (*Dioscorea alata*). On a montré que les patates (pommes de terre douces) et les pâtures de manioc peuvent remplacer partiellement le mais jaune dans les rations pour les poulets rôtis. Cependant les deux pâtures étaient inférieures au mais pour favoriser le gain de poids. Ces résultats sont similaires avec ceux obtenus ailleurs dans des travaux antérieurs, dans lesquels on avait observé que le mais pouvait être remplacé seulement par des pâtures de pommes de terre douces (patates), jusqu'à la limite de 50–57%.

Les oiseaux nourris avec des pâtures d'ubi, pongapong et gabi réagirent mal avec des proportions beaucoup plus réduites d'efficacité du gain de poids et de l'alimentation. Cependant, la pâture de gabi était très agréable au goût et savourée par les oiseaux.

Resumen

Harina de tubérculos como fuentes de hidratos de carbono en las raciones de los pollos de freir

Se llevaron a cabo seis pruebas sobre el uso de tubérculos producidos localmente como fuentes de energía para pollos de freir. Los tubérculos usados fueron las batatas (*Impomea batatas*), cazabe (*Manihot esculenta*), gabi (*Colocasia esculentum*), pongapong (*Amorphallus campanulatus*) y ubi (*Dioscorea alata*). Se demostró que las harinas de batatas y de cazabe pueden reemplazar parcialmente el maíz amarillo en las raciones para los pollos de freir. Sin embargo, las dos harinas eran inferiores al maíz para estimular el aumento de peso. Estos hallazgos están de acuerdo con los de los primeros trabajos llevados a cabo en otras partes, en los que se observó que el maíz podía reemplazarse únicamente por harina de batata hasta el 50–57%.

Se obtuvieron malos resultados con las aves alimentadas con harinas de ubi, pongapong y gabi, con índices significativamente más bajos de eficacia en la ganancia de peso y la alimentación. Sin embargo, la harina de gabi era muy agradable y del gusto de las aves.

Limitations imposed by scarcity of maize and by
competition with human consumption have forced
many farmers, especially those engaged in small scale
poultry raising, to look for other sources of energy as
substitutes for this cereal.

Root crops or tubers have been used for this purpose
as dehydrated meals, after roasting, or even raw. Of
the tubers, the sweet potato (*Ipomoea batatas*) as an
energy source for poultry is, perhaps, the most studied
(Tillman & Davis, 1943; Rosenberg & Sen, 1952;
Squibb, 1955; Yoshida & Norimoto, 1958, 1959;
Yoshida, *et al.*, 1961; Magay, 1972; and Saure, 1972).
Almost all these workers agreed that sweet potato (or
camote as it is locally known) cannot completely
substitute for maize, depressed growth resulting when
levels exceeded at most 50–75% of the maize in the
ration. Squibb (1955) claimed that sweet potatoes
could satisfactorily substitute all of the maize in his
ration (30%). Even then he obtained decreased growth
in the chicks, although the difference was not
statistically significant.

Studies using dehydrated cassava meal for poultry
have also been conducted abroad (Vogt & Stute,
1964; Enriquez and Ross, 1972). Their results were
promising but their work was largely on older birds.

The trials described here used day-old broiler chicks
and also explored the possibility of using other tubers
as energy sources besides the sweet potato or camote
and cassava.

Materials and methods

In all the feeding trials, newly harvested tubers were
used. These were washed free of soil, chopped or
sliced raw (except for pongapong which was first
boiled), dried at 80°C and ground to a fine meal.
All ration ingredients were analysed for crude protein
content and the values used in formulating the
different rations. Rations were isonitrogenous and
isocaloric with the control, (either normal yellow or
white corn) containing 22–22.5% protein and 2,800
kcal metabolisable energy.

Tubers used were:
1. Sweet potato (camote), *Ipomoea batatas*
2. Cassava, yellow and/or white variety, *Manihot
   utillissima*
3. Gabi, *Colocasia esculenta*
4. Pongapong, *Amorphophallus campanulatus*
5. Ubi, *Dioscorea alata*

In each of the trials, treatments were replicated three
or four times with ten one day-old broiler chicks per
replicate. Feeding was ad libitum for four or five
weeks, water being provided at all times. Initial and
weekly or biweekly weighings were done in all trials,
and records were kept of feed consumption, weights,
and mortality. Total protein consumed was calculated
using the actual analysis of the ration as determined
in the laboratory. Multiple co-variance analysis and
Duncan's multiple range test (Steel & Torrie, 1960)
were used to test differences in treatment means.

Average daily gains were adjusted for initial weights
and protein consumed.

Results and discussion

A summary of the results in all trials conducted is
presented in Table 1.

Trial 1

Results of the study using Cobbs broiler chicks
showed that rations containing as high as 50%
cassava meal (100% replacement of corn in the
ration) can be fed from day-old to five weeks without any deleterious effects. Average daily weight
gains adjusted for initial weight and protein consumed
showed no significant differences between the two
rations. Reduction of the cassava meal to 37.5% in
the ration (75% replacement of maize) slightly
increased weight gains. Differences among the three
treatments were not statistically significant. Values
obtained were 17.94 g for yellow corn, 19.15 g for
50% cassava and 20.33 g for 37.5% cassava. Feed
efficiencies were also not significant although the
cassava meal appeared more efficient than the other
two rations with 1.97 versus 2.58 and 2.29 for corn
and 37.5% cassava, respectively.

Trial 2

In this trial two tuber meals, camote and pongapong,
were compared with the maize control ration (50%
levels) using Peterson broiler chicks. Feed efficiency
and adjusted weight gains of camote meal, though
slightly lower, did not differ significantly from the
white maize control. However, the birds fed pongapong meal performed very poorly with an adjusted
weight gain of only 12.93 g and feed/gain ratio of
2.72. These values were significantly lower (P⧸0.05)
compared to white maize (19.67, 1.93) and camote
meal rations (18.36, 2.13), respectively. A total of
seven birds died in the pongapong lots compared to
one for maize and three for camote out of a total of
30.

Trial 3

No significant treatment differences for weight gains
were obtained in this trial. Adjusted weight gains
were about equal for the maize and camote meal
rations (17.16 versus 17.17 g). The cassava gave
14.89 g gain, lower than but not significantly different
from the maize and camote, and higher than but not
significantly different from the gabi meal ration. The
weight gain of 11.78 for the gabi was the lowest
among all rations. Treatment differences in feed
efficiencies showed the value for gabi meal of 3.87
was significantly lower (P⧸0.01) compared to maize,
cassava and camote meal. The differences among the
latter three treatments were, however, not significant.
It was also observed that gabi meal was very much

TABLE 1. *Performance of broiler chicks fed tuber meals as sources of carbohydrates*

| | Av. daily gain (g) | | Feed, 2 |
	Unadjusted	Adjusted 1, 2	gain
Trial 1 (Cobbs Str.) March–April, 1971			
50% yellow maize	17.30	17.94[a]	2.58[a]
50% cassava meal	19.37	19.15[a]	1.97[a]
37.5% cassava, 12.5% yellow maize	20.79	20.33[a]	2.29[a]
Trial 2 (Peterson Str.) June–August, 1971			
50% white maize	20.32	19.67[a]	1.93[a]
50% camote meal	17.90	18.36[a]	2.13[a]
50% pongapong meal	12.80	12.93[a]	2.72[b]
Trial 3 (Peterson Str.) September–November, 1971			
50% yellow maize	17.70	16.16[a]	2.09[a]
50% cassava meal	13.47	14.89[ab]	2.46[a]
50% camote meal	15.91	17.17[a]	2.35[a]
50% gabi meal	13.77	11.78[b]	3.87[b]
Trial 4 (Peterson Str.) February–March, 1972			
50% yellow maize	24.92	22.79[a]	1.87[a]
50% camote meal	16.05	17.27[b]	2.44[a]
50% ubi meal	14.36	15.81[b]	3.35[b]
50% white cassava meal	13.72	13.76[b]	2.35[a]
50% yellow cassava meal	15.28	14.39[b]	2.46[ab]
Trial 5 (Peterson Str.) June–August, 1973			
50% yellow maize	19.43	20.74[a]	2.23[a]
50% camote meal	16.30	14.99[a]	2.42[a]
Trial 6 (Peterson Str.) December, 1972–January, 1974			
50% yellow maize	22.74	21.71[a]	2.58[a]
50% cassava meal	20.91	21.93[a]	2.41[a]

1. Adjusted for initial weight and protein consumed.
2. Figures with the same superscript in the same column are not significantly different at 5% level.

relished by the birds and had no toxic effect. Not a single mortality occurred in the four weeks collection of data, but growth was poor.

Trial 4

Camote, ubi and two varieties of cassava (white and yellow) were compared with the yellow maize control ration (50% levels) and fed for four weeks to Peterson broiler chicks. Significant differences in adjusted mean weight gains were obtained (P$\angle$0.100). Partitioning of treatment degrees of freedom showed significant differences (P$\angle$0.05) between yellow maize and the tuber meal rations. No significant differences were obtained among the tuber meals studied, although the camote appeared to be superior to ubi, white cassava and yellow cassava with an adjusted daily gain of 17.27 g against 15.81, 13.75 and 14.39 g, respectively. This substantiates results obtained in Trial 3.

Significant differences in feed efficiency were obtained (P$\angle$0.05). The yellow maize ration with a feed efficiency of 1.87 was the most efficient, but was not significantly different from camote (2.44), white cassava (2.35) and yellow cassava (2.46). The least efficient was the ubi meal ration, requiring 3.35 kg to produce 1 kg gain in weight. Actual weight gains were used to calculate feed efficiency ratios in all trials.

Trial 5

While the F-test failed to show significant differences in treatment means, weight gains adjusted for protein intake showed that the yellow maize control ration performed better than the camote meal ration (20.74 g versus 14.99). This may be due to the few number of observations which required a very large difference to show significance. Feed efficiencies of the two rations were not statistically significant. Weight gains indicate that camote may replace yellow maize partially, but not wholly.

Trial 6

Results of this study comparing cassava meal with yellow maize show a favorable response of the Peterson broilers to cassava. The adjusted daily gains and feed efficiencies of the tuber meal-fed chicks compared favorably with those of the cereal-fed lots, the differences obtained not being statistically significant. Feed/ gain ratios were 2.41 for the cassava and 2.58 for the yellow maize while average adjusted daily gains were 21.93 and 21.71 g for the cassava and yellow maize respectively.

References

Enriquez, F. G. & Ross, E. (1972) Cassava root meal in grower and layer diets. *Poultry Sci.* **51**: 228.

154   **Magay, E. J.** (1972) Utilisation of carotene from yellow sweet potato by broilers and digestibility of sweet potato rations by swine. Unpublished Master's Thesis.

Rosenberg, N. M. & Sen, J. (1952) Sweet potato root meal vs. yellow corn meal in chick rations. *World Poult. Science J.* 8(2): 93.

Saure, R. V. (1972) Sweet potato meal as replacement for corn in isonitrogenous and isocaloric broiler and swine rations. Unpublished Master's Thesis.

Squibb, R. L. (1955) The value of sweet potato and achiole meal in rations for chicks. *World Poult. Sci. J.* 11(4): 343.

Steel, R. G. D. & Torrie, J. H. (1960) Principles and Procedures of Statistics. New York. McGraw-Hill Book Co., Inc.

Tillman, A. D. & Davis, H. J. (1943) Studies on the use of dehydrated sweet potato meal in chick rations. *Louisiana Expt. Sta. Bull.* 258.

Vogt, H. & Stute, K. (1964) Testing tapioca pellets in complete broiler feed. *Nut. Abst. and Rev.* 35: 541.

Yoshida, M. & Morimoto, H. (1958) The nutritive value of sweet potato as carbohydrate source of poultry feeds. *World Poult. Sci. J.* 14(3): 246.

Yoshida, M. & Morimoto, H. (1959) Nutritive value of sweet potato as carbohydrate source in poultry feeds. 2. Effect of sweet potato on day-old chicks. *World Poult. Sc. J.* 1(2): 226.

Yoshida, M. H., Joshi, Q. & Morimoto, H. (1961) Nutritive value of sweet potato as carbohydrate source of poultry feed. 2. Effect of vitamin A supplementation on chick growth. *World Poult. Sci. J.* 17(3): 391.

Feeding and replacement value of calcium hydroxide treated maize cobs for dairy cows

A. Darwish and A. Gh. Galal

Assiut University, Assiut, Arab Republic of Egypt

Summary

Maize cobs were finely ground and treated with calcium hydroxide to increase digestibility. The nutritive value of the treated cobs was assessed for sheep and lactating cows.

The nutritive value of the treated cobs was almost double that of the untreated material, with a starch value of 46.7% which approached that of coarse wheat bran. The inclusion of treated cobs in the ration of the lactating cows in place of the wheat bran caused a small decrease in milk yield and a small increase in fat yield. However, neither of these differences was statistically significant. Treatment of maize cobs with calcium hydroxide could help to increase milk production in the Arab Republic of Egypt.

Résumé

Les pâtures et la valeur du remplacement par des chaumes de mais traités par l'hydroxyde de calcium pour les vaches laitières

Les chaumes de mais ont été finement moulus et traités avec de l'hydroxyde de calcium pour augmenter la digestibilité. La valeur nutritive des chaumes traités a été évaluée pour les moutons et les vaches latières.

La valeur nutritive des chaumes traités était presque le double du matériel non-traité, avec une teneur en amidon de 46.7%, qui approchait celle du son de blé à gros grains. L'inclusion des chaumes de mais traités, dans la ration des vaches latières au lieu du son de blé, produisit une petite diminution de la production de lait et une augmentation infime de la production de graisse. Cependant, aucune de ces différences n'était significative au point de vue statistique. Le traitement des chaumes de mais avec l'hydroxyde de calcium pourrait aider à augmenter la production de lait dans la République Arabe de l'Egypte.

Resumen

Alimentación y valor de sustitución de las mazorcas de maíz tratadas con hidróxido de calcio para las vacas de leche

Se molieron finamente las mazorcas de maíz y se trataron con hidróxido de calcio para aumentar su digestibilidad. Se estimó el valor nutritivo de las mazorcas tratadas para corderos y vacas de leche.

El valor nutritivo de las mazorcas tratadas fué casi el doble que el del material no tratado, con un valor de almidón del 46,7% que se aproximó al del salvado de trigo bruto. La inclusión de las mazorcas tratadas en la ración de las vacas de leche en lugar del salvado de trigo originó un pequeño descenso en la producción de leche y un pequeño aumento en la producción de grasa. Sin embargo, ninguna de estas diferencias fué significativa estadísticamente. El tratamiento de las mazorcas de maíz con hidroxido de calcio podría contribuir a aumentar la producción de leche en la República Arabe de Egipto.

Introduction

Shortage of feed is considered to be the major problem in animal production in Egypt. Efforts are being made to find new sources of feed for animals, or to increase the feeding value of some of the poor roughages from farm residues and agricultural by-products which are presently available, by chemical treatment. Several other investigations on these lines have been reported elsewhere (Abou-Raya *et al.*, 1966; Backman & Sitzbar, 1919; Ferguson, 1943; Godden, 1920; Honcamp & Banman, 1921; Woodman & Evans, 1947).

Maize cobs have a feeding value approaching that of wheat straw, ie a starch value of between 20.6 and

27.6. It was found that the best method of chemical treatment to increase feeding value was to soak them in a mixture of calcium hydroxide $(Ca(OH)_2)$ and water without subsequent washing. This treatment raised the starch value (SV) to as much as 53.33.

The present work was initiated to study how $Ca(OH)_2$ soaking affected the feeding value of maize cobs and also what effect the soaked cobs would have on the milk and fat yields of dairy cattle when fed as a replacement for coarse wheat bran.

Experimental methods

Soaking method of $Ca(OH)_2 : H_2O$ mixture

The mixture for soaking was prepared by adding 1.5 parts of $Ca(OH)_2$ to 100 parts of water. The ground maize cobs were soaked in the solution for 24 h at room temperature, then drained without washing and dried at room temperature. The method was similar in many ways to that described by Ghoneim et al. (1966).

Digestibility trials

Five digestibility trials were performed with clover hay being used as the basal ration, treated maize cobs, wheat bran, concentrate mixture I (25% wheat bran, 50% decorticated cotton seed meal, 10% maize and 12% rice bran) and concentrate mixture II (5% wheat bran, 20% treated maize cobs, 53% decorticated cottonseed meal, 7% maize and 12% rice bran). The two mixtures were supplemented by calcium carbonate and sodium chloride at a rate of 2.0% and 1.0% respectively.

Feeding trial

One experiment based on the 'swing over' method (Kellner, 1926), was performed including a control ration (concentrate mixture I) and one tested ration (concentrate mixture II). Four Jersey crosses with native cattle were selected during their high lactation, shortly after the peak. The experimental period lasted for 78 days and was divided into a transition period of 11 days preceding the initial control ration, which lasted for 15 days, which was followed by an 11-day transition period to the tested ration, which continued for 15 days. The control ration was finally fed in the same manner as initially used. The daily requirements of SV and digestible protein were calculated as described by Darwish (1963) from the knowledge of the average body weight, milk yield and fat percentage during the week before the experiment. The SV of concentrate mixtures I and II were 63.10 and 62.50 respectively. The corresponding digestible protein contents were 22.50% and 21.36%. During the control periods as well as the test one, each cow received 4 kg wheat straw in addition to the concentrate mixture.

Milk samples

Individual daily milk samples were taken from each animal during the test and two control periods. Proportionate samples were taken from the evening and morning milk.

Analytical methods

The analysis of feeding-stuffs and faecal material was carried out by conventional methods (Association of Official Agriculture Chemists, 1960) using duplicate samples of 2–3 g each. The milk fat was determined by the method of Gerber. Statistical analysis was carried out by the procedures described by Snedecor (1968).

Results and discussion

Data in Table I concerning the chemical analysis of treated cobs, show that they had a very low content

TABLE 1. *Analysis, digestion coefficients and feeding value of feeding-stuffs (%)*

| | Dry matter as offered | Crude protein | Composition of dry matter | | | |
			Crude fat	Crude fibre	NFE	Ash
Clover hay	92.00	15.56	2.81	22.69	41.27	17.67
Wheat bran	91.52	5.41	1.48	21.46	60.02	3.15
Maize cobs (treated)	93.00	1.01	0.00	34.98	49.32	14.79
Mixture I	93.00	26.00	5.36	12.83	44.61	11.20
Mixture II	89.00	24.99	3.67	13.86	46.14	11.34

| | Digestion coefficients | | | | Feeding value | |
	Crude protein	Crude fat	Crude fibre	NFE	Starch value	TDN
Clover hay	52.58	62.36	57.39	75.86	38.82	42.20
Wheat bran	80.21	69.43	26.36	71.43	48.44	55.15
Maize cobs* (treated)	89.87	–	90.00	82.50	46.69	67.09
Mixture I	86.53	90.71	62.69	72.23	63.10	68.53
Mixture II	85.48	85.92	75.39	80.56	62.50	67.65

*Crude fibre deduction = 0.58 SV: unit crude fibre.

of crude protein and were devoid of crude fat. The major contents were the nitrogen-free extract (NFE) and crude fibre. The relative high content of ash was to be expected after treatment with $Ca(OH)_2 : H_2O$ mixture. Except for the ash content with composition in general was similar to that published by Ghoneim *et al.* (1966) when the same method of soaking without washing was used.

The proximate analysis indicated that the control ration (Mixture I) and the tested one (Mixture II) were very similar. The digestion coefficients of the treated maize cobs were of a high order being over 82%.

Treatment with $Ca(OH)_2$ increased the digestion coefficients very noticeably and clearly indicated that the cobs were very much affected by the alkali treatment. This was noticeably reflected in their feeding value, the SV of air dried maize cobs approaching that of the coarse wheat bran. The total digestible nutrients (TDN) of treated maize cobs were in fact 21.7% higher than those of wheat bran, while their SV was almost double that of untreated cobs. The 370,000 t of maize cobs which are estimated to be available in Egypt have a feeding value of about 65,000 t of SV but by the alkali treatment this could be increased to about 173,000 t (Ghoneim *et al.*, 1966).

Feeding trial

The average daily milk yields with the control ration (containing 25% wheat bran) were 8.71, 10.71, 13.08 and 8.97 kg, while the corresponding average daily milk yields with the tested ration were 8.44, 9.61, 12.35 and 8.70 kg when feeding the animals on the same level of SV but replacing the wheat bran with the treated maize cobs. The average milk yields were 8.46, 10.03, 10.78 and 8.29 kg respectively when the control ration was fed back after the tested ration.

The percentage differences in the calculated milk yields of the tested ration from the initial control of the four cows were: 1.38, 6.53, 4.66 and 1.34% respectively, the average differences − 0.478% ± 2.15, not being statistically significant.

It may be concluded that using the treated corn cobs instead of wheat bran in the ration (provided there was no reduction in protein) causes an insignificant decrease in milk yield of only 0.478%. There is, therefore, an advantage in feeding calcium hydroxide treated corn cobs to lactating dairy cows, thereby replacing a noticeable part of the expensive concentrate components and thereby reducing the demand for feeding stuffs which are in short supply.

Regarding the change in fat yield with the tested ration, the average percentage increase in fat yield was 0.62% ± 3.77, which was not statistically significant. Similar results have been recorded by several workers (Ghoneim *et al.* 1966; Clark *et al.,* 1968; Firestone, 1970; Giyonko & Gleyn, 1969).

Applying the fat corrected milk (FCM) system, the percentage differences in FCM yield of the tested ration from the initial control of the four cows were −1.50, −8.26, 7.61 and 1.16% respectively, the average difference (−0.99% ± 2.84) not being statistically significant.

It may therefore be concluded that the milk yield as well as the fat yield and the FCM were not significantly affected by feeding the treated maize cobs to dairy cows as a replacement for wheat bran. If it were feasible to treat all the maize cobs in Egypt with $Ca(OH)_2$, around 100,000 additional tons of SV would become available and this could provide energy needed to produce 400,000 t of fat corrected milk.

References

Abou-Raya, A. K., Ghoneim, A. & Abou-El-Hassan, A. (1966) *Proc. 2nd Animal Prod. Conf., Cairo, 1963,* Vol. II, 593.

Association of Official Agriculture Chemists (1960) Official Methods of Analysis, 9th Ed., A.O.A.C.

Backman, E. & Sitzbar (1919) *Acad-Wicc,* 275.

Clark, J., Preston, T. R., Willis, M. B. & Valman, I. (1968) Revista Cubana de Ciencia Agricola, English Ed., 2(2), 195.

Darwish, A. (1963) Ph.D. Thesis, Faculty of Agriculture, University of Cairo, A.R. of Egypt.

Ferguson, W. S. (1943) *J. Agric.,* 33, 174.

Firestone, E. E. (1970) *Dairy Science Abstracts* 32(7).

Ghoneim, A., Abou-Raya, A. K. & Abou-ElOHassan, A. (1966), *Proc. of 2nd Animal Prod. Conf. Cairo, 1963,* Vol. II, 607.

Giyonko, G. M. & Kleyn, D. H. (1969) *J. Dairy Science,* 52, 1379.

Godden, W. (1920) *J. Agric. Sci.,* 10, 437.

Honcamp, F. & Bauman, F. (1921) *III-Fandw. Fers. Sta.,* 98, 46.

Kellner, O. (1926) The Scientific Feeding of Animals, 2nd Ed. by Goodwin, W., Duckworth, London.

Snedecor, G. W. (1968) Statistical Methods, 6th Ed. Iona State College Press, Iona, USA.

Woodman, H. E. & Evans, R. E. (1947) *J. Agric. Sci.,* 37, 202.

The use of some by-products in feed mixtures for animals

A. K. Abou-Raya, E. R. M. Abou-Hussein, A. Abou-El Hassan and M. M. El-Shinnawy

Animal Production Department, Faculty of Agriculture,
University of Cairo, Arab Republic of Egypt

Summary

Trials were carried out in which mechanically treated maize stalks and rice hulls were included in pelleted rations for cattle and sheep. Other ingredients of the pelleted rations were undecorticated cotton seed cake, urea and molasses. It was found that they could be used successfully in practice, and it was considered that they could be of great value in extending supplies of animal feed in Egypt.

Résumé

L'utilisation de quelques sous-produits dans les pâtures mixtes pour les animaux

Des essais ont été entrepris dans lesquels des chaumes de mais et des cosses de riz traitées mécaniquement ont été incluses en boulettes dans des rations pour le bétail et les moutons. Autres ingrédients des rations de boulettes étaient les tourteaux de graines de coton, l'urée et les mélasses. On a trouvé que les boulettes pouvaient être utilisées dans la pratique et on a estimé qu'elles pourraient être de grande valeur pour développer les pâtures pour les animaux en Egypte.

Resumen

El uso de algunos productos secundarios en los piensos compuestos para animales

Se llevaron a cabo pruebas en las que los tallos de maíz y las cáscaras del arroz tratados mecánicamente se incluyeron en raciones granuladas para ganado vacuno y corderos. Otros ingredientes de las raciones granuladas fueron tortas de orujo de algodón no descortezado, urea y melazas. Se halló que podían usarse con éxito en la práctica, y se consideró que podrían ser de gran valor para aumentar los suministros de piensos para animales en Egipto.

Owing to the feed shortage in Arab Republic of Egypt, two important roughages, maize stalks and rice hulls were mechanically treated, side supplemented and introduced into suitable pelleted feed mixtures. Twenty seven trials were carried out to find the most suitable proportions for successful pelleting using local pelleting plants. Mixtures which could be successfully pelleted and which had suitable qualities and feeding values were produced, on a part or full production scale. They were subjected to metabolism trials, using mature rams, for nitrogen balance and digestibility of nutrients and energy. The feeding experiments made use of 32 Aussimy lambs, 38 Barki lambs, 99 steers, 33 buffalo bullocks, 14 Friesian lactating cows and 7 crosses of Friesians with local cows.

The results indicated that successful pelleting needed the addition of 40% of undecorticated cotton seed cake to both roughages, and adding the roughage increased the amount of molasses which could be used in the feed mixture. Moreover, pelleting increased the feeding value and the bulk density. Metabolism trials showed that the starch equivalents of the two mixtures including maize stalks, PM_1 and PM_2, and that including rice hulls, PM_3, were 40.42, 42.85 and 35.67% respectively, the digestible protein equivalents being 9.79, 12.71 and 9.23% respectively. Pelleting mixtures including poor roughages seemed to make their feeding value greater than would be expected from that of the individual ingredients. When pelleting an all concentrate mixture, PM_4, a reduction in the feeding value appeared to occur.

Feeding the pelleted mixture alone resulted in a satisfactory intake by rams, covering maintenance requirements with a surplus for production. Practically the same results were obtained when PM_1 replaced a unified feed mixture (UM) by up to 75% in the ration of fattening steers. PM_2 could be used for fattening steers by gradual replacing up to 75% of the concentrate mixture PM_4 (58.2% starch equivalent and 15.5% digestible crude protein), and up to 50% with buffalo bullocks, without affecting the rate of gain.

Replacement of PM_4 with up to 50% of PM_2 in the ration of lactating Friesian cows resulted in an insignicant increase in milk yield. Results with PM_3 for fattening local steers, indicated that it could be used with success as a substitute for good quality hay, a part of the concentrate mixture or a suitable proportion of them both. It could replace a part of the concentrate mixture and wheat straw in the ration of fattening lambs shortly after weaning, without affecting the rate of gain. Introducing 4 kg of PM_3 into the daily ration of lactating cows (Friesian crosses) to replace a part of the roughage and a part of the concentrate, resulted in similar yields of butter fat.

The introduction of these roughages into pelleted mixed feeds also containing undercorticated cotton seed cake, urea and molasses results in feeds which can be used successfully in practice. These mixed feeds could be of great value in extending supplies of available feeds in the Arab Republic of Egypt, and further studies on the lines described above deserve the fullest support on a national scale to assist in solving the problem of feed shortage.

The nutritional value of conophor seed (*Tetracarpidium conophorum* **Welw.**)

V. A. Oyenuga

Department of Animal Science, University of Ibadan, Nigeria

Summary

Conophor seeds are widely consumed in West and Central Africa. They are obtained from the pod-like fruits of a perennial climbing plant which grows in forest areas. Conophor seeds contain around 23% of crude protein and 56% of a highly unsaturated oil the main fatty acid of which is linolenic acid (72%). Ash content is around 3.5%, and the seed is rich in Mg, Ca, P, K and Fe.

Amino acid analysis showed conophor seed protein to be deficient in lysine (3.74 g/16 g N) but with a good content of methionine + cystine (5.33 g/16 g N), tryptophan (4.45 g/16 g N) and threonine 5.30 g/16 g N). Rat feeding trials showed the biological value of conophor seed protein to be 49.9. Autoclaving marginally increased protein biological value to 51.6 and markedly increased true digestibility. It is concluded that despite certain deficiencies conophor seeds are a valuable source of protein in high consuming areas during the season of production.

Résumé

La valeur nutritive des graines de conophore (*Tetracarpidium conophorum* Welw.)

Les graines de conophore sont largement consumées dans l'Afrique Centrale et de l'Ouest. Les graines sont obtenues des fruits en gousse d'une plante grimpante qui croît dans les régions forestières. Les graines de conophore contiennent environ 23% de protéines brutes et 56% d'une huile fortement non-saturée, dont le principal acide gras est l'acide linolénique (72%). Le contenu en cendres est d'environ 3.5% et les graines sont riches en Mg, Ca, P, K et Fe.

L'analyse des aminoacides montra que la protéine des graines de conophore est pauvre en lysine (3.74 g/16 g N) mais ayant une bonne teneur de méthionine +

cystine (5.33 g/16 g N), tryptophane (4.45 g/16 g N) et thréonine (5.30 g/16 g N). Des essais de nourrir les rats avec ce produit montrèrent des valeurs biologiques des protéines des graines de conophore de 49.9. La mise à l'autoclave augmenta faiblement la valeur biologique des protéines à 51.6, mais augmenta d'une façon marquée la vraie digestibilité. On conclut qu'en dépit de certaines insuffisances, les graines de conophore sont une source précieuse de protéines dans les régions de grande consommation, pendant la saison de production.

Resumen

El valor nutritivo de la semilla de "conophor" (*Tetracarpidium conophorum* Welw.)

Las semillas de "conophor" se consumen mucho en Africa Occidental y Central. Se obtienen de los frutos de forma de vaina de una planta trepadora perenne que crece en áreas de bosques. Las semillas de "conophor" contienen alrededor del 23% de proteína bruta y el 56% de un aceite altamente no saturado, y cuyo principal ácido graso es el ácido linolénico (72%). El contenido de ceniza es de alrededor del 3,5%, y la semilla es rica en Mg, Ca, P, K, y Fe.

El análisis de los aminoácidos demostró que la proteína de la semilla de "conophor" era deficiente en lisina, (3,74 g/16 g N) pero con un buen contenido de metionina + cistina (5,53 g/16 g N), triptófano (4,45 g/16 g N) y treonina (5,30 g/16 g N). Las pruebas de alimentación de ratas demostraron que el valor biológico de la proteína de la semilla de "conophor" era de 49,9. En el autoclave, el valor biológico de la proteína aumentó ligeramente a 51,6 y aumentó marcadamente la digestibilidad verdadera. La conclusión es que, a pesar de ciertas deficiencias, las semillas de "conophor" son una fuente valiosa de proteína en áreas de consumo alto durante la estación de producción.

Introduction

In a general survey of the biochemical and nutritive qualities of the foods and feeds of West Africa, consideration was given to the conophor fruit (*Tetracarpidium conophorum* welw.), or (*Awusa nut* (yoruba), which is widely distributed and consumed, particularly by the inhabitants of the Guinean zone of West and Central Africa.

Description and characteristics of Awusa nut

The tree bearing this pod-like fruit is a forest rambling perennial climber, commonly growing wild or sometimes cultivated as a subsidiary crop in a cocoa farm. It climbs round a supporting tree, sometimes attaining the height of 30–31 m until the crown is able to obtain enough light for its photosynthetic activities. The stem may reach some 12 cm in diameter at the base after 10 to 12 years growth. The plant produces flowers around March, sets its fruits in April and May which mature and are harvested between June and September, harvesting of prolonged but lighter crops sometimes extending to October and November. The cultivation of Awusa nuts for their oil was encouraged during and after the 1939–45 war when linseed oil was scarce.

The fully developed fruit of *T. conophorum* contains four nuts, hence the generic name, tetracarpidium. However, all the four nuts are not always developed and fruits with three, two or even one nut are not uncommon. In a random sample of some 360 pods, those containing four nuts occurred most frequently followed by those with three and two nuts respectively. Only approximately one out of 20 of the pods observed contained one nut.

Materials and methods

Proportion of pods to nuts, etc.

Some 360 pods were randomly chosen and from these the average weight of pods, proportion of pods to nuts, and other components of nuts were determined. The pods made up the bulk of the dry weight of the fruit; pods with one nut, on the average, accounted for approximately 84% of the total weight of the raw fruit. In the case of fruits containing four nuts, the total weight of the four nuts accounted for approximately 64% of the fruit, the empty pod being 36% (Table 1). In the nut itself, the kernel (endosperm) accounted for approximately 73% the cotyledon 0.8% and the empty shell approximately 26% of the total weight of the dry nut. The proportions of the different components of the nut were not appreciably altered by cooking carried out at over 100°C for some 150 minutes.

Chemical analysis

After removing the nuts from the pod, and with the shell also removed, chemical analyses were carried out on the raw and cooked kernels with and without the cotyledon (embryo) and on the cotyledon only respectively, employing AOAC (1970) methods. Proximate chemical analyses were also carried out on ground meals obtained from partial oil extraction by both the hydraulic pressure and solvent extraction methods respectively. The mineral content were estimated by atomic absorbtion spectrophotometry using a Perkin-Elmer 290 instrument after wet ashing with a nitric acid/perchloric acid/sulphuric acid mix. Phosphorus was determined by the phosphovanado-molybdate method (AOAC 1970). The fatty acid components of the extracted methylated oil were separated by gas chromatography (Pye, model 204). Column temperature was 170°C, rate of flow of argon, 56 ml/min, attenuation was 1×10^3 and the sample size was $1\mu l$.

Amino acid analysis

The amino acid constituents of the oil extracted protein were determined using an automated Hitachi-Perkin-Elmer analyser—Model KLA-3B after hydrolysis of the material (100 mg) with 6N HCl (10 ml) at 110°C in an atmosphere of nitrogen for 24 hours. Tryptophan was chemically estimated by the method of Miller (1967). Energy was determined with a Gallenkamp oxygen ballistic bomb calorimeter.

Rat bioassay

Weanling male albino rats of Wister strain, 28–30 days old with average weight of between 52 and 56 g were used to estimate protein efficiency ratio (PER), net protein retention (NPR), net protein utilization (NPU), biological value (BV), apparent and true digestibility (AD & TD) and the other parameters as described by

TABLE 1. *Average proportion of pod constituents of conophor fruit (% dry fruit and nut)*

	Raw Samples		Cooked Samples	
	Fresh	*Dried*	*Wet*	*Dried*
Weight of whole fruit (g)	96.53	44.86	–	–
Pod without nut	65.22	26.81	–	–
One nut to the pod	11.62	16.05	–	–
4 nuts to the pod	46.48	64.20	–	–
Shell per nut	29.11	25.86	26.09	26.72
Kernel (endosperm) per nut	71.23	72.12	74.20	72.42
Cotyledon (embryo) per kernel	1.02	0.81	0.92	0.82
Cotyledon per nut	0.72	0.60	0.63	0.59

TABLE 2. *Chemical composition of the components of the conophor seed*

| | Uncooked material | | | Cooked material | |
	Kernels with cotyledons	Kernels without cotyledons	Cotyledons only	Kernels with cotyledons	Kernels without cotyledons
Dry matter (%)	95.1	95.8	94.9	94.7	95.8
Crude protein (%)	22.7	20.8	9.9	21.2	21.4
Crude fibre (%)	3.7	3.5	0.05	4.3	4.5
Ether extract (%)	56.0	57.7	21.0	47.6	46.6
Silica-free ash (%)	3.6	3.5	4.7	3.6	3.4
Nitrogen-free extract (%) (by difference)	14.0	14.5	64.4	23.5	24.1
Gross energy (Kcal 100 g)	686.3	703.7	896.5	713.5	667.2

TABLE 3. *Chemical composition of the residual meal of the conophor kernel after partial oil extraction, (%)*

	Extracted cooked conophor meal (hydraulic press)	Extracted cooked conophor meal (solvent extracted)	Extracted autoclaved conophor meal (solvent extracted)
Dry matter	96.5	96.8	93.3
Crude protein	31.5	42.6	43.0
Crude fibre	5.9	6.4	4.4
Ether extract	27.2	17.9	16.7
Silica-free ash	4.7	5.3	5.6
Nitrogen-free extract	43.3	23.5	22.8

the NAS/NRC (1963). Eight rats were used per test diet and an additional eight for the nitrogen-free diet in each of the four experiments conducted, the means of which are reported in the tables of results.

Results and discussion

Chemical composition

The proximate chemical composition of conophor nuts is shown in Table 2. Raw *T. conophorum* contains approximately 23% of crude protein (N x 6.25) and a somewhat lower amount (21%) when cooked suggesting that a small proportion of the protein may be present in the form of hot water soluble amides. The amount of protein in the cotyledon is, however, substantially lower than that contained in the endosperm. When the inedible cotyledon (embryo) is removed, the kernel (endosperm) contains a somewhat lower amount of protein (21%).

The level of protein in conophor seed, therefore, compares well with a number of other well known protein rich seeds. Crude protein content is in the same range with Bambara groundnut (21.1%), cashew nut (21.2%), pigeon pea (23.8%) but lower than in lima bean (24.1%), groundnut (24.8%), cow pea (24.7%) green gram (28.5%), the African locust bean (30.4%) and the soya bean (44%) (Oyenuga, 1968). Protein content is also higher than that of palm kernel meal. The extracted meal has a crude protein content varying between 31.5 and 43% depending on the degree of extraction (Table 3).

Per capita daily consumption of conophor seed during peak production may average between 4 and 6 nuts, amounting in dry weight to between 21 and 32 g (each dried kernel without cotyledon weighing 5.28 g). This gives between 4–7 g of daily crude protein intake. This will meet about 25% of the daily protein require-

ment per person in Nigeria during the peak production season of conophor seed.

The amount of ether extract (oil) in the raw kernel is high (56%) and it is still higher with the cotyledon removed (58%). The amount drops, however, to approximately 58% in the cooked kernels, suggesting that some of the oil fraction is volatile and there were some losses in the cooking water. Conophor seed, therefore, ranks with other well known tropical oil-rich seeds. It compares well in oil content with water melon seed (*citrullus vulgaris* Schrad 57.06%), lower than coconut seed (65.9%), but higher than decorticated ground-nut (50.9%), cocoa beans (42.8%), sunflower seed (33.4%), African locust bean (30.3%), soya beans (19.1%) and cotton seed (14%)(4).

Table 4 shows that conophor seed oil is highly unsaturated, being a predominantly linolenic acid oil (72%) with oleic acid and linoleic acid accounting for 12.6 and 10% respectively of the total weight of the oil. The saturated acids, palmitic and stearic (2.4 and 3.0% respectively), occur in smaller proportion. The seed oil contains about two-thirds of tri-unsaturated glycerides, the rest being almost wholly di-unsaturated. The oil is therefore liquid at ordinary temperature, and

TABLE 4. *Fatty acid composition and other characteristics of conophor seed oil*

Palmitic acid, (% total fat)	2.4
Stearic acid (% total fat)	3.0
Oleic acid, (% total fat)	12.6
Linoleic acid, (% total fat)	10.0
Linolenic acid, (% total fat)	72.0
Iodine value	153.8
Saponification value	200.3
Hydroxyl value	1.052
Acid value	14.8
Acetyl value	0.000784
Refractive index	1.4753
Specific gravity	0.899

is characterised by a high iodine value (153.8) and a fairly high saponification number. With its high linolenic acid value, conophor seed oil is superior to linseed oil in technical application for paints, linoleum etc. (Table 5.) With the two essential fatty acids, linolenic and linoleic acids present, it is a highly nutritious oil.

TABLE 5. *Component fatty acids of linseed oil and conophor seed oil (% molecular weight)*

	Linseed oil	*Conophor seed oil*
Saturated (total)	15.7	12.3
Oleic	13.7	10.8
Linoleic	14.4	11.2
Linolenic	56.2	65.7
Iodine value	182.0	201.6

Source: Hilditch & Seavell (1950).

The silica-free ash accounts for 3.5% of the cotyledon-free kernel in the raw state, being somewhat higher in the kernel containing the cotyledon and noticeably higher in the cotyledon (Table 2). The amount did not vary very much when the seeds were cooked. When the oils were partially removed (Table 3) the amount of silica-free ash in the residual meal rose to between 4.7 and 5.6%.

Conophor seed is rich in Mg, Ca, P, K and Fe (Table 6). The proportion of these elements in the endosperm is high but relatively low in the embryo except P and Fe which are higher in the embryo than in the kernel. The Ca : P ratio in the different parts of the kernel is also noteworthy.

Ca : P *Ratio in conophor seed*

Whole kernel	444 : 331 = 1.1 : 1
Kernel only	529 : 389 = 1.4 : 1
Cotyledon only	154 : 448 = 1 : 2.9

TABLE 6. *The mineral content of the conophor seed meal mg/100 gm*

	Whole kernel	Kernel without	Cotyledon
Calcium	446.1	528.9	154.0
Phosphorus	330.7	389.2	448.0
Magnesium	309.3	353.38	426.1
Potassium	960.8	1,046.1	955.4
Sodium	3.95	4.6	9.0
Chloride	179.0	160.6	372.9
Iron	26.3	9.00	25.2
Manganese	2.77	1.75	2.24
Copper	1.61	1.70	1.80
Zinc	4.5	5.5	14.1
Cobalt	0.34	0.50	–

The Ca : P ratio in the edible portions (whole kernel and split kernel) occurs in the approximated ratio of 1.4 : 1 whereas in the embryo, it occurs in the reverse order 1 : 2.9. Of the trace minerals, the seed is rich in iron, zinc, cobalt and copper. It is also high in chloride though somewhat low in sodium. It is a rich source of magnesium. Conophor seed is as rich or even richer in Ca, P and Fe than any other tropical concentrate seed and it is a valuable source of these and other minerals.

Amino acid pattern

Table 7 compares the amino acid components of the conophor seed protein with those of soya bean meal and groundnut meal; that of the whole hen's egg is included as a reference. The protein of *T. conophorum* is distinctly higher in methionine, cystine, threonine and tryptophan and it contains as much valine and nearly as much histidine as soya bean meal. Its content of these amino acids are higher than those of groundnut meal. It however contains lower amounts

TABLE 7. *Comparative amino acid composition of conophor seed meal, soya bean meal, groundnut meal and whole hen's egg (g per 16gN)*

	Conophor seed meal	Soya bean meal	Groundnut meal	Whole hen's egg
Arginine	9.45	8.24	11.07	6.10
Histidine	2.17	2.40	2.13	2.43
Isoleucine	4.41	7.60	3.12	6.29
Leucine	6.79	7.63	6.30	8.82
Lysine	3.74	6.22	3.84	6.98
Methionine	2.62	1.14	0.85	3.36
Cystine	2.71	1.80	1.20	2.43
Methionine + Cystine	5.33	2.94	2.05	5.79
Phenylalanine	3.03	5.76	4.69	5.63
Tyrosine	4.84	4.05	4.16	4.16
Phenylalanine + Tyrosine	7.87	9.81	8.85	9.79
Threonine	5.30	3.62	2.58	5.12
Tryptophan	4.45	1.37	1.24	1.62
Valine	4.94	4.93	4.00	6.85
Alanine	3.23	4.44	3.54	5.92
Aspartic acid	11.34	12.94	11.95	9.02
Glutamic acid	11.64	20.32	19.50	12.74
Proline	4.74	3.30	2.46	4.16
Serine	5.89	4.83	4.90	7.65
Glycine	8.16	4.46	6.10	3.31
Total amino acids (mg/gN)	6,217	6,567	5,852	6,411

of isoleucine, leucine, and phenylalane than soya bean meal, but is higher in tyrosine. Its content of leucine and isoleucine is higher than those of groundnut meal, but it contains a lower amount of phenylalanine than groundnut meal and of the total aromatic amino acids than either soya bean meal or groundnut meal. The total amino acids expressed in mg per gN is higher than that of groundnut, though lower than that of soya bean meal.

The amino acid pattern expressed as milligrams of amino acids per gram of total essential amino acid (A/E), the total essential amino acids, the ratio of the total essential amino acids to the total amino acids (E/N) (all expressed in mg/gN), the proportion of the total nitrogen derived from essential amino acids (E/T ratio) and the chemical score of the three proteins are all shown in Table 8. Table 8, therefore, shows not only the comparative contributions of the essential amino acids, but also the supply of total nitrogen that the proteins of soya bean meal, groundnut meal and conophor seed meal can make to the body. The total essential amino acids from conophor seed are considerably higher than those of groundnut meal and somewhat lower that those of soya bean meal and of egg. Similarly the E/T (g/gN) ratio of conophor seed meal protein is distinctly higher than that of groundnut, and similar to that of soya bean. The ratio of essential to total amino acids (E/N) is higher than that of groundnut and similar to that of soya bean meal. The chemical score of 64 is also superior to either of the other two meals (59 for soya bean and 57 for groundnut).

The protein of conophor seed meal with its high content of the sulphur-containing amino acids, methionine and cystine, and of threonine and tryptophan will complement very well the protein of soya bean meal and those of many other legumes and oilseeds which although low in these amino acids are more adequate in lysine, leucine and phenylalanine in which conophor seed protein is marginal (Oyenuga,

1966). It will also complement the protein of some of the cereals.

Rat studies

Table 9 compares the protein quality of the conophor seed meal with those of soya bean meal, groundnut meal and whole hen's egg, all fed at the critical level of 10% protein diet. The conophor seed meal was fed 'cooked' and 'autoclaved' to the rat. They showed similar apparent and true digestibility values with heat treated soya bean meal and with groundnut meal, all of which appear less readily digestible than the hen's egg. The protein of the 'autoclaved' sample of the conophor was apparently better digested than the cooked samples.

Statistical analysis, using Duncan's Multiple Range Test, showed that the mean differences in apparent digestibility between the whole hen's egg and the 'autoclaved' conophor seed meal were not significant, but the apparent digestibility of the 'autoclaved' conophor seed meal by the rat was significantly superior to those of the soya bean meal, groundnut meal and 'cooked' conophor seed meal, while the differences observed between the meals of soya bean, groundnut, 'cooked' conophor and the egg were not statistically significant.

The picture shown by the true digestibility data is, however, somewhat different. In this case, the 'autoclaved' conophor meal was not statistically superior to soya bean and ground-nut meals, but statistically superior to 'cooked' conophor meal and inferior to whole hen's egg. The differences between the egg and the soya bean, and between the soya bean and the groundnut meal were not significant although the egg was significantly superior to the groundnut meal.

The egg diet resulted in significantly superior liveweight gains of the rat than the soya bean, groundnut and 'autoclaved' conophor meal diets. But the differences which occurred between the diets containing the soya

TABLE 8. *Comparative essential amino acid pattern of conophor seed meal, soya bean meal, groundnut meal and whole hen's egg expressed as mg amino acid per gram of total essential amino acids (A/E)*

	Conophor seed meal		Soya bean meal		Groundnut meal		Whole hen's egg	
Isoleucine	103		172		98		123	
Leucine	158		173		197		172	
Lysine	87		141		120		136	
Total aromatic amino acid	184		222		277		191	
Phenylalanine		71		131		147		110
Tyrosine		113		92		130		81
Total sulphur amino acids	124		67		64		113	
Methionine		61		28		27		66
Cystine		63		41		38		47
Threonine	124		82		81		100	
Tryptophan	104		31		39		32	
Valine	115		112		125		134	
Total essential amino acids (mg/gN)	2,677		2,758		1,999		3,203	
E/T Ratio	2.68		2.76		2.00		3.20	
E/N Ratio (% total essential amino acid over total amino acid)	43		42		34		50	
Chemical Score (A/E)*	64		59		57		–	

*$\frac{A/Ex}{A/Ee}$ % (where A/Ex represents the ratio of each essential amino acid in the food protein to the total essential amino acid in that protein; and A/Ee represents similar ratio for whole hen's egg as reference protein).

TABLE 9. *Comparative protein quality of the conophor seed meal, soya bean meal, groundnut meal and whole hen's egg when fed in a 10% protein diet to weanling albino rats*

| | Conophor seed meal | | Soya bean meal* | Groundnut meal | Whole hen's egg |
	Cooked	Autoclaved	(heat treated)		
Weight gains at 10 days (g)	(−2.2)d	6.43b	10.60b	9.05b	35.40a
Protein intake (g)	4.86c	5.85bc	6.12bc	6.85ab	8.16a
Carcass moisture (%)	68.6	67.40	65.20	62.60	62.40
Carcass N (% of liveweight)	2.83b	2.80b	2.64b	2.60b	2.37a
Carcass N (% dry weight)	9.02c	8.63bc	7.57ab	6.90a	7.00a
$H_2O/_N$ ratio	24.40b	24.10b	24.70b	24.60b	26.3a
PER	(−0.34)c	1.10b	1.73b	1.32b	4.10a
NPR	2.42b	3.28b	2.61b	2.90b	5.47a
NPU	38.40c	46.20c	55.70b	46.00c	85.70a
BV	49.90b	51.60b	60.30b	59.50b	87.50a
Apparent digestibility (%)	76.40c	87.60a	79.90bc	81.60b	84.20ab
True digestibility (%)	77.1c	89.90b	92.40ab	91.20b	95.70a

*The soya bean meal used in this studies was prepared by the method of Coates, Hewitt and Golob (1970) and then autoclaved for 20 minutes at 120°.

a, b, c, d, ... For the different parameters, values in the same row followed by the same suffix are not significantly different (P ⟨ 0.05).

bean meal, groundnut meal and 'autoclaved' conophor meal were not statistically significant. The rats fed on the cooked meal of conophor lost weight all through when fed at this critical 10% level. Gain in weight did not strictly correlate with protein intake in all cases although protein intake by rats fed on 'cooked' conophor were statistically inferior to those on groundnut and egg but not to those on 'autoclaved' conophor and heated soya bean meal. The differences which existed between the BV of soya bean meal, groundnut meal and the conophor meals were not statistically significant, although those of the egg were statistically superior to all the protein meals from plant sources. The NPU of the soya bean was also superior to those of the other plant protein meals, none of which was superior to the other. The egg was of course significantly superior to all of them.

A study of these biological data and the amino acid pattern suggested that Awusa nut protein would be a valuable supplement to a cereal based diet. For this reason the conophor protein was fed to the rat as protein supplement in a maize diet at total protein levels of 20 and 24% respectively with an egg supplemented maize diet at a level of 20% protein as standard (Table 10). The mean liveweight gain, BV, NPU, EFC and the N_2 retained all showed good levels for a vegetable protein although they were significantly inferior to the whole egg supplemented diet.

In conclusion, conophor seed is therefore a good source of protein and will supply high amounts of

total nitrogen required for the synthesis of the dietary dispensable amino acids; in addition, it will supply the sulphur-containing amino acids, tryptophan and threonine as well as glycine in high amounts; valine in adequate quantities and lysine, leucine and isoleucine at marginal levels. It is, however, low in the aromatic amino acids which tend to limit its growth promoting ability in the rat, reduces its otherwise balanced amino acid pattern and thus depresses its NPU, BV and other protein evaluation parameters in this test animal. It obviously makes a valuable contribution to human diet during its peak production season in the areas of production and high consumption.

References

Association of Official Analytical Chemists (1970) Official methods of analysis 11th ed. Washington, DC.

Coates, M. E., Hewitt, D. & Golob, P. (1970) A comparison of the effects of raw and heated soya bean meal in diets for germ-free and conventional chicks *Brit. J. Nutrition,* 24, 213.

Hilditch, T. P., & Seavell, A. J. (1950) *J. Oil Chem. Assocn.* 33, 24.

Miller, E. L. (1967) Determination of the tryptophan content of feedingstuffs with particular reference to cereals. *J. Sci. Fd. Agric.* 18, 381–386.

National Academy of Sciences/National Research Council (1963) Evaluation of protein quality. *Publs. Natn. Acad. Sci. Natn. Res. Coun.* Washington DC, No. 1100.

Oyenuga, V. A. (1966) Improvement of Nutritional Status in Developing Countries by Improved Food Production Legumes. *Proc. VIIth Int. Congr. Nutrition,* Hamburg, 3, 35–60.

Oyenuga, V. A. (1968) Nigeria's Foods and Feedingstuffs. Their chemistry and nutritive value. Ibadan Univ. Press 99pp.

TABLE 10. *Nutritive value of conophor seedcake to the rat*

| | Level of protein fed from | | |
| | Awusa nuts | | Egg |
	20%	24%	20%
Nitrogen retained (g)	8.5	9.7	13.0
Efficiency of food conversion (EFT)	4.1	3.7	2.5
Biological value (BV)	63.8	62.1	80.7
Net protein utilisation (NPU)	51.5	51.6	72.3
Net protein value (NPV)	10.3	12.4	14.5
Protein efficiency ratio (PER)	1.24	1.14	2.03
Mean liveweight gain/day (g)	2.4	2.5	4.1

Discussion

Mr Muller: I would like to draw Dr Babatunde's attention to the high levels of calcium in his pig diets. Pigs are extremely sensitive to calcium and in these diets calcium came up to almost lethal levels, and even supplementation with zinc would not avoid parakeretosis in these circumstances. High calcium levels were probably why good performance was not obtained under such high levels of meat and bone meal.

Dr Babatunde: I agree that high calcium was perhaps a factor leading to our poor results with meat and bone meal. Our objective was of course to see if meat and bone meal could be used as a major source of protein for pig rations and our experiments show quite clearly that it cannot.

Mr Ola: I refer to Professor Oyenuga's paper on conophor seed. If this were partially diverted to animal consumption from its existing food use might not human beings suffer from lack of it?

Professor Oyenuga: The studies on conophor seed were part of a general survey being conducted on West African foods. You will see from my paper that in fact the interpretation of the results was directed towards its use as a food for humans rather than as an animal feed. If the appropriate research and development inputs were provided to increase the availability of conophor seed, it could have a very big future both as a human food and an animal food. I should remind you that the soya bean was originally only cultivated as a human food.

Fifth Session

Use of oilseed cakes and meals as animal feed

Thursday 4th April
Morning

Chairman
Dr R. Bressani
Institute of Nutrition of Central
America and Panama (INCAP)
Guatemala

Nutritive value of oilseed cakes and meals

R. Roberts

Bibby Agriculture Ltd., Liverpool, England

Summary

Oilseeds represent genera of plants where metabolic activity has resulted in reserve food substances in the form of fats rather than the more common food reserve, starch.

The major oil seeds are of tropical or subtropical origin but some can be grown in more temperate regions and 14 examples have been selected to cover the major seeds and some of lesser importance in the world scene. Their feeding characteristics, in terms of both nutrients and anti-nutrients, are different but only the former characteristic is dealt with, making reference not only to cakes and meals but also to the possibilities of feeding whole seeds.

The nature of the oil in the original seed, and therefore remaining in the pressed cake, ranges from oils with drying characteristics, such as linseed and soya oil, to hard vegetable fats, such as coconut and palm kernel fat. An equation relating composition and metabolisable energy is described in the context of acid oils.

Oil seed residues are mainly used as a source of protein and it is in this area where the amino acid matrix of the materials is described and quantified on a comparative basis. The value of these characteristics will vary with the class of stock under consideration and may well be important in the selection of any particular livestock enterprise.

A total evaluation of raw materials is important for both tactical and strategic purposes. A simple exercise using the price of maize and soya will partly satisfy this requirement and a more complex approach involving a number of nutritional factors is described.

Résumé

La valeur nutritive des tourteaux de graines oléagineuses et des pâtures

Les graines oléagineuses représent des types de plantes où l'activité métabolique a résulté dans des substances alimentaires de réserve sous forme de graisses plutôt que la réserve alimentaire plus habituelle, l'amidon.

Les graines oléagineuses importantes sont d'origine tropicale ou sous-tropicale, mais quelques unes peuvent pousser dans des régions plus tempérées et 14 exemples ont été sélectionnés pour couvrir les graines importantes et quelques unes de moindre importance sur la scène mondiale. Les caractéristiques en termes nutritifs et anti-nutritifs sont différentes, mais ici on s'occupe seulement des caractéristiques précédentes, ayant rapport pas seulement avec les tourteaux et les pâtures, mais aussi avec les probabilités des pâtures de graines intégrales.

La nature de l'huile dans les graines originales et par conséquent restant dans les tourteaux pressés, s'étend des huiles avec des caractéristiques siccatives comme les graines de lin et l'huile de soja jusqu'aux graisses végétales dures comme les graisses des noix de coco et les graines intégrales de palmier. Une équation se rapportant à la composition et à l'énergie métabolisable est décrite dans le contexte des huiles acides. Les résidus des graines oléagineuses sont utilisés surtout comme une source de protéines et c'est dans ce domaine, où la matrice d'aminoacides des matériels est décrite et mesurée, sur une base comparative. La valeur de ces caractéristiques varie avec la catégorie du stock en question et elle pourrait être bien importante dans la sélection de n'importe quelle entreprise de bétail.

Une évaluation totale des matières premières est importante pour les buts tactique et stratégique. Un simple exercice utilisant le prix du mais et de la soja satisfera partiellement cette exigence; on décrit une approche plus complexe impliquant un nombre de facteurs alimentaires.

Resumen

Valor nutritivo de las tortas de orujo y harinas de semillas oleaginosas

Las semillas oleaginosas representan géneros de plantas en las que la actividad metabólica ha producido la reserva de sustancias alimenticias en forma de grasas más bien que en la más común reserva de alimentos, el almidón.

Las semillas oleaginosas principales son de origen tropical o subtropical, pero algunas pueden cultivarse en regiones más templadas y se han seleccionado 14 ejemplos para abarcar las semillas principales y algunas de importancia menor en la escena mundial. Sus características alimenticias, tanto respecto a los elementos nutritivos como antinutritivos, son diferentes pero se trata solamente de las primeras haciendo referencia no solamente a las tortas de orujo y harinas, sino también a las posibilidades con el alimento completo de semillas.

La naturaleza del aceite en la semilla original, que permanece por consiguiente en la torta prensada, varía desde aceites con características de secado, tales como el aceite de linaza y el aceite de soja, hasta grasas vegetales sólidas, tales como la grasa de coco y la de palma. En el contexto de los aceites ácidos se describe una ecuación en relación con la composición y la energía metabolizable.

Los residuos de las semillas oleaginosas se usan principalmente como fuente de proteínas y es en esta zona donde el aminoácido matriz de los materiales se describe y valora sobre una base comparativa. El valor de estas características variará con la clase de ganado que se esté considerando y puede ser muy importante en la selección de cualquier empresa ganadera particular.

Es importante una evaluación total de las materias primas, tanto para fines tácticos como estratégicos. Un ejercicio sencillo usando el precio del maíz y el de la soja satisfará en parte esta exigencia y se describe una forma más compleja de abordar el asunto implicando cierto número de factores nutritivos.

In many genera of plants metabolic activity leads to the production of reserve food substances in the form of oils or fats, rather than the more common food reserve, starch. These are built up from carbohydrates synthesized in the leaves and then transported to the storage region, usually the fruit and developing seed. The fats are built up from various saturated and unsaturated fatty acids in association with glycerol and remain in that form until required for use by the plant. Two types of oil are produced by plants, fixed non-volatile oils and essential or volatile oils. The non-volatile oils are the more important commercially and oils and fats of vegetable origin are in great demand in world markets. Oilseed cakes and meals can be considered a by-product of this demand and have become a very important group as a source of protein for both man and animals.

For this paper I have selected 14 examples which include materials of both great and lesser importance in the world protein scene. We have heard of the 'World oilcake and meal supplies' in a previous paper (see p. 49), but Table 1 will remind us of the relative significance of the selected materials. Nearly 90% of the total under consideration is provided by five oilseeds, soya, cotton, groundnut, sunflower and rape. The addition of three more materials, linseed, coconut and sesame brings the total to over 95% and amounts to an overall world tonnage of approximately 115 mt. The significance of this figure is the amount

TABLE 1. *World production of oilseeds*

Oilseed		% of total world* production	Total world production* (million tonne)
Soya	*Glycine soja*	40.3	
Cotton	*Gossypium spp.*	18.3	
Groundnut	*Arachis hypogaea*	15.4	
Sunflower	*Helianthus annuus*	8.1	
Rape	*Brassica napus/campestris*	6.6	
	Total	88.7	106.4
Linseed	*Linum usitatissimum*	3.0	
Coconut	*Cocos nucifera*	2.7	
Sesame	*Sesamum orientale*	1.7	
	Total	96.1	115.3
Castor	*Ricinus communis*		
Palm kernel	*Elaeis guineensis*		
Safflower	*Carthamus tinctorius*		
Mustard	*Sinapis alba/nigra*		
Niger	*Guizotia abyssinica*		
Crambe	*Crambe abyssinica*		
	Total		Approx. 120

Source: FAO (1971)

of protein represented and this would amount to approximately 30 mt which can be compared with the world production of fish meal protein at approximately 2 mt. In a study of nutritive value of such vast protein supplies one is impressed by the considerable variation in yields that are achieved in various producing countries, for example, the yield of soya beans varies between 690 kg/ha in the USSR to a mean yield of 1850 kg/ha in the USA. Within the limits of climate and expertise there must be room for hope towards improved world supplies by improving yields. It is often difficult to establish priority needs for improvements in energy, protein or amino acid balance in a particular material, but an overall increase in the yield of the existing package cannot be criticised.

Having stated that the oilseed cake or meal is a by-product of the oils and fats industry it is relevant to consider the different fat make up in individual seeds because of the residual fat that certainly remains in a pressed cake. These fats are classically described as drying, semi-drying, or non-drying oils and fats that are normally solid at room temperature. The drying oils have a great affinity for oxygen and form a thin elastic film on exposure to air. The nutritional significance of these categories lies in the fatty acid composition of the material (see Table 2), and in the

TABLE 2. *Major fatty acids of oilseeds*

Fats	
(50% C_{14})	Coconut
	Palm kernel
Non-drying oils	
(mainly $C_{18:1}$)	Groundnut
	Sesame
	Castor (Ricinoleic)
(mainly $C_{22:1}$)	Rape
	Mustard
	Crambe
Semi-drying oils	
(40–60% $C_{18:2}$)	Cotton
	Sunflower
	Soya
Drying oils	
(65–70% $C_{18:2}$)	Safflower
	Niger
(40–65% $C_{18:3}$)	Linseed

Source: Hilditch and Williams (1964)

examples chosen we have linseed oil containing 55% linolenic acid which represents a drying oil, whereas palm kernel and coconut oils have approximately 50% of the saturated C_{14} + C_{12} acids, myristic and lauric. Apart from the non-drying oils, groundnut, sesame and castor, the remainder lie in the drying or semi-drying categories with high proportions of linolenic acid $C_{18:2}$. While there is a need for essential fatty acids in livestock feeds, the majority of cereals are well supplied and there is seldom a deficiency. However, in the specific context of egg size in poultry the level can be significant and the energy level of such fats will be high, but oxidative rancidity may be a problem in some circumstances in the ground material. Energy level is not the major criterion in dairy feeds, however,

where high levels of unsaturated fats has been shown to depress butter fat content in the milk and more saturated fats with high levels of oleic, eg groundnut, and especially coconut and palm kernel with high levels of C_{12} and C_{14} can be advantageous. Such aspects are important where butter fat is included, in the definition of quality for payment, but the fundamental rightness of such an approach can be criticised.

There are four oilseeds containing somewhat unique fatty acids, with erucic acid $C_{22:1}$ making up approximately 50% of rape, mustard and crambe oils and hydroxy $C_{18:1}$ ricinoleic acid being approximately 90% of castor oil. Erucic acid appears to have rather poor availability in itself but due to unfavourable reports about its effect on human health, strains of rape have been developed which contain little or no erucic acid. Castor oil has a variety of uses from medicine, through lubricants to plastics and because of the toxicity of ricinoleic acid castor meal has little significance as a feed ingredient.

Oil refineries remove free fatty acids present in the oil by saponification, and acidification of these soaps produces a material composed of free fatty acids and glycerides. Such material will reflect the pattern of production in terms of fatty acid composition and it can be a useful energy source for animal feeds. The energy level of such a material will depend on many factors such as the proportion of glycerol and glycerides, the structure of such glycerides and nature of the free fatty acids, non-saponifiable material and water level. From the results of some twenty trials with chicks a formula has been derived by which the energy level can be estimated, see Table 3, but while 92.5% of the variation was explained the sample cannot be considered representative of the group of materials considered in this paper.

TABLE 3. *Metabolisable energy (chicks) in relation to fatty acid composition*

Metabolisable Energy

$$MJ/kg = 34.69 + 0.51 x + 0.08 y - 1.46 a - 0.35 b - 0.19 c - 0.04 d$$

x = % unsaturated acids C_{17} or less
y = % unsaturated acids C_{18} or more
a = % non saponifiable material
b = % saturated acids C_{18} or more
c = % saturated acids C_{16} or less
d = % free fatty acids

Source: Bibby data

The major significance of oilseed residues in animal feeds lies in their protein content and as this can be influenced by the degree of oil extraction, energy contents of pressed cake or meal can vary. The nature of the original material will also affect the energy level for different species of livestock because of indigestible residues. As protein levels may vary between 20 and 50% and crude fibre levels between 5 and 15% both because of the nature of the seed and also the

174 degree of dehulling that has taken place, there is little point in presenting data in this particular area. Mention should however be made of the possibility of feeding whole seeds as ingredients in livestock feeds. Soya beans are suitable for this operation because of the relatively low oil content and some form of cooking, through micronisation or extruders, is essential to control the anti-nutrients and produce a physical form suitable for use in the feed plant. Good control of heating is required to optimise this, and micronisation provides such control and may allow exploitation of bean prices which are at times favourable compared with extract prices. Figure 1 shows the changes in price for groundnut cake, whole nuts and oil during 1972 and is produced to suggest that the seeds are geared to oil prices and not extract prices. Soya would have been chosen but there were considerable gaps in the data during 1973 because of the USA action on exports.

Extracts of various publications have provided the data in Table 4 but the degree of extraction and level of dehulling is pertinent. The level of protein and aspects relating to 'anti-nutrients', which will be the subject of a later paper, suggest some of the selected materials as more suitable for the ruminant animal and digestibility data apply to such species. Some data is presented for protein digestibility with pigs but figures for safflower, castor and crambe were not found. The protein digestibility for cattle varies from 76% with palm kernel to 90% for groundnut and with pigs the protein digestibility in soya, sunflower and sesame are all greater than 90%.

Following digestibility of the protein, the component amino acids in the protein are particularly important with special reference to the simple stomached animal. However one is reminded of efforts to treat proteins in such a way as will allow minimal breakdown in the rumen where bacterial activity would downgrade the quality, and amino acid composition may become increasingly important with ruminant animals in the future along with solubility and speed of deamination. There is a mass of data available on amino acid com-

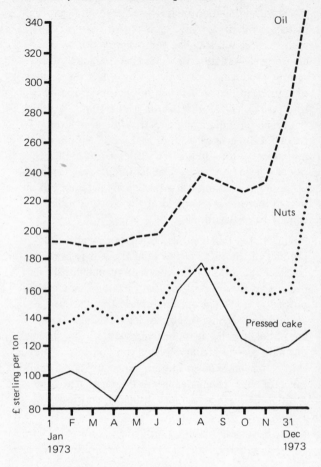

Figure 1
Groundnut prices in the UK during 1973

Source: The Public Ledger 1973–74

position but a minimum of sources has been selected because of the influence of origin, variety and analyst and therefore the figures in Table 5 may not be fully representative. However, when all the seeds are combined, one is intially impressed by their similarity in composition with the coefficient of variation (see Table 6). It will be seen that for valine, isoleucine, threonine and histidine it is below 15%, while those for lysine, methionine + cystine, tryptophan and leucine are 26, 27, 20 and 17% respectively. Some of

TABLE 4. *Content and digestibility of protein*

	Crude protein %	Digestibility %	
		Cattle	Pigs
Soya	51.5	–	91
Cotton	44.8	80	–
Groundnut	51.5	90	–
Sunflower	50.3	–	90
Rape	44.0	83	–
Linseed	38.6	83	–
Coconut	23.1	84	–
Sesame	51.5	–	94
Castor	34.0	–	–
Palm kernel	17.3	76	–
Safflower	49.1	–	–
Mustard	37.3	86	–
Niger	34.0	90	–
Crambe	49.0	–	–

Sources: United States – Canadian Tables of Feed Composition (1969); Schneider (1947); Altschul (1958); Miller *et al.* (1962); Williams (1966).

Oilseed	Lysine	Methionine + Cystine	Valine	Tryptophan	Isoleucine	Threonine	Histidine	Leucine	Oil Content	Protein Content (N x 6.25)
	g/16 g N								%	
Soya	6.3	2.8	5.2	1.3	5.5	3.7	2.4	7.4	20	42
Cottonseed	4.2	3.7	5.0	1.6	3.9	3.5	2.7	6.1	20	20
Groundnut	4.9	2.1	5.9	1.1	4.2	3.2	2.5	7.8	49	28
Sunflower	3.6	4.7	4.9	1.1	4.5	3.2	2.1	5.6	20	23
Rape	4.4	3.4	4.8	0.9	3.8	4.1	2.4	3.8	40	20
Linseed	3.3	3.2	4.9	1.5	5.5	3.6	2.1	5.4	38	24
Coconut	3.0	2.8	4.6	0.9	3.1	3.1	2.6	7.0	65	10
Sesame	2.7	4.2	5.0	1.6	4.4	3.3	2.3	7.1	50	25
Castor	3.0	–	5.4	1.1	4.6	3.2	1.7	5.6	50	18
Palm Kernel	3.4	2.1	5.4	1.0	4.0	3.1	1.6	6.4	49	9
Safflower	3.1	3.2	5.5	1.4	4.1	3.2	2.4	5.5	30	16
Mustard	5.4	4.2	4.7	1.1	3.8	4.0	2.6	6.3	–	–
Niger	4.4	3.2	3.6	–	4.2	3.9	2.4	6.6	45	18
Crambe	5.0	4.4	4.9	1.5	3.9	4.4	2.4	5.4	35	32

References: United States – Canadian Tables of Feed Composition (1969); Gates (1972/73); FAO (1970); Harvey (1970).

TABLE 6. *Amino acid composition of oilseed proteins: mean coefficient of variation*

	Mean ± SD g/16gN	Coefficient of variation, %
Lysine	4.1 ± 1.07	26
Methionine + Cystine	3.4 ± 0.83	27
Valine	5.0 ± 0.53	11
Tryptophan	1.2 ± 0.26	20
Isoleucine	4.2 ± 0.63	15
Threonine	3.5 ± 0.42	12
Histidine	2.3 ± 0.32	14
Leucine	6.1 ± 1.05	17

this variation could be due to difficulties of assay in the case of methionine + cystine and tryptophan but it is of benefit to look at these acids and suggest materials that appear to be good or bad sources of particular essential amino acids.

Lysine or methionine + cystine will be the first limiting amino acids in the majority of circumstances for the monogastric. Soya stands out as league leader for lysine with mustard, rape, crambe, niger, groundnut and cotton, forming an average group. The remainder lie below the mean with sesame at the bottom of the table. For methionine + cystine we find sunflower as the leader together with crambe, mustard and sesame in the top group. A group of average composition would include cotton, rape, linseed, safflower and niger, and poor sources of the sulphur acids would include soya, coconut, groundnut and palm kernel.

It is impossible to predict the second and third limiting amino acids because of the variety of purposes to be satisfied in terms of species, age of stock, and composition and density of the feed. However, the high coefficient for tryptophan and leucine deserve examination as these acids may be significant in some circumstances, especially for maize based diets and trypto-

phan. Sesame is a good source of tryptophan and together with cotton, crambe, linseed, safflower and soya, could be looked upon as supplements for maize. The remaining oilseeds have a tryptophan content similar to that of maize and must be considered poor sources of this amino acid. Leucine will not be limiting in the majority of circumstances and because of the high level of this amino acid in cereals (between 7 and 11 g/16g nitrogen for barley, wheat and maize, and 15g for sorghum), there will usually be a considerable surplus to requirement. Groundnut, soya, sesame and coconut, contain high levels of leucine whereas castor, sunflower, safflower, linseed, crambe and rape, are below the mean group for this amino acid.

It is stimulating to consider the oilseeds that constitute the majority of world output and their variety of inhibitory or toxic factors that have been overcome or accommodated by carefully controlled processing conditions and objective restriction of use to certain species and circumstances. There must therefore be a future for oilseeds that are currently considered of minor importance in the world scene because one can confidently predict that man's ingenuity through breeding, and processing, will overcome the antinutrient properties that will be discussed in a later paper (see Liener, pp. 179). Of the minor oilseeds safflower, niger and palm kernel can be considered bland feed ingredients but castor, mustard, and crambe will demand considerable treatment before they can be accepted as routine materials. Mustard and crambe can be considered in the same category as rape but more exhaustive treatment is necessary and a feeding trial with detoxified mustard with pigs gave promising results as indicated in Table 7. Crambe can be given a soda ash treatment and this together with treatment with ammonia have produced an improvement in feeding value as reported by Mustakas *et al* (1968). Castor however presents more difficult problems but because

TABLE 7. *Treated mustard extract for pigs*

Treatment	Control	Treated* mustard	Raw* mustard
77 lb Slaughter			
Liveweight gain (kg)	0.69	0.68	0.65
Feed conversion ratio	3.08	3.14	3.19
120 lb Slaughter			
Liveweight gain (kg)	0.73	0.71	0.73
Feed conversion ratio	3.33	3.42	3.27

*Material ex Colman's, Norwich.

of the unique properties of the oil in the field of plastics and nylon, the seed residue will remain a potential source of protein and while breeding for the elimination of ricin appears to have failed, treatment of the material may prove successful. Ricin and allergens in castor bean are discussed by Spies *et al.* (1962) as being sensitive to aqueous calcium hydroxide treatment, but residual castor oil in the meal may have harmful physiological effects if fed over extended periods.

Turning now to utilisation and market price, mention must be made of a feature that has not been considered. An acceptable intake of feed is necessary for satisfactory performance in livestock and therefore palatability of the feed is important. Palatability can be considered as a complex of texture, odour and taste and one should beware of using human acceptability as a yardstick. Some of the materials discussed have undesirable characteristics in this area with particular reference to cattle and for pigs. Palm kernel can be considered unpalatable because of its gritty nature, as are also rape, mustard and crambe due to their content of mustard oils, while safflower is not readily accepted by dairy cows. However the nature of the basal feed, in which such materials are incorporated, has a considerable bearing on the overall acceptance of the feed as a whole and upper limits of inclusion must be established for particular purposes. These considerations are, of course, in addition to any consideration of potential toxicity that must also be applied.

In establishing a value for any particular feed ingredient many aspects of that material must be assessed. However the major elements for consideration are protein and energy content and these must be weighed in the current climate of availability and offered price. For a producing country the first decision must be to either retain or export, and in either case whether to retain or export the whole or part of the seed, or the whole or part of the cake, extract and oil. Following such a decision the local feedstuff manufacturer must look upon the home grown material in competition with other protein sources that may be imported or with suitable alternative crops. The end result will be a market price that must either be accepted or rejected in favour of alternatives. Coombes and Romaser (1961) offered an approach to establishing such relative values using a principle of 'Partial Nutrient Worth' and provided an equation of the form,

$$\text{Nutritional Worth} = AX + BY + C$$

in which A relates to the protein content, B the energy content and C represents the notional added value.

The values for A, B and C are specific to an individual raw material but X and Y are extracted from a table relating to the currently pertaining price for yellow corn and dehulled soya extract. Such an approach has obvious merit for those with minimal computer facilities because the values given covered a considerable range in the price for maize and soya and could therefore be used to satisfy a particular market situation. Since the time of this publication however we have all experienced inflation to various degrees and one can look back with sadness on the days when the maximum price for soya extract and yellow maize were considered to be approximately £40/t and £25/t respectively.

For those who have a more comprehensive computer approach to formulation it is possible to evaluate particular ingredients on the basis of constraint costs. It is in such circumstances that the difference in value of the same raw material for different feeds is readily seen and examples of two feeds, a pig grower and a medium energy dairy cake, are given in Table 8. The

TABLE 8. *Constraint costs*

	Pig grower (10–30 kg liveweight)	Dairy cake (medium energy)
Crude protein	− 0.017	− 1.446
Oil	− 0.916	− 0.460
Lysine	− 27.347	−
Calcium	+ 1.319	+ 0.236
Phosphorus	− 3.886	− 4.209
Salt	+ 0.436	+ 0.018
Crude fibre	+ 1.819	+ 0.001
Total	− 0.538	− 0.120

contribution to value varies considerable in the case of protein and amino acids for the two feeds, and the element of 'naked total' is very significant in the grower feed. These differences are reflected in the value of a particular protein source for a specific feed and as an example, groundnut cake is worth £132/t for the dairy feed but only £109/t for the grower feed. This is a result of the high value of naked crude protein in the first case where the relatively poor amino acid matrix is no detriment. However the value of soya extract in the two selected cases gives the reversed picture being worth only £108/t for the dairy feed but £144/t in the context of pig growers.

The correct use of a particular raw material therefore depends on its nature and the purpose for which it will be used and the decision to 'buy' or 'not to buy' can only be satisfactorily made on a complex series of calculations. Such calculations for the multi-brand operator must include constraint costs for each brand, their relative tonnage, the stock position and correct forecasting of sales and replacement costs and I am sure that there are many solutions to this complex matter reflected in the minds of the present audience. However a solution to a group of brands can be achieved by stepwise multiple regression using the analyses of raw materials to explain price and such a

TABLE 9. *Raw material value and analysis (layer feeds)*

Material value

$$£/ton = 1.524 \times \text{Crude protein (\%)}$$
$$+ 1.00 \times \text{Oil (\%)}$$
$$+ 0.98 \times \text{Energy (MJ/kg)}$$
$$- 0.018 \times \text{Xanthophylls (mg/kg)}$$
$$+ 23.563$$

solution derived for laying feeds is given in Table 9. This prediction of value is only pertinent to one specific raw material price climate and is therefore not as flexible as the soya:maize approach. However the formula resulted in a predicted price for barley of £53.8/t and for soya of £102.5/t whereas the actual prices were £53.0 and £102.5/t respectively. Neither of the systems for evaluating a raw material can be considered perfect, and one's application to the problem will depend on one's ingenuity and available facilities, but in some way the feed formulator must optimise the particular aspects of each raw material towards the mutual satisfaction of both the feed compounder and livestock producer.

We will be hearing of the toxic factors associated with oilseeds in a later paper and many of the major materials in use to-day suffer from endogenous toxics and antinutrients. They supply a vast amount of protein into the world pool because such factors have either been eliminated, minimised, or tolerated by controlled use. I am sure that the barriers of the fourth Ice Age and the Sahara Desert are no longer significant and many of the lesser used oilseeds, with some undesirable qualities, will be managed, through breeding, processing and understanding, and will become significant supplements to the world protein need. This will only be so, however, if due regard is made to growing the material in the right place to satisfy the right need and at the right price.

References

Altschul A. M. (1958) Processed Plant Protein Foodstuffs. New York: Academic Press Inc.

Coombes G. F. & Romaser G. L. (1961) Partial nutrient worth of seed ingredients for poultry rations. *Miscellaneous Publications 158*. University of Maryland, USA.

FAO (1970) Nutritional Studies. No. 24. Rome. FAO.

FAO (1971) Production Yearbook. Rome: FAO.

Gates G. (1972/73) Feedstuffs Yearbook. **44**. No. 38. Minneapolis.

Harvey D. (1970) Tables of the Amino Acids in Foods and Feedingstuffs, Commonwealth Agricultural Bureau, Technical Communication No. 19, 2nd Ed.

Hilditch T. P. & Williams P. N. (1964) The Chemical Constitution of Natural Fats. 4th Ed. London. Chapman & Hall.

Miller R., Van Etten, C. H., McGlew, Clara, Wolff I. A. & Jones Q. (1962) Amino acid composition of seed meals from forty one species of *Cruciferae. J. Agric. Fd. Chem.* **10**, 426.

Mustakas G. C., Kirk L. D. & Griffin, E. L. (1968) Crambe seed processing. Improved seed meal by soda ash treatment. *J. Am. Oil Chem. Soc.* **45**, 53.

The Public Ledger (1973/74) London: UK Publication Ltd. Jan 1973–Jan 1974.

Schneider G. T. (1947) Feeds of the World. Morgantown: West Virginia University. Agricultural Experiment Station.

Spies J. R., Coulston E. J., Bernton H. S., Wells P. A. & Stevens H. (1962) The chemistry of allergens. Inactivation of the castor bean allergens and ricin by heating with aqueous calcium hydroxide. *J. agric. Fd. Chem.* **10**, 140.

United States-Canadian Tables of Feed Composition. National Academy of Sciences (1969) Publication 1684. Washington D.C.

Williams K. A. (1966) Oils Fats and Fatty Foods. 4th Ed. London: J & A Churchill Ltd.

Endogenous toxic factors in oilseed residues

I. E. Liener
University of Minnesota, USA

Summary

The protein-rich residue that remains after removal of the oil from many plant materials is known to contain a wide variety of substances which, if not destroyed by proper processing, may adversely affect the nutritive properties of the protein. Some of these may be protein in nature as exemplified by the trypsin inhibitors and phytohemagglutinins, which are found mainly in soya beans and other leguminous seeds. Oilseed residues may also contain glycosides which, upon enzymatic hydrolysis, release substances which can be toxic. Examples of such glycosides are the goitrogens of rapeseed, the cyanogens of linseed, and the saponins and estrogens of soya beans. Gossypol is a well-known example of a toxic substance which seems to occur exclusively in the cottonseed. Lesser known, and still rather poorly characterised, are factors in oilseed proteins which interfere with the utilisation of certain micronutrients such as metals and vitamins. In this paper particular emphasis is placed on the practical importance of the processing conditions which are necessary to destroy these toxic factors so that oilseed residues may provide dietary sources of protein which do not pose a hazard to the animals to which they are fed.

Résumé

Les facteurs toxiques endogènes dans les résidus des graines oléagineuses
Le résidu riche en protéines qui reste après l'enlèvement de l'huile de nombreux matériels des plantes contient comme on le sait, une grande variété de substances qui, si elle ne sont pas détruites par un procédé convenable, peut affecter défavorablement les propriétés nutritives des protéines. Certaines peuvent être des protéines en nature, exemplifiées par les inhibiteurs trypsiniques et les phytohémagglutinines qui sont trouvées surtout dans les graines de soja et autres graines légumineuses. Les résidus des graines oléagineuses peuvent contenir aussi des glycosides qui par hydrolyse enzymatique relâchent des substances qui peuvent être toxiques.

Exemples de tels glycosides sont les goitrogenes des graines de colza, les cyanogènes des graines de lin, les saponines et les oestrogènes des graines de soja. Le gossypol est un exemple bien connu d'une substance toxique qui paraît se produire exclusivement dans les graines de coton. Moins connus et plutôt insuffisamment caractérisés, sont les facteurs dans les protéines des graines oléagineuses qui entravent l'utilisation de certains micronutriments comme les métaux et les vitamines. Dans ce rapport on accentue particulièrement l'importance pratique des conditions de traitement qui sont nécessaires pour détruire ces facteurs toxiques, de sorte que les résidus des graines oléagineuses puissent procurer des sources diététiques de protéines qui ne présentent pas de risque pour les animaux, nourris avec ces substances.

Resumen

Factores tóxicos endogenos en residuos de semillas oleaginosas
Los residuos ricos en proteínas que quedan después de la extracción del aceite de muchos materiales de plantas se sabe que contienen una amplia variedad de sustancias que, si no se destruyen mediante el tratamiento adecuado, pueden afectar de manera adversa las propiedades nutritivas de la proteína. Algunas de éstas pueden ser proteínas en la naturaleza como los ejemplificads por los inhibidores de la tripsina y fitohemoaglutininas, que se encuentran principalmente en la soja y otras semillas de leguminosas. Los residuos de semillas oleaginosas pueden contener también glucósidos que, por hidrólisis enzímica, dejan en libertad sustancias que pueden ser tóxicas. Ejemplos de tales glucósidos son los goiterógenos de la colza, los cianógenos de la linaza y las saponinas y estrógenos de la soja. El gosipol es un ejemplo bien conocido de sustancia tóxica que parece encontrarse exclusivamente en la semilla del algodón. Menos conocidos y todavía mal caracterizados, son los factores en las proteínas de las semillas oleaginosas que se interponen en la utilización de ciertos elementos nutritivos microscópicos tales como

180 los metales y las vitaminas. En este documento se concede una atención particular a la importancia práctica de las condiciones de tratamiento que son necesarias para destruir estos factores tóxicos, a fin de que los resíduos de las semillas oleaginosas puedan facilitar fuentes dietéticas de proteínas que no pongan en peligro a los animales que se están alimentado con ellos.

Introduction

Nutritionists are agreed that oilseed residues can and do constitute a valuable source of protein for feeding animals, but they are likewise cognisant of the fact that this protein cannot be utilised to the fullest degree unless these products are 'properly' processed. The main theme of this paper will centre on the latter part of this statement which is based on the fact that there is present in most oilseed residues a number of naturally occurring or endogenous factors which, unless destroyed or inactivated, can detract quite markedly from the full nutritional potential of the protein which they contain.

Although poisonous substances are known to be widely distributed in the plant kingdom, including many of the forage crops used for feeding animals, we shall be concerned here with only those oil-bearing seeds which have present or potential value as a source of protein for animal feed, such as the soya bean, groundnut, cotton seed, and, of somewhat lesser importance perhaps, the cruciferous oilseeds. Wherever possible an indication will be given of the various ways and means whereby these toxic components can be effectively eliminated so that the residue remaining after the extraction of the oil no longer poses a hazard to the animal to which it is fed.

It is convenient to classify these substances by their chemical structure (where known) according to the scheme shown in Figure 1.

Proteins

Trypsin inhibitors

Perhaps the best known of the antinutritional factors shown in Figure 1 are the trypsin inhibitors which come under the category of proteins. These are widely distributed in the plant kingdom, particularly among the legumes. These inhibitors have attracted the attention of animal nutritionists because of the possible role which these substances might play in determining the nutritive value of plant proteins.

It was not long after soya beans were introduced into the USA during the early part of this century that Osborne and Mendel (1917) observed that unless cooked for several hours soya beans would not support the growth of rats. This observation concerning the beneficial effect of heat treatment was subsequently found to hold true for most animal species. The reason for this effect of heat became evident when it was discovered that raw soya beans contained a substance capable of inhibiting trypsin (Read & Haas, 1938; Ham & Sandstedt, 1944; Bowman, 1944). This

Figure 1

Classification of endogenous toxic factors according to chemical properties

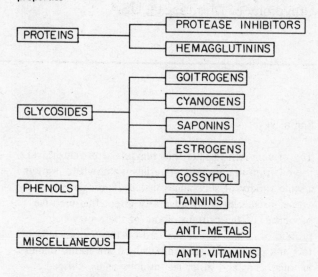

substance was subsequently crystallized and characterized as a protein by Kunitz in 1945 (Kunitz, 1945). Although this so-called Kunitz inhibitor is the main inhibitor found in soya beans, several others have also been found to be present to a lesser extent (Rackis & Anderson, 1964).

In general the extent to which the tryspin inhibitors are destroyed by heat is a function of the temperature, duration of heating, particle size, and moisture conditions; variables which must be closely controlled in the processing of soya bean meal in order to obtain a product having maximum nutritive value. As shown in Figure 2 the destruction of over 95% of the trypsin

Figure 2

Effect of autoclaving on protein efficiency and trypsin inhibitor activity of raw soya bean meal

Conditions: live steam at atmospheric pressure at either 5% or 19% moisture content prior to autoclaving

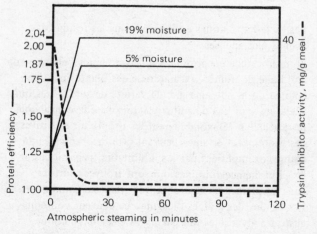

Source: Rackis (1965)

inhibitor activity by atmospheric steaming, in the presence of 5 or 19% moisture, is accompanied by a marked improvement in the nutritive value of the protein. Protein isolates and concentrates, unless heat processed, generally contain sufficient residual levels of trypsin inhibitor to cause some inhibition of rat growth (Rackis, 1965; Rackis et al. 1963). This residual activity is most likely due to incomplete separation of the whey protein, in which the trypsin inhibitors reside, from the curd which makes up the bulk of the soya bean protein.

The mechanism whereby the trypsin inhibitors cause growth inhibition has been the object of much study (see review by Liener and Kakade, 1969), but we still do not have all the answers. There is little doubt that hypertrophy of the pancreas constitutes one of the primary physiological effects produced by the inhibitor, an effect which is accompanied by a stimulation of its secretory activity. This results in an endogenous loss of the pancreatic enzymes, trypsin and chymotrypsin, which are rich in the sulphur-containing amino acids. It is believed that this endogenous loss of sulphur-containing amino acids further accentuates a methionine deficiency already created by the fact that this is the limiting amino acid of soya bean protein.

There is ample evidence to indicate that the trypsin inhibitors alone cannot account for all the growth inhibition observed on a diet in which the protein is provided by unheated soya beans. It has recently been shown (Kakade et al., 1973) that if you remove the trypsin inhibitors from a crude extract of raw soya beans, the increase in protein efficiency ratio (PER) is only about 40% of that obtained by heat treatment (Table 1). It is interesting to note that the removal of the trypsin inhibitor without the application of heat did not restore the pancreas to its normal size, but again accounted for about 40% of the total effect of unheated soya bean protein. Therefore, it is necessary to conclude that more than half of the growth inhibition and pancreatic hypertrophy observed with rats on diets containing raw soya bean must be ascribed to heat-labile factor or factors other than trypsin inhibitors.

TABLE 1. *Contribution of trypsin inhibitors to the growth inhibition and pancreatic hypertrophy induced in rats by diets containing unheated soya bean protein*

Source of protein in diet	PER	Wt of pancreas g/100g body wt
Soya flour extract, unheated	1.4	0.71
Soya flour extract, heated	2.7	0.57
Soya flour extract minus inhibitor*	1.9	0.65
Change due to removal of trypsin inhibitors (%)	+38	−43

Source: Kakade et al. (1973).
*Trypsin inhibitors were removed by passing a crude aqueous extract of soya bean flour through a column of Sepharose-bound trypsin.

The only other oilseed residue that is known to contain a trypsin inhibitor is the groundnut which has about 20% of the activity of raw soya beans (Anatharaman, & Carpenter, 1969). The presence of this inhibitor in the groundnut, however, is of doubtful nutritional significance since heat effects little if any improvement

in the nutritive quality of the protein, and there is no correlation between nutritive value and trypsin inhibitor activity (Borchers & Ackerson, 1950; Woodham & Dawson, 1968).

Phytohemagglutinins

Since, as indicated earlier, the poor nutritive value of raw soya beans cannot be fully accounted for by trypsin inhibitors, efforts have been made to uncover the presence of other toxic factors. One of the group of substances which merit consideration in this respect are the so-called phytohemagglutinins which are present in most legumes (see review by Jaffe, 1969). These proteins exhibit the unique phenomenon of being able to combine with glycoprotein components of the cell membrane, and in the case of red blood cells, this is accompanied by agglutination of the cells. Some of these are extremely toxic, such as *ricin* from the castor bean; others, such as those isolated from kidney beans (*Phaseolus vulgaris* Linn.) while not as toxic as ricin, are nevertheless capable of retarding the growth of experimental animals. The ability of the hemagglutinin isolated from the soya bean to inhibit the growth of rats is demonstrated by the data shown in Table 2. It is apparent from these data that the soya bean hemagglutinin accounts for almost 50% of the growth inhibition observed on raw soya flour. Thus the combined effects of the trypsin inhibitor and the hemagglutinins would seem to account for close to 90% of the growth inhibitory properties of raw soya beans.

TABLE 2. *Contribution of the soya bean hemagglutinin (SBH) to the growth inhibition induced in rats by diets containing soya bean protein*

Source of protein in diet	wt gain g/2 wk	growth inhibition (%)
Heated soya bean meal	60.0	0
Unheated soya bean meal	28.0	43
Heated soya bean meal + o.8% SBH*	45.0	26

Source: Liener (1953).
*This level of SBH provides the same level of hemagglutinating activity as the unheated soya bean meal used in this study.

The exact mechanism whereby the phytohemagglutinins inhibit growth is not known, but it is most likely related to the ability of the hemagglutinins to interact with components of the cell membrane. The absorptive capacity of the cells lining the intestinal mucosa may thus be impaired resulting in the inefficient utilisation of ingested nutrients.

Glycosides

Goitrogens

Goitrogenic agents in the form of glucosides are found predominately in the cruciferae family which includes such oilbearing seeds as the rapeseed (*Brassica napus* L.),

182　mustard seed (*B. nigra* Koch) and crambe (*Crambe abyssinica* Hochst. ex R. E. Fries) (see review by van Etten, 1969). The rapeseed in particular has received a great deal of attention as a potential source of protein because of its high protein content (40–45%) and balanced amino acid composition (Miller *et al.*, 1962; Ballester *et al.*, 1970). Its use as animal feed, however, has been restricted by the presence of thioglycosides or glucosinolates which, upon hydrolysis, release products which are goitrogenic and growth inhibitory. The predominant glucosinolate of rapeseed is *progoitrin* which, upon enzymatic hydrolysis, releases an isothiocyanate and 5-vinyloxazolidine-2-thione (see Figure 3). The latter two compounds are antithyroid agents which inhibit the organic binding of iodine in the thyroid. Their goitrogenic effect is not prevented by the addition of iodine to the diet.

Detoxification of cruciferous oilseed meals can be effected by

(i)　destroying the thioglucosidases which accompany the glucosinolates,

(ii)　removal of the glucosinolates, or

(iii)　removal of the goitrogenic products.

Although the enzyme is readily inactivated by moist heat treatment (Eapen *et al.*, 1968), there is the everpresent possibility that, if the thioglycosides are allowed to remain in the meal, subsequent hydrolysis by the bacteria in the intestinal tract may release the goitrogenic agents. The glucosinolates may be removed from the meal by extraction with hot water (Eapen *et al.*, 1969; Agren & Eklund, 1972), dilute alkali (Kozloska *et al.*, 1972) or acetone (Van Etten *et al.*, 1965) or by decomposition with iron salts (Kirk *et al.*, 1971) or soda ash (Mustakas *et al.*, 1968). If the meal is moistened so that the enzyme is allowed to hydrolyse the glucosinolates, the goitrogenic products may be removed by extraction with acetone (Tookey *et al.*, 1965) or water (Ballester *et al.*, 1970) or by steamstripping of the volatile isothiocyanates (Mustakas *et al.*, 1962). In general, cruciferous meals which have been effectively detoxified by any of the above techniques have a nutritive value which is essentially equivalent to properly processed soya bean meal.

Efforts to breed strains of rapeseed with low glucosinolate content have met with some degree of success (Downey *et al.*, 1969; Josefson, 1970). One such variety (Bronowski) was found to promote better growth and efficiency of feed utilisation than commercial rapeseed meal (Lo & Hill, 1971; Josefson & Munck, 1972) although there is some indication of a high-molecular weight constituent which still inhibits growth and is unrelated to the glucosinolates (Josefson & Munck, 1972).

Several procedures have been described for the preparation of protein isolates from rapeseed meal which are relatively free of glucosinolates (Lo & Hill, 1971; Owen *et al.*, 1971; Eklund *et al.*, 1971). These

Figure 3

Reactions depicting the enzymatic hydrolysis of progoitrin to produce goiterogenic end-products

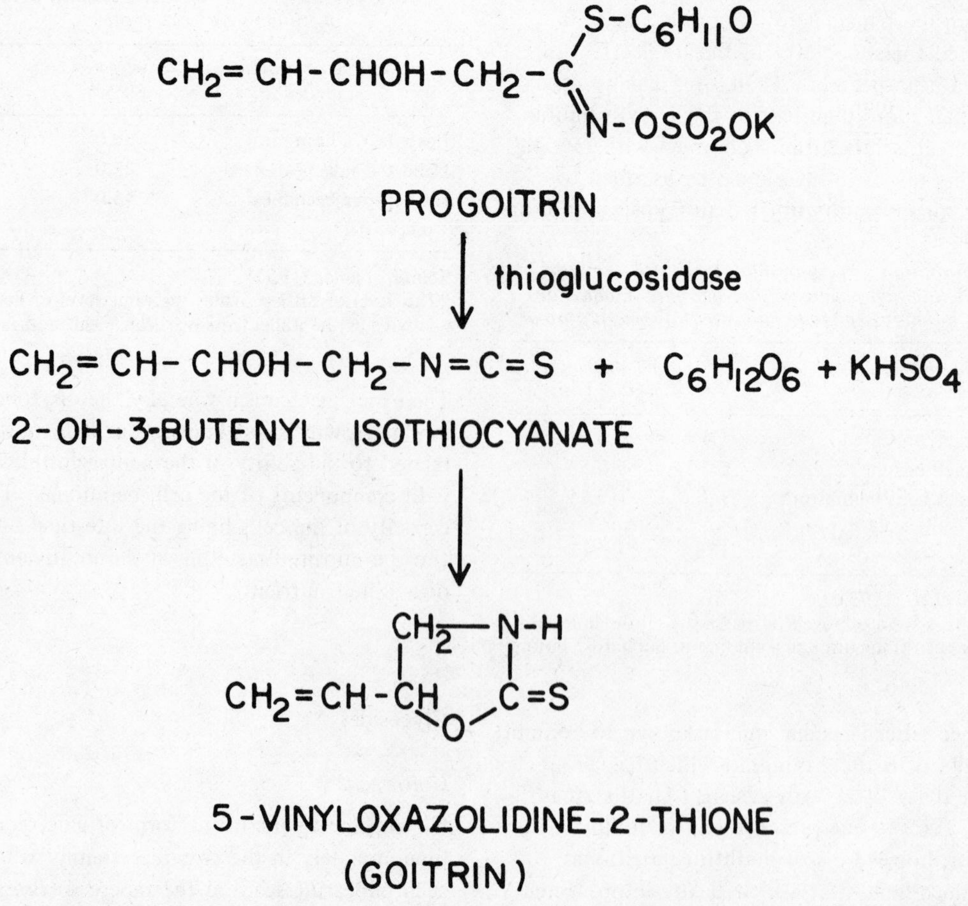

$$CH_2=CH-CHOH-CH_2-C\underset{N-OSO_2OK}{\overset{S-C_6H_{11}O}{\big<}}$$

PROGOITRIN

thioglucosidase

$$CH_2=CH-CHOH-CH_2-N=C=S \quad + \quad C_6H_{12}O_6 + KHSO_4$$

2-OH-3-BUTENYL ISOTHIOCYANATE

5-VINYLOXAZOLIDINE-2-THIONE

(GOITRIN)

protein isolates have been found to be essentially non-toxic and to have a biological quality equivalent to such protein sources as casein or soya bean meal (Agren & Eklund, 1972; Lo & Hill, 1971; Eklund *et al.*, 1971).

The feeding of unheated soya beans to rats and chicks produces a marked enlargement of the thyroid gland (Patton *et al.*, 1939), but in this instance the goitrogenic principle does not appear to be a glucosinolate but rather a low molecular weight oligopeptide composed of two or three amino acids or one or two amino acids and a sugar residue (Konijn *et al.*, 1973). Its mechamism of action appears to be similar to that of the glucosinolates since it interferes with the incorporation of iodine into the thyroid gland. Rats fed groundnuts also develop an enlarged thyroid gland, an effect which has been attributed to a phenolic glycoside which resides in the skin (Srinivasan *et al.*, 1957). It was suggested that the phenolic metabolites found from this glycoside are preferentially iodinated and thereby deprive the thyroid of available iodine.

Cyanogens

It has long been recognised that a wide variety of plants are potentially toxic because they contain glycosides which release HCN upon hydrolysis (see review by Conn, 1973). Plants such as sorghum, New Zealand clover, and certain varieties of lima beans have frequently been responsible for livestock poisoning. Among the oilseeds, however, few are known to be cyanogenic with the exception of linseed meal. Linen flax is known to contain a glycoside, linamarin, which releases HCN when hydrolysed by the enzyme, linamarase (a β-glucosidase) and oxynitrilase as shown in Figure 4. Since linseed meal is usually processed under conditions which almost certainly cause the inactivation of these enzymes, it is highly unlikely that the cyanogenic glycosides pose any real threat to the health of animals consuming this oilseed residue.

Saponins

Saponins are foam-producing glycosides in which the non-sugar residue (sapogenin) is a triterpenoid alcohol. They are usually bitter-tasting, but those of soya beans are non-bitter tasting. A number of different saponins are found in soya beans and groundnuts (see review by Birk, 1969), and, at one time, these were thought to be one of the factors contributing to the poor nutritive value of unheated soya bean protein (Potter & Kummerow, 1954). Birk *et al.* (1963), however, have shown that the hemolytic activity of the soya bean saponins is unaffected by the heat treatment necessary to produce the optimum nutritive value of soya bean meal. Although high levels of soya bean saponins inhibit the proteolytic action of trypsin and chymotrypsin, this effect is simply the manifestation of a non-specific interaction of soya bean saponins with proteins (Ishaaya & Birk, 1969). Finally, soya bean saponins do not impair the growth of rats or chicks when added at five times the concentration found in soya bean meal (Ishaaya *et al.*, 1969); there is in fact no evidence that they are even absorbed into the body following ingestion (Gestetner *et al.*, 1968).

Estrogens

A number of isoflavones having estrogenic activity have been reported to be present in soya beans (see review by Stob, 1973). One of these, genistein (4',5,7-trihydroxyisoflavone), in addition to having estrogenic activity, when fed to male rats at a level of 0.5% of the diet, inhibited their growth and produced elevated levels of zinc in the liver and bones and increased the deposition of calcium, phosphorus, and manganese in the bones (Magee, 1963). Since hexane-extracted soya beans contain only 0.1% of the parent glycoside, genistin, (Walter, 1941) it is highly unlikely that there are sufficient amounts of this estrogen in soya bean meal to cause any problems in feeding animals.

Figure 4
The enzymatic hydrolysis of linamarin, the cyanogenetic glycoside of flax seed

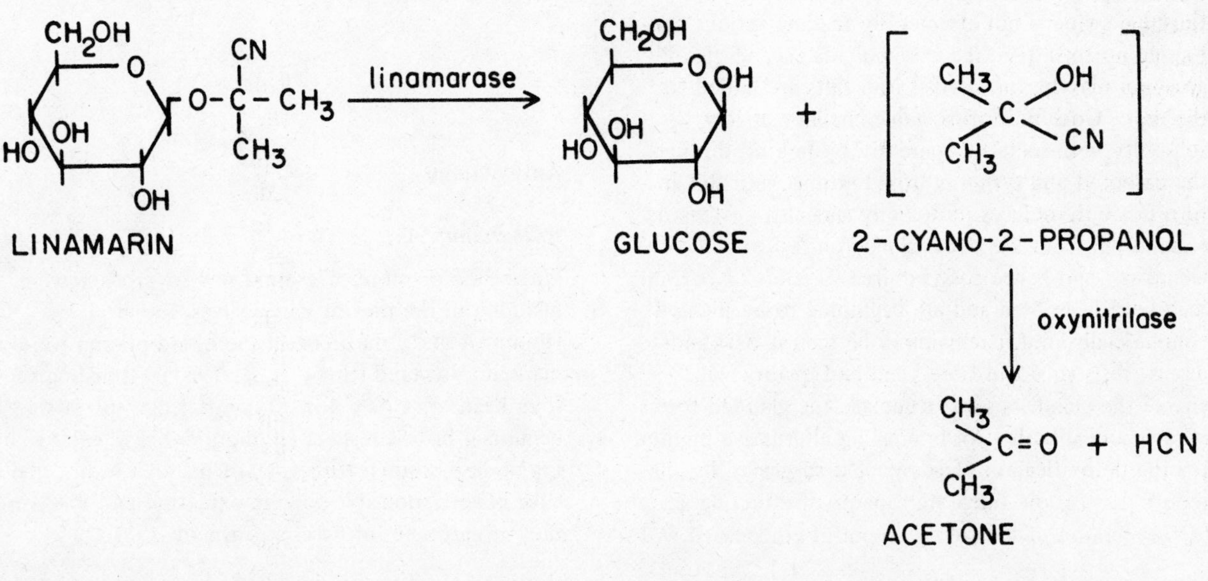

Gossypol

One of the more familiar naturally-occurring plant toxins is gossypol, a phenol-like compound which has the structure shown in Figure 5. This substance is peculiar to the pigment glands of the cotton seed and constitutes about 20 to 40% of the weight of the glands. Although gossypol is toxic to non-ruminants, it is much less so to ruminants presumably because the gossypol becomes bound to protein while in the rumen (Reiser & Fu, 1962). The heat generated during the commercial production of cotton seed meal serves to cause 80 to 90% of the gossypol to become bound to the protein rendering it non-toxic. The amount of gossypol inactivated in this manner depends on the type of commercial processing involved (see review by Berardi and Goldblatt, 1969). Unfortunately the binding of gossypol to protein reduces the availability of the lysine because of the interaction of the aldehyde groups of gossypol with the epsilon amino group of lysine.

Figure 5
Structure of gossypol

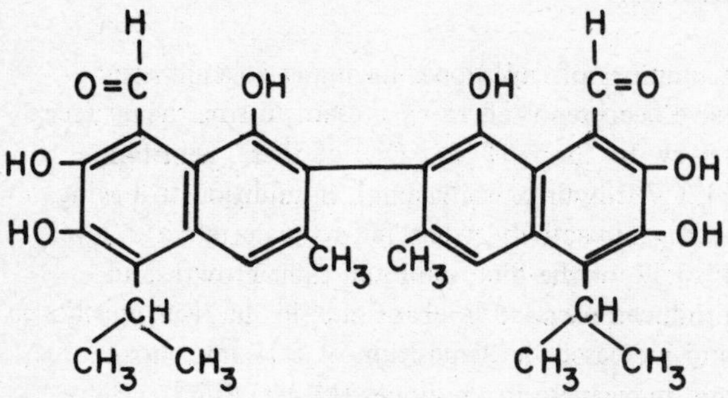

The level of free gossypol that remains after processing is an important consideration in the use of cotton seed meal for feeding animals (Smyth, 1970). Dietary levels of 0.015% or less free gossypol are believed to be safe when cotton seed meal is used as a protein supplement in balanced diets for poultry, although levels of gossypol in excess of 0.005% cause egg yolk discoloration in laying hens. The performance of growth-finishing swine is not affected by feeding rations containing up to 0.01% free gossypol. Higher levels of gossypol may be tolerated if iron salts are added to the diet. Gossypol forms stable chelates of low solubility with metals, apparently by linking through the carbonyl and *ortho*-hydroxyl groups, and this in turn not only reduces its toxicity but also hastens its elimination from the body (Singleton & Kratzer, 1973). Glandless, and hence gossypol-free varieties of cotton seed have been bred and are beginning to be planted commercially, but it remains to be seen if pest and disease resistance and fibre yield and quality will enable the glandless seed to replace the glanded forms on a practical basis. An interesting alternative method for the detoxification of gossypol is suggested by the report that certain fungi are capable of effecting a 90% reduction of the free gossypol of cottonseed with concomitant elimination of its toxicity to rats and chicks (Baugher & Campbell, 1969).

Although another group of phenolic compounds, the tannins, is widely distributed in nature, the only oil-seed residue which appears to contain significant amounts of these substances is rapeseed meal. Limited studies (Yapar & Clandinin, 1972) have shown that rapeseed meal from which the tannins had been extracted had a significantly higher metabolisable energy value for chicks than the unextracted meal. Adding the extracted tannins to soya bean meal resulted in a reduction in its metabolisable energy value. Rather surprisingly, however, the adverse effect of tannins did not seem to be due to its effect on nitrogen absorption; the possibility that the tannins might have an adverse effect on certain enzyme systems was suggested (Yapar & Clandinin, 1972).

Miscellaneous

Metal-binding constituents

Growth experiments with chicks and turkeys have shown that the dietary requirement for such metals as zinc, manganese, and calcium is significantly increased when isolated soya bean protein is the main source of protein in the ration (see review by Liener, 1969). Autoclaving the soya bean protein or adding ethylenediamine tetraacetate (EDTA) to the diet reversed this increased requirement for metals. *In vitro* experiments have provided direct evidence that isolated soya bean protein can bind zinc. The specific component responsible for the binding of zinc is most likely phytic acid since soya bean protein from which the phytic acid had been removed was no longer capable of binding this metal. Phytic acid may also be responsible for the unavailability of calcium and zinc from sesame meal and cotton seed meal reported by several investigators in studies with chicks (Pensack *et al.*, 1958; Lease & Williams, 1967). A marked reduction in the dietary requirement of chicks for zinc can be achieved by treating cottonseed meal with the enzyme phytase prepared from *Aspergillus ficcum* (Rojas & Scott, 1969).

Anti-vitamins

Anti-vitamin A

When 30% or more of ground raw soya beans were included in the diet of dairy calves, the levels of vitamin A and β-carotene in the blood plasma were markedly lowered (Shaw *et al.*, 1951). Roasting the soya beans at 100°C for 30 minutes did not prove beneficial in this respect (Ellmore & Shaw, 1954). Lipoxidase, known to be present in soya beans, may have been responsible for the oxidation and inactivation of carotene in these experiments.

Anti-vitamin D

Unheated soya bean meal, or the protein isolated therefrom can produce rickets in turkey poults (Carlson *et al.*, 1964), chicks (Jensen & Mraz, 1966), and swine (Miller *et al.*, 1965) but this rachitogenic effect is abolished by autoclaving of the soya bean protein or by supplementation with vitamin D. Whether this rachitogenic effect is simply due to the binding of calcium by phytic acid which results in an interference with the absorption of calcium is still not certain.

Anti-vitamin E

Although an antagonist to vitamin E was first reported to be present in kidney beans (*Phaseolus vulgaris* Linn.) (Hintz & Hogne, 1964), anti-vitamin E activity has also been recently noted in isolated soya bean protein (Fisher *et al.*, 1969). Unheated soya bean protein isolate was found to increase the chicks requirement for a-tocopherol as measured by growth, mortality, exudative diathesis, and encephalomalacia. The identity of the factor responsible for anti-vitamin E activity is not known at present.

Anti-pyridoxine

Although flax seed (*Linum usitatissimum* L.) is considered a poor source of protein for the chick, considerable improvement can be effected by extraction with water, autoclaving, or by supplementation with pyridoxine (Kratzer, 1947; Kratzer & Williams, 1948). Klosterman *et al.* (1967) have isolated an antagonist of pyridoxine from flax seed which they identified as a γ-glutamyl derivative of 1-amino-D-proline as shown in Figure 6. This peptide was given the name of linatine although the anti-vitamin itself is 1-amino-D-proline. Its antagonistic action against pyridoxine can be explained by the fact that it forms a complex with pyridoxal phosphate.

Anti-vitamin B_{12}

Since soya beans are devoid of vitamin B_{12}, it is not surprising that diets containing soya bean protein require supplementation with this vitamin in order to support normal growth of rats. Vitamin B_{12} supplementation, however, improves the growth of animals receiving raw soya beans to a greater extent than similar supplementation of diets containing heated soya beans (Baliga & Rajagopalan, 1954). Recent studies by Edelstein and Guggenheim (1970a, 1970b) would suggest a decreased availability of the vitamin B_{12} produced by the microflora when raw soya is included in the diet as well as the presence of a heat labile substance which somehow accentuates the requirement for vitamin B_{12}.

Conclusions

It should be apparent that, although we can cite numerous examples of naturally occurring toxic factors in oilseed residues, these products can and do provide a valuable source of protein for feeding animals. This can be attributed in large part to the fact that, having once recognized the presence of such substances in feed materials, man has been able to apply his technological skills to the development of processes which eliminate them or at least reduce them to non-toxic levels. Nevertheless there is the ever present possibility that the prolonged consumption of a particular oilseed residue which has been improperly processed could bring to the surface toxic effects which would not otherwise be manifest. It is also conceivable that the efforts of plant breeders to develop plants having a better balance of amino acids could inadvertently lead to the acquisition of strains having the genetic capacity to synthesise heretofore unrecognized toxins. All of those who are concerned with the important problem of animal production should at least be cognisant of such possibilities and be prepared to apply their knowledge and skill to meet this challenge should it arise.

Figure 6
Anti-pyridoxine factor in linseed meal

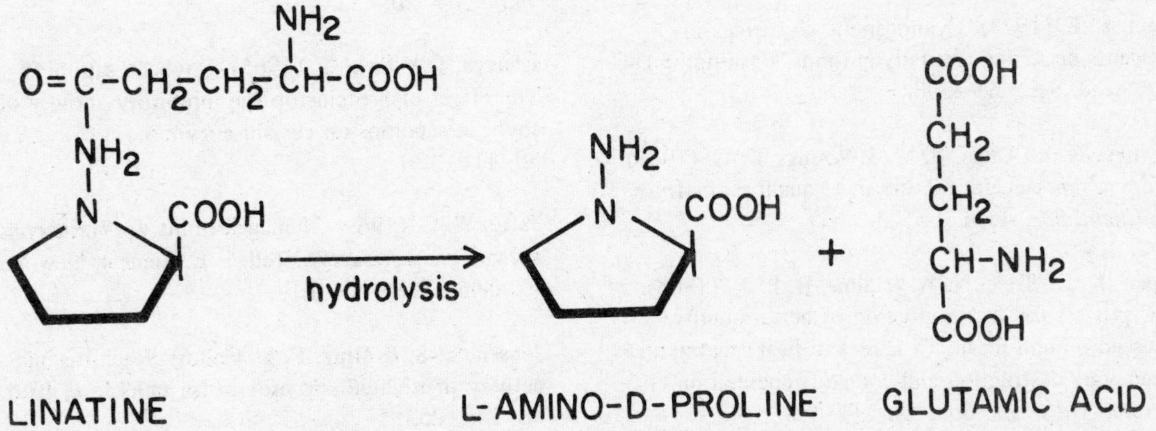

LINATINE   L-AMINO-D-PROLINE   GLUTAMIC ACID

Agren, G. & Eklund, A. (1972) The nutritive value of detoxified protein concentrate prepared from rapeseed by hydraulic pressing. *J. Sci Food Agr.* **23**, 1457–1462.

Anantharaman, K. & Carpenter, K. J. (1969) Effects of heat processing on the nutritional value of groundnut products. I. Protein quality of groundnut cotyledons for rats. *J. Sci. Food Agr.* **20**, 703–708.

Baliga, B. R. & Rajagopalan, R. (1954) Influence of vitamin B_{12} on the biological value of raw soya bean. *Current Sci.* (India) **23**, 51–52.

Ballester, D., Rodrigo, R., Nakouzi, L., Chichester, C.O., Yañez, E. & Mönckeberg, F. (1970) Rapeseed meal. II. Chemical composition and biological quality of the protein. *J. Sci. Food Agr.* **21**, 140–142.

Ballester, D., Rodrigo, R., Nakouzi, L., Chichester, C.O., Yañez, E. & Mönckeberg, F. (1970) Rapeseed meal. III. A simple method for detoxification. *J. Sci. Food Agr.* **21**, 143–144.

Baugher, N. L. & Campbell, T. C. (1969) Gossypol detoxification by fungi. *Science* **164**, 1526–1427.

Berardi, L. C. & Goldblatt, L. A. (1960) Gossypol. In *Toxic constituents of plant foodstuffs* (ed. I. E. Liener), New York; Academic Press, 211–266.

Birk, Y., Bondi, A., Gestetner, B. & Ishaaya, I. (1963) A thermo-stable hemolytic factor in soybeans. *Nature* **197**, 1089–1090.

Birk, Y. (1969) Saponins. In *Toxic constituents of plant foodstuffs* (ed. I. E. Liener), New York: Academic Press, 169–210.

Borchers, R. & Ackerson, C. W. (1950) The nutritive value of legume seeds. X. Effect of autoclaving and the trypsin inhibitor test for 17 species. *J. Nutr.* **41**, 339–345.

Bowman, D. E. (1944) Fractions derived from soybeans and navy beans which retard the tryptic digestion of casein. *Proc. Soc. Exp. Biol. Med.* **57**, 139–140.

Carlson, C. W., Saxena, H. C., Jensen, L. S. & McGinnis, J. (1964) Rachitogenic activity of soybean fractions. *J. Nutr.* **82**, 507–511.

Conn, E. E. (1973) Cyanogenetic glycosides in toxicants occurring naturally in foods, Washington D.C.: Nat. Acad. Sci., 299–308.

Downey, R. K., Craig, B. M. & Youngs, C. G. (1969) Breeding rapeseed for oil and meal quality. *J. Amer. Oil. Chem. Soc.* **46**, 121–123.

Eapen, K. E., Tape, N. W. & Sims, R. P. A. (1968) New process for the production of better quality rapeseed oil and meal. I. Effect of heat treatments on enzyme destruction and color of rapeseed oil. *J. Amer. Oil Chem. Soc.* **45**, 194–196.

Eapen, K. E., Tape, N. W. & Sims, R. P. A. (1969) New process for the production of better quality rapeseed oil and meal. II. Detoxification and dehulling of rapeseeds – feasibility study. *J. Amer. Oil. Chem. Soc.* **46**, 52–55.

Edelstein, S. & Guggenheim, K. (1970) Causes of the increased requirement for vitamin B_{12} in rats subsisting on an unheated soybean flour diet. *J. Nutr.* **100**, 1377–1382.

Edelstein, S. & Guggenheim, K. (1970) Changes in the metabolism of vitamin B_{12} and methionine in rats fed unheated soya bean flour. *Brit. J. Nutr.* **24**, 735–740.

Eklund, A., Agren, G. & Langler, T. (1971) Rapeseed protein fractions. I. Preparation of a detoxified lipid-protein concentrate from rapeseed (*Brassica napus* L.) by a water-ethanol extraction method. *J. Sci. Food Agr.* **22**, 650–652.

Eklund, A., Agren, G., Langler, T., Stenram, U. & Nordgren, H. (1971) Rapeseed protein fractions. II. Chemical composition and biological quality of a lipid-protein concentrate from rapeseed (*Brassica napus* L.), *J. Sci. Food Agr.* **22**, 653–657.

Ellmore, M. F., & Shaw, J. C. (1954) The effect of feeding soybeans on blood plasma carotene and vitamin A of dairy calves. *J. Dairy Sci.* **37**, 1269–1272.

Fisher, H., Griminger, P. & Budowski, P. (1969) Antivitamin E activity of isolated soybean protein for the chick. *Z. Ernaehrungswiss* **9**, 271–278.

Gestetner, B., Birk, Y. & Tencer, Y. (1968) Soybean saponins. Fare of ingested soybean saponins and the physiological aspect of their hemolytic activity. *J. Agr. Food Chem.* **16**, 1031–1035.

Ham, W. E. & Sandstedt, R. M. (1944) A proteolytic inhibitory substance in the extract from unheated soybean meal. *J. Biol. Chem.* **154**, 505–506.

Hintz, H. F. & Hogue, D. E. (1964) Kidney beans (*Phaseolus vulgaris*) and the effectiveness of vitamin E for prevention of muscular dystrophy in the chick. *J. Nutr.* **84**, 283–287.

Ishaaya, I., Birk, Y., Bondi, A. & Tencer, Y. (1969) Soybean saponins. IX. Studies of their effect on birds, mammals, and cold-blooded organisms. *J. Sci. Food Agr.* **20**, 433–436.

Ishaaya, I. & Birk, Y. (1965) Soybean saponins. IV. The effect of proteins on the inhibitory activity of soybean saponins on certain enzymes. *J. Food Sci.* **30**, 118–120.

Jaffé, W. G. (1969) Hemagglutinins. In *Toxic constituents of plant foodstuffs* (ed. I. E. Liener), New York: Academic Press, 69–101.

Jensen, L. S. & Mraz, F. R. (1966) Rachitogenic activity of isolated soy protein for chicks. *J. Nutr.* **88**, 249–253.

Josefson, E. (1970) Glucosinolate content and amino acid composition of rapeseed (*Brassica napus* L.) as affected by sulfur and nitrogen nutrition. *J. Sci. Food Agr.* **21**, 98–103.

Josefson, E. & Munck, L. (1972) Influence of glucosinolates and a tentative high molecular detrimental factor on the nutritional value of rapeseed meal. *J. Sci. Food Agr.* **23**, 861–869.

Kakade, M. L., Hoffa, D. E. & Liener, I. E. (1973) Contribution of trypsin inhibitors to the deleterious effects of unheated soybeans fed to rats. *J. Nutr.* **103**, 1772–1778.

Kirk, L. D., Mustakas, G. C., Griffin, E. L., Jr. & Booth, A. N. (1971) Crambe seed processing. Decomposition of glucosinolates (thioglycosides) with chemical additives. *J. Amer. Oil. Chem. Soc.* **48**, 845–850.

Klotsterman, H. J., Lamoureux, G. L. & Parsons, J. L. (1967) Isolation, characterization, and synthesis of linatine, a vitamin B_6 antagonist from flax seed (*Linum usitatissimum*). *Biochemistry* **6**, 170–177.

Konijn, A. M., Gersham, B. & Guggenheim, K. (1973) Further purification and mode of action of a goitrogenic material from soybean flour. *J. Nutr.* **103**, 378–383.

Kozlowska, H., Sosulski, F. W. & Youngs, C. G. (1972) Extraction of glucosinolates from rapeseed. *Can. J. Food Sci. Tech.* **5**, 149–154.

Kratzer, F. H. (1947) Effect of duration of water treatment on the nutritive value of linseed meal. *Poultry Sci.* **26**, 90–91.

Kratzer, F. H. & Williams, D. E. (1948) The relation of pyridoxine to the growth of chicks fed rations containing linseed oil meal. *J. Nutr.* **36**, 297–305.

Kunitz, M. (1945) Crystallisation of trypsin inhibitor from soybeans. *Science* **101**, 668–669.

Lease, J. G. & Williams, W. P. Jr. (1967) Availability of zinc and comparison of *in vitro* and *in vivo* zinc uptake of certain oil seed meals. *Poultry Sci.* **46**, 233–241.

Liener, I. E. (1953) Soyin, a toxic protein from the soybean. I. Inhibition of rat growth. *J. Nutr.* **49**, 527–540.

Liener, I. E. (1969) Miscellaneous toxic factors. In *Toxic constituents of plant foodstuffs* (ed I. E. Liener), New York: Academic Press, 431–432.

Liener, I. E. & Kakade, M. L. (1969) Protease inhibitors. In *Toxic constituents of plant foodstuffs* (ed. I. E. Liener), New York: Academic Press, 8–68.

Lo, M. T. & Hill, D. C. (1971) Evaluation of protein concentrates prepared from rapeseed meal. *J. Sci. Food Agr.* **22**, 128–130.

Magee, A. D. (1963) Biological responses of young rats fed diets containing genistin and genistein. *J. Nutr.* **80**, 151–156.

Miller, E. R., Ullrey, D. E., Zutant, C. J., Hoefer, S. H. & Luecke, R. L. (1965) Comparison of casein and soy proteins upon mineral balance and vitamin D_2 requirements of the baby pig. *J. Nutr.* **85**, 347–354.

Miller, R. W., van Etten, C. H., McGrew, C., Wolff, I. A. & Jones, Q. T. (1962) Amino acid composition of seed meals from forty-one species of cruciferae. *J. Agr. Food Chem.* **10**, 426–430.

Mustakas, G. C., Kirk, L. D. & Griffin, E. L. Jr. (1962) Mustard seed processing: bland protein meal, bland oil, and allyisothiocyanate as a by-product. *J. Amer. Oil Chem. Soc.* **39**, 372–377.

Mustakas, G. C., Kirk, L. D., Griffin, E. L. Jr. & Clanton, D. C. (1968) Crambe seed processing. Improved feed meal by soda ash treatment. *J. Amer. Oil. Chem. Soc.* **45**, 53–57.

Osborne, T. B. & Mendel, L. B. (1917) The use of soybean as food. *J. Biol. Chem.* **32**, 369–387.

Owen, D. F., Chichester, C. O., Granadino, J. C. & Mönckeberg, F. B. (1971) A process for producing nontoxic rapeseed protein isolate and an acceptable feed by-product. *Cereal Chem.* **48**, 91–96.

Patton, A. R., Wilgus, H. S. Jr. & Harshfield, G. S. (1939) The production of goiter in chickens. *Science* **89**, 162–163.

Pensack, J. M., Henson, J. N. & Bogdonoff, P. D. (1958) The effects of calcium and phosphorus on the zinc requirements of growing chicks. *Poultry Sci.* **37**, 1232.

Potter, G. C. & Kummerow, F. A. (1954) Chemical similarity and biological activity of the saponins isolated from alfalfa and soybeans. *Science* **120**, 224–225.

Rackis, J. J., Smith, A. K., Nash, A. M., Robbins, D. J. & Booth, A. N. (1963) Feeding studies on soybeans. Growth and pancreatic hypertrophy in rats fed soybean meal fractions. *Cereal Chem.* **40**, 531–538.

Rackis, J. J. & Anderson, R. L. (1964) Isolation of four soybean trypsin inhibitors by DEAE – cellulose chromatography. *Biochem. Biophys. Res. Commun.* **15**, 230–235.

Rackis, J. J. (1965) Physiological properties of soybean trypsin inhibitors and their relationship to pancreatic hypertrophy and growth inhibition of rats. *Fed. Proc.* **24**, 1488–1493.

Read, J. W. & Haas, L. W. (1938) Studies on the baking quality of flour as affected by certain enzyme actions. V. Further studies concerning potassium bromate and enzyme activity. *Cereal Chem.* **15**, 59–68.

Reiser, R. & Fu, H. C. (1962) The mechanism of gossypol detoxification by ruminant animals. *J. Nutr.* **76**, 215–218.

Rojas, S. W. & Scott, M. L. (1969) Factors affecting the nutritive value of cottonseed meal as a protein source in chick diets. *Poultry Sci.* **48**, 819–835.

188 Shaw, J. C., Moore, L. A. & Sykes, J. F. (1951) The effect of raw soybeans on blood plasma carotene and vitamin A and liver vitamin A of calves. *J. Dairy Sci.* **34**, 176–180.

Singleton, V. L. & Kratzer, F. H. (1973) Plant phenolics in toxicants occurring naturally in foods. Washington, D. C.: Nat. Acad. Sci., 309–345.

Smith, K. J. (1970) Practical significance of gossypol in feed formulation. *J. Amer. Oil Chem. Soc.* **47**, 448–450.

Srinivasan, V., Mougdal, N. R. & Sarma, P. S. (1957) Studies on goitrogenic agents in food. I. Goitrogenic action of groundnuts. *J. Nutr.* **61**, 87–95.

Stob, M. (1973) Estrogens in foods in toxicants occurring naturally in foods. Washington, D. C.: Nat. Acad. Sci., 550–557.

Tookey, H. L., van Etten, C. H., Peters, J. E. & Wolff, I. A. (1965) Evaluation of enzyme-modified solvent-extracted crambe seed meal by chemical analyses and rat feeding. *Cereal Chem.* **42**, 507–514.

Van Etten, C. H., Daxenbichler, M. E., Peters, J. E., Wolff, I. A. & Booth, A. N. (1965) Seed meal detoxification, seed meal from *Crambe abyssinica* *J. Agr. Food Chem.* **13**, 24–27.

Van Etten, C. H. (1969) Goitrogens. In *Toxic constituents of plant food stuffs* (ed. I. E. Liener), New York: Academic Press, 103–142.

Walter, E. D. (1941) Genistin (an isoflavone glucoside) and its aglycone, genistein, from soybeans. *J. Amer. Chem. Soc.* **63**, 3273–3276.

Woodham, A. A. & Dawson, R. (1968) The nutritive value of groundnut protein. 1. Some effects of heat upon nutritive value, protein composition and enzyme inhibitory activity. *Brit. J. Nutr.* **22**, 589–600.

Yapar, Z. & Clandinin, D. R. (1972) Effect of tannins in rapeseed meal on its nutritional value for chicks. *Poultry Sci.* **51**, 222–228.

Discussion

Professor Göğüs: No mention has been made by either Dr Roberts or Professor Liener of hempseed or poppyseed. Are there any objections to their use as feed ingredients, and has Dr Liener any information on possible narcotic effects due to the presence of opium in poppy seed cake?

Dr Roberts: We do not in fact use hemp or poppyseed cake. I have no direct information on possible toxic factors but would certainly anticipate little problem with hempseed.

Professor Liener: I can only echo the comments of Dr Roberts. I am not aware of any reported experimental results on the use of poppyseed. Does Professor Göğüs have any information on this?

Professor Göğüs: There is discussion in Turkey at present as to whether poppyseed cake should be used in animal rations. Poppyseed is rich in protein but it is not considered desirable to use the cake at inclusion rates of more than 5–10%.

Dr Babatunde: In the paper by Dr Roberts some factors were used in the calculation of 'Material Values' for feed materials. How were these factors derived and are they dependent on other factors which are not controllable by man? We are very interested in the possibility of using palm kernel cake for mongastric feeding in Nigeria, but experiments we have conducted in this regard have yielded disappointing results. Does Dr Roberts have any information on the use of palm kernel cake for monogastric feeding in other parts of the world.

Dr Roberts: Palm kernel cake is mainly used for ruminants in Britain. We have no experience of using palm kernel cake in pig and poultry rations but would anticipate that its digestibility would be lower for these than for ruminants. Regarding your other question, value is a rather complex issue. The table in my paper which you mentioned was the result of a solution of many values in particular specific situations. The price of protein in a chick feed would be different to that in a layer feed or a pig feed, but it is possible to produce an average price or value for protein by taking a selected group of feeds and the table in my paper was this in the context of layer feeds.

Mr Clermont-Scott: In the feeding trial with raw and treated mustard mentioned in Dr Robert's paper, what were the levels fed to the pigs, and has it also been fed to other animals? Since the toxicity appeared to be low does this mean that allyl isothiocyanate might not be as toxic as is thought?

Dr Roberts: Two levels of mustard were fed. We started at 7.5% but as this was quite unacceptable the inclusion rate was reduced to 2% at which it was readily accepted. The material came from a commercial mustard producer in Britain, and I am not sure of its precise botanical description. The treated material had neither isothiocyanate nor oxazolidine in it.

Mr Clermont-Scott: This question is directed to Professor Liener. Since properly processed rapeseed is not goitrogenic and can in fact only be after the hydrolysis of the glucosinolates followed by treatment with a polar solvent, what evidence is there that goitrogens are produced in the stomachs of animals, and is their so-called presence not significant but merely a mask for other toxic factors?

Professor Liener: I do not consider that oxazolidinethiones are the whole answer to the poor growth frequently observed on some cruciferous meals. There is an increasing amount of evidence to indicate that there must be other factors present in the meals which are causing some degree of toxicity or growth inhibition. A recent paper by Monk from Sweden has pointed out that even those varieties of rape or mustard seed which contain little if any oxazolidin-ethiones produce growth below that expected and a factor which is not a glucosinolate has been isolated which inhibits growth.

Mr Najar: I would like to ask Dr Roberts if he has any information on the use of olive cake as an animal feed.

Dr Roberts: I am sorry I have no information on olive cake to offer.

Dr Andrews: Would Professor Liener care to comment on the presence of a toxic factor in rapeseed causing high mortality in laying hens which has been reported in the literature by several workers. This does not appear to relate in any way to goitrogenic activity but causes massive liver haemorrhage particularly in laying hens.

Professor Liener: There is still a lot to learn about rapeseeds and if they are to be used more extensively as feed materials considerable efforts must be made to identify these non-goitrogenic toxic substances.

Professor Abou-Raya: I agree with Dr Roberts that the basis for the commercial evaluation of a feed depends on the species of animals which are to utilise it. However, it might be misleading to base the evaluation on percent digestibility of protein. Perhaps it might be better to base feed value on true rather than apparent digestibility.

Dr Roberts: We do not in fact use digestibility figures in assessing the value of feedstuffs. I personally would like to use availability or some other angle, but am not convinced that the data are sufficiently acceptable under all circumstances.

The production, marketing and utilisation of oilseed cakes and meals in Turkey

A. K. Göğüs

Department of Animal Nutrition, University of Ankara, Turkey

Summary

Production of oilseeds in Turkey is now approaching 1.5 m t/a. The oilseeds produced are mainly cotton seed and sunflower seed together with small quantities of sesame, groundnut, soya and linseed. Turkey is the fourth largest producer of sunflower seed and the ninth largest producer of cotton seed. There is considerable potential for increasing the production of oilseeds in Turkey.

Turkey has the installed capacity to process all the oilseeds it produces and oilcake production exceeds 500,000 t/a. However, total consumption of oilcakes in Turkey during 1972 was only 164,000 t, the remainder being exported. Despite the large numbers of livestock kept in Turkey, demand for concentrate feeds is low due to lack of application of scientifically based methods of feeding, and Turkey will continue to be a major exporter of oilcakes despite a desire to increase local consumption by some Government authorities.

There are many technical problems associated with the production, marketing and utilisation of Turkish oilcakes. These include inadequate storage facilities, toxic factors such as gossypol and transportation difficulties. Also Western European countries importing Turkish oilcakes have standards and regulations which have to be met.

Résumé

La production, la commercialisation et l'utilisation des tourteaux de graines oléagineuses et des pâtures en Turquie

La production de graines oléagineuses en Turquie approche maintenant 1.5 million de tonnes par an. Les graines oléagineuses produites sont surtout des graines de coton et de tournesol ensemble avec des petites quantités de sésame, arachide, soja et graines de lin. La Turquie est le quatrième des pays producteurs de graines de tournesol et le neuvième pour les graines de coton. Il y a un potentiel considérable pour augmenter la production de graines oléagineuses en Turquie.

La Turquie a un nombre suffisant d'installations pour traiter toutes les graines oléagineuses produites et la production de tourteaux dépasse 500,000 t/a. Cependant, la consommation totale de tourteaux d'huile pendant 1972 était seulement de 164,000 t, le reste étant exporté. En dépit du grand nombre de bétail élevé en Turquie, la demande pour les pâtures concentrées est réduite à cause du manque d'application des méthodes de nourrissage sur une base scientifique et la Turquie continuera d'être un exportateur majeur de tourteaux d'huile, en dépit du désir de certaines autorités gouvernementales d'augmenter la consommation locale.

Il y a des nombreux problèmes techniques associés avec la production, la commercialisation et l'utilisation des tourteaux d'huile de Turquie. Ces-ci incluent les facilités inadéquates de stockage, les facteurs toxiques comme le gossypol et les difficultés de transport. Egalement, les pays de l'Europe Occidentale important les tourteaux d'huile de Turquie ont des standards et réglements qui doivent être satisfaits.

Resumen

La producción, comercialización y utilización de tortas de orujo y harinas de semillas oleaginosas en Turquía

La producción de semillas oleaginosas en Turquía se está aproximando ahora a 1,5 milliones de toneladas anuales. Las semillas oleaginosas que se producen son principalmente la semilla del algodón, la de girasol junto con pequeñas cantidades de sésamo, cacahuete, soja y linaza. Turquía es el cuarto productor mundial de semilla de girasol y el noveno productor mundial de semilla de algodón. Hay un potencial considerable en Turquía para aumentar la producción de semillas oleaginosas.

Turquía tiene la capacidad instalada para tratar todas las semillas oleaginosas que produce y la producción de tortas de orujo excede de las 500.000 toneladas por año. Sin embargo, el consumo total de tortas de orujo en Turquía durante 1972 fué únicamente de 164.000 toneladas, siendo exportadas las restantes. A pesar de la gran cantidad de ganado que hay en Turquía, la demanda de piensos concentrados es baja debido a la falta de aplicación de métodos de alimentación basados científicamente, y Turquía continuará siendo un exportador principal de tortas de orujo a pesar del deseo de aumentar el consumo local por algunas autoridades del gobierno.

Hay muchos problemas técnicos asociados con la producción, comercialización y la utilización de tortas de orujo turcas.

Incluyen éstos los medios de almacenamiento inadecuados, factores tóxicos tales como el gosipol y dificultades de transporte. También, los países de Europa occidental que importan las tortas de orujo turcas tienen niveles y reglamentos que tienen que cumplirse.

There are three aspects to be covered in this paper, the first being the extent to which Turkey produces oilcakes and the extent to which they are used in the country. The second is the quality of oilcakes and some of the factors affecting quality, and the last is concerned with oilcake standards, regulations and marketing possibilities.

It is a well known fact that Turkey is one of the less developed countries. Although most of the population and farm animals consume inadequate quantities of protein, protein concentrates are still exported from Turkey on a large scale. Almost 3–4% of the total national income of Turkey comes from oilcake exports. The total national income from oilseed cake exports reached US $22,083,000 in 1971. Indeed, some of the scientific and government authorities are now asking 'Why do we export oilcakes while we need protein concentrates and while our farm animals suffer from protein deficiency,' and some others say in answer 'We must export oilcakes because we cannot consume all of the product.' As a result of these different viewpoints, the Turkish government stopped oilcake exportation for a while, but then gave permission for it to be resumed. So, the question is 'Shall Turkey export oilcakes or should oilcake exportation be stopped?' To answer this question, we must have an idea about oilcake production, oilcake quality, oilcake preservation possibilities in Turkey and to what extent oilcakes can be used as ingredients of compound feeds in Turkey. Also can Turkish animal husbandrymen use oilcakes in their own mixed feeds, and can Turkish villagers be taught methods of feeding oilcakes?

I must confess that we do not have reliable figures on oilcake production in Turkey, but we do have nearly reliable estimates of oilseed production (see Table 1). As can be seen in Table 1, cotton seed production has doubled and sunflower seed production increased 2.2 times over the last 20 years, while poppy seed and linseed production has decreased to a large extent. Soya bean production is very small.

The main oilseed producing countries and their production are shown in Table 2, from which it will be seen that Turkey takes an important place among the main producer countries. Almost 7.8% of hempseed, 2.24% of sesame seed, 3.1% of cotton seed, and 2.9% of sunflower seed produced in the world is produced in Turkey. In addition to these oilseeds, Turkey also has feed grade hazel nuts for oilcake, and poppy seed, Syrian scabious seed and safflower seed in appreciable amounts.

Average chemical composition of oilcakes produced in Turkey is shown in Table 3, from which it will be seen that there is no significant difference in composition between those of the Deutsche Landwirtschafts-Gesellschaft (DLG) and Turkey. But, it seems to me that Turkish oil cakes contain a little less protein and a little more oil than oilcakes of European origin. The crude protein content of Turkish cotton seed cakes ranges in between 38.2 to 49.4%, and oil content ranges between 3.4 to 10.6%.

According to the national standards regulations of Turkey undecorticated expeller processed cotton seed cakes should not contain more than 9.0% of oil and not less than 19% of crude protein. Partially decorticated expeller processed cotton seed cakes should contain a maximum of 9.0% of oil and a minimum of 32.0% of protein. Partially decorticated extraction processed

TABLE 1. *Oilseed production in Turkey (tonnes)*

Year	Cotton seed	Sunflower seed	Sesame seed	Poppy seed	Linseed	Hempseed	Peanut	Soya bean	Safflower seed	Rapeseed	Total
1951	308,000	107,512	27,857	20,000	25,000	4,000	9,167	1,877	715	3,213	507,341
1955	285,000	138,000	51,000	20,000	18,000	4,300	16,200	4,000	320	1,730	538,550
1960	305,700	123,000	44,000	24,500	22,300	6,000	16,800	6,000	540	4,000	552,040
1965	527,000	160,000	34,000	11,000	14,000	3,500	30,000	5,000	90	7,500	792,900
1970	640,000	375,000	36,000	7,500	6,700	2,500	37,000	12,000	900	3,100	1,120,700
1971	845,000	465,000	–	–	–	–	–	–	–	–	–

TABLE 2. *Oilseed production of main producer countries, thousands of metric tons*

Cotton seed		Unshelled groundnuts		Rapeseed		Linseed	
USA	4,028	India	5,410	India	1,491	USA	678
USSR	4,026	China (Rep.)	2,353	China (Rep.)	1,012	Canada	641
China (Rep.)	2,870	Nigeria	1,227	Canada	985	Argentina	551
		USA	1,170	Poland	493	USSR	456
Brazil	1,235	Senegal	864	France	465	India	398
Pakistan	997	Brazil	775	Pakistan	352	Poland	63
Egypt	851	Indonesia	452	Sweden	215	Ethiopia	61
Mexico	808	Burma	438	Germany (D. Rep.)	211	Uruguay	55
Turkey	643	S. Africa	328	Germany (Fed.)	161	Romania	35
		Sudan	306	Czechoslovakia	71	France	28
World total	20,978	Argentina	305	Japan	71	Czechoslovakia	15
Sunflower seed		Niger	240	Chile	59	Mexico	13
		Uganda	219	Australia	51	Hungary	12
USSR	6,094	Zaïre	183	Denmark	32	Pakistan	12
Argentina	922	Cameroun	182	Hungary	24	Turkey	11
Romania	729	Malawi	171	Switzerland	12		
Bulgaria	444	Thailand	161	Netherlands	10	World total	3,192
Turkey	276	Mali	135	Finland	8	*Soya beans*	
		Mozambique	127	Turkey	6	USA	28,201
World total	9,438	Japan	125	World total	4,286	China (Rep.)	11,041
Sesame seed		Taiwan	119	*Hempseed*		Brazil	1,085
		Gambia	115			USSR	519
India	465	Chad	114	USSR	287	Indonesia	416
China (Rep.)	363	Pakistan	105	Turkey	37	Canada	236
Sudan	228	Rhodesia	87			Korea (Rep.)	217
Mexico	164	Mexico	80			Mexico	181
Burma	122	Zambia	64			Japan	174
Venezuela	88	Dominican Republic	61			Iran	26
Ethiopia	59	Ghana	59			Turkey	9
Turkey	41	Dahomey	41				
		Egypt	40				
		Turkey	34				
World total	1,827	World total	16,899	World total	467	World total	42,752

Figures represent average production of 10 years (1961–1971)
Source: FAO Production Yearbooks

cotton seed cakes should contain a maximum of 1.5% of oil and a minimum of 35.0% of protein. The moisture content of all types of cotton seed cakes and meals should not be more than 10%.

The average amino acid content of cotton seed cake is 3.9% lysine, 1.7% methionine, 1.6% cystine, 6.7% leucine, 3.9% isoleucine, 6.0% valine, 3.6% threonine, 6.3% phenylalanine, 1.3% tryptophan, 2.5% histidine and 7.4% arginine. Total essential amino acid content of a representative sample of cotton seed protein is 44.9%; NPU value is around 50%, digestibility 75% and biological value 60%. Total nonessential amino acid content is around 51.4%.

Free gossypol content of cotton seed cakes produced in western Anatolia by the extraction process ranges from 0.044 to 0.097% while expeller process cotton seed cakes contain 0.026 to 0.066% free gossypol. Cotton seed cake which is produced in southern Anatolia contains 0.09 to 0.12% free gossypol. According to our research work cotton seed cakes which contain the lesser quantities of gossypol quoted above, can be used in poultry diets at inclusion rates of up to 45%. But we do not recommend using cotton seed cake at inclusion rates of more than 10% in any poultry feed. We also use this oilcake in rations for dairy cattle,

sheep and beef cattle, but at not more than 10 to 20% of the total ration. Very few villagers use this meal in their animal rations. The only concerns which use cotton seed cake in Turkey are the national and private feed industries.

Turkey has 22 oil mills in Cukurova (eastern part of southern Anatolia), of which two use solvent extraction methods and the others use expeller methods. The capacity of these mills is around 671,000 t per year, while the cotton seed production of this region is only around 420,000 t per year. Therefore there is a surplus in processing capacity of 37%. In Antalya region (western part of southern Anatolia), there are two oil mills with a capacity of around 114,000 t per year. The amount of processed cotton seed was around 82,000 t and that of sunflower seed around 10,000 t in 1971, indicating some 19% of spare capacity. In western Anatolia (in Ege region), there are 11 cotton seed oil extraction units and seven sunflower oil extraction units, the extraction capacities of these plants being around 407,000 t per year. Almost 310,000 t of cotton seed are produced in this region per year, so there is also 34% of spare capacity in this region for cotton seed processing. Sunflower seed production in Ege region is around 69,000 t per year, and the

TABLE 3. *Average chemical composition of some oilcakes (%)*

		Moisture	Protein	Fat	Fibre	N-Free Extract	Mineral Matter	No. of analyses
Cotton seed oilcake (Expeller)*	Turkey	6.7	42.1	7.7	11.0	25.7	6.8	(23)
Cotton seed oilcake (Extraction)*	Turkey	10.6	43.7	2.6	6.8	28.1	8.2	(10)
Cotton seed oilcake (Expeller)*	DLG**	10.3	46.0	9.4	6.0	22.2	6.1	(10)
Cotton seed oilcake (Extraction)*	DLG	11.5	43.2	0.7	9.2	27.9	7.7	(11)
Sunflower seed oilcake (Expeller)	Turkey	6.7	33.4	7.9	24.5	20.7	6.8	(22)
Sunflower seed oilcake (Expeller)	DLG	8.0	29.9	9.9	23.5	22.7	6.0	(2)
Sunflower seed oilcake (Extraction)	DLG	9.7	35.6	1.7	22.1	25.0	5.9	(11)
Sesame seed oilcake (Expeller)	Turkey	9.1	40.1	10.0	11.0	20.0	9.0	(1)
Sesame seed oilcake (Expeller)	DLG	14.1	36.8	11.4	8.1	21.8	7.8	(2)
Sesame seed oilcake (Extraction)	DLG	9.5	44.7	1.1	6.9	25.0	12.8	(8)
Linseed oilcake (Expeller)	Turkey	7.8	29.0	8.3	12.3	35.0	7.6	(4)
Linseed oilcake (Expeller)	DLG	9.8	33.6	6.3	10.0	34.3	6.0	(122)
Linseed oilcake (Extraction)	DLG	11.1	34.8	1.6	9.5	36.7	6.3	(54)
Groundnut oilcake (Expeller)	Turkey	8.9	50.8	7.0	4.1	24.3	4.9	(2)
Groundnut oilcake (Expeller)	DLG	9.5	44.3	6.0	10.9	23.3	6.0	(33)
Groundnut oilcake (Extraction)	DLG	9.3	46.2	1.1	9.4	28.3	5.7	(44)
Soya bean oilcake (Expeller)	Turkey	9.5	40.8	7.0	6.7	28.6	7.4	(4)
Soya bean oilcake (Extraction)	Turkey	13.0	43.4	2.1	6.1	30.0	6.0	(3)
Soya bean oilcake (Expeller)	DLG	12.2	43.4	5.7	4.8	29.6	5.3	(10)
Soya bean oilcake (Extraction)	DLG	13.1	45.1	0.9	5.9	30.1	5.9	(499)
Rapeseed oilcake (Expeller)	Turkey	10.0	33.3	10.8	10.5	26.5	8.9	(1)
Rapeseed oilcake (Expeller)	DLG	10.7	32.9	5.4	13.7	29.5	7.8	(27)
Rapeseed oilcake (Extraction)	DLG	12.0	35.2	1.7	12.9	31.4	7.6	(65)
Poppy seed oilcake (Expeller)	Turkey	8.6	37.4	6.0	16.8	21.0	10.2	(5)
Poppy seed oilcake (Expeller)	DLG	10.6	36.9	6.2	13.0	20.1	13.2	(4)
Poppy seed oilcake (Extraction)	DLG	11.0	36.1	1.2	15.1	23.8	12.8	(11)

*From decorticated seed
**DLG = Deutsche Landwirtschafts-Gesellschaft (1968)

capacity of the oil extraction plants 46,000 t per year, indicating that processing capacity is not enough for all sunflower seed produced in this region.

There are 94 expeller and four extraction oil plants in the region of Marmara and Trakya, the total capacity of which being around 732,000 t per year. But only sunflower seed production is produced in this region, almost 345,000 t of sunflower seed being produced each year. So, there is 47% of spare processing capacity in Marmara and Trakya regions.

On average oilcake production reaches nearly 550,000 t per year in Turkey, of which around 200,000 t is consumed within Turkey. This means that around 350,000 t of oilcakes are exported from Turkey each year. If government were to stop cake exportation, the local prices of oilcakes would decrease. As for instance, while one kilogram of barley is around 220 kuruş, one kilogram of corn 235 kuruş, one kilogram cotton seed oilcake is around 155 kuruş, and sunflower seed oilcake was around 139 kuruş in 1974, in fact, animals in Turkey do not suffer so much from a shortage of oilcakes, but more from a lack of carbohydrate-rich feeds.

Turkey does not have suitable storage facilities for keeping oilcakes safely, and state and private feed manufacturers are unable to use all oilcake produced in Turkey.

Turkey has nearly 36,471,000 sheep, 19,487,000 goats, 12,756,000 head of cattle, 5,000,000 dairy cows and 3,000,000 horses, mules and donkeys. Poultry numbers are around 34,600,000. The amount of compound feed produced in national and private feed manufacturing plants was around 353,000 t in 1972. Almost half of this amount was poultry feeds. This would indicate a compound feed consumption per animal per year at around 2.4 kg for large animals, and 5.1 kg for poultry. The average oilcake content of compound feed of a large animal is around 27.5%, and the average oilcake content of mixed feed of poultry is around 22.5%, and according to the best estimation, national and private feed manufacturing plants can currently only consume some 25% of Turkey's oilcake production.

The relation between consumed and exported amounts of oilcakes is shown in Table 4. As is shown in this table, the consumption rate of oilcakes in Turkey has greatly increased in five years. The amount of oilcakes consumed by the feed industry has also increased. In spite of these important increases, exports have not decreased, indicating that oilcake production in Turkey is increasing more and more as time goes on.

TABLE 4. *Consumption and exportation rates of oilcakes (t), in Turkey*

Years	Consumed by domestic market	Consumed by feed industries	Consumed by private use	Exported amount
1967	68,047	21,692	46,354	301,329
1968	109,195	32,181	77,014	272,703
1969	112,773	34,340	78,843	274,000
1970	120,552	36,837	83,715	269,842
1971	215,788	N/A	N/A	273,421
1972	163,922	88,450	75,472	349,652

The main countries to which oilcakes are exported are Denmark, the Federal Republic of Germany, France, Switzerland, Sweden and Norway; cotton seed oilcake and sunflower seed oilcake being the main exports. The countries are listed in accordance with the amounts exported to them, Denmark being the greatest and Norway the smallest. Some difficulties are encountered in exporting oilcakes, such as if oilcakes are not exported promptly fermentation takes place and fires may occur in harbour depots. Shortages of shipping and port capacity sometimes cause problems, Turkey having only a limited number of ports. The second difficulty is that oilcakes are not produced in accordance with the minimal quality requirements of Turkish standards. Almost no quality control takes place in Turkey, and in any case foreign countries do not accept the results of quality control analysis. This is because some of the manufacturers mix olive oilcake, sunflower shells, and cotton seed hulls with oilcakes greatly reducing the quality of the oilcakes. There is no regular and continuous communication between Turkey and foreign countries because of lack of telex facilities, which leaves exporters ignorant of sudden fluctuations in world market prices.

There are also some difficulties in the domestic marketing of oilcakes. There are often no transportation facilities, no stores and no oilcake selling centres. The oilcake prices in domestic markets are too low, while those in foreign markets are too high. All oilcakes should be sold according to quality standards. Exporters should establish new storage facilities at ports. Two or three oilcake hexane extraction units are also needed for extracting excess oil in oilcakes. There must be more feed industries for using oilcakes in compound feeds, and I believe that we must continue to export oilcakes surplus to domestic requirements until Turkey's long term problems have been solved.

References

Deutsche Landwirtschafts–Gesellschaft (1968) Futterwerttabelle Für Wiederkaver Arbeiten der DLG, Band 17, DLG–Verlags–GmbH, Frankfurt am Main, Rüsterstrasse 13.

FAO (1961–71) Production Yearbooks

The supply and utilisation of oilseed cakes in the Sudan

A. M. S. Mukhtar

University of Khartoum, Democratic Republic of the Sudan

Summary

Animal feeds, especially concentrates, are a major factor in the development of the livestock industry in the Sudan. As feeds represent up to 70% of the total input costs of any commercial production enterprise, adequately balanced feeds based on least cost rations are the major criterion for a profitable livestock and poultry industry. In the Sudan the most important concentrates used in animal feeding are oilseed cakes. The production of oilseed cakes in the Sudan amounts to several hundred thousand tons per annum, but they are mostly exported. On the other hand production of cereals in the Sudan does not at times even meet demands for human use.

The most expensive oilseed cake available is that of sesame, then groundnut cake, the cheapest being cotton seed cake. Trials have been carried out by the private industrial sector for the production of pelleted feed containing 20-40% of oilseed cakes, but these are expensive compared with the oilseed cakes themselves.

The trend in production of sesame, groundnut and cotton seed cakes in the Sudan is reviewed together with their pattern of utilisation. Figures for proximate composition and amino acid composition of the proteins are presented, and the use of oilseed cakes in the Sudan discussed.

Résumé

Les provisions et l'utilisation des tourteaux de graines oléagineuses dans le Sudan

Les pâtures pour les animaux, surtout les concentrés sont un facteur majeur dans le développement de l'industrie du bétail dans le Sudan. Comme les pâtures représentent jusqu'à 70% du coût de l'investissement total de n'importe quelle entreprise de production commerciale, des pâtures convenablement équilibrées, basées sur des rations d'un coût minime, sont le critère majeur pour un industrie profitable du bétail et de la volaille.

Dans le Sudan, les plus importants concentrés utilisés dans les pâtures des animaux sont les tourteaux de graines oléagineuses. La production de tourteaux de graines oléagineuses dans le Sudan s'élève à quelques centaines de mille de tonnes par an, mais la plupart est exportée. D'un autre côté, la production de céréales dans le Sudan, quelquefois ne peut pas satisfaire même les demandes pour l'usage humain.

Le plus coûteux tourteau de graines loéagineuses disponible est celui de sésame, puis les tourteaux d'arachide, le moins coûteux étant le tourteau de graines de coton. Des essais ont été entrepris par le secteur industriel privé pour la production des pâtures en boulettes, contenant 20-40% de tourteaux de graines oléagineuses, mais ces-ci sont chers comparées avec les propres tourteaux de graines oléagineuses.

La tendance de production du sésame, des cacahuètes et des tourteaux de graines de coton dans le Sudan est passée en revue, en même temps que le mode d'utilisation. On présente des chiffres de la composition approximative et de la composition en aminoacides des protéines et on discute l'utilisation des tourteaux de graines oléagineuses dans le Sudan.

Resumen

El suministro y utilización de tortas de orujo en el Sudán

Los piensos para animales, especialmente concentrados, son un factor principal en el desarrollo de la industria ganadera en el Sudán. Como los piensos representan hasta el 70% de los costes de entrada totales de cualquier empresa de producción comercial, piensos equilibrados adecuados basados en las raciones de coste menor son el criterio principal para una industria provechosa de ganadería y aves de corral. En el Sudán los concentrados más importantes usados en la alimentación de los animales son las tortas de orujo de semillas oleaginosas. La producción de tortas de orujo de semillas oleaginosas en el Sudán alcanza varios cientos de miles de toneladas por año, pero se exporta la mayor parte. Por otra parte, la producción de cereales en el

Sudán en ocasiones no satisface siquiera las demandas para uso humano.

La torta de orujo de semillas oleaginosas más cara de que se dispone es la del sésamo, luego la torta de orujo de cacahuete, siendo la más barata la torta de orujo de la semilla de algodón. Se han llevado a cabo pruebas por el sector industrial privado para la producción de pienso granulado con un contenido det 20–40% de tortas de orujo de semillas oleaginosas, pero éste es caro en comparación con las mismas tortas de orujo de semillas oleaginosas.

Se revisa la tendencia en la producción de tortas de orujo de sésamo, cacahuete y semillas de algodón en el Sudán, junto con su tipo de utilización. Se presentan cifras de la composición aproximada y de la composición de los aminoácidos de las proteínas y se discute el uso de las tortas de orujo de semillas oleaginosas en el Sudán.

The Sudan is mainly an agricultural country, the most important cash crops being cotton, groundnuts and sesame seed.

Cotton is the backbone of the economy of the Sudan, and the total area under cotton cultivation both by irrigation and rainfall is now approaching 2m. feddans (one feddan = 1.038 ac.). Cotton is ginned locally with the production of several hundred thousand tons of seeds annually. The oil is expressed from undecorticated seed leaving the cake as a by-product. Most of the cotton seed cake is exported the remainder being used locally for feeding animals of economic importance.

For example in 1968 it was estimated that 201,590 t of undecorticated cotton seed cake were produced in the Sudan of which only some 33,805 t was used for feeding animals within the Sudan (Kirby & Halliday, 1971). There is evidence to suggest that the local consumption of cotton seed cake is, however, gradually increasing.

Cotton seed cake

Cotton seed cake is more readily available and less costly than groundnut and sesame cakes. In 1972 the purchase price of a tonne of cotton seed cake was seventeen Sudanese pounds, but today the price has increased to forty Sudanese pounds.

The proximate composition of a sample of Sudanese cotton seed cake analysed at the Tropical Products Institute was found to be as follows:

Moisture (%)	3.0
Crude protein (N x 6.25: %)	25.9
Oil (%)	6.5
Crude fibre (%)	24.1
Ash (%)	5.0
Nitrogen - free extract (%)	35.5

This is in line with typical values reported in the literature for cotton seed cake as given by Morrison (1956). Sudanese cotton seed cake is low in moisture content, and provided it is properly stored it does not deteriorate in quality during storage.

The large export trade in Sudanese cotton seed cake has a great effect on local prices, and even availability at certain times. This has proved a considerable handicap to the operations of local feed manufacturing enterprises.

The animal feed manufacturing industry in the Sudan is relatively recent. Five years ago a small feed manufacturing plant was established by the private sector to produce pelleted feeds for cattle, but it did not prove a commercial success. The price of the pelleted product, which contained 25% of cotton seed cake and 20% of groundnut shell in addition to other ingredients was similar to that of cotton seed cake alone, which provided little incentive to purchase the new product. However now that the price of cotton seed cake has doubled, the prospects for pelleted feed for cattle would appear to be better. Moreover small-scale feedlot operations are now being established by the private sector and beef carcasses are being exported by air to neighbouring countries. At present the feed ingredients are bought separately and only crudely mixed by hand, and the use of well mixed manufactured feeds of subsequent nutritive value for such feedlot operations would be a great advantage. Cotton seed production is still increasing in the Sudan and it is likely that more cotton seed cake will become available.

Cotton seed cake has been used in ruminant feeding in Sudan especially for lactating cows and in fattening calves. For lactating cows cotton seed cake constitutes one third of the concentrate mixture, while for fattening calves it is fed at levels varying between 20 – 25% of the total ration. Other ingredients in the fattening ration include wheat bran, hay, molasses and other materials.

In practice many cattle merchants use large amounts of cotton seed cake (together with some straw) for feeding export sheep and cattle. This practice is wasteful in that the animals are getting more nutrients than are necessary and it would be expected that digestibility would be reduced. El Shafie (1966) fed zebu cattle a ration containing 20% each of cotton seed cake, sorghum grain, wheat bran and 13% each of sorghum hulls, lucerne and molasses, the mineral and salt content being 1%. The proximate composition of the ration was as follows:

Moisture (%)	7.4
Crude protein (N x 6.25: %)	13.8
Oil (%)	2.8
Crude fibre (%)	13.1
Ash (%)	2.5
Nitrogen - free extract (%)	60.4

Average daily gains on this ration were 0.83 kg. Mukhtar and Shafie (1972) fed a similar ration to zebu calves obtaining an average feed conversion of 5.4 kg of feed for 1 kg liveweight gain. Cotton seed cake has also been used at inclusion rates of 7–10% in poultry rations (the same inclusion rates used for groundnut and sesame cakes). However, it is known that a high level of cotton seed cake in layer rations can lead to discoloration of egg yolks due to the effect of gossypol.

Groundnut cake

Groundnuts are grown in most parts of the Sudan by both rain-fed and irrigated agricultural systems. It has been estimated that the total area devoted to groundnut production is more than 1m. feddans.

Sudanese groundnuts are partly exported as kernels or consumed as edible nuts, the remainder being crushed in the Sudan to produce oil and cake. Both decorticated kernels and undecorticated nuts are crushed. The main proximate constituents of Sudanese decorticated and undecorticated groundnut cakes are given below:

	Decorticated	Undecorticated
Moisture (%)	4.4	3.7
Crude protein (N x 6.25%)	36.3	22.8
Oil (%)	11.9	14.0
Crude fibre (%)	10.3	24.5

The oil content of both types of groundnut cake is generally high and may effect the keeping quality. Table 1 gives statistics for the production of groundnuts in the Sudan, together with exports of seed and cake during the period 1960/61 to 1970/71. The amount of groundnut cake consumed locally is very

TABLE 1. *Production and export of groundnut seeds and cake for the period 1960/61 to 1970/71 (tonnes)*

Year	Seed production	Seed export	Cake export
1960/61	192,430	118,776	2,642
1961/62	264,750	166,607	17,255
1962/63	308,630	164,863	32,608
1963/64	379,069	217,376	38,714
1964/65	328,610	217,371	18,140
1965/66	304,601	141,791	21,406
1966/67	341,254	149,626	47,886
1967/68	297,366	117,199	43,335
1968/69	246,962	106,765	27,182
1969/70	384,650	91,318	36,514
1970/71	338,618	167,151	24,880

small compared with the amount exported, and has been estimated to be only some 5% of total cake production. Groundnut cake is as would be expected rather more expensive than cotton seed cake.

Sesame cake

Sesame is grown in the Sudan by rain-fed agriculture and about 1m. feddans are cultivated annually. Sesame is mainly exported as the unprocessed seed but some is processed locally into oil and cake. Sesame cake has a higher nutritive value than cotton seed and groundnut cake and this is reflected in its price which is about double that of cotton seed cake. The main proximate constituents of a sample of Sudanese sesame cake analysed at the Tropical Products Institute were as follows:

Moisture (%)	3.8
Crude protein (N x 6.25: %)	46.2
Oil (%)	11.9
Crude fibre (%)	7.6

Oil content is high due to poor methods of processing and this may effect keeping quality. Sesame protein is particularly rich in methionine compared with other oilseed proteins such as cotton seed and groundnut. Statistics for the production and export of sesame seed and the export of sesame cake are given in Table 2.

TABLE 2. *Production and export of sesame seeds and cake export figures for the period 1960/61 to 1970/71 (tonnes)*

Year	Seed production	Seed export	Cake export
1960/61	146,651	63,124	33,278
1961/62	132,190	77,389	17,285
1962/63	141/970	69,649	33,550
1963/64	173,871	101,410	22,336
1964/65	183,699	70,588	24,256
1965/66	160,088	74,573	25,199
1966/67	133,975	75,498	8,811
1967/68	186,368	84,725	12,127
1968/69	192,550	112,602	28,841
1969/70	174,473	81,890	16,631
1970/71	296,967	86,260	10,647

Conclusions

The oilcakes of cotton seed, groundnut and sesame are the main oilcakes produced in the Sudan. Practically all the current annual production of 250–300,000 t of cotton seed is crushed locally as is also a large proportion of groundnut and sesame seed production. One of the main limitations of Sudanese oilcakes is high oil content which may adversely affect keeping quality. This is due to the primitive nature of much oilseed processing in the Sudan, and the situation can only be improved by modernisation of the industry.

The Sudan currently produces 50% of its requirements for sugar and wheat, and it is hoped to achieve self-sufficiency in these by 1976/77. The consequent production of molasses and wheat milling offals will be of great assistance in establishing a local feed industry. The current animal population numbers in the Sudan include 12 m. head of cattle, 10m. sheep, 9m. goats and 2.5 m. camels, all of which have a need for good quality feedingstuffs. In the world today there is a shortage of grains as well as animal protein and it is essential that the requirement for animal feed be met as far as possible by materials not consumed directly by humans.

For years to come food production will continue to be a serious problem in the tropics. In this age of technology it will be the technical knowhow that will pave the way for production of high quality feeds which will in turn lead to an increase in supplies of animal products.

References

El Shafie, S. A. (1966). Further observations on fattening of Sudan zebu cattle. *Sudan J. vet. Sci. and Anim. Husb.* 7, 1 22–27.

Kirby, R. H. & Halliday, D. (1971) Report on feasibility of setting up an animal feed industry in the Democratic Republic of the Sudan, London, Confidential Report of Tropical Products Institute.

Morrison, F. B. (1956) Feeds and Feeding, 22nd Ed. Morrison Publishing Company. Ithaca, New York, USA.

Mukhtar, A. M. S. & S. A. Shafie (1972) Feedlot performance and carcass characteristics of individually fed zebu calves. *World Rev. of Anim. Prod.* 8, 4.

The feedingstuff potential of cashewnut scrap kernel meal

B. L. Fetuga, G. M. Babatunde and V. A. Oyenuga
Department of Animal Science, University of Ibadan, Nigeria

Summary

The proximate composition, amino acids and protein quality of meals prepared from good grade and discarded cashewnut kernels were studied. In another study, the comparative value of cashewnut scrap kernel meal and groundnut meal as supplements in growing and fattening pig diets were assessed.

The crude protein content of both sources was high and after oil extraction they produced meals with protein levels comparable to those of groundnut and soya bean meals. The trytophan and sulphur amino acids were higher than in soya bean and groundnut meals, while the lysine levels were lower than in soya bean meal, but comparable to that in groundnut.

The good grade meal was better digested and superior in quality to the scrap kernel meal. Rats fed the good grade meal did not respond to methionine supplementation, but showed slight responses to lysine, indicating an adequacy of the sulphur amino acids and slightly low available lysine level. Lysine, but not methionine, supplementation significantly improved the quality of the scrap kernel meal.

When evaluated at a critical protein level with growing pigs, the scrap kernel meal was superior to groundnut meal. Also in practical type diets for fattening pigs, the scrap kernel meal diets were superior to the groundnut meal diets in terms of growth and efficiency of feed utilisation but the carcass values were identical.

Résumé

Le potentiel nutritif des pâtures de déchets des noyaux des noix d'acajou

On a étudié las composition approximative et la qualité des aminoacides et des protéines des pâtures faites de noyaux des noix d'acajou de bonne qualité ou rebutées. Dans une autre étude on a évalué la valeur comparative des pâtures de noyaux des noix d'acajou et des pâtures de cacahuètes comme suppléments dans les diètes pour les cochons en développement ou mis à engraisser.

Le contenu en protéine brute de deux sources était élevé et après l'extraction de l'huile, elles produirent des pâtures avec des niveaux de protéine comparables à celles des pâtures d'arachides et de graines de soja. Le tryptophane et les aminoacides sulfurés étaient plus élevés que dans les pâtures de graines de soja et de cacahuètes, tandis que les niveaux de lysine étaient plus bas que dans les pâtures de graines de soja mais comparables à celles des pâtures de cacahuètes.

Les pâtures de bonne qualité étaient mieux digérées et supérieures en qualité aux pâtures de restes de noyaux. Les rats nourris avec des pâtures de bonne qualité ne réagirent pas aux suppléments de méthionine, mais montrèrent une légère réaction à la lysine, indiquant une juste proportion d'aminoacides sulfurés et un niveau légèrement bas de lysine disponible. Les suppléments de lysine, mais non de méthionine, améliorèrent significativement la qualité des pâtures de restes de noyaux.

Quand évaluée à un niveau critique protéinique, avec des cochons en voie de croissance, la pâture de restes de noyaux était supérieure à la pâture de graines de cacahuètes. Quoique dans les diètes de type pratique pour l'engraissement des cochons, les diètes de pâtures de restes de noyaux étaient supérieures aux diètes de pâtures de graines d'arachides, en termes de croissance et efficacité de l'utilisation des pâtures, les valeurs des carcasses étaient identiques.

Resumen

El potencial como pienso de la harina de semillas anacardo desechadas

Se estudiaron la composición aproximada, la calidad de los aminoácidos y proteína de harinas preparadas de semillas de anacardo de buena calidad y desechadas. En otro estudio se estimaron el valor comparativo de la harina de semilla de anacardo desechada y la harina de cacahuete como suplementos en las dietas de crecimiento y engorde de cerdos.

El contenido de proteína bruta de ambas fuentes fué alto y después de la extracción del aceite, produjeron harinas con niveles de proteína comparables a los de las

202 harinas de cacahuete y soja. El triptófano y los sulfuraminoácidos eran más altos que en las harinas de soja y de cacahuete, al tiempo que los niveles de lisina eran más bajos que en la harina de soja, pero comparables a los de la de cacahuete.

La harina de buen grado era mejor digerida y superior en calidad que la de la harina de semillas de desecho. Las ratas alimentadas con la harina de buen grado no respondieron al suplemento de metionina, pero mostraron respuestas ligeras a la lisina, indicando una suficiencia de los sulfuraminoácidos y nivel de lisina

disponible ligeramente bajo. El suplemento de lisina, pero no de metionina, mejoró significativamente la calidad de la harina de semillas de desecho.

Cuando se valoró a un nivel de proteína crítico con cerdos en proceso de crecimiento, la harina de semillas desechadas fué superior a la harina de cacahuete. También, en dietas de tipo práctico para cerdos en engorde, las dietas de harina de desecho fueron superiores a las dietas de harina de cacahuete, en relación al crecimiento y eficacia de la utilización del alimento, pero los valores de la carcasa fueron idénticos.

Introduction

The feeding of pigs and poultry in many developing countries continues to pose many problems, because of a lack of understanding of the nutritional requirements of these animals under the local conditions and a lack of knowledge of the nutritive value of locally available feedingstuffs. Furthermore the increase in the prices of tested conventional sources of protein like soya meal, groundnut cake and fishmeal which has occurred in recent times has increased the interest in using alternative indigenous materials, especially if increased dependence on imported concentrate feeds is to be avoided.

TABLE 1. *Price fluctuations of some commonly used ingredients of animal diets in Nigeria*

Ingredients	Prices per ton (N)*		Increase
	1970	1973	%
Maize	90.00	150.00	66.7
Guinea corn	90.00	150.00	66.7
Groundnut cake	70.00	130.00	85.7
Palm kernel cake	75.00	75.00	0.00
Fishmeal	160.00	490.00	206.3
Bloodmeal	70.00	70.00	0.00
Meat and bonemeal	40.00	50.00	25.00
Luru	90.00	110.00	22.2
Wheat offals	20.00	54.00	170.0
Rice bran/chaff	10.00	12.00	0.00
Brewers' dried yeast	60.00	60.00	0.00
Oyster shell	38.00	48.00	26.3
Bonemeal	56.00	70.00	25.0
Palm oil (per drum of 200C)	34.00	60.00	76.5
Cane molasses	6.00	7.00	16.7

N1 = £ sterling 1.5

Cashewnuts (*Anarcardium occidentale*, L.) have been produced in Nigeria for many years, and although they are obviously mainly consumed as edible nuts, there is potential for their use to some extent for animal feeding. This is because only 60–65% of cashewnuts which are produced are suitable for edible purposes and of these a further 35–40% are discarded as broken kernels or due to scorching in the roasting process. The studies reported here were conducted to evaluate

the discarded broken and scorched nuts in comparison with the good grade nuts, bearing in mind that while the good grade nuts and their extraction meal may find use in human nutrition, the scrap kernel meal may find use in monogastric animal diets as a cheap protein supplement.

Materials and methods

Meals

Good grade and scrap kernels were obtained from the Western Nigeria Development Corporation (WNDC) processing factory in Ibadan, Western State. Processing treatment at the factory had involved roasting of the whole nuts for 90 seconds in cashewnut shell liquid at 185°C., separation into good grade and scrap kernels being carried out after this stage. Further treatment in the laboratory involved coarse grinding and extraction to exhaustion with petroleum ether of 40–60°C. boiling range. After oil extraction samples were further ground to pass through a 30 mm mesh sieve, and left in the open to allow the evaporation of residual petroleum spirit. The dry oil-free material was stored in screw-capped bottles at −5°C. until used.

Scrap kernel meals used in the pig studies were prepared by crushing scrap kernels in an expeller plant at the premises of Vegetable Oils Nigeria Ltd., Ikeja, Lagos. The groundnut cake used for comparison was obtained from a commercial oil mill in Kano and had been produced by the expeller process.

Chemical analysis

The proximate composition of the samples was determined on the air dried samples by the methods prescribed by the AOAC (1970). Gross energy determinations were carried out with an oxygen ballistic bomb calorimeter, while the mineral constituents were determined by atomic absorption spectrophotometry after wet ashing. Phosphorus was determined colorimetrically by the phosphovanadomolybdate method (AOAC, 1970).

Amino acid composition of the proteins was determined using an automated Hitachi-Perkin Elmer amino acid analyser (model KLA-3B), after hydrolysis of the samples with 6N hydrochloric acid for 24 hours in an atmosphere of nitrogen. Tryptophan was determined by the method of Miller (1967).

Rat feeding trials

Protein efficiency ration (PER) net protein retention (NPR), biological value (BV), net protein utilisation (NPU) and apparent and true digestibility (AD and TD) were determined with weanling albino rats of the Wistar strain weighing on the average 50–55 gm and about 28 days old, as described by the National Academy of Sciences/National Research Council, (1963), for two samples of the cashewnut scrap meal (CSM), a sample of the cashewnut good grade meal (CGM), heat treated soya bean meal (SBM) and freeze dried ether extracted whole hen's egg (HE). There were eight rats per test diet and an additional group of eight rats on a nitrogen free diet making a total of 48 rats.

The basal diet contained corn starch 65%, glucose 5%, sucrose 10%, non-nutritive cellulose 5%, groundnut oil 10%, mineral supplement (Miller, 1963) 4% and vitamin mixture (Miller, 1963) 1%. The protein sources were included in the basal diet at the expense of corn starch such that they provided 10% of crude protein in the final diet.

In a second trial 64 weanling male albino rats were used to study the effect on the protein quality, of adding either 0.15% lysine and/or 0.20% DL-methionine to either of the two grades of cashewnut meal, using techniques already described.

Pig studies

Two experiments were conducted with pigs, one with growing pigs and the other with fattening pigs. The first experiment employed 24 Landrace x Large White pigs weighing initially 15 kg to compare the growth promoting ability of CSM, GNM and fishmeal (FM) as well as the relative nitrogen retention on these protein sources. The diets fed consisted of a basal maize diet to which was added the protein concentrates in such a way that they contributed 10% crude protein, the total crude protein content of the test diets being about 15%. Diets were made isocaloric by the inclusion of lard and corn starch in varying ratios (Table 6). Feeding was four times daily in individual stalls. The experiment lasted eight weeks and weight gains and feed intake records were kept on a weekly basis.

The total protein efficiency ratio (TPER) was calculated as:—

$$\frac{\text{Weight gain of pig}}{\text{Total protein (cereal + supplement) consumed per pig}}$$

Nitrogen metabolism studies lasting seven days were carried out on four pigs per diet three weeks after commencement of the experiment in special metabolic cages with facilities for separate collection of urine and faeces (Oyenuga, 1961) as previously described by Babatunde *et al.* (1971).

The second experiment, which employed 40 Landrace x Large White pigs weighing initially 51 kg divided into five groups of eight pigs each, was to compare the effects of CSM and GNM at two levels of inclusion on the growth performance and carcass characteristics of growing-fattening pigs. The composition of the diets used is shown in Table 6. Diet 1, the control, was the one commonly used for fatteners at the University of Ibadan piggery. Diets 2 and 3 containing 15.38% and 17.98% crude protein respectively were compounded such that GNM at inclusion rates of 17.14% and 22.85% furnished approximately 50% and 58%

TABLE 6. *Composition of experimental diets (pigs, experiments I and II)*

Ingredients	Experiment I (Growers) Diets			Experiment II (Fatteners) Diets				
	1	2	3	1	2	3	4	5
Yellow maize	47.60	47.60	47.60	66.50	71.36	65.65	61.52	54.65
Fishmeal (FM)	15.43	–	–	1.50	–	–	–	–
Groundnut meal (GNM)	–	20.00	–	16.00	17.14	22.85	–	–
Bloodmeal	–	–	–	5.00	–	–	–	–
Cashewnut scrap meal (CSM)	–	–	27.94	–	–	–	26.98	33.85
Rice bran	2.40	2.40	2.40	7.50	7.50	7.50	7.50	7.50
Bonemeal	2.00	2.00	2.00	1.50	1.50	1.50	1.50	1.50
Oyster shell	0.80	0.80	0.80	0.50	0.50	0.50	0.50	0.50
Lard	3.72	3.72	2.54	1.25	1.25	1.25	1.25	1.25
*Vit. – Min. – Antibiotic mix	0.25	0.25	0.25	0.25	0.25	0.25	0.25	0.25
Maize starch	27.80	23.22	16.47	–	–	–	–	–
Salt (NaCl)	–	–	–	0.5	0.5	0.5	0.5	0.5
**Zinc	+	+	+	–	–	–	–	–
+BHT	+	+	+	+	+	+	+	+
	100.0	100.0	100.0	100.0	100.0	100.0	100.0	100.0

*A Pfizer livestock feeds product supplying the following per kg of finished diet, Vit. A, 98331 I.U; Vit.D$_3$, 1965 I.U.; Vit E, 69 I.U.; Vit K, 20 mg; Vit. B$_{12}$, 10 mg; Riboflavin, 41 mg; Nicotinic acid, 246 mg; Pantothenic acid, 96 mg; Folic acid, 10 mg; Cobalt, 5 mg; Copper, 244 mg; Iodine, 20 mg; Manganese, 341 mg; Zinc, 100 mg; Iron 100 mg and oxytetracycline hydrochloride 20 mg/t
**Zinc oxide was added to the growers diet to provide 100ppm Zinc
+ Butylated hydroxy-toluene (an antioxidant)

respectively of the total dietary protein, while diets 4 and 5 were the equivalents of diets 2 and 3, except that CSM replaced GNM and contained approximately the same amount of crude protein (15.48 and 17.76% respectively). Feed and body weight gain records were kept on a weekly basis. As the individual pigs attained approximately 92 kg body weight they were slaughtered after 18 hours fasting and conventional carcass data were collected.

Statistical analysis

All data were subjected to analysis of variance (Steel & Torrie, 1960) and treatment mean were compared by the multiple range test of Duncan (1955).

Results and discussion

Proximate and amino acid composition

The proximate composition and minerals in both grades of cashew nut are shown in Table 2. The extracted meals had crude protein contents of 40–45% depending on the efficiency of oil extraction which are similar to those of extracted SBM and GNM. Phosphorus and potassium were the most abundant mineral constituents, being higher than values reported for GNM (Miller, 1963). Calcium and zinc contents were however lower than in either SBM or GNM.

Table 3 compares the amino acid compositions of the proteins of both the CSM and CGM to those of GNM, SBM and HE. The lysine and threonine contents of

TABLE 2. *Composition of cashewnut (good grade) kernel meal (CGM) and cashewnut scrap meal (SCM), (%)*

| | Cashewnut meals | | | |
| | Good grade kernel | | Discarded kernel | |
Item	Undefatted	Defatted	Undefatted	Defatted
Residual moisture	5.5	7.4	4.4	6.5
Crude protein	21.2	40.9	21.6	42.8
Ether extract	48.1	1.3	45.5	1.3
Crude fibre	0.8	1.5	2.3	4.1
Silica-free ash	3.3	5.3	3.8	6.8
Calcium	0.04	0.06	0.03	0.06
Phosphorus	0.88	1.72	0.84	1.64
Sodium	0.005	0.02	0.016	0.03
Potassium	0.57	1.42	0.52	0.98
Magnesium	0.28	0.54	0.24	0.48
Iron	0.008	0.01	0.006	0.009
Copper	0.002	0.006	0.002	0.007
Zinc	0.004	0.009	0.003	0.007
Manganese	0.002	0.004	0.001	0.003
Gross energy, kcal/g	7.76	4.28	7.32	4.14

TABLE 3. *Amino acid compositions of cashewnut (good grade) kernel meal (CGM), cashewnut scrap kernel meal (CSM), groundnut meal (GNM), soya bean meal (SBM) and whole hen's egg (HE) expressed as g per 16 g total nitrogen.*

| Amino acids | Cashewnut meals | | Groundnut meal* (GNM) | Soya bean meal* (SBM) | Whole hen's egg* (HE) |
	Good grade (CGM)	Discards (CSM)			
Arginine	10.70	9.87	12.30	7.57	6.10
Histidine	2.06	1.96	3.04	2.68	2.43
Isoleucine	3.86	3.79	3.58	4.58	6.29
Leucine	6.51	6.63	7.09	7.94	8.82
Lysine	4.04	3.86	3.90	6.12	6.98
Methionine	1.40	1.38	0.91	1.25	3.36
Cystine	1.78	1.68	1.14	1.64	2.43
Methionine + Cystine	3.18	3.06	2.05	2.89	5.79
Phenylalanine	3.89	3.74	5.60	5.68	5.63
Tyrosine	2.37	2.68	4.34	4.32	4.16
Phenylalanine + Tyrosine	6.26	6.42	9.94	10.00	9.79
Threonine	3.10	3.09	3.04	3.82	5.12
Tryptophan	1.37	1.34	1.24	1.26	1.62
Valine	5.80	5.23	4.27	5.51	6.85
Alanine	3.70	3.77	4.19	4.83	5.92
Aspartic acid	9.20	9.13	11.82	11.46	9.02
Glutamic acid	18.74	19.42	21.12	17.70	12.74
Proline	3.72	3.46	5.06	5.10	4.16
Serine	4.76	4.34	5.33	5.32	7.65
Glycine	4.60	4.16	6.30	4.63	3.31

*Values reported for these feedingstuffs were also obtained at Ibadan using similar techniques as described for the cashewnut meals.

the CSM and CGM proteins were lower than in HE and SBM protein but comparable to the levels in GNM.

The CGM protein had a higher sulphur amino acid (methionine + cystine) content than those of GNM and SBM, while both the CGM and CSM proteins had higher contents of tryptophan than those of SBM and GNM. The other essential amino acids appear to be well represented in the cashewnut protein.

Rat studies

Table 4 summarises results of biological evaluation with rats. The egg diets showed significantly superior PER, NPR, NPU and BV to all other diets. The two samples of CSM showed significantly (P<0.05) inferior protein quality indices compared to the SBM and CGM. The CGM was in turn found to be superior to SBM in these respects.

TABLE 4. *Protein quality of cashewnut (good grade meal (CGM), cashewnut scrap meal (CSM), heat-treated soya bean meal (SBM) and freeze-dried ether-extracted hen's egg): biological data*

	Good grade meal	Scrap meal I	Scrap meal II	Soya bean meal	Hen's egg
PER	2.01b	1.12c	0.76d	1.74b	3.94a
NPR	4.01b	3.13c	2.86c	3.94b	6.04a
NPU	63.0b	46.7d	41.3e	58.9c	94.0a
BV	68.6d	46.7d	48.9d	64.8c	98.4a
Apparent digestibility (%)	83.8b	77.9c	77.4c	82.4b	93.7a
True digestibility (%)	91.8b	84.6c	84.3c	92.0b	98.4a

a, b, c, d, e. All mean in the same row with the same suffix are not significantly different (P<0.05).

The apparent and true digestibility values (Table 4) for HE were significantly (P<0.05) higher than for CGM and SBM which were in turn significantly (P<0.05) higher than for the two samples of CSM compared with those for CGM. The lower protein quality indices observed for CSM which showed identical chemical and amino acid composition to CGM might be attributable to heat damage. Such damage could lead to a reduced digestibility as evidenced in the observed digestibility values (84.6% for CSM as against 91.8% for CGM) and hence a reduced availability of the essential amino acids. Work by Miller *et al.* (1965), Ford (1965) and Neishem & Carpenter (1967) have all shown heat damage to result in reduced protein digestibility, as well as an impairment of the protein quality. That the scrap kernel meal designated II was significantly (P<0.05) inferior to that designated I may also indicate that the scrap kernels coming from the factory are not of constant quality and that the extent of damage to kernels may differ from batch to batch. This is important particularly where they are intended for use as supplements because such variations could result in considerable variations in the level of available amino acids in mixed diets.

Addition of synthetic DL methionine to CSM (Table 5) resulted in slight but non-significant improvement in quality. The CGM responded even less to additional methionine. Piva *et al.* (1971) had also reported non-response of rats to methionine supplementation of Tanzanian cashewnut extraction meals, and it is unlikely that methionine is a limiting amino acid in cashewnut protein.

Moreover, its cystine content (1.68g/16g N) is high compared to most other oilseeds and this along with the methionine present might be sufficient to meet the requirements of rats for the sulphur amino acids.

Addition of lysine alone or lysine and methionine to CSM resulted in a remarkable improvement in all protein quality indices, while CGM showed smaller but also significant (P<0.05) improvement in protein quality indices. This seems to suggest that available lysine is a limiting factor in the cashewnut meals for rat growth. The differences in the degree of response to lysine and/or methionine supplementation tends

TABLE 5. *Amino acid supplementation of cashewnut scrap kernel (CSM) and good grade kernel (CGM) extraction meals: protein quality indices*

	Weight gain at 10 days g	Protein intake g	PER	NPR	PRE	NPU	BV	AD %	TD %
Cashewnut scrap kernel meal + no amino acid	7.2a	6.36	1.12a	3.13a	50.9a	46.0a	54.5a	78.2a	83.9a
Cashewnut scrap kernel meal + 0.15% Lysine HCL	21.3d	8.14	2.51ef	4.10b	65.9b	65.6bc	74.0c	76.4a	84.4a
Cashewnut scrap kernel meal + 0.15% lysine HCL + 0.20% DL methionine	17.3bc	7.30	2.39de	4.15b	66.4b	65.0b	77.4c	76.1a	84.9a
Cashewnut scrap kernel meal + 0.20% DL methionine	8.3a	6.39	1.31b	3.30a	52.9a	50.0a	57.9b	76.9a	86.2a
Cashewnut good grade extraction meal + no amino acids	15.3b	7.30	2.11c	4.06b	64.8b	65.9bc	79.8cd	83.8b	91.4b
Cashewnut good grade extraction meal + 0.15% lysine HCL	20.6cd	7.60	2.71g	4.42cd	71.8cd	68.4bc	83.2ac	82.7b	90.8b
Cashewnut good grade extraction meal + 0.15% lysine HCL + 0.20% DL methionine	18.3bcd	6.80	2.67fg	4.57d	73.1d	70.4c	84.8e	84.6b	92.1b
Cashewnut good grade extraction meal + 0.20% DL methionine	15.1b	6.70	2.23cd	4.25bc	68.1bc	66.4bc	81.6de	84.9b	91.9b

a, b, c, d, e, – All means in the same column with the same suffix are not significantly different (P 0.05)

to reflect the more severe heat treatment and damage to CSM compared to CGM.

Pig studies

Results of experiment I (growing pigs) and experiment II (fattening pigs) are presented in Tables 7 and 8 respectively. In the first experiment, all pigs receiving the FM diet grew at a highly significant (P<0.005) faster rate and consumed significantly less feed per kg gain than those receiving CSM and GNM diets. Those receiving the CSM diet also grew significantly (P<0.05) faster than those on GNM diets. The total protein efficiency ratio (TPER) followed a similar trend. Nitrogen retention by pigs on the FM diet was significantly higher than by those on CSM or GNM diets, both of which did not show significantly different nitrogen retention value.

In the second experiment (fattening pigs) analysis of variance showed significant (P<0.05) treatment differences for average daily gain. However, multiple comparison of the treatment means showed that only pigs on treatment 2 (17.4% groundnut meal) grew at a significantly (P<0.05) slower rate than those on diet 1 (control) and diet 5, the higher protein level CSM diet, but not than those on diets 3 and 4, while growth rates on diets 1, 3, 4 and 5 were also not significantly different.

The most efficient utilisers of feed were the pigs on the control group (diet 1) followed by the pigs on diet 5. Compared at each of two levels pigs on the CSM diets utilised feed more efficiently than those on the GNM diets, but the differences were not significant.

The lower level of groundnut inclusion (diet 2) showed the poorest conversion efficiency which was significantly (P<0.05) lower than for the control group (diet 1).

For all the carcass parameters studied, there were no significant treatment differences either due to the level of inclusion or to the source of protein fed. The observed superiority of the CSM to GNM when fed at a fairly low level of protein may be due to higher content of the essential amino acids methionine and cystine and tryptophan. Similarly in the fattener's experiment, the CSM was superior to the GNM diets (though not significantly so) at both levels of inclusion, for the same reason as above. The non-significance of the difference in the fattener's experiment may probably be associated with the fact that at heavier weights, the pigs' requirements for amino acids become less critical, but the CSM diets still showed marginal superiority because the efficiency of utilisation of any source improves as the pattern of amino

TABLE 7. *Mean liveweight gain, and efficiency of feed utilisation and nitrogen retention of pig feed either fishmeal, groundnut meal or scrap cashewnut meal (experiment 1—pigs)*

	Dietary treatments			SE of mean
	1	2	3	
Average initial liveweight kg	13.80	13.94	14.00	–
Liveweight (kg)	38.30	23.54	25.66	–
Average daily liveweight gain (g)	439.0a	149.00b	206.5c	17.42
Average daily feed intake (kg)	0.90a	0.59b	0.72c	0.11
Feed conversion efficiency (kg feed/kg gain)	1.98a	4.01b	3.50c	0.21
Total protein efficiency ratio	3.13a	1.63b	1.80c	0.08
Nitrogen retained (g/day)	38.81a	21.26b	26.14b	1.24
Ingested nitrogen retained (%)	63.40a	46.80b	48.20b	1.46

Row values not bearing the same superscript are significantly different (P<0.05)

TABLE 8. *Mean liveweight gains, feed intake, feed efficiency and carcass measurements of market-weight pigs either of two levels of cashew scrap meal or groundnut meal*

	Dietary Treatments					SE of means (±)
Average initial liveweight (kg)	51.40	50.80	51.20	50.90	51.30	–
Average final liveweight (kg)	92.54	92.41	92.40	92.54	92.62	–
Average daily gain (kg)	0.63	0.50	0.57	0.54	0.61	0.04*
Average daily feed consumed (kg)	2.26	2.17	2.31	2.22	2.43	0.34 (NS)
Feed conversion efficiency (kg feed/kg gain)	3.58	4.26	4.05	4.11	3.98	0.21*
Average number of days to reach slaughter weight	65.3	81.59	72.3	77.1	67.8	–
Average backfat thickness (cm)	3.72	3.82	3.79	3.74	3.68	0.24 (NS)
Average carcass length (cm)	79.84	78.96	79.44	79.56	78.95	2.18 (NS)
Lean meat in carcass (%)	54.82	54.34	54.68	54.82	54.91	1.08 (NS)
Bone in carcass (%)	11.87	11.21	11.94	11.88	11.76	0.18 (NS)
Fat in carcass (%)	31.48	31.52	31.28	31.14	30.98	0.68 (NS)
Skin in chilled carcass (%)	14.38	14.94	14.56	14.51	14.58	0.46 (NS)
Ham in chilled carcass (%)	24.86	24.84	24.79	25.20	25.38	0.64 (NS)
Shoulder in chilled carcass (%)	21.04	21.14	21.23	20.98	21.54	0.53 (NS)
Fat cuts in chilled carcass (%)	22.98	23.01	22.94	23.11	22.86	0.68(NS)
Four lean cuts in chilled carcass (%)	59.68	59.44	59.38	59.40	59.56	1.42(NS)
Average loin eye area (cm2)	23.15	22.98	23.21	23.16	23.44	0.94(NS)

* Significant treatment differences among treatment mean (P 0.05). Details of those significantly different treatments are given in text.

NS = No significant treatment differences among the means.

acids in the diet approaches that of the essential amino acid requirement of the animal. There appears to be no report in the literature on the value of CSM in pig diets, with which comparisons of results could be made. Results with rats, however, for the commercial grade meal (Piva *et al.* 1971, and Fetuga *et al.* 1973) show clearly the superiority of this source over soya bean and groundnut meal. That the scrap kernel meal is less effective, however, has been suggested to be associated with lower levels of available amino acids compared to the good grade meal.

In conclusion, it is suggested that CSM may find considerable use in Nigeria and other countries producing cashewnut as an effective supplement in cereal-based diets for growing and fattening pig diets, particularly in combination with other sources high in available lysine, due to the high tryptophan and fairly high content of the sulphur amino acids (groundnut meal which is available as an alternative to CSM in Nigeria has lower levels of these amino acids). This is particularly significant because an earlier study by Fetuga (1972) of a range of protein sources available in Nigeria, had shown the sulphur amino acids, methionine and cystine, as well as tryptophan to be the amino acids in shortest supply.

It must be stressed however, that cashewnuts are not at present processed into CSM and oil on a commercial basis, and that cashewnut production in Nigeria is presently very low. For these reasons the use of CSM for animal feed must be regarded as a future rather than a present possibility.

References

Association of Official Analytical Chemists (1970) Official methods of analysis. 11th ed. Washington, D.C.: Association of Official Analytical Chemists.

Babatunde, G. M., Fetuga, B. L. & Oyenuga, V. A. (1971) The effects of varying the dietary calorie: protein ratios on the performance characteristics and carcass quality of growing pigs in the tropics. *Anim Prod.*, 13, 675–702.

Duncan, D. B. (1955) Multiple range and multiple F. test. *Biometrics*, 11, 1–42.

Fetuga, B. L. (1972) Assessment of the protein quality of certain Nigerian foods and feedstuffs in the nutrition of the pig and the rat. Ibaday: Ph.D Thesis, University of Ibadan, Nigeria.

Fetuga, B. L., Babatunde, G. M. & Oyenuga, V. A. (1973) Protein quality of some Nigerian feedstuffs. II. Biological evaluation of protein quality. *J. Sci. Fd. Agric.*, 24, 1515–1523

Ford, J. E. (1965) A microbiological method for assessing the nutrition value of proteins. 4. Analysis of enzymatically digested food protein by sephadex-gel filtration. *Brit. J. Nutr.*, 19, 277–289

Miller, D. S. (1963) In Evaluation of protein quality *Publs. Natn. Acad. Sci.–Natn. Res. Council*, Washington DC No. 1100, 34–36

Miller, E. L., Carpenter, K. J. & Milner, C. K. (1965) Availability of sulphur amino acids in protein foods. 3. Chemical and nutritional changes in heated cod muscle. *Brit. J. Nutr.*, 19, 547–564.

Miller, E. L. (1967) Determination of the tryptophan content of feedingstuffs with particular reterence to cereals. *J. Sci. Fd. Agric.*, 18, 381–386

National Academy of Sciences/National Research Council (1963). Evaluation of protein quality. *Publs. Natn. Acad. Sci. – Natn. Res. Coun.* Washington DC No. 1100

Neishem, M. C. & Carpenter, K. J. (1967) The digestion of heated damaged protein. *Brit. J. Nutr.*, 21, 399–411.

Oyenuga, V. A. (1961) Nutritive value of cereal and cassava diets for growing and fattening pigs in Nigeria. *Brit. J. Nutr.* 15, 327–338.

Piva, G., Santi, E. & Ekpenyong, T. E. (1971) Nutritive value of cashewnut extraction meal. *J. Sci. Fd. Agric.*, 22, 22–23.

Steel, R. G. D. & Torrie, J. H. (1960) Principles and Procedures of Statistics New York, Toronto and London: McGraw-Hill Book Company, Inc., p.99

Woofroof, J. G. (1969) Composition and use of peanuts in the diet. *Wld. Rev. Nutr. Diet.* 11, 142–169

Discussion

Mr Ola: I would like to ask Professor Göğüs whether the relative economics of solvent extraction and expeller processing of oilseeds have been considered in Turkey. With solvent extraction solvent residues may be a problem while residual oil may go rancid in expeller cakes.

Professor Göğüs: With modern expellers, oil content of cakes is only 5—7% and if they are not stored too long rancidity is no problem. However, in Turkey our expeller plants mostly produce cake containing 9—10% of oil. There is only one plant in Turkey (in the Ege region) using solvent extraction methods.

Dr Traore: I would like to comment on some of the matters raised in Dr Mukhtar's paper. Cotton seed oil-cakes produced in Mali do not contain aflatoxin and problems in their use are associated rather with gossypol. We are just starting to grow glandless varieties of cotton seed in Mali on an experimental basis. We are feeding whole cotton seed to cattle at the rate of 2—3 kg/day without any negative effects. Molasses is also being fed with good results in combination with whole cotton seed.

Dr Mukhtar: Gossypol does not have any effect on ruminants. It only seems to affect laying hens when inclusion rates exceed 20%. In the Sudan eggs produced in villages are darker because hens are fed on cotton seed cake. We have also had good results in the Sudan from the feeding of combinations of cotton seed cake and molasses to fattening cattle, but when a good quality roughage was also fed, molasses had no effect at inclusion rates of higher than 30%. Mixed rations containing cotton seed cake, molasses and groundnut shell have also been fed to fattening cattle with good results.

Chairman: We have been using cotton seed flour for the past 12 years in Central America to feed children. This cotton seed flour is produced by pre-press solvent extraction using hexane as the solvent. The specifications for this cotton seed flour are minima of 50% protein and 3.6g of available lysine per 100g of protein, and maxima of 0.006 and 0.95% of free and total gossypol respectively. This same type of flour is now being used to develop milk replacers for calves. In producing such edible-grade cotton seed flours it is important to control moisture content and temperature during processing. If the seed is of good quality there are likely to be no problems with regard to rancidity of fat in the meal or flour.

Mr Eme: I refer to the paper presented by Dr Fetuga on cashew nut scrap meals. Cashewnuts are an important crop in the East Central State of Nigeria and there is a programme for the expansion of production. Does Dr Fetuga have any information on the feeding value of cashewnut apples as well as the scrap kernel meal?

Dt Fetuga: Cashewnut apples are commonly eaten by people in Nigeria. On the basis of proximate composition they are not likely to be of any great value to monogastic animals. It would perhaps not be a good thing to divert their use from humans to animals. The cashewnut shells could perhaps be considered for ruminant feeding but these are used as a source of 'Cashewnut shell liquid'.

Sixth Session

Assessment and standardisation of feed quality

**Thursday 4th April
Afternoon**

Chairman
Professor V. A. Oyenuga
University of Ibadan, Nigeria

The basis for feed quality assessment

K. J. Carpenter
University of Cambridge, England

Summary

The feedingstuffs in international trade are valued primarily as concentrated sources of digestible energy and protein. In addition, the buyer wants his purchases to be palatable and non-toxic.

None of these things is measured directly by the traditional analytical measurements—the proximate constituents—but they characterise samples in a rough way, ie provide a type of simple 'fingerprint' that, in combination with visual and microscopic examination, gives buyers confidence in their genuineness and acceptability. They can also be used to classify a sample within a range of variable materials; for example, with meat-and-bone meals, according to their 'crude protein: ash' ratio.

The above is true for classes of materials of which there has been long experience. A new product cannot be accepted just on the basis of chemical analysis however thorough. Extensive feeding trials and measurements of digestibility are also required, plus some experience as to whether the material has any risk attached to it as a carrier of unwanted contaminants such as salmonella, aflatoxins and nitrosamines. There is no avoiding the use of materials in which aflatoxins are liable to occur but it is possible to use simple screening tests to check that they are not present in significant quantities.

High protein feeds also command a particularly high price but their quality as sources of available amino acids can depend on their processing and subsequent conditions of storage as well as on their raw material. Methods used as indicators of protein quality include 'solubility of nitrogen after treatment with dilute pepsin', 'lysine reactive with fluorodinitrobenzene' and 'protein binding with Orange 12 dye'. Each of the tests has its advantages and disadvantages and the results may need careful interpretation.

Vitamins and trace minerals are vital for the production of balanced animal diets, but analytical values for these do not greatly influence the international trade in feedingstuffs since pre-mixed vitamin-mineral concentrates are normally required in any case, to complete compound diets, and are relatively inexpensive.

Résumé

La base de l'évaluation de la qualité des pâtures
Les pâtures dans le commerce international sont appréciées premièrement comme sources concentrées d'énergie digestible et de protéines. En plus, l'acheteur désire que ses achats soient agréables au goût et non-toxiques. Aucun de ces éléments n'est pas mesuré directement par des calculs analytiques traditionnels -les constituants approximatifs-mais ils caractérisent des échantillons, de façon approximative, par ex: ils procurent un type de simple 'empreinte digitale' qui en combinaison avec les examens visuels et microscopiques, donne aux acheteurs une confiance dans leur autenticité et acceptabilité. Ils peuvent être utilisés aussi pour classifier un échantillon dans une rangée de matériels variables; par ex: avec des pâtures de viande et d'os, suivant leur 'protéine brute': la proportion 'des cendres'.

Ceci est vrai pour les catégories de matériels dont il y a eu une longue expérience. Un nouveau produit ne peut pas être accepté justement sur la base de l'analyse chimique, n'importe combien précise qu'elle soit. Des essais étendus des pâtures et des calculs de la digestibilité sont exigés et en plus une certaine expérience pour savoir si le matériel présente un risque quelconque attaché, comme étant un porteur de contaminants indésirables comme la salmonelle, les aflatoxines et les nitrosamines. Il n'y a pas moyen d'éviter l'usage des matériels dans lesquels les aflatoxines sont passibles d'être trouvées mais il est possible d'utiliser un simple test pour se rendre compte qu'elles ne sont pas présentes en quantités significatives.

Les pâtures riches en protéines également règlent des prix particulièrement élevés, mais leur qualité comme source d'aminoacides disponibles peut dépendre de leur traitement et des conditions subséquentes de stockage aussi bien que de leur matière première. Les méthodes utilisées comme indicatrices de la qualité des protéines incluent 'la solubilité de l'azote après le traitement avec pepsine diluée', 'la réaction de la lysine avec le fluorodinitrobenzène' et 'l'union des protéines avec le colorant Orange 12'. Chacun de ces tests a ces avantages et ses désavantages et les résultats pourraient nécessiter une interprétation rigoureuse.

Les vitamines et les traces minérales sont vitales pour la production des diètes équilibrées pour les animaux mais les valeurs analytiques pour ces substances n'influencent pas beaucoup le commerce international de pâtures puisque les concentrés vitamines-minéraux pré-mélangés, en tout cas sont normalement exigés, pour complétér les diètes combinées et sont rélativement peu coûteux.

Resumen

La base de la estimación de la calidad del alimento
En el comercio internacional se valoran los piensos, en primer lugar, como fuentes concentradas de energía digestible y proteína. Además, el comprador quiere que sus adquisiciones sean agradables y no tóxicas.

Ninguna de estas cosas se mide directamente por las mediciones analíticas tradicionales-los constituyentes aproximados-sino que caracterizan muestras de una manera tosca, es decir, suministra un tipo de "huella dactilar" sencilla que, en combinación con el examen visual y microscópico, da a los compradores confianza en su autenticidad y aceptabilidad. Pueden usarse también para clasificar una muestra dentro de una gama de materiales variables; por ejemplo, con harinas de carne y de huesos, de acuerdo con su proporción "proteína bruta: cenizas".

Lo precedente es verdad para clases de materiales de los que ha habido larga experiencia. Un nuevo producto no puede aceptarse solamente sobre la base del análisis químico aunque sea completo. Se requieren también pruebas de alimentación amplias y medidas de digestibilidad, más alguna experiencia referente a si el material tiene algún riesgo especial como portador de contaminantes no deseados, tales como la "salmonella", "aflatoxinas" y "nitrosaminas". No se puede evitar el uso de materiales expuestos a que se encuentren en ellos las "aflatoxinas" pero es posible usar pruebas de examen sencillas para comprobar que no están presentes en cantidades significativas.

Los piensos ricos en proteínas tienen también un precio alto pero su calidad como fuentes de amino-ácidos disponibles puede depender de su preparación y condiciones subsiguientes de almacenamiento, así como en su materia prima. Los métodos usados como indicadores de la calidad de la proteína incluyen 'solubilidad de nitrógeno después del tratamiento con pepsina diluída', 'reactivo de lisina con fluorodinitrobenzeno' y 'ligado de proteína con tinte naranja 12'. Cada una de las pruebas tiene sus ventajas y desventajas y los resultados pueden necesitar una interpretación cuidadosa.

Las vitaminas y oligominerales son vitales para la producción de dietas para animales equilibradas, pero los valores analíticos de éstos no influyen en gran manera en el comercio internacional de piensos, puesto que en cada caso se requieren normalmente concentrados de vitamina-mineral previamente mezclados, para completar las dietas compuestas, y son relativamente baratos.

We are primarily concerned here with feeds of importance in international trade. In practice this means the 'concentrated' feeds, rather than roughages, and most come into the three categories of cereals, oilseeds and animal products. The greater portion goes to the feeding of poultry and pigs, though the third use is for ruminants (and pre-ruminants) as supplements to roughages.

The first requirement is, of course, that they should not be harmful. Bindloss (1974: see below pp 00) will discuss the commercial implications of standards being legally set up for feedingstuffs. Obviously the subject is extremely complicated. We cannot put all materials into two clear cut pigeon holes: 'toxic' or 'non-toxic'. Many have characteristics which put limits on the safe upper levels at which they can be used in different types of diet, but below these levels they make valuable contributions to the total supply of balanced feeds.

Any assessment must therefore consider both positive and negative aspects but it will be more convenient in the present paper to take the positive aspects first.

Vitamins and minerals.

A partial deficiency of *any* one nutrient in a diet can cause inefficient and uneconomic production. Nevertheless, it may be quite reasonable for a feed formulator to be indifferent as to the vitamins and trace minerals contributed by a particular material. This is usually because he knows that he is going to have to add some vitamins and minerals to balance his formulae and, since the amounts of each micronutrient needed will vary from one formula to the next, it is economic to have a limited number of pre-mixes which are sufficiently inclusive and general in context to cover all contingencies. This can be economic because their cost is only a small fraction of the total cost of a balanced ration, and it acts as a precaution against relying on ingredient analyses which are expensive and sometimes still imprecise.

Some ingredients are, of course valued for their contribution of UGF (unidentified growth factors), particularly in poultry production where it is common practice to fix a minimum level of usage of fishmeal and fermentation products. By the nature of things there is no chemical procedure for this type of activity, and it has been suggested that the response may be more analagous to that obtained with antibiotics than the response to an essential nutrient.

Energy

Energy is still the most expensive nutrient to supply in a balanced animal diet. For each type of material there are published 'metabolisable energy' (ME) values for both pigs and poultry (National Academy of Sciences, 1971; Allen, 1973). Usually formulators are content to use ME values but there is reasonable evidence that ME from fat has a significantly higher net (or productive) value than the same quantity of ME from carbohydrate (de Groote, 1974). This can be simply allowed for if one has a fair estimate of the digestible lipid content of the material.

Determination of ME has the great advantage over any of the net energy systems (such as starch equivalents) in that one does not need to employ animal calorimeters or respiratory chambers. Bomb calorimetry is now quite a simple operation and ME determination with even a small number of replicates seems to be satisfactorily reproducible from one laboratory to another (Carpenter & Clegg, 1956; Hill et al. 1960). It is important, however, when comparing values to check that they have been calculated in the same way:– eg on a dry matter basis or corrected to some standard moisture content. Also, in the case of values determined with chicks that have a high rate of growth (and thus of nitrogen retention) it has become standard to correct ME values to N equilibrium, (ie to subtract from the classical values an allowance for the amount of uric acid gross energy that would appear in the urine if the retained protein were to be metabolised as an energy source, (Hill et al. 1960). In practice, this usually means about a 3% deduction. Before any new product is put on the market, their ME value for pigs and poultry should certainly be determined and this need not be expensive compared with an ordinary feeding trial.

We now come to the crunch: the problem of how the buyer is to check whether the value of a particular batch of material is the same as that of the average values for material listed under the same name in published tables of feedingstuff values. Gross energy (ie the value obtained by complete combustion in a bomb calorimeter) is almost useless because most of the differences between feeds is in the digestibility of the gross energy which, itself remains fairly constant (Carpenter & Clegg, 1956).

Proximate analysis

Before turning to newer chemical analyses, what can we learn from the traditional Weende or proximate analysis which has been the basis of legislation about feedstuff composition? It is easy enough to make fun of the system, for example by pointing out that a mixture of 'vaseline, sand, old socks and oxalic acid' can give the same analysis as soya bean meal, but the buyer works with his eyes open and would probably spot this one even before he had brought out his microscope!

In practice, the values for crude protein, ether extract, crude fibre and ash do provide a crude kind of 'finger print' or profile of a sample. The overwhelming proportion of the energy in animals feeds comes from carbohydrates and the main cause of lower-than-expected ME in a batch of material is that an unduly high proportion of the carbohydrate is 'structural' rather than 'storage' in character. This is encountered particularly with oilseed meals where there is incomplete removal of husk or cortex at one stage, for example with sunflower or sesame meals.

This, fortunately, is also automatically reflected in an increased crude fibre value in such samples. Then, provided that one has knowledge of determined ME values of samples of different crude fibre content, one can interpolate a reasonable estimate of ME for the sample in question.

What one must *not* do is to assume that crude fibre corresponds to the total indigestible carbohydrate and that NFE (Nitrogen-free extractives, by difference) is entirely digestible. There is a variable, but sometimes quite important fraction of the carbohydrate which is indigestible to poultry though brought into solution by the treatments with acid and then alkali that constitute the crude fibre determination (Carpenter & Clegg, 1956; Bolton, 1954; Carpenter, 1961). Results with different grades of wheat by-products provide an example (Table 1). This makes proximate analysis of only limited value for predicting the ME value of mixtures, or of materials of which one has little or no background experience.

Workers interested mainly in the utilisation of herbage products have investigated many modifications of the 'fibre' determination in attempts to obtain measures that correspond more closely to the fraction indigestible by ruminants. For the rather different problem of evaluating concentrate feeds, and mainly for use with non-ruminants, it seems more useful to obtain a direct estimate of the useful carbohydrates instead of leaving them to a 'residue by difference'. We have found the anthrone reagent to give a simple colorimetric

TABLE 1 *The gross and metabolisable energy (ME) for poultry of wheat products in relation to their carbohydrate analysis (Carpenter & Clegg, 1956)*

Material (at 90% dry matter)	Energy, kcal/g		Crude fibre, %	Nitrogen-free extractives (by difference)	Starch + sugars
	gross	ME			
Whole wheat	3.93	3.32	1.6	72.9	63.4
Milling offals:					
a. Fine parings	4.13	2.65	5.5	60.7	42.9
b. French pollards	4.12	2.00	7.6	56.6	29.1
c. Coarse bran	4.12	1.32	10.6	54.6	18.1

measure that includes both sugars and starch, and that the formula:

$$0.059 + 0.038 (1.1 \times \% \text{ starch} + \% \text{ sugar} + \% \text{ crude protein} + \% \text{ ether extract})$$

gave a prediction of classical ME values (k cal/g) for poultry with a standard error of approximately ± 0.21 (Carpenter & Clegg, 1956). In a further study of this approach in Canada (Sibbold et al. 1965). the formula:

$$10.01 (4.1 \times \% \text{ starch} + 3.55 \times \% \text{ sugar} + 3.52 \times \% \text{ crude protein} + 7.85 \times \% \text{ ether extract})$$

gave a slightly closer prediction of N-corrected ME values (kcal/g) for a larger series of feeds and diets of all types.

Fat

Fats are another significant source of energy in animal diets. The ME of *digested* fat is almost constant regardless of source and the problem is to know how digestible a particular material will be. For relatively small quantities naturally present in conventional 'concentrate' feeds, 'ether extract' gives a reasonable measure of fat content and the digestibility seems generally high. Commercially available 'fats' as such are commonly blends of different origin. The problem from an energy point of view rises from observations that highly saturated fats are not so well digested if fed at levels of 5% or more, as a result of reactions giving insoluble calcium stearate (Sibbold et al. 1961). Iodine values and solidifying temperatures give a simple characterisation of commercial-grade fats.

A dispute about methodology has arisen over materials such as fishmeals that contain highly unsaturated fat. A proportion of this may autoxidise during storage and become insoluble in petrol ether (and to a lesser extent in diethyl ether). The ether extract of a batch of material may therefore show a decrease on successive analyses (Almquist, 1956). This is frustrating to someone who likes stability in his analyses, even if the real world is changing. The EEC is therefore ordering standardised fat analyses with a preliminary acid-hydrolysis stage to solubilise any oxidised fat. Is the extra work involved justified? My own view is that we want analyses to reflect nutritional value as far as possible and that since oxidised lipids insoluble in ether, are also apparently indigestible (March *et al* 1965), one loses more than one gains by changing from the traditional 'ether extract' procedure.

It is only in recent years that the positive contribution of fat has been emphasised. In the past it was thought of primarily as a source of trouble, and I will touch on this aspect of feed quality also. Dietary fatty acids are all deposited to some extent in animal tissues, though their subsequent turnover is quite rapid. Animals receiving fish lipids up to (or close to) the time of slaughter will therefore contain some of the 20-24 carbon, polyunsaturated acids which seem to oxidise during the cooking of meat to give a characteristically fishy or 'kippery' taste if present in large enough quantities (Opstvedt, 1971). However, it seems to be a safe rule to feed diets containing no more than 0.8% fish lipid during a 'finishing' period.

The presence of oxidised fat in feedstuffs themselves is quite a different matter. As stated above they are generally indigestible, but it was a practice in some quarters to use peroxide (ie hydroperoxide) values as an index of 'rancidity' and to condemn materials as 'unfit for feeding' if the value exceeded a particular figure. Research workers have been unable to find evidence of significant adverse effects of hydroperoxides at the level at which they might be encountered in practical diets, *nor* from free fatty acids that may occur in feeds from hydrolysis of lipids (ie hydrolytic rancidity (Carpenter *et al*, 1966). My view, discussed at length elsewhere (Carpenter, 1968) is that with the general use of stabilised fat-soluble vitamins, rancidity is not a direct problem, but that provided one knows the norms for fair, average samples it may serve as an indirect indicator of excessive staleness or mould infestation in particular classes of material.

Protein

The second big item in the cost of balanced animal diets comes from the provision of adequate protein. Proximate analysis tells us the quantity of 'crude protein' (N x 6.25) in a feed and, in general, the great majority of this is truly protein. However, there have always been exceptions and, recently, there has been discussion of the high level of nucleic acid in single cell protein preparations, so that a significant proportion of the total nitrogen is in 'non-protein' and non-nutritive' form. The earliest analytical refinements were to subtract from the total nitrogen either 'non-protein N' (soluble after treatment with alkali and copper ions), 'ammonia N' or even 'hot-water soluble N'. There may still be situations where these measurements serve as relatively cheap and rapid indicators of quality in a particular product. But, in general, the major differences in quality of the crude protein from one feed to another arises because of differences in the digestibility and the amino acid composition of the true protein.

As an academic exercise we can construct experimental diets in which the limiting factor is simple ammonium ions or non-specific, non essential amino acids. But in practice, the cost-limiting factors in pig and poultry diets are the essential amino acids. Further it has generally been found that if adequate levels of lysine and of methionine + cystine have been provided in a diet from the range of conventional feeds, the other amino acids are almost always present in adequate supply. This follows from it being either lysine or the sulphur-containing amino acids that are first limiting in the proteins of nearly all the common feedstuffs (McLanghlan et al. 1959). Where a portion of the supply of these amino acids is provided as the synthetic chemical then one has to check that tryptophan, isoleucine and arginine are also in adequate supply, but this is of much less importance at the present time.

There are now a number of published tables of mean values for the amino acid composition of all the common concentrate feeds (National Academy of Sciences, 1971; FAO, 1970; Combs & Nott, 1967). Some of the older tables include serious errors because some materials contain factors that we now realise interfered with the microbiological assay procedures used for the earlier studies, so that care is necessary (Atkinson & Carpenter, 1970). The figures in the standard tables are sometimes referred to as 'total' amino acid values, because they are derived from *total* hydrolysis of the proteins by refluxing in strong acid (or in alkali for the estimation of tryptophan). They are thus comparable to gross energy values obtained by *total* combustion of samples within a bomb calorimeter. In both cases the nutritionist is really concerned with the proportion of those amino acids that are biologically available.

By and large it is probably fair to accept that 'available', as applied to amino acids, means no more than 'digestible', but it is uncertain whether or not one can measure digestibility satisfactorily as 'whatever goes in at the mouth end of an animal and cannot be recovered by analysis of its faeces'. Some amino acid units may remain undigested in the small intestine and then be fermented by bacteria in the large intestine. Such molecules will thus have been apparently digested even though they were of no value to the animal. There is also the point that some derivatives of lysine, that may be encountered in processed foods, are absorbed but unutilisable and re-excreted in the urine. Until the practical effect of these two possibilities is shown to be negligible, research workers are continuing to carry out biological assays for the potency of feeds as sources of the key amino acids, using the growth rate of rats and chicks as their measure of response. Combs & Nott (1967) have reported such availability factors for the amino acids of the majority of poultry feedstuffs, and they

range of 98% down to 65%. Some minor modifications have been suggested in another paper (Carpenter, 1971).

With pigs the only relevant published data are for the digestibility of individual amino acids (Dammers, 1964; Eggum, 1973). However, it seems likely that if there were any problems in utilisation of a particular type of protein young chicks would be at least as sensitive as pigs in showing it up.

For a new type of high protein material it is important that amino acid analyses and availability studies should be organised by its producers. If formulators are not fully informed of the characteristics of a material they will, inevitably, be reluctant to place much reliance on what it can contribute to a balanced diet. However, the compounder's second problem, which applies to all materials, is judging whether the sample offered to him has the same protein content and quality as the 'fair, average' values in published tables.

Many of the protein rich feeds are processed materials: fish and meat meals have had to be dried and oil seeds meals fat-extracted with processes that usually involve heat at some stage. So that, in addition to the variability in the quality of raw materials, there is also the possibility of damage to the protein during manufacture. Proteins can undergo a number of reactions leading to loss of nutritional quality and in certain cases these can occur even during storage. Fishmeals containing particularly reactive oils were liable, prior to their stabilisation with anti-oxidants, to heat up when stored in bulk as a result of the exothermic autoxidation of the oil, and in extreme cases the digestibility of the protein was seriously reduced and a portion of the cystine destroyed. In stored milk powders where the moisture level is above 5% there can occur a series of Maillard reactions between the aldehyde group of the lactose and the epi-amino group of lysine units in the protein; this results in the lysine becoming nutritionally unavailable even though this is not fully detected by analyses for total amino acids (Carpenter & Booth, 1973).

Proximate analysis gives a first chance of picking out a 'rogue' sample. Obviously a low crude protein value is a basis for rejection or adjustment of price. Gross overheating of a sample is also easily picked out by smell and appearance to an experienced user. However, colour can be misleading, for example with meat meals where depth of colour may reflect the inclusion of blood in the raw material rather than overheating in manufacture. Also, in the case of fish meals, colour may reflect pigments in the particular plankton on which the fish were feeding when they were caught, and be no indicator of quality. Also it has been surprising that so-called flame-dried materials, are in general, of no lower quality than corresponding steam-dried products even though they may be rather darker. The former process, really drying in a current of very hot air, involves drying of small crumbs of tumbling material in a cloud of their own steam, so that they do not over-heat unless they remain in the current whilst already dry.

This is the case only for a small proportion of finer crumbs that do look burnt. In the latter process heating is from pipes containing circulating steam, and it takes considerably longer. Damage is a function of both time and temperature, and most materials are produced with a high coefficient of availability; the exceptions are feather meal and blood meal.

In practice, total amino acid analysis is too expensive and too slow for use in quality control. The most common single amino acid analysis that has been used for occasional quality control has been the determination of FDNB-reactive lysine. Fluorodinitrobenzene (FDNB) will react with an intact protein at those lysine units that have not engaged in reactions (eg Maillard reactions) that reduce the value of the protein. After a preliminary treatment of the feed with this reagent, hydrolysis with acid yields a 'labelled' lysine derivative that can be measured colorimetrically. There are many possible procedures for separating the lysine derivative from possible interfering colours as reviewed elsewhere (Carpenter & Booth, 1973) but it can be done reasonably well in most cases without special equipment (Carpenter, 1960). A high value in this test indicates both that the raw material was of high quality *and* that processing and storage have not resulted in significant damage.

For routine use one really needs even simpler tests. Dye-binding with an azo dye such as Acid Orange 12 is certainly rapid and some results have shown quite close correlations between dye-binding capacity and FDNB-reactive lysine for different samples of fishmeal. However, others have not confirmed this (Sandler, 1972;) this may be explained by different levels of arginine and histidine in the raw materials since all three di-basic amino acids bind with the dye. However, there seems to be considerable scope of the development of this type of procedure, which has so far been used commercially only for estimating protein 'quality' in materials where the protein is of almost constant character, as in milk for example. (National Academy of Sciences, 1971).

Another approach has been to measure 'pepsin digestibility' of the nitrogen in feeds. This is a misleading name for the test which really measures the solubility of the nitrogen in a feed after *in vitro* digestion with pepsin. As originally operated, with a relatively strong solution of pepsin, only extremely indigestible material such as 'hoof and horn' were left insoluble (Gehrt *et al.*, 1955). Nearly all other protein was at least split into peptides that were sufficiently small to be water-soluble. However, it was realised that heat damage which considerably reduced the nutritional value of materials had little effect on its solubility with pepsin under these conditions.

An alternative procedure was developed using much more dilute pepsin, and measuring just the extra nitrogen brought into solution as compared with that soluble in water without pepsin. This proved to reflect nutritional damage to heated samples with considerable sensitivity and showed promising correlations with nutritive value for quite a range of materials (Lovern *et al.*, 1964). However, it has been found not to be applicable to all materials. In particular it appears that the presence of unsaturated fat in samples may hinder the effect of pepsin *in vitro*, and thus give low 'solubility' values while the actual nutritional value of the samples is high (Contreras & Komo, 1965). The discrepancy is presumably explained by the animals' gut having lipases as well as protein-digesting enzymes, so that the film of fat is first removed from particles and then the protein is hydrolysed.

Where a single type of material is being subjected to quality control, excessive heat during manufacture may be detected by changes in the solubility of the nitrogen in saline under standardised conditions (Lyman *et al.*, 1953). However, denaturation of proteins by heat does not *per se* reduce their nutritional value, so that any standard for a solubility test must be worked out empirically in relation to materials of the same type that have been subjected to actual nutritional tests. As will be discussed below, some materials actually benefit from a certain amount of cooking.

For the sulphur-containing amino acids there are, unfortunately, no really satisfactory procedures that can be used by a general, routine laboratory. If one is able to use microbiological assay procedures, the proteolytic organism *Streptococcus zymogenes* has been found to give values for 'available methionine' that agree well with the results of biological assays (Ford, 1962; Miller *et al.*, 1965). However, this still leaves cystine unmeasured and this is a particularly labile amino acid.

Attempts have been made to use 'total sulphur' as an indirect estimate of the sum of methionine and cystine (Miller & Naismith, 1958, Porter, 1972; Boulter *et al.*, 1973). When the results are expressed as 'percent of sample' and the samples tested have also had a considerable range of protein content, quite high correlations have been found, but it is doubtful whether it is a reliable method for assessing the quality of a given amount of crude protein as a source of these two amino acids (FAO, 1970). Certainly we did not find it to be so with a range of animal by-product materials.

Undesirable factors

These are so various that it is difficult to say much in a general review. Broadly we can put them into three categories:

1. Those always expected to be present in a particular type of material, so that it needs only correct classification of a particular batch for it to be restricted to 'safe' uses.
2. Those always present in a particular class of raw material but which can be removed or inactivated by suitable processing methods before being offered for sale.

3. Those which are not naturally present in a material but which experience has led us to believe may get there in a proportion of the samples of particular manufactured products.

A mild example in the first category would be the pigments in yellow maize. In the UK they are undesirable in broiler diets because broiler producers aim to sell white-skinned chickens, and maize pigments are deposited there to some extent and make them less 'white'. Beauty, as ever, is in the eye of the beholder, and the same pigments transferred to the yolk of hen's eggs are, of course, looked on with pleasure.

In the second category are the growth inhibitors in raw soya, and it is normal practice to add a toasting stage to the treatment of the meal after it has been solvent extracted. That this has been adequate can be tested empirically by the 'urease' procedure (Croston et al. 1955). It seems that the amount of heat needed to reduce the level of this enzyme (naturally present in raw soya) to low levels is also sufficient to inactivate the growth inhibitors (trypsin inhibitors, haemagglutenins etc., (Liener, 1974), and assay of the enzyme can be a simple procedure. Other examples would be the level of free gossypol in cotton seed meals and of active goitrogens in rapeseed meals (Liener, 1969).

In the last category we have the micro-organisms and their toxic metabolites. The aflatoxin problem is considered in another paper at this Conference (Jones, 1970; see pp below). Fortunately there are rapid chemical procedures for its extraction and detection that do not require expensive equipment. Others will be speaking on possible problems with other mycotoxins. Recent history has shown how feeds can be contaminated, as a result of quite unpredicted chances and interactions with toxic chemicals such as nitrosamines and polychlorinated biphenyls. Looking for such problems is obviously a highly specialised matter outside the range of ordinary feed assessment. But this makes it all the more important, and in their own interest, for producers of particular materials to call in specialist help as soon as there is suspicion that inexplicable bad results are being obtained with their product.

Similarly the analyst must resist the attempt to impress his clients by appearing all-knowing. The methods we have (and, in particular, the methods that we can afford to use with a limited expenditure of time and money) can only serve as indicators. As in any scientific endeavour we may with confidence assert the 'negative', in this instance 'that a particular batch of material does *not* correspond to certain standards'. But we can never assert dogmatically from analysis alone that a batch of material is safe and of high nutritive value.

References

Allen, R. D. (1973) Ingredient Analysis Tables. Feedstuffs Yearbook. Minneapolis: Miller, p.24.

Almquist, H. J. (1956). *J. Agr. Fd Chem.* 4, 638.

Atkinson, J & Carpenter, K. J. (1970) *J. Sci. Fd Agric.* 21, 366.

Bindloss, A. A. (1974). Current trends in international feed standards, *Conf. on Animal Feeds of Trop. and Sub-Trop. Origin.* London, April.

Bolton, W. (1954) *Proc. 10th World Poultry Congr. (Edinburgh),* p. 94.

Boulter, D. *et al.* (1973) in *Nutritional improvement of food legumes by breeding* [Ed. M. Milner]. New York. Protein Advisory Gp., UN.

Carpenter, K. J. & Clegg, K. M. (1956) *J. Sci. Fd Agric.* 7, 45.

Carpenter, K. J. (1960) *Biochem. J.* 77, 604.

Carpenter, K. J. (1961) in *Nutrition of Pigs & Poultry* (Ed. J. T. Morgan & D. Lewis), London: Butterworth, p. 29.

Carpenter, K. J. L'Estrange, J. L. & Lea, C. H. (1966) *Proc. Nutr. Soc.* 25, 25.

Carpenter, K. J. (1968) *Proc. 2nd Nutr. Conf. Feed Mnfctrs.* 54.

Carpenter, K. J. (1971) *Feedstuffs* 43, August 7, p. 31.

Carpenter, K. J. & Booth, V. H. (1973) *Nutr. Abstr. Rev.* 43, 423.

Cole, E. R. (1969) *Rev. Pure Appl. Chem.* 19, 109.

Combs, G. F. & Nott, H. (1967) *Feedstuffs,* 39, October 21, 36.

Contreras, E. & Romo, C. (1965) *Inst. Fomento Perquuere, Santiago,* Chile, Publ. No. 12.

Croston, C. B., Smith, A. K. & Cowan, J. C. (1955) *J. Amer. Oil. Chem. Soc.* 32, 279.

Dammers, J. (1964) Digestibility in the pig. Factors influencing the digestion of components of the feed and the digestibility of the amino acids. Hoorn: Inst. Veevoedingsonderzoek, Netherlands.

De Groote, G. (1974) *Brit. Poult. Sci.* 15, 75.

Eggum, B. O. (1973) A study of certain factors influencing protein utilisation in rats and pigs. *Benetn Forsogslab. Copenhagen.* No. 406.

FAO (1970) Amino Acid Contents of Foods and Biological Data on Proteins, *FAO Nutr. Studies,* No. 24.

Ford, J. E. (1962) *Brit. J. Nutr.* 16, 409.

Gehrt, A. J., Caldwell, M. J. & Elmslie, W. P. (1955) *J. Agr. Fd Chem.* 3, 159.

Hill, F. W., Anderson, D. L., Renner, R. & Carew, L. B.; (1960) *Poult. Sci.* 39, 573.

Jacobsen, E. E. *et al.* (1972) Evaluation of the dye-binding method as a tool for the practical check of fish meal quality. Hillerød; A/S N. Foss Electric Ltd, Denmark.

220   Jones, B. D. (1974) Aflatoxin in feedingstuffs—its incidence, significance, and control, *Conf. on Animal Feeds of Trop. and Sub-Trop. Origin,* London, April.

Liener, I. E. (1969) Toxic Constituents of Plant Foodstuffs. New York: Academic Press.

Liener, I. E. (1974) *J. Agr. Fd Chem.* **22**, 17.

Lovern, J. A., Olley, J. & Pirie, R. (1964) *Fishing News International,* **3**, 310.

Lyman, C. M., Chang, W. Y. & Couch, J. R. (1953) *J. Nutr.* **49**, 678.

March, B. E. *et al.* (1965) *Poult, Sci.* **44**, 697.

McLaughlan, J. M. *et al.* (1959) *Can. J. Biochem. Physiol.* **37**, 1293.

Miller, D. S. & Naismith, D. J. (1958) *Nature.* Lond. **182**, 1786.

Miller, E. L. & Carpenter, K. J. (1964) *J. Sci. Fd Agric.* **15**, 810.

Miller, E. L., Carpenter, K. J., Morgan, C. B. & Boyne, A. W. (1965) *Brit. J. Nutr.* **19**, 249.

National Academy of Sciences (1971) Atlas of Nutritional Data on United States and Candian feeds. Washington, D. C.: National Academy of Sciences.

Opstvedt, J., Nygard, E. & Olson, S. (1971) *Acta Agric. Scand.* **21**, 125.

Porter, W. M. (1972) *Genetic control of protein and sulphur contents in dry bean,* Phaseolus vulgaris. Ph.D. thesis, Purdue University.

Sandler, L. (1972) *Fish. Ind. Res. Inst. Univ. of Cape Town, 26th Ann. Rept.,* p. 32.

Sibbold, I. R., Slinger, S. J. & Ashton, G. C. (1961) *Poult. Sci.* **40**, 303.

Sibbold, I. R., Czarnocki, J., Slinger, S. J. & Ashton, G. C. (1963) *Poult. Sci.* **42**, 486.

Current trends in international feed standards

A. A. Bindloss

Animal feeds co-ordination, Unilever Ltd., London, England

Summary

Objectives at national and farm levels are to optimise livestock conversions, both physical and economic, of raw material resource inputs. Maintaining production, by ensuring that diet standards are met at least cost, whilst varying formulations as ingredient prices and availabilities change, demands maximum access and choices between types and sources of wholesome ingredient materials.

Consequent features of commodities, understood between buyers and sellers, formally written into contracts and dealt with outside contracts, are discussed. These, orginally concerned with nutrient features, begin to include impurity and contaminant features associated with natural materials, but of potential hazard to consuming livestock, or to man as consumer of livestock products.

Increasing demands that safety and quality of public environment are preserved create pressures on technical and administrative authorities, national and international, to study further potentially hazardous contaminants of natural nutrient ingredients. The list of such features, of concern to buyers and sellers, is likely to be extended.

Current forms of such studies are discussed as to possible eventual forms of controls imposed by legislative or other means at national or international levels.

Possible conflicts that may arise, between choice and availability of raw materials needed as nutrients to assure availability of livestock food products, and need to safe-guard public health and safety, are examined with reference to future technical and commercial relationships between buyers and sellers of internationally exchanged ingredient commodities.

Résumé

Tendances actuelles dans les standards internationaux des pâtures

Les objectifs, aux niveaux nationaux et des fermes, sont d'obtenir le maximum de conversion du bétail, tant physiquement qu'économiquement, de l'ingestion des ressources de matières premières. Maintenir la production, en s'assurant que les diètes standard sont obtenues avec le moindre coût, même qu'en variant les formules à cause de changements des prix des ingrédients et des disponibilités, demande un maximum d'accès et de choix entre les types et les sources de matériels constituants sains.

On discute les points concernant les denrées, agrées entre les acheteurs et les vendeurs, écrits formellement dans les contrats et-négociés aussi en dehors des contrats. Ces denrées, quant aux caractéristiques nutritives, commencent à inclure des impurités et des contaminants associés avec les produits naturels, présentant un risque potentiel pour le bétail on pour l'homme comme consommateur des produits de bétail.

Des demandes de plus en plus grandes pour que la sûreté et la qualité de l'environnement public soient conservées, exercent une pression sur les autorités administratives, nationales et internationales, d'étudier d'autres contaminants virtuellement hasardeux des ingrédients nutritifs naturels. La liste de telles caractéristiques qui préoccupent les acheteurs et les vendeurs sera probablement élargie.

Les formes actuelles de telles études sont discutées concernant les formes possibles de vérification imposées par la loi ou par d'autres moyens, au niveau national ou international.

On examine les conflits possibles qui pourraient arriver entre le choix et la disponibilité des matières premières nécessaires comme substances nutritives, pour assurer la disponibilité des produits alimentaires pour le bétail et le besoin de sauvegarder la santé publique et la sécurité, au sujet des rapports futures techniques et commerciaux entre les acheteurs et les vendeurs d'ingrédients échangés sur une échelle internationale.

Resumen

Tendencias actuales en las normas de piensos internacionales

222 Los objectivos, a niveles nacionales y de granja, son mejorar al máximo las conversiones de la ganadería, tanto físicas como económicas, de las entradas de recursos de materias primas. Mantener la producción, garantizando que las normas de las dietas se satisfacen al coste mínimo, al tiempo que se varían las fórmulas cuando los precios y las disponibilidades de los ingredientes cambian, exige acceso y opciones máximos entre tipos y fuentes de materiales ingredientes sanos.

Se discuten aspectos consiguientes de los artículos, entendido entre compradores y vendedores, escrito formalmente en contratos y negociado con contratos de fuera. Estos, interesados originalmente en los aspectos nutritivos, empiezan a incluir la impureza y características contaminantes asociadas a los materiales naturales, pero de peligro potencial para el ganado que los consume o para el hombre como consumidor de productos de la ganadería.

Las demandas crecientes de que la seguridad y la calidad del medio ambiente público se preserven, crea presión sobre las autoridades técnicas y administrativas, nacionales e internacionales para estudiar más contaminantes peligrosos potencialmente de los ingredientes nutritivos naturales. La lista de tales aspectos, que atañen a comprobadores y vendedores, es probable que aumente.

Se discuten las formas actuales de tales estudios como formas eventuales posibles de controles impuestos por medios legislativos u otros a niveles nacionales o internacionales.

Los conflictos posibles que puedan surgir, entre la elección y la disponibiladad de las materias primas requeridas como elementos nutritivos para garantizar la disponibilidad de productos alimenticios de la ganadería, y la necesidad de salvaguardar la salud y seguridad públicas, se examinan con referencia a las futuras relaciones técnicas y comerciales entre comprobadores y vendedores de productos para ingredientes intercambiados internacionalmente.

Introduction

Food is the absolutely basic necessity of man, and in modern complex society, the price of food to the consuming public is a highly critical economic and political issue. All those concerned in the entire food-conversion chain work against the background of these two simple statements, and any matters affecting food volumes, and food economics, are of concern to all of us.

Resource conversion

Livestock farming exists to provide meat, milk and eggs for human diets; the task of the animal is to convert feeds into these products.

On individual farms, or on the aggregate of these which is the national livestock farm of any country, two features of this conversion have to be optimised concurrently. Efficiency of physical conversion must be secured — that is, the lowest possible consumption of feed in order to produce a given quantity of eggs, meat or milk. This physical efficiency, however. must be itself achieved at the highest possible efficiency in economic terms — that is, cost of paying for the feeds eaten must be as low as possible in relation to the value of the resulting animal products. Only when these two objectives are pursued together can conversion of total resources be optimised.

Least-cost formulation

Simultaneous pursuit of these twin targets has led to the development of sophisticated techniques for drawing up planned diets for livestock. Science has established, but continually elaborates and refines, the levels of total nutrients which animals need daily for various productive purposes; so much energy, so much protein and constituent amino-acids, so much of the many necessary minerals and vitamins. These patterns for different purposes and species, the Diet Standards, are the master-features determining productive efficiency. So long as they are provided in diets, they can satisfactorily be built up from any number of contributions from individual materials in the diet formula.

If the content of various nutrient features, and the cost, of all available feed ingredients are known, therefore, it is possible to find a mix of these which at any one time satisfies the Diet Standard as to nutrient pattern of levels and balances, at the lowest feasible cost of doing so: this is a least-cost formula. Choices of which ingredients are included, and of their levels of indicated inclusion, thus are made both on their nutrient contents and their prices; price is a major influence on the solution. The technique is in almost universal use to guide purchase of ingredient commodities. Its use is primarily directed to the purely *nutrient* objective, least-cost formulation to Diet Standards, so as to support optimum physical conversion by consuming stock at minimum feasible cost of feeding for that production.

Any factor which effects the price of an ingredient commodity, whether or not it has any bearing on the material *as a source of nutrients,* will affect whether or not the material features in a least-cost diet formula, and if it does, will affect its level of indicated use.

International transactions in feed materials take place under commercial contracts between buyers and sellers. These have to reflect descriptions, features, and specifications desired by the purchaser, and agreed by the seller as reasonable, in that he can locate and supply a commodity meeting these yardsticks. A price has then to be agreed between the parties for the commodity meeting the technical yardsticks specified, when the contract as a whole can be completed between them.

Contractual features of descriptions and specifications can only be included, for verification of which satisfactory methods exist, since on these depends proof that contracts have been honoured; or that minor variations exist in the material as received (when price adjustment provisions of the contract operate); or that major variations exist, when decisions on physical and financial procedures are made by independent arbitration procedure. It is well-recognised by buyers and sellers, therefore, that many features about commodities, important to the technical uses for which they are purchased, cannot necessarily feature in commercial contracts, but are nevertheless understood by both parties to be taken into account in purchasing decisions. Such matters are indeed discussed and worked upon by both parties, and very important improvements have resulted in matters such as the qualities of proteins rather than their total contents, the hygienic status of commodities, and the freedoms of materials from moulds and associated mycotoxins, whether or not features such as these could be dealt with in commercial contracts.

In this discussive paper, the term International Feed Standards is taken to embrace all matters which are understood between sellers and buyers as relevant to the qualities of commodities, to volumes than can be offered meeting these qualities, and to the contractual prices that will be put upon offers of these volumes.

Impact of legislation on transactions

If the law of a nation, or of a group of nations such as the EEC working under harmonised law binding upon its members, lays down specific and quantified features of feed materials which *must* be complied with, it follows that buyers in those countries have to insist on the inclusion of these features in their specifications; because compliance is matter of the law, these will be written into commercial contracts.

Since these contracts have to be accepted by sellers, who thus undertake to meet these features in materials supplied, it is clear that methods of verification must first exist which can be applied under the practical operational conditions of trading, both to the selection by sellers of material capable of meeting contracts, and to the checking of contracted material on receipt by the buyer. Unless such methods exist, such laws, however clearly stated, cannot be effective because they cannot be brought to bear at operational levels.

Remembering always that feed commodities are, in terms of resource conversion objectives, sold and bought as sources of convertible nutrients, it is nevertheless already true that buyers and sellers take into account various non-nutrient features of materials. For many years, both have been aware of the restraints on effective economic use *as sources of nutrients,* that may be imposed by, for instance, contaminations of natural materials with potentially pathogenic organisms such as *Salmonella;* presence, absence, or levels present of mycotoxins such as aflatoxin, due to moulds which establish themselves on the commodity concerned; or, in mineral materials, presence and level of presence of fluorine as a contaminant in phosphate sources.

When, in the light of experience in use, it is shown to be necessary to control matters like these in the diets of livestock, new restraints on levels tolerable in diets lead to new and additional features in commodities taken into account by the buyer. They have little or nothing to do with the purpose for which the commodity is basically purchased, but because of their *association* with the commodity and its nutrient content, they form a new, and added constraint on its potential permissible use in prac- tical feeding. Whether imposed on technical grounds because of knowledge of the uses proposed, or whether imposed by legislation, buyers add further features to their specifications, and sellers have to consider whether they can be met, and if so, in what volumes they can meet the new more stringent specifications, and at what prices.

At the same time, and it is most important to emphasise this, the least-cost diet formulation technique is itself modified to take note of the newly imposed restraints. The task now set to the computer becomes a multiple one in that from the simple concept set about *nutrient requirements* in diet standards met at lowest feasible cost, we now begin to demand formulae which:

(a) fulfil this optimum solution, whilst

(b) keeping the diet formula such that additional, *non- nutrient* restraints on, say, total fluorine or total aflatoxin levels are not exceeded.

Now consider this position from the seller's point of view. He is asked to undertake new and added commitments as to qualities, to guarantee which means further quality controls to guide his selection of contractable material. This in turn may show him he can only offer a reduced volume of material meeting the widened, more stringent, specification. He will probably react in one of two ways. He may seek to continue to sell to such markets, in which case in the price he asks he will seek to reflect both the added costs of quality controlling, and any lessening of the 'selected' volume he can offer meeting the new more exacting specification. Alternatively, he may decide to abandon the attempt to sell to this market, and turn to other markets whose specifications are less stringent, and who may thus be capable of taking his available volume at acceptable prices.

Consequently, the pattern emerges that the addition of non-nutrient restraints, especially by legislative means necessitating inclusions in contractual specifications, is likely to lead to diminished volume of offers to those countries operating them, and to increased prices asked for those volumes which are offered.

Basis for non-nutrient restraints

Whereas, looking back over the years, non-nutrient restraints such as those on hygienic features (salmonella), on aflatoxin, or on fluorine were first introduced into buyer and seller standards because of their potential hazard to consuming livestock, the current position on such restraints has taken on a new, and very important aspect. Governments, medical authorities, scientists and commercial feed and food processors, and the consuming public themselves, are increasingly pre-occupied with such features in relation to real or potential hazards to the consuming public.

Pesticides, ions of heavy metals, or aflatoxins or other mycotoxins may be associated with feed materials, and hence with formulated mixed feeds. If on feeding, residues of these are present in milk, meat or eggs in amounts which can be dangerous to those eating these in normal quantities, then public safety requires control on such features. The need for and means of establishing such controls is under active study, and as maximum permissible levels are identified, such controls will add further features to specifications and feed standards understood between buyers and sellers.

Impact of more extensive, more stringent feed standards

We have seen that each new feature added to specifications, and the stringency of control specified, tends towards reduction of volumes on offer, to increased price for these reduced volumes, and to a reduction of buying and formulating choice between available ingredients. These ingredients are needed and purchased as resources for the conversion of their nutrient contents.

This brings into play two inevitable results. Though the basic purpose, conversion of nutrients, is not affected, the fact that supply and price are modified causes the direct impact to fall on the economics of livestock farming. Cost of feeding goes up, supply of feed may become restricted, whilst efficiency of food conversion is still maintained. The total impact of course is the aggregate of impacts made by each control considered necessary and imposed, but all fall on the livestock producer.

Rising feed cost will lead to demand by the farmer to be paid more for his products. Inevitably, therefore, measures to underwrite the safety of the public as consumers of animals products have got to be paid for in higher prices of eggs, meat and milk. Since the cost of food is itself a major public issue, authorities are thus faced with the probability that the consumer whose safety they seek to guarantee will, when the impact of safety-measures becomes evident, protest as strongly about food prices, and press for adjustment of purchasing incomes to compensate for increases.

Public safety and health considerations

No one involved in the food chain ending in the human consumer would contest that, where truly necessity exists to control matters affecting human health and safety, such controls should be devised. Where these concern livestock products, they concern all those involved in feeds, and in the nutrient commodities which are the raw materials of the livestock industry on whom their impact will fall.

Because that impact will be determined by the scope of controls (the number of such features) and by the stringency of the maximum levels of permitted presence laid down for each, it is appropriate to consider questions relevant to both of these.

Decision-making on essential non-nutrient controls

Clearly, those by whose decisions specific figures in the schedules of control laws finally are adopted, have to draw upon areas of expertise which are extremely diverse.

That controls of this nature will come is not questioned. Because the impacts on the feed industry and on the commodities involved in feeding are clear, and may materially affect food prices, it is desirable and sensible for these to be confined to those *essential to* consumer safety, and within these, *no more stringent than is essential* to fulfil their purpose.

Fact finding: what controls are essential, and at what levels?

These questions are suggested as critical:
(1) From the aspect of human health and safety, is attempt made to consider the priorities between known hazards based on clinical evidence of ill-effects, and potential hazards postulated by toxicological and biochemical knowledge only?
(2) Is attempt made to assess priorities according to the clinical significance of hazards? For instance, between aflatoxins as demonstrated carcinogens, and other substances whose significance is of minor, or less well-established clinical importance?

(3) Are toxicological data based on 'acceptable daily intake' figures, based on realistic daily intakes of individual articles of human diet which will be affected?

(4) Are figures proposed by control authorities based on these considerations, or are they first subjected to modifications by factors representing additional margins of safety? If so, what is the order of these margins and is there certainty that they are essential?

(5) As to features, and levels of these, on which is it proposed to make controls compulsory, is it first ascertained that methods of sampling and of detection and estimation exist, to enable controls to be applied under practical operating conditions at acceptable costs?

(6) Lastly, before adopting such controls as law, are the impacts on supply and pricing of feeds and of foods first fully evaluated, and the decision to enforce them taken on acceptance of these impacts against the benefits of underwriting the safety of the human consumer?

The demand of man for protection against hostile elements of the environment he has himself created is likely, in the context of livestock farming and the feeds involved in it, to have important impacts on supplies and costs of food commodities. Such controls cannot be approached unrelated to these impacts.

If decisions on controls are taken without first carefully balancing their benefits to the consumer against their impacts on the consumer, that is if in extent and stringency such controls are in excess of what is absolutely necessary for protection of public health and safety, it is perfectly possible that they may if enforced lead to actual shortages of foods. Under various degrees of privation, it is a matter of conjecture whether the consumer-population would derive much comfort from the knowledge that this arose because their diet had been not merely rendered safe, but 'safer-than-safe'.

Such decisions on non-nutrient standards are therefore of the greatest importance to all who are, from whatever basis of activity, concerned in the provisioning and the production of livestock.

Discussion

Mr de Speville: Mauritius is a small over-populated island, and most of the land available for agriculture is under sugar cane, leaving very little for crops such as maize and soya beans. We would very much like to use sugar as a source of energy in animal feeds especially those for pigs and poultry. I would be grateful if Dr Carpenter could inform me as to the best source of protein to compensate for the lack of protein in sugar? Also has sugar ever been used on an economic basis in pig and poultry rations and would mixed feeds containing sugar have poor physical properties in humid climates eg would they stick in automatic feeders.

Dr Carpenter: I cannot answer your questions myself. However much work has been carried out on the use of sugar and sugar by-products in poultry rations in Hawaii, which also has a large sugar industry. I do not think this work has been published, but I am sure that if contact were established with the Hawaian Agricultural Experiment Station some useful information would be forthcoming.

Professor Fuller: I would like to refer to the comment by Dr Carpenter to the effect that rancidity of fats is not a direct problem. I agree that peroxide value or rancidity is not a good measure of rancidity by itself. Fats which have "gone rancid" may not have serious adverse effects if the diet is properly fortified with fat soluble vitamins and an antioxidant. It is in the early stages of oxidation of the fatty acids before the peroxide value has risen greatly (during free radical formation) that it is so destructive of sensitive nutrients in the food. We feel that the "active oxygen method" test gives a more reliable indication of the potential danger of unstabilised fats. In any event fat should be stabilised before addition to the feed. Unstable fats oxidise quite readily upon being mixed with feeds because of increased aeration, contact with trace elements etc.

Dr Carpenter: I agree that fat stability is a very complicated subject, and it cannot be said that a tallow which has started to oxidise is bad and one which has not yet oxidised at the time of mixing is good. It may be the one that seems good which is going to cause the most trouble, because it will go through the oxidation period in the diet. I agree that it is important to ensure that vitamin A is in stabilised form and that enough vitamin E is present. As Professor Fuller mentioned it is possible to precipitate vitamin E deficiency and this is thought not necessarily to be due to oxidation in the diet but just to the increased levels of highly unsaturated fatty acid in the tissue.

Professor Abou-Raya: I refer to the paper presented by Dr Carpenter. We have found the Weende proximate analysis (crude fibre and nitrogen-free extract) to be just as useful for the evaluation of poultry feeds as the more detailed breakdown of the carbohydrate fraction. However, detailed analysis is very important when the physiology of the digestion of a specific component is being investigated. We have found pentosans in cotton seed cake, rice bran and wheat bran to be appreciably digested by poultry.

Dr Carpenter: I have no comments to make on what was said by Professor Abou-Raya except to say that there has been very little work on the digestion of pentosans and that the work he mentioned is important and I would like to see it published.

International feed nomenclature

L. E. Harris
Utah State University, Logan, USA

H. Haendler
University of Hohenheim, Stuttgart, Germany

L. R. McDowell
University of Florida, Gainesville, USA

Summary

Because of the confusion associated with the naming of feedstuffs throughout the United States and Canada, the United States National Academy of Science Committee on Animal Nutrition recognised the need for a systematic feed nomenclature. This would describe feed accurately, be adaptable for coding of feed names and data on electronic computers, be adaptable for retrieving data for specific tabulations, and be useful internationally.

To meet these objectives, a new system was devised which makes it possible to know the contents and other characteristics of a feed from its name. This system is known as the 'International Feed Nomenclature.'

A complete 'International Feed Name' consists of nine component terms, and to the extent that information is available, it gives a quantitative description of the feed.

Feeds have also been categorised into eight different classes. This is done according to the origin of the feed product.

An 'International Reference Number' is assigned to each feed. The first digit of this number is the class of feed and the remaining digits are assigned consecutively. The reference number is used to identify the feed in the computer for calculating diets. It is also used for summarisation of the data and for printing feed composition tables, which may be printed directly from the computer.

Through this project, an international network of feed information centres has been developed. At the present time, cooperating centres are located in Australia, Canada, Germany, the United States, and FAO (Rome). The International Feedstuffs Institute at Utah State University correlates the technical phases of the work while FAO handles the non-technical phase.

Résumé

La nomenclature internationale des pâtures

A cause de la confusion associée avec les dénominations des pâtures d'un bout à l'autre des Etats-Unis et du Canada, le Comité de l'Académie Nationale des Sciences des Etats-Unis pour la nutrition des animaux, a reconnu le besoin d'une nomenclature systématique des pâtures. Celle-ci décrira avec précision, sera adaptable pour coder les noms des pâtures et les données dans des ordinateurs, sera adaptable pour recouvrer ces données pour des classifications spécifiques et sera utile internationalement.

Pour satisfaire ces objectifs, on a imaginé un nouveau système qui rend possible de savoir en partant de son nom, les contenus et les autres caractéristiques d'une pâture. Ce système est connu comme la 'Nomenclature Internationale des Pâtures'.

Une Dénomination Internationale complète des Pâtures consiste en neuf termes composants et jusqu'à la limite ou l'information est disponible, elle donne une description quantitative de la pâture.

Les pâtures ont été aussi catégorisées dans huit classes différentes. Ceci est fait selon l'origine du produit alimentaire.

Un 'Numéro International de Référence' est assigné à chaque pâture. Le premier chiffre de ce numéro représente la catégorie de la pâture et les chiffres restants sont assignés consécutivement. Le numéro de référence est utilisé pour identifier la pâture dans l'ordinateur pour calculer les diètes. Il est aussi utilisé pour résumer les données et pour imprimer les tables de composition des pâtures, qui peuvent être imprimées directement de l'ordinateur.

Avec ce projet, on a développé un réseau international de centres d'information des pâtures. A présent des centres coopérants se trouvent en Australia, au Canada, en Allemagne, aux Etats-Unis et à FAO (Rome). L'Institut International des Pâtures de l'Université d'Etat de Utah met en corrélation les phases techniques du travail tandis que la FAO exécute les phases non-techniques.

Resumen

Nomenclatura internacional de piensos

A causa de la confusión asociada con la denominación de piensos a través de los Estados Unidos y Canadá, el Comité de la Academia Nacional de Ciencias de los Estados Unidos sobre Nutrición Animal reconoció la necesidad de una nomenclatura sistemática de piensos. Esta describiría el pienso exactamente, sería adaptable para la codificación de los nombres de los piensos y datos en computadores electrónicos, sería adaptable para la recuperación de datos por tabulaciones específicas, y sería útil internacionalmente.

Para cumplir estos objetivos, se ideó un nuevo sistema que hace posible conocer los contenidos y otras características de un pienso por el nombre que tiene. Este sistema es conocido como la "Nomenclatura Internacional de Piensos".

Un "Nombre de Pienso Internacional" completo consta de nueve términos componentes, y hasta el punto en que se puede disponer de información, da una descripción cuantitativa del pienso.

También se han clasificado los piensos en ocho categorías diferentes. Esto se hace de acuerdo con el origen del tipo de pienso.

Se asigna un "Número de Referencia Internacional" a cada pienso. El primer dígito de este número es la clase de pienso y los dígitos restantes se asignan consecutivamente. El número de referencia se usa para identificar el pienso en el computador para calcular las dietas. Se usa también para resumir los datos y para imprimir tablas de composición de los piensos, que pueden ser impresas directamente por el computador.

Mediante este proyecto, se ha desarrollado una red internacional de centros de información sobre los piensos. Actualmente, existen centros que cooperan en Australia, Canadá, Alemania, los Estados Unidos y la FAO (Roma). El Instituto Internacional de Piensos en la Universidad del Estado de Utah correlaciona las fases técnicas del trabajo mientras que la FAO se ocupa de la fase no técnica.

Introduction

Data on feed analyses and animal trials with feeds started more than one hundred years ago. Since the beginning of our century, there has been a tradition of compiling feed composition tables as Kellner (1905) did in Germany and Henry (1898) did in the USA.

Research that is effectively determining the nutrient requirements of animals on the one hand and the concomitantly increasing need for more accurate information about composition and nutritive values of feeds, requires specialised methods of processing large amounts of data.

It seems to be more than mere accident that in the home countries of the afore-mentioned pioneers in compiling of feed composition tables, Germany and USA, two separate centres of feed data documentation were built up. The German documentation began in 1949 (Haendler, 1963; Haendler and Jager, 1971) and the documentation in Utah (USA) began in 1952 (Harris, et al. 1968).

International network of feed information centres (INFIC)

Although there was some contact between Utah and Hohenheim for several years, it was not possible to combine or adapt the two systems to each other. Personnel at the Utah centre contacted the FAO concerning the need for world co-operation. The FAO in turn sent a consultant to visit various centres compiling data on feed composition (Alderman, 1971).

This resulted in two informal meetings which were held at FAO headquarters in Rome in 1971 and 1972 (Harris and Christiansen, 1972; Haendler and Harris, 1973).

Representatives attended these meetings from the Australian Feeds Information Centre, Canberra, Australia; the Canada Feed Information Centre, Ottawa, Canada; the Dokumentationsstelle der Universität Hohenheim, Stuttgart-Hohenheim, Federal Republic of Germany; the Florida Feeds Information Centre, US-AID for Latin America, Gainesville, Florida, USA; the FAO Feed Information Centre, Rome, Italy; and the Agency for International Development (AID), Washington, DC, USA. It was decided by this group that an 'International Network of Feed Information Centres' (INFIC) should be established (Figure 1). It should be noted that others may join.

In 1973, another informal meeting of the INFIC group was held in Hohenheim, Germany, to finalise the organisation. Some of the most important decisions reached during these meetings were:
(1) To adopt the 'International Nomenclature' for naming feeds (Harris et al. 1968).
(2) To combine the best parts of the Utah and Hohenheim classifications and coding systems into an International system (this should be completed in 1974).
(3) To develop 'International Definitions' of nutrition terms.
(4) To adopt an 'International Record Keeping System' for recording feed composition data (Harris, 1970) which needs to be enlarged to record other substances including those with negative effects such as pesticides, mycotoxins and others.

Figure 1
Organisation of the international network of feed information centres

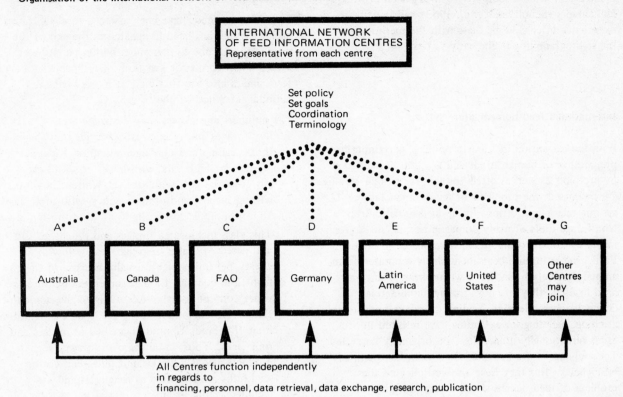

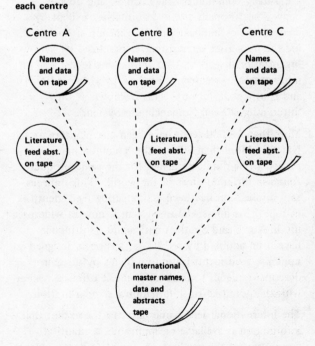

(5) To adopt summarising procedures so data from all centres can be dovetailed by a computer.
(6) To exchange data.
(7) To work co-operatively with the USA Agency for International Development and similar European institutions.
(8) To work with the International Union of Nutritional Scientists (UINS) to standardise terminology and methods for analysing feed samples.

It was suggested that different centres be responsible for various regions of the world. With the help of the FAO the two data processing centres, Hohenheim and Utah, would extend their activities to the developing countries. Hohenheim would be responsible for the Near East including the African countries bordering the Mediterranean. The Federal Agency for Economic Co-operation would support laboratory work in these countries, and data would be sent to Hohenheim. Utah would receive data through the FAO from the other African countries, while data from Australia, New Zealand, and the Southwest Pacific would be summarised at Canberra. Canada would collect data and send it to Utah.

Utah would collect data from the USA and Latin America. (The Animal Science Department at the University of Florida has been collecting data from Latin America but this responsibility will be transferred to Utah on July 1, 1974). The Hohenheim Centre would furthermore extract data from literature as it did at present and would intensify its activity in European countries. The International Feedingstuffs Institute at Logan, Utah would dovetail data from all centres, make a master tape, and send copies of the tape to all centres (Figure 2). The system of combining data is illustrated in the '*Atlas of Nutritional Data*

Figure 2
International feed names and data and abstracts of feed literature will be dove-tailed at the Utah International Feedstuffs Institute into a master tape which will be sent periodically to each centre

on *United States and Canadian Feeds*', National Academy of Sciences (1971) and the '*Latin American Tables of Feed Composition*' prepared by McDowell *et al.* (1974a; 1974b).

Eventually, the 'International Data Bank' will also contain summary and literature feed abstracts (Figure 2), which are presently already available in certain centres. Thus, Hohenheim has been preparing such abstracts regularly since 12 years. The ultimate use of such data will be in conjunction with animal nutrient requirements to calculate diets and feed mixtures manually or with a computer. Among such centres Hohenheim (Scholtyssek *et al.* 1963) had the first experience in calculating diets with the method of linear programming in the early sixties.

International feed nomenclature system

With the expansion of food technology pertaining to preparation of human foods and as more human food by-products as well as offals suitable for animal feeding are processed, the problems of feed nomenclature assume major proportions for applied nutritionists. Non-forage diets of many domestic animals no longer primarily consist of unaltered farm-produced grains. Feed production has been increasingly separated from animal production; and even when entire grains are used in diets, they often go through commercial channels where cleaning, grading, blending, and possibly grinding, pelleting, or extruding have resulted in products nutritionally unlike the crop originally harvested. These changes have increased diets values and animal efficiencies. But they have also complicated the problem of feed names. Feed control officials of both the USA and Canada therefore approved names for the feeds regulated by feed legislation in their countries and described the processes involved and permitted in their manufacture. Such names, however, are usually common or trade names, and do not convey nutritionally useful information, except by coincidence or implication. As a means of correcting the numerous inconsistencies practiced by various organizational workers who are naming feeds, a new international system was proposed by Harris, (1963) and Harris *et al.* (1968). This is known as the 'International Feed Nomenclature System'.

More than 18,000 feeds have been given 'International Names'. More than 20% of the 'common names' proved to be duplicate names for the same product found in different areas of the world. This unnecessary duplication had been complicating feed identification. This nomenclature system is now in widespread use in North and South America and its principles have been adopted by INFIC. It has been designed to minimise feed identification problems by assigning descriptive names to feedstuffs, and it offers a system with the potential for international standardisation.

The international feed names give, to the extent that information is available or applicable, a quantitative description of the product. A complete name consists of nine component terms:

(1) Scientific name (genus and species)
(2) Origin (or parent material)
(3) Common species, variety, or kind
(4) Part actually eaten
(5) Process(es) and treatment(s) to which the original material or part eaten has been subjected before being fed to the animal.
(6) Stage of maturity (applicable primarily to forages)
(7) Cutting or crop (applicable primarily to forages)
(8) Grade, quality designations, and guarantees
(9) Classification (according to nutritional characteristics).

The different feeds have been assigned to eight classes each of which has been designated at the end of the international name by a number within parentheses. This number forms the last term of the name of a feed, and is also the first digit of its six-digit international reference number.

The numbers and classes they designate are:

1. Dry forages and roughages: subdivided into hay (both legume and non-legume), straw, fodder (aerial part with ears, with husks or with heads), stover (aerial part without ears, without husks or without heads) and other products with more than 18% fibre, hulls and shells.

   This class includes all forages and roughages, cut and cured. Forages or roughages are low in net energy per unit weight, usually because of their high fibre content, though sometimes because the water content is high. According to the nomenclature, products that in the dry state contain more than 18% crude fibre are classified as forages and roughages. Thus, in addition to forages, such products as oat hulls, peanut hulls, and cotton seed hulls are classified as roughages and are included in this group.

2. Pasture, range plants and forages fed green: included in this group are all forage feeds not cut (pasture), all feeds cut and fed green, and feeds cured on the stem, such as dormant range plants. The term 'fresh' is used as a process term for most of these feeds although they may be dry and matured when consumed.

3. Silages: subdivided into maize, legume and grass.

4. Energy feeds: subdivided into cereal grains low in cellulose, cereal grains high in cellulose, cereal milling by-products low in cellulose, cereal milling by-products high in cellulose, fruits, nuts and roots. Products with less than 20% protein and less than 18% crude fibre are classified as energy feeds.

5. Protein supplements: subdivided into those of animal, avian, marine, milk and plant origin. Products which contain 20% or more protein.

6. Mineral supplements.

7. Vitamin supplements

8. Additives: antibiotics, colouring materials, flavouring, hormones, medicants.

The guidelines for classification are approximate and there is some overlapping. Each class used in this system has certain properties that are considered in balancing diets.

Description of components of international feed names

Scientific names

The orderly classification of plants and animals according to their presumed natural relationships forms a basic biological discipline. In the nomenclature, the scientific name(s) consists of the genus and species.

Origin

The origin indicates the parent material from which the feed originates (ie, plant, animal or mineral). For most plants, this source is designated by the common name such as 'timothy' or 'wheat'. Whenever such a name is not the only one used, cross references are used so it will be recognised internationally. An example of this is 'alfalfa'. In some countries this plant is known as 'lucerne'. In view of this, 'alfalfa' is used as the international origin, but there is a cross reference: 'lucerne see alfalfa'.

For non-plant feeds, the name of the animal or bird is used as the origin term, such as 'cattle', 'crab', 'chicken', 'horse', 'sheep', 'turkey', or 'whale'. 'Fish' is used as the origin term of all fishes and the species, or variety, follows, ie, 'fish, cod'; or 'fish, salmon'. When the specific origin is unknown, the words 'animal', 'fish', or 'poultry' are employed as origins.

Proposing any of the common names of a feedstuff as an international standard does not imply that its other common names are not as good, but merely means that one name must be selected in order to minimise confusion.

Common species, variety, or kind

If the nutritive value of a feed has been influenced by the species or kind of product comprising it, this description is included. Examples would be 'cattle, holstein', 'fish, salmon', 'maize, dent yellow', or 'copper sulfate'. When species or varietal information is lacking, the species term is omitted.

Part actually eaten

The third component of a feed name is the actual part of the parent material consumed. To the layman, the edible parts of plants or animals are usually considered to be the obvious structures as leaves, stems, seeds, milk, or bone. But today food technologists are fractionating natural foods and reconstituting their parts into new 'processed' foods. For example, many modern cereals are blends of parts of numerous cereals with or without added synthetic or purified minerals, vitamins, amino acids, antioxidants, and flavours. This extensive processing of plant seeds and manufacturing of new products provides innumerable by-products that find use as feeds for animals, as do the by-products from the preparation of table meats and fish.

The parts of the feed names have been defined (Harris *et al.* 1968; National Academy of Sciences, 1971). Examples of various parts and their definitions are as follows:

Bran: the pericarp of cereal grains. There is also a feed known as pineapple bran, but as the definition of 'bran' does not fit this feed the name was changed to pineapple cannery residue. However, pineapple bran is carried as another name.

Cob: the fibrous inner portion of the ear of maize from which the kernals have been removed.

Germ: the embryo found in seeds and frequently separated from the starch endosperm during milling.

Gluten: the tough, viscid, nitrogenous substance that remains after the flour or wheat or other grain have been washed to remove the starch.

Husks: leaves enveloping an ear of maize; the outer coverings of kernels or seeds, especially when dry and membranous.

Process and treatments

A feed may be modified by a food technologist or a feed manufacturer by a variety of processes, some of which may greatly alter the feeding value of the products treated (National Academy of Sciences, 1973). While heat may damage some components, such as the amino acids and vitamins, it may make other nutrients more available, such as the carbohydrate in potatoes or beans.

Mink may suffer from thiamine deficiency and die when fed certain types of raw fish that contain the enzyme thiaminase. This enzyme destroys the thiamine present in a mixed ration. Since thiaminase is heat labile, cooking the causative fish before mixing the remaining diet ingredients is a satisfactory means of avoiding this problem.

Grinding may change the completeness of digestion of cellulose and protein, while grinding or pelleting of forages alters the proportions between the volatile fatty acids produced by rumen microflora of milking cows and hence affects the relation between the fat and non-fat solids in the milk produced. For this reason, pelleted forages are not recommended for dairy cows.

It is important, then, that persons who feed animals be aware of the processes to which a given feed has been subjected. Examples of representative processes are listed and defined below:
(1) Condensed: reduced to denser form by removal of moisture.
(2) Cracked: particle size reduced by a combined breaking and crushing action.
(3) Dehydrated: having been freed of moisture content by thermal means.
(4) Dry-rendered: residues of animal tissues cooked in open steam-jacketed vessels until the water has evaporated. Fat is removed by draining and pressing the solid residue.

Stages of maturity

The stage of maturity may be unimportant or inapplicable to many feeds, but is probably the most important factor in determining the nutritive value of a forage. Alfalfa is a good example of this. If it matures beyond the one-tenth bloom stage, its lignin content increases markedly and this lowers its digestibility and the voluntary intake of the harvested hay. Table 1 lists the terms used in the international feed names to denote stages of maturity.

TABLE 1. *Stage of maturity terms used in international feed names*

Blooming plants	Non-blooming plants[a]
Germinated	1 to 14 days' growth
Early vegetative	15 to 28 days' growth
Late vegetative	29 to 42 days' growth
Early bloom	etc.
Midbloom	
Full bloom	
Late bloom	
Milk stage	
Dough stage	
Mature	
Post ripe	
Stem cured	
Regrowth early vegetative	
Regrowth late vegetative	

[a]These classes are for species that remain vegetative for long periods and apply primarily to plants that grow in the tropics. When the name of a feed is developed, the age class forms part of the name (eg, Pangolagrass, aerial part fresh 150 to 28 days' growth). For plants growing longer than 14 days, the interval is increased by increments of 14 days.

Cutting or crop

The cutting refers to the sequence during a specific growing season, designated as cut 1, cut 2, cut 3, etc. This pertains largely to the hays.

Grade or quality designation

This information is useful in describing the nutritional value of the feed as well as the limits set by law for control purposes. The maximum and minimum quantities of various constituents in the international feed name will have to be established by the countries or

TABLE 2. *Variability in composition in relation to description of a feed*

Name of feed and nutrient	Protein dry basis %	Coefficient of variation %
Alfalfa, hay, sun cured		
Protein	18.2	16
Calcium	1.32	42
Alfalfa, hay, sun cured, cut 2		
Protein	18.4	11
Calcium	1.25	36
Alfalfa, hay, sun cured, early bloom, cut 2		
Protein	17.6	4
Calcium	1.27	20

international agencies dealing with the particular specifications permissible. Standards for feedstuffs processed and fed within the territorial boundaries of a country should be established by officials of that country. Where feedstuffs are being exported to foreign markets, quality standards may be established by an international agency or by specifications established by the importing nation.

Using the 'component' system, it is possible to describe and essentially reproduce a feed ingredient. The exactness of a given name depends upon the accuracy and the amount of information available. A feed described in detail is a much more consistent entity as far as the coefficient of variability is concerned than is one for which less information is presented (Table 2).

A high consistency in description is only possible if the component terms are unambiguous and if the vocabulary is free from synonyms. This is why at Hohenheim a team supported by a grant from the

TABLE 3. *International feed names broken down into components*

Component			Name components	
Description number		Example 1	Example 2	Example 3
International names				
Genus	025	*Medicago*	*Bos*	*Linum*
Species	030	*sativa*	*spp*	*usitatissimum*
Origin	155	Alfalfa	Cattle	Flax
Part	215	aerial part	whey	seeds
Process	245	dehydrated ground pelleted	condensed	mechanical extracted ground
Maturity	275	early bloom	—	—
Cutting	300	cut 2	—	—
Grade	325	—	min. solids declared	max 10% fibre
Class		forage 1	energy feed 4	protein supplement 5
Official and other names for the feeds above				
AAFCO	425	—	Condensed whey	Linseed meal, mechanical extracted
CFA	430	—	—	Linseed meal
Other	435	Lucerne dehydrated, ground, pelleted, early bloom cut 2	whey, condensed	Linseed oil meal, expeller extracted
Other	440	—	whey, evaporated	Linseed oil meal, hydraulic extracted
Other	445	—	whey, semisolid	Linseed oil meal, old process
International feed reference no.		1−07−733	4−01−180	5−02−045

Institute für Dokumentationswesen, Frankfurt, is working to revise and complete (and translate into German and French) a vocabulary for describing feedstuffs. This work is benefiting from long experience with the use of the Hohenheim classification system and the intense study of documentation languages. The results of this work were submitted to a committee established at the INFIC meeting in Hohenheim in September 1973, in order to arrive at a controlled vocabulary for international use.

The international system of naming may be illustrated by three examples (Table 3). In normal linear form, these international names are written as follows:

1. Alfalfa, aerial part, dehydrated ground pelleted, early bloom, cut 2, 1.
   Lucerne
   International feed reference number (IFR)
   1—01—733

2. Cattle, whey, condensed, minimum solids declared, 4.
   condensed whey (AAFCO)
   whey, condensed
   whey, evaporated
   whey, semisolid
   IFR 4—01—180

3. Flaxseeds, mechanical extracted ground, maximum 10% fibre, 5.
   linseed meal, mechanical extracted (AAFCO)
   linseed meal (CFA)
   linseed oil meal, expeller extracted
   linseed oil meal, old process
   IFR 5—02—045

The six-digit IFR number listed after each name may be used as the 'numerical name' of a feed when using an electronic computer to calculate diets to obtain maximum profit. Note the first digit of the reference number is its class designation.

In the linear form, each component is in a certain order without commas so the component terms may be recognized. This order follows a logical sequence. The examples illustrate the principle that the full international name of a feed consists of all the nine components minus only those inapplicable to the feed. At the present time, over 18,000 international feed names are on magnetic tape in accordance with the component system.

American Association of Feed Control Officials (1974) names are designated by 'AAFCO' and Canada Feed Act (1967) names by 'CFA'. Their official names cannot usually be used as international names because individual names are either incomplete or do not begin with the 'origin' or parent material. For example, if the scientific name *Bos spp.* is to be included in a table of feed composition data, then the term 'whey' must be preceded by 'cattle' since whey in this instance comes from cattle milk. The AAFCO name begins with 'condensed' and does not show origin.

Official names from other countries will be added as they become known. For Europe the official names will be determined and defined by the competent authority of the European Communities, a draft of which has been in existance for several years.

Figure 3

Composition of Latin American feeds

NAME OR NUTRIENT	UNIT	AS FED	DRY	CV	NO
AVENA SATIVA, OATS					
AVENA SATIVA, GRAIN, (4)					
REF NO 4-03-309					
DRY MATTER	%	88.9	100.0	7	23
ORGANIC MATTER	%	85.0	95.6		
ASH	%	3.9	4.4	50	35
SHEEP	DIG COEF %	27.	27.		
CRUDE FIBER	%	12.6	14.2	31	38
HORSES	DIG COEF %	44.	44.		1
SHEEP	DIG COEF %	32.	32.		1
SWINE	DIG COEF %	14.	14.		1
ETHER EXTRACT	%	4.7	5.3	28	39
HORSES	DIG COEF %	80.	80.		1
SHEEP	DIG COEF %	80.	80.		1
SWINE	DIG COEF %	79.	79.		2
NITROGEN FREE EXTRACT	%	57.8	65.0	8	27
HORSES	DIG COEF %	79.	79.		1
SHEEP	DIG COEF %	78.	78.		1
SWINE	DIG COEF %	76.	76.		2
PROTEIN	%	9.9	11.2	23	51
HORSES	DIG COEF %	79.	79.		1
SHEEP	DIG COEF %	78.	78.		1
SWINE	DIG COEF %	81.	81.		2
CATTLE	DIG PROT % *	5.6	6.3		
GOATS	DIG PROT % *	6.6	7.5		
HORSES	DIG PROT %	7.8	8.8		
SHEEP	DIG PROT %	7.7	8.7		
SWINE	DIG PROT %	8.0	9.0		
ENERGY	GE KCAL/KG	4075.	4584.		1
CATTLE	DE MCAL/KG *	2.69	3.03		
SHEEP	DE MCAL/KG *	2.88	3.24		
SWINE	DE KCAL/KG *	2735.	3077.		
CATTLE	ME MCAL/KG *	2.21	2.48		
CHICKENS	ME-N KCAL/KG	1262.	1420.		2
SHEEP	ME MCAL/KG *	2.36	2.65		
SWINE	ME KCAL/KG *	2564.	2884.		
CATTLE	NE-M MCAL/KG	1.34	1.51		
CATTLE	NE-GAIN MCAL/KG *	0.83	0.94		
CATTLE	NE-LACTATION MCAL/KG *	(1.47)	(1.65)		
CHICKENS PRODUCTIVE	KCAL/KG	1651.	1858.		1
CATTLE	TDN %	61.0	68.6		
HORSES	TDN %	67.5	75.9		1
SHEEP	TDN %	65.3	73.4		1
SWINE	TDN %	62.0	69.8		2
CALCIUM	%	(0.20)	(0.23)	98	14
CHLORINE	%	(0.10)	(0.11)		1
COBALT	MG/KG	(0.026)	(0.029)		1
IODINE	MG/KG	0.064	0.072		1
IRON	%	0.005	0.005	15	10
PHOSPHORUS	%	0.27	0.30	21	14
POTASSIUM	%	(0.42)	(0.47)		2
SODIUM	%	(0.06)	(0.07)		1
SULPHUR	%	0.20	0.23		1
ASCORBIC ACID	MG/KG	(3.2)	(3.6)	25	10
BIOTIN	MG/KG	(0.26)	(0.32)		1
CHOLINE	MG/KG	(945.)	(1063.)		2
FOLIC ACID	MG/KG	0.22	0.24		1
NIACIN	MG/KG	(16.7)	(18.7)	28	10
PANTOTHENIC ACID	MG/KG	13.1	14.8	3	5
RIBOFLAVIN	MG/KG	1.3	1.5	11	10
THIAMINE	MG/KG	(2.5)	(2.8)	10	10
ARGININE	%	(0.64)	(0.72)	30	4
CYSTINE	%	(0.17)	(0.19)		3
GLYCINE	%	(0.49)	(0.55)		1
HISTIDINE	%	(0.16)	(0.18)		3
ISOLEUCINE	%	(0.58)	(0.65)		2
LEUCINE	%	(0.86)	(0.96)		2
LYSINE	%	(0.37)	(0.42)	36	4
METHIONINE	%	(0.17)	(0.19)		3
PHENYLALANINE	%	(0.60)	(0.68)		2
THREONINE	%	(0.39)	(0.44)		2
TRYPTOPHAN	%	(0.14)	(0.16)	21	5
TYROSINE	%	(0.73)	(0.82)		2
VALINE	%	0.65	0.73		2

Atlas computer format is useful for recording a large amount of data about a feed (about 450 entities are being recorded). Data are from Latin America. A feed there is known by its scientific name—hence this is listed as 'origin'. The common name is shown after the scientific name. 'CV' is the coefficient of variation and 'NO' is the number of analyses. A '*' indicates the values were estimated using a formula: parentheses indicate the values were taken from US data, which makes a more complete table. (Taken from McDowell et al., 1974a).

International record system

An 'International Record System' to provide a unified method for recording laboratory analytical data on an 'international source form' has been developed (Harris, 1970). This allows for direct transfer of data to an electronic computer system and the addition of data to a master file on cards or tape. Data are summarised by and master copies for tables of feed composition can be made directly from the computer. Tables are photographed and printed by the off-set method (Figures 3 and 4) which makes it possible to publish a feed composition table in about four weeks (National Academy of Sciences, 1970; National Academy of Sciences, 1971; McDowell et al., 1974b).

Tablas de composición de America Latina

Linea No.	Especie generica nombre comun / Nombre Internacional del alimento / Otro nombre	Alimento Internacional Referencia No.	Materia seca %	Ceniza %	Fibra cruda %	Extracto Etereo %	Proteina %	ED Bovinos Mcal/kg	ED Ovinos Mcal/kg	ED Suinos kcal/kg	EM$_n$ Pollos kcal/kg
01	ZARZA DE CAMPECHE										
02	-SEE LEUCAENA LEUCOCEPHALA, PARTE										
03	AEREA, FRESCO, (2)										
04	ZEA MAYS, MAIZ										
05	-OLOTES, MOL, (1)	1-02-782	87.6	4.2	31.5	2.7	2.9	2.19*	1.95*	-	-
06			100.0	4.8	36.0	3.0	3.3	2.50*	2.23*	-	-
07											
08	-PARTE AEREA, CURADO AL SOL (C-S), (1)	1-02-775	82.8	6.4	23.3	1.4	7.9	2.01*	2.09	-	-
09			100.0	7.7	28.2	1.7	9.5	2.43*	2.52	-	-
10											
11	-PARTE AEREA, DESH, (1)	1-02-768	94.9	6.4	32.1	2.1	6.9	2.28*	2.21*	-	-
12			100.0	6.7	33.9	2.2	7.3	2.40*	2.33*	-	-
13											
14	-PARTE AEREA SIN MAZORCAS Y SIN TUZA,	1-13-325	88.2	9.7	32.0	1.7	4.8	1.89*	1.87*	-	-
15	PICADO DESH Y MOL, (1)		100.0	11.0	36.3	1.9	5.5	2.14*	2.13*	-	-
16											
17	-PARTE AEREA, FRESCO, (2)	2-02-806	28.2	2.4	7.8	0.8	2.0	0.79*	0.76*	-	-
18			100.0	8.7	27.8	2.8	7.3	2.81*	2.68*	-	-
19											
20	-PARTE AEREA, FRESCO, PROXIMO A LA	2-02-803	-	-	-	-	-	-	-	-	-
21	MADUREZ, (2)		100.0	8.1	29.2	1.6	8.8	2.61*	2.66*	-	-
22											
23	-PARTE AEREA, FRESCO, MITAD DE FLORACION, (2)	2-13-763	-	-	-	-	-	-	-	-	-
24			100.0	10.7	23.5	0.9	9.8	2.79*	2.75*	-	-
25											
26	-PARTE AEREA, FRESCO PICADO, (2)	2-10-357	22.3	1.8	6.6	0.4	1.6	0.57*	0.59*	-	-
27			100.0	7.9	29.6	2.0	7.0	2.57*	2.64*	-	-
28											
29	-TALLOS, FRESCO, (2)	2-02-814	18.2	0.9	6.2	0.3	0.6	0.41*	0.46*	-	-
30			100.0	5.1	33.9	1.7	3.3	2.23*	2.54*	-	-
31											
32	-PARTE AEREA, ENSILADO, (3)	3-02-822	27.5	2.2	7.6	0.9	2.3	0.80*	0.80*	-	-
33			100.0	8.1	27.8	3.3	8.5	2.90*	2.90*	-	-
34											
35	-PARTE AEREA, ENSILADO, ESTADO DE LECHE, (3)	3-02-818	23.4	1.6	7.2	0.6	1.5	0.64	0.63*	-	-
36			100.0	7.0	30.9	2.5	6.2	2.75	2.68*	-	-
37											
38	-PARTE AEREA, ENSILADO, PROXIMO A LA	3-02-819	25.5	1.3	7.4	0.8	2.6	0.69	0.74*	-	-
39	MADUREZ, (3)		100.0	5.2	28.9	3.2	10.0	2.68	2.91*	-	-
40											
41	-PARTE AEREA, ENSILADO EN SILO DE	3-02-824	-	-	-	-	-	-	-	-	-
42	TRINCHERA, (3)		100.0	6.4	30.5	6.3	7.3	2.52*	2.75*	-	-
43											
44	-GRANO, (4)	4-02-879	89.1	1.6	2.2	4.6	9.9	3.24*	3.62*	3344.*	-
45			100.0	1.8	2.5	5.2	11.1	3.63*	4.07*	3752.*	-
46											
47	-GRANO, FERTILIZADO, (4)	4-10-359	86.0	1.3	2.1	1.7	8.7	3.10*	3.30*	3293.*	-
48			100.0	1.6	2.5	2.0	10.1	3.60*	3.84*	3831.*	-
49											
50	-GRANO, MOL, (4)	4-10-422	88.8	2.2	2.5	4.0	8.9	3.08*	3.36*	3310.*	-
51			100.0	2.4	2.8	4.5	10.0	3.47*	3.79*	3727.*	-
52											
53	-MAZORCAS, MOL, (4)	4-02-849	86.2	1.9	7.9	3.4	7.3	2.83*	2.98*	3040.*	2810.*
54			100.0	2.2	9.2	3.9	8.5	3.28*	3.46*	3526.*	3259.*
55											
56	-SUB-PRODUCTO DE LA FABRICACION DE	4-02-887	89.1	5.7	5.0	9.8	16.4	3.71*	3.49*	3492.*	1952.*
57	SEMOLA, MN 5% GRASA, (4)		100.0	6.4	5.6	11.0	18.4	4.16*	3.91*	3918.*	2190.*
58											
59	-AFRECHO CON GERMEN, MOL, (5)	5-13-529	89.3	4.3	9.4	8.6	14.5	-	3.48*	-	-
60			100.0	4.8	10.5	9.6	16.3	-	3.90*	-	-
61											
62	-GLUTEN CON AFRECHO, MOLINERIA HUMEDA	5-02-903	90.4	5.7	9.3	2.7	24.6	3.28*	3.31*	3395.*	1688.*
63	Y DESH, (5)		100.0	6.3	10.3	2.9	27.3	3.63*	3.66*	3758.*	1868.*
64											
65	-GLUTEN, MOLINERIA HUMEDA Y DESH, (5)	5-02-900	87.6	6.5	5.3	5.9	40.8	3.13*	3.40*	-	2427.*
66			100.0	7.4	6.1	6.7	46.6	3.58*	3.88*	-	2772.*
67											
68	-GRANOS DE DESTILERIA CON SOLUBLES,	5-02-843	92.6	4.7	9.1	9.7	27.2	3.15*	3.02*	-	2479.*
69	DESH, MN 75% SOLIDOS ORIGINALES, (5)		100.0	5.1	9.8	10.5	29.4	3.40*	3.26*	-	2677.*
70											
71	-SOLUBLES DE DESTILERIA, DESH, (5)	5-02-844	92.2	7.7	5.0	8.0	27.2	3.27*	3.46*	3332.*	2928.*
72			100.0	8.3	5.4	8.7	29.5	3.55*	3.75*	3614.*	3176.*
73	ZEA MAYS, MAIZ, AMARILLO										
74	-GRANO, (4)	4-07-911	88.2	1.5	2.3	4.2	9.0	3.16*	3.37*	3307.*	-
75			100.0	1.7	2.6	4.8	10.2	3.58*	3.82*	3748.*	-
76	ZEA MAYS, MAIZ, BLANCO										
77	-GRANO, (4)	4-09-907	88.3	1.5	2.9	4.1	9.4	3.17*	3.35*	3314.*	-
78			100.0	1.7	3.2	4.7	10.7	3.59*	3.80*	3752.*	-
79	ZEA MAYS, MAIZ, OPACO 2										
80	-TALLOS, FRESCO, PROXIMO A LA MADUREZ, (2)	2-13-506	-	-	-	-	-	-	-	-	-
81			100.0	7.1	35.8	0.5	5.7	-	2.40*	-	-
82											
83	-GRANO, (4)	4-11-445	86.3	1.7	2.5	4.4	9.7	3.11*	3.28*	3235.*	-
84			100.0	2.0	2.9	5.1	11.2	3.61*	3.80*	3748.*	-
85	ZEA MAYS, INDENTATA, MAIZ, AMARILLO DENTADO										
86	-INDENTATA AMARILLO DENTADO, GRANO, (4)	4-02-935	87.0	1.5	2.9	3.5	8.8	3.08*	3.30*	3489.*	3329.*
87			100.0	1.7	3.3	4.0	10.1	3.54*	3.80*	4008.*	3825.*
88	ZEA MAYS, INDENTATA, MAIZ, BLANCO DENTADO										
89	-INDENTATA BLANCO DENTADO, GRANO, (4)	4-02-928	87.5	-	-	-	-	-	-	-	-
90			100.0	-	-	-	-	-	-	-	-
91											

The computer output in long format. This has been used in all the USA National Research Council (1970) nutrient requirement series. It is useful when only a small amount of data is to be reported on each feed. The 'origin' is the scientific name, and the remainder of the names are in Spanish. (Components are translated from English to a different language and fed into the computer, which will then print out the translated names). (McDowell et al., 1974b).

	En base a tal como ofrecido y en base seca																		
Linea no.	NDT Bovinos %	NDT Ovinos %	NDT Suinos %	Calcio %	Cobalto mg/kg	Cobre mg/kg	Manganesio mg/kg	Fosforo %	Potasio %	Zinc mg/kg	Colina mg/kg	Niacina mg/kg	Acido Pantotenico mg/kg	Pro Vitamina A mg/kg	Riboflavina mg/kg	Cistina %	Lisina %	Metionina %	Triptofano %
	49.6*	44.2	-	0.26	-	-	-	0.17	0.79	-	-	-	-	-	-	-	-	-	-
	56.6*	50.5	-	0.30	-	-	-	0.19	0.91	-	-	-	-	-	-	-	-	-	-
	45.6*	36.0	-	0.31	-	-	-	0.14	-	-	-	-	-	-	-	-	-	-	-
	55.1*	43.5	-	0.38	-	-	-	0.17	-	-	-	-	-	-	-	-	-	-	-
	51.7*	50.0*	-	1.86	-	-	-	0.14	-	-	-	-	-	-	-	-	-	-	-
	54.5*	52.8*	-	1.96	-	-	-	0.14	-	-	-	-	-	-	-	-	-	-	-
	42.8*	42.5*	-	0.41	-	-	-	0.14	-	-	-	-	-	-	-	-	-	-	-
	48.5*	48.2*	-	0.47	-	-	-	0.16	-	-	-	-	-	-	-	-	-	-	-
	18.0*	17.1*	-	0.11	-	-	-	0.01	-	-	-	-	-	-	-	-	-	-	-
	63.8*	60.8*	-	0.39	-	-	-	0.04	-	-	-	-	-	-	-	-	-	-	-
	-	-	-	-	-	-	-	-	-	-	-	-	-	-	-	-	-	-	-
	59.1*	60.2*	-	-	-	-	-	-	-	-	-	-	-	-	-	-	-	-	-
	63.2*	62.4*	-	-	-	-	-	-	-	-	-	-	-	-	-	-	-	-	-
	13.0*	13.4*	-	0.04	-	6.2	11.1	0.03	-	-	-	-	-	-	-	-	-	-	-
	58.4*	60.0*	-	0.19	-	27.9	49.8	0.13	-	-	-	-	-	-	-	-	-	-	-
	9.2*	10.5*	-	-	-	-	-	-	-	-	-	-	-	-	-	-	-	-	-
	50.6*	57.6*	-	-	-	-	-	-	-	-	-	-	-	-	-	-	-	-	-
	18.1	18.1	-	0.16	-	0.5	13.2	0.07	0.32	4.3	-	5.4	-	-	-	-	-	-	-
	65.8	65.8	-	0.57	-	1.7	48.2	0.24	1.18	15.8	-	19.5	-	-	-	-	-	-	-
	16.1	14.2*	-	0.07	-	-	-	0.06	0.37	-	-	-	-	-	-	-	-	-	-
	69.0	60.9*	-	0.28	-	-	-	0.24	1.57	-	-	-	-	-	-	-	-	-	-
	17.1	16.8*	-	0.27	-	3.4	8.8	0.44	1.03	5.4	-	-	-	-	-	-	-	-	-
	67.0	65.9*	-	1.06	-	13.3	34.3	1.71	4.02	21.0	-	-	-	-	-	-	-	-	-
	-	-	-	-	-	-	-	-	-	-	-	-	-	-	-	-	-	-	-
	57.1*	62.5*	-	-	-	-	-	-	-	-	-	-	-	-	-	-	-	-	-
	73.4*	82.2	75.8*	0.07	-	19.8	58.4	0.33	0.71	30.7	-	27.0	28.2	-	-	0.11	0.27	0.14	0.16
	82.4*	92.2	85.1*	0.04	-	22.3	65.5	0.37	0.80	34.5	-	30.3	31.6	-	-	0.13	0.30	0.15	0.18
	70.3*	75.0*	74.7*	0.02	-	2.8	1.7	0.26	0.22	17.2	-	-	-	-	-	0.17	0.23	0.13	0.40
	81.8*	87.2*	86.9*	0.02	-	3.3	2.0	0.31	0.25	20.0	-	-	-	-	-	0.20	0.27	0.15	0.47
	69.9*	76.2*	75.1*	0.12	-	17.3	7.7	0.36	0.25	24.0	-	18.1	-	-	-	0.25	0.33	0.17	0.11
	78.7*	85.9*	84.5*	0.13	-	19.5	8.6	0.41	0.28	27.0	-	20.4	-	-	-	0.28	0.38	0.20	0.12
	64.2*	67.6	69.0	0.09	0.133	3.0	4.5	0.23	0.46	8.9	398.	17.6	4.4	-	0.9	0.14	0.18	0.14	0.07
	74.5*	78.4	80.0	0.10	0.154	3.5	5.2	0.26	0.53	10.3	462.	20.4	5.1	-	1.1	0.16	0.21	0.17	0.08
	84.2	79.1	79.2*	0.44	0.059	13.2	14.4	1.16	0.45	-	964.	59.6	7.8	-	2.2	0.19	0.45	0.18	0.12
	94.5	88.7	88.9*	0.47	0.066	14.8	16.1	1.31	0.51	-	1081.	66.9	8.8	-	2.4	0.21	0.51	0.20	0.13
	-	79.0*	-	0.84	-	35.7	13.4	0.88	0.97	-	-	-	-	-	-	-	-	-	-
	-	88.6*	-	0.94	-	40.0	15.0	0.99	1.09	-	-	-	-	-	-	-	-	-	-
	74.5	75.1*	77.0	0.33	0.084	45.8	23.1	0.71	0.58	-	1495.	70.0	12.9	-	4.2	0.54	1.02	0.55	0.26
	82.4	83.1*	85.2	0.37	0.093	50.7	25.6	0.79	0.64	-	1654.	77.5	14.3	-	4.6	0.60	1.13	0.61	0.28
	71.1*	77.0*	-	0.29	0.068	27.2	8.6	0.67	0.29	-	890.	42.3	9.5	-	1.6	0.63	0.83	0.92	0.20
	81.2*	88.0*	-	0.34	0.078	31.1	9.8	0.76	0.34	-	1016.	48.3	10.9	-	1.8	0.71	0.95	1.05	0.23
	71.4*	68.5	-	0.17	0.111	56.5	23.9	0.73	0.65	-	2637.	-	-	-	-	0.34	0.75	0.50	0.18
	77.1*	73.9	-	0.18	0.120	61.1	25.8	0.79	0.70	-	2848.	-	-	-	-	0.37	0.81	0.54	0.20
	74.2*	78.5*	75.6	0.31	0.194	82.7	73.7	1.24	1.75	84.3	4798.	-	-	-	-	0.44	0.94	0.55	0.21
	80.5*	85.1*	82.0	0.34	0.211	89.7	79.9	1.35	1.90	91.4	5203.	-	-	-	-	0.47	1.02	0.60	0.23
	71.6*	76.4*	75.0*	0.05	-	53.8	4.4	0.30	0.44	-	-	19.5	-	-	790.4	0.08	0.20	0.14	0.06
	81.1*	86.6*	85.0*	0.06	-	61.0	5.0	0.34	0.50	-	-	22.1	-	-	895.8	0.09	0.27	0.16	0.07
	71.9*	76.1*	75.2*	0.01	-	-	-	0.27	-	-	-	18.6	-	-	450.7	-	0.29	-	-
	81.4*	86.1*	85.1*	0.02	-	-	-	0.31	-	-	-	21.1	-	-	510.3	-	0.33	-	-
	-	-	-	-	-	-	-	-	-	-	-	-	-	-	-	-	-	-	-
	-	54.4*	-	-	-	-	-	-	-	-	-	-	-	-	-	-	-	-	-
	70.6*	74.4*	73.4*	0.12	-	-	-	0.27	-	-	-	-	-	-	-	0.18	0.36	0.12	0.06
	81.8*	86.2*	85.0*	0.14	-	-	-	0.31	-	-	-	-	-	-	-	0.21	0.42	0.13	0.07
	69.9*	75.0*	79.1	0.01	-	3.4	4.1	0.25	0.33	10.3	435.	21.3	5.6	-	1.1	0.11	0.20	0.17	0.09
	80.4*	86.1*	90.9	0.01	-	3.9	4.7	0.28	0.38	11.8	500.	24.4	6.5	-	1.3	0.12	0.23	0.20	0.10
	-	-	-	0.04	0.027	2.6	3.9	0.26	-	-	-	6.8	1.8	-	0.6	0.09	0.26	0.09	0.09
	-	-	-	0.04	0.031	3.0	4.4	0.30	-	-	-	7.8	2.0	-	0.7	0.10	0.30	0.10	0.10

Members of the international network of feed information centres (INFIC) plan to publish feed composition tables for given regions or countries. These tables and animal nutrient requirement tables (National Academy of Sciences, 1970; Agricultural Research Council, 1965) are of most use to developing countries or persons without access to a computer or chemical laboratory, since diets, rations, or feed mixtures may be calculated manually using data from the tables.

Figure 5
Requirements of animals should be calculated to fit a group of animals in a given environment. Utilisation of the diet should be calculated by regression equations. Balance the diet using amount of nutrient per 1,000 kcal Metabolisable Energy and regression equations to estimate utilisation of ME. Introduce cost to make up the most profitable diet.

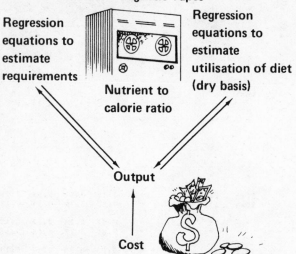

Figure 6

A portable terminal may be plugged into an electrical outlet on a farm or at a feed mill. An operator connects the terminal to the central computer at Utah State University by phone. The operator types in the costs of the feeds and codes to designate to the computer the animal nutrient requirements and feeds to be used. The central computer calculates the animal diet and the portable terminal prints out the most profitable one.

Dr Lorin E. Harris is typing the data on the portable terminal. Leonard C. Kearl is holding a 720-page *Atlas of Nutritional Data on United States and Canadian Feeds.* Scientists from Utah State University and Dr E. W. Crampton of McDonald College in Quebec prepared the atlas for the USA National Academy of Sciences. It is used as a 'bible' in its field.

Ultimately, the user's feed will be analyzed for critical nutrients. Other feed analyses and nutrient requirements data will be kept on magnetic tapes or discs in a computer. Using the analysed data, stored feed analyses, and nutrient requirement data, plus equations, a more precise diet, ration, or feed mixture can be calculated by the computer to obtain maximum profit (Harris *et al.* 1972; Harris *et al.* 1974).

A remote terminal may be set up in a farmer's kitchen, for example, or in a feed mill office. The terminal is connected to a telephone, signals are fed into the computer, and the diet or feed mixture is printed by the terminal (Figure 6).

The proposed 'international bank' of chemical and biological feed data will allow rapid provision of answers to a wide range of inquiries on feeds or methods of feeding.

Presently the Hohenheim centre answers questions concerning feedstuffs and feeding, and supplies data on request. In the future it will be possible to obtain data stored in a common data bank from the Documentation centre in Stuttgart-Hohenheim, Germany, or from the International Feedstuffs Institute at Logan, Utah, USA (Figure 2), if desired.

References

Agricultural Research Council. A.R.C. (1965) Nutrient requirements of farm livestock. No. 2 ruminants. London, United Kingdom.

Alderman, G. (1971) Proposals for the establishment of a feeds information system for the Food Agricultural Organization. FAO AGA/MISC/71/28.

Association of American Feed Control Officials (USA) (1974) Feed Control Official Publication. Ernest A. Epps, Jr. Division of Agri. Chemistry, P.O. Box 16390—A, Baton Rouge, Louisiana 70803.

Canada Feed Act and Feed Regulations (1967) Roger Duhamel, Queen's Printer and Controller of Stationery, Ottawa, Canada.

Haendler, H. (1963) Bedeutung und Aufgaben des Archivs für Futtermittel Hohenheim. *Kraftfutter* **46** (11): 555—556.

Haendler, H. & Jager, F. (1971) Stand und Entwicklung der Befunddokumentation Futtermittel. *Mitteilungen der Gesellschaft für Bibliothekswesen und Dokumentation des Landbaues* H., **15**, 23–31.

Haendler, H. & Harris, L. E. (1973) Data documentation in the field of feed analysis and the international co-operation it includes. *Quarterly Bull. Interna. Assoc. Agr. Librarians and Documentalists* **18** (2): 87–94.

Harris, L. E. (1963) A system for naming and describing feedstuffs, energy terminology, and the use of such information in calculating diets. *J. Animal Sci,* **22**: 535–547.

Harris, L. E. (1970) Nutrition research techniques for domestic and wild animals Vol. 1. Logan, Utah: L. E. Harris.

Harris, L. E., Asplund, J. M. & Crampton, E. W. (1968) An international feed nomenclature and methods for summarizing and using feed data to calculate diets. *Utah Agr. Exp. Sta. Bul.* **479**, Logan, Utah, USA.

Harris, O. E. & Christiansen, W. C. (1972) International network of feed information centres (INFIC). Special publication from Mitteilungen der Gesellschaft fur Bibliothekswesen and Documentation des Landbaues. **H 17**: 93–100.

Harris, L. E., Kearl, L. C. & Fonnesbeck, P. V. (1972) Use of regression equations in predicting availability of protein and energy. *J. Animal Sci.,* **35** (3): 658–680.

Harris, L. E., Kearl, L. C., Stenquist, N. J. & Barnard J. J (1974). Maximising profits by use of a computer to calculate diets for beef and dairy cattle and sheep. Utah Ext. Service Cir. Logan, Utah, USA.

Henry, W. A. (1898) Feeds and feeding. 1st ed. Madison, Wisconsin, U.S.A.

Kellner, O. J. (1905) Die Ernährung der Landwirtschaftlichen Nutztiere. Verlagsbuchhandlung. Berlin: Paul Parey. Germany.

McDowell, L. R., Conrad, J. H., Thomas, Jenny E. & Harris, L. E. (1974a) Latin American tables of feed composition. University of Florida: Gainesville, Florida, USA.

McDowell, L. R., Conrad, J. H., Thomas, Jenny E. & Harris, L. E. (1974b) Latin American tables of feed composition, abridged, Spanish ed. University of Florida: Gainesville, Florida, USA.

National Academy of Sciences (1970) Nutrient requirements of domestic animals – No. 4, Nutrient requirements of beef cattle, 4th rev. ed. W. Burroughs *et al.,* Printing and Publishing Office, National Academy of Sciences, Washington, DC, 20418.

National Academy of Sciences (1971) Atlas of nutritional data on United States and Canadian Feeds (prepared by E. W. Crampton & L. E. Harris). Printing and Publishing Office, National Academy of Sciences, Washington DC, 20418.

National Academy of Sciences (1973) Effect of processing on the nutritional value of feeds. Printing and Publishing Office, National Academy of Sciences, Washington, DC, 20418.

Scholtyssek, S., Lorenz, G. & Haendler, H. (1963) Legehennen Futter und Lineare Programmierung, *Kraftfutter,* **46** (3), 122.

The standardisation and control of quality of animal feeds

H. Bhagwan

Director (Agriculture and Food), Indian Standards Institution

Summary

Production of compounded feeds in India started only in 1959, but by 1972 has already reached an annual rate of production of 371,500 t. Of this 183,000 t was for poultry and approximately the same amount for cattle. Demand for compounded feeds in India is expected to grow considerably over the next few years.

The Animal Feeds Sectional Committee (AFDC 15) of the Indian Standards Institution has made an important technical contribution to the development of the Indian feed industry since it was first established in 1959. Standard specifications for both poultry and cattle feeds have been prepared by the Committee on the basis of research carried out at the Indian Veterinary Research Institute, Izatnagar and the National Dairy Research Institute, Karnal respectively. Standards have also been prepared for the various ingredients of compounded feeds available to manufacturers in India, and this work is continuing. The Committee which is composed of experts from the industry, research institutions, animal husbandry and animal breeding farms, also acts as a clearing house for all major technical problems facing the Indian feed industry. For example, the Committee recently studied the practical use of urea for cattle feeds under tropical and sub-tropical conditions.

It is desirable that the minimum limits prescribed in the various standards are followed by all feed manufacturers in India, so that poultry and dairy farmers can be assured of obtaining feeds of a reasonable nutritive value. However only a few manufacturers have so far come forward to market their products under the ISI Certification Mark. The growth and development of ISI certification is reviewed and suggestions for its better operation invited.

atteint déjà un degré de production annuelle de 371,500 t. De cette production 183,000 t. étaient pour la volaille et environ la même quantité pour le bétail. On compte que la demande pour pâtures en l'Inde augmentera considérablement dans les années suivantes. Le Comité Régional pour les Pâtures des Animaux (AFDC 15) de l'Etablissement des Standards de l'Inde, a fait une importante contribution technique au développement de l'industrie Indienne de pâture depuis sa création en 1959. Les spécifications standard pour les pâtures de la volaille et le bétail ont été préparées par le Comité sur la base de recherches entreprises à l'Institut Indien de Recherches Vétérinaires, Izatnagar et à l'Institut National de Recherches Fermières, à Karnal. On a préparé aussi des standards pour les différents ingrédients des pâtures combinées, disponibles pour les fabricants en l'Inde et ce travial continue. Le Comité qui est composé d'experts de l'industrie, des instituts de recherches, de l'élevage du bétail et des fermes pour la réproduction des animaux, agit aussi comme un bureau de clarification des problèmes techniques majeurs que l'industrie Indienne de pâtures doit affronter. Par exemple, le Comité a étudié récemment l'utilisation pratique de l'urée dans les pâtures pour le bétail, sous des conditions tropicales et sous-tropicales.

Il est désirable que les limites minimum préscrites dans les différents standards soient suivies par tous les fabricants de pâtures en l'Inde, pour que les fermiers de volaille et de bétail soient assurés d'obtenir des pâtures d'une valeur nutritive raisonnable. Pourtant, seulement quelques fabricants se sont présenté pour commercialiser leurs produits sous la marque de certification ISI. On passe en revue la croissance et le développement de la certification ISI et on demande à faire des suggestions pour un meiller fonctionnement.

Résumé

La standardisation et la vérification de la qualité des pâtures pour les animaux

La production de pâtures combinées en l'Inde a commencée seulement en 1959, mais en 1972 elle a

Resumen

La normalización y el control de calidad de los piensos para animales

La producción de piensos compuestos en la India se inició solamente en 1959, pero para 1972 había

alcanzado ya un índice anual de producción de
371.500 t. De éstas, 183.000 t. eran para las aves
de corral y approximadamente la misma cantidad para
la ganadería. Se espera que la demanda de piensos
compuestos en la India aumente considerablemente
durante los próximos pocos años.

El Comité de Sección de Piensos para Animales
(AFDC 15) de la Institución de Normas Indias ha
hecho una contribución técnica importante al desarrollo
de la industria de piensos india desde que se estableció
por primera vez en 1959. Especificaciones patrón,
tanto para piensos para aves de corral como para
ganado, se han preparado por el Comité a base de la
investigación llevada a cabo en el Instituto de
Investigación Veterinaria India, Izatnagar y el Instituto
de Investigación Lechera Nacional, Karnal, respectivamente. Se han preparado también patrones para los
varios ingredientes de piensos compeustos disponibles
a los fabricantes en la India, y este trabajo está
continuando. El Comité, que está compuesto de

expertos procedentes de la industria, las instituciones
de investigación, granjas de animales, y granjas para la
crianza de animales, también actúa como un centro
de estudio de los problemas técnicos principales a que
hace frente la industria de piensos india. Por ejemplo,
el Comité estudió recientemente el uso práctico de la
urea para piensos para el ganado bajo condiciones
tropicales y subtropicales.

Es deseable que los límites mínimos prescritos en los
varios patrones sean seguidos por todos los fabricantes
de piensos de la India, a fin de que los avicultores y
los granjeros dedicados a la producción de leche
puedan tener la seguridad de obtener piensos de valor
nutritivo razonable. Sin embargo, hasta ahora solamente unos pocos fabricantes se han lanzado a
comercializar sus productos bajo la Marca de Certi-
ficación ISI. Se revisa el aumento y el desarrollo de
la certificatión ISI y se invita a que se envíen
sugerencias para su mejor funcionamiento.

Introduction

Standards serve the purposes of communication,
comparison and measurement. Standardisation, by and
large the process of formulating standards, can therefore be developed at various levels, such as company,
association (industry), national or international.
Sen (1969) expressed it well when he said: 'The
process of development of national standards is
basically the process of co-ordinating the three interests
represented by producers, consumers and technologists.
The technical significance of this statement is that
national standards should be based on the current
state of scientific knowledge on the subject, permit
production at economical level and serve the need of
consumers in general. The crucial point is to secure
the delicate balance between the three'.

While securing this delicate balance, one tends to
acquire knowledge and to form an integrated and
balanced view of research, industry and marketing.
It is this acquired knowlege which is presented in this
paper along with the Indian experience of standardisation and quality control in the field of animal feeds.

Beginning of standardisation in India

In India the need for compiling standards in animal
feeds was felt as early as 1956 and this resulted in the
setting up of the 'Animal Feed Sectional Committee'
(AFDC 15) of the Indian Standards Institution. The
objective of this committee, comprising experts drawn
from research institutions, industry, government departments and large size farms, was:
(a) to lay down Indian standards for poultry feeds;
     and

(b) to investigate the need for the formulation of
     standards for cattle feeds, and submit appropriate
     recommendations.

It will be clear from these initial terms of reference
that a point seemed to have been reached in scientific
research for poultry feeds at which standards could be
laid down but with respect to cattle feeds a similar
exercise was considered slightly premature.

It is interesting to recall that even at the first meeting
of AFDC 15 held in July 1957 it was emphasised that
'while augmenting food resources of the country, the
tendency hitherto appears to have been to consider
the needs of human population only, leaving out the
animals to fend for themselves on grazing or such
by-products of crops which are grown mainly for
human consumption. Under these circumstances, it is
not possible to produce high grade stock either for
milk or for work without provision of adequate or
balanced feed for them'.

Leading scientists in India had already recommended
that plants should be set up in the country for producing balanced feeds. It was in the search for a
standard on this balanced feed for mass production
that it was decided at the first meeting of AFDC 15
that detailed information should be collected from all
concerned on availability of raw materials and their
nutritional value, nutritional deficiencies of cattle and
poultry, and availability of waste products from
slaughter houses, farms, etc. This information was
collected and with this, work was started on compiling
standards for poultry and cattle feeds.

Poultry feeds

A proposed draft standard on poultry feeds covering
requirements for growing poultry only, was submitted

| | Requirements for | | |
Characteristic	Poultry starter feed	Poultry grower feed	Poultry layer feed
Moisture (% weight, max)	10	10	10
Crude protein (% weight)	20 to 25	18 to 23	15 to 20
Crude fat or ether extract (% weight)	3 to 6	3 to 6	3 to 6
Crude fibre (% weight, max)	7	8	8
Total ash (% weight, max)	10	10.5	11
Acid insoluble ash (% weight, max)	3.5	3.5	3.5
Calcium (% weight)	1.0 to 1.3	1.0 to 1.3	2.0 to 2.3
Phosphorus, (% weight)	0.6 to 0.8	0.6 to 0.8	0.8 to 1.0
Growth assay:			
Combined average weight in g of male and female chicks at the age of 8 weeks (min):			
White Leghorn chicks	380	–	–
Rhode Island Red chicks	400	–	–

TABLE 2. *Requirements for poultry feed as prescribed in IS:1374–1968*

Characteristic	Poultry starter feed	Poultry grower feed	Poultry layer feed
i Moisture (% weight, max)	10	10	10
ii Crude protein (% weight min)	20	16	15
iii Crude fibre (% weight, max)	7	8	10
iv Acid insoluble ash (% weight, max)	4.0	4.0	4.0

Note: Requirements for the chacteristics given in (ii) to (iv) are on moisture-free basis.

by an expert of the Indian Veterinary Research Institute (IVRI) in early 1958. The draft standard with the addition of requirements for laying poultry feeds was widely circulated for comment. In the light of comments received, the standard was finalised for publication by the middle of 1959. The finalised version of this standard included requirements for poultry starter feeds also. Table 1 shows detailed requirements for the three types of feeds. Other important features of this standard were:

(1) A comprehensive list of permissible ingredients was included.
(2) Requirements for minerals and vitamins for in-process inspection were specified separately.
(3) A growth test with 25 day-old chicks (White Leghorn or Rhode Island Red) and having a duration of about 8 weeks was specified for assessing the quality of the starting poultry feeds.
(4) A few feed formulae were also included. These formulae which were intended to guide the feed manufacturers had been found to give satisfactory performance in various research investigations.

Soon after the publication of IS:1375–1959 Specification for Poultry Feeds, the Indian Council of Agricultural Research (ICAR) advised all the states to purchase poultry feeds conforming to this standard. This directive included a further critical re-appraisal of this standard which was carried out on the following lines:

(1) The mineral and vitamin requirements were modified and the list of recommended ingredients enlarged to include cheaper ingredients.

(2) A few more feed formulae as suggested by various workers were included.
(3) The duration of the growth test was reduced from 8 weeks to 4 weeks. Further, it was decided that if a manufacturer used any of the formulae included in the standard, it should not be necessary for him to carry out the growth test.

The first revision of the standard incorporating these modifications was brought out in 1964.

Despite this revision, the industry continued to press that the growth test was not practical for adoption as a routine quality control measure. Further, it was emphasised that for lot to lot examination for the purpose of ISI certification, the characteristics to be checked should be minimised so that the cost of testing does not greatly increase product cost. Accordingly, for routine quality control purposes, Table 2 was finalised in 1968. In this table only four characteristics were retained: moisture, crude protein, crude fibre and acid insoluble ash. Crude protein was retained to ensure the minimum quantity of the most vital nutrient, crude fibre and acid insoluble ash limits to ensure the use of good quality and the right type of ingredients, and the moisture limit to ensure better shelf life. Simultaneously, it was recognised that for the characteristics given in Table 3, manufacturers themselves should adopt additional quality control measures and make their records open for inspection by the certifying agency, whenever necessary.

By and large, the 1968 version of IS:1374 remains acceptable to the industry. The only important

TABLE 3. *Additional requirements (for quality control by manufacturers) for poultry feeds as prescribed in IS:1374—1968*

Characteristic	Poultry starter feed	Poultry grower feed	Poultry layer feed
Crude fat (% weight, min)	3	3	3
Calcium (%weight, min)	1	1	2
Available phosphorus (% weight, min)	0.5	0.5	0.5
Lysine (% weight, min)	0.9	0.7	0.5
Methionine (% weight, min)	0.40	0.35	0.30
ME/kg, min.	2,640	2,530	2,740

Note: The energy values may also be declared as productive energy values; 1.6 calories of metabolisable energy is equal to 1 calorie productive energy.

TABLE 4. *Requirements for broiler feeds as prescribed in IS:4018—1967*

Characteristic	Requirement for	
	Broiler starter feed	Broiler finisher feed
i Moisture	10	10
ii Crude protein (nitrogen x 6.25) (% weight, min)	22	22
iii Crude fibre (% weight, max)	6	6
iv Acid insoluble ash (% weight, max)	3.5	3.5
v Calcium (% weight)	1.0 to 1.2	0.8 to 1.0
vi Available phosphorus (% weight, min)	0.4	0.4

Note: The values specified for the characteristics at SI No. (ii) to (vi) are on moisture-free basis.

amendment which has been carried out is that quality control of the particle size of the ready mixed feed has also been left to the manufacturer who is required to ensure that the feed is palatable and eaten by the birds.

It was felt in 1963 that a specification for broiler feeds should also be formulated in addition to ISI:1374—1959. The expert from IVRI submitted a proposed draft in early 1965 which was then thoroughly scrutinised by AFDC 15 and circulated widely for comment. Then a special panel was constituted to prepare a final version of the standard in March 1967. This version was again circulated to leading experts in the country and their comments were re-examined in IVRI to arrive at the specification given in Table 4. This standard, namely, ISI:4018—1967, also includes a method for determination of metabolisable energy-to-protein ratio with a view to helping the manufacturers to assess the quality before marketing. This ratio is specified as 130 to 135 cal/kg:1 in broiler starter feeds and 155 to 165 cal/kg:1 in broiler finisher feeds. These values are required to be declared by the manufacturer. Simultaneously, the importance of the limits for lysine and methionine were emphasized and their minimum limits were specified as 0.90 and 0.50% by weight respectively.

Cattle feeds

Work on the specification for cattle feeds was also started in 1957. It took more than three years to crystallise the scientific opinion in the country. Salient features of the first series of decisions on this subject are as follows:

(1) A national standard for balanced feed mixtures based on the requirements of Indian cattle would definitely serve as a guide to the manufacturers.

(2) The balanced feed mixtures are intended to supplement the basal ration comprising roughages or green feeds and coarse fodder.

(3) In computing balanced feed mixtures, it was essential to know the nutritional requirements for various classes of livestock for work and production as well as nutritive value of the available ingredients. The scientific data available at the National Dairy Research Institute (NDRI) and U.P. College of Veterinary Sciences and Animal Husbandry, along with the ICAR Bulletin number 25 (2) could be utilised for the purpose.

(4) For the guidance of the manufacturers a few feed formulae found satisfactory at the research institutes were also included in the standard. But it was also provided that the manufacturers could make suitable modifications in the feed formulae according to the locality, season and availability of the ingredients.

(5) The basis for computation of feed mixtures was that each feed formula should provide 14 to 16% of digestible crude protein (DCP) and 68—74% of total digestible nutrients (TDN). One kg of any of these mixtures would be required per 2.5 kg of milk in cows or 2 kg of milk in buffaloes. In the case of bullocks, 2—3 kg of any of these mixtures would be adequate while for hard work it should be raised to 3—4 kg.

(6) It would also be necessary to supplement the ration with mineral mixtures.

In May 1961 a draft specification for balanced feed mixtures for cattle was widely circulated for comment. At the time of finalisation of this standard, it was strongly pleaded that as the industry was in its infancy, more time should be allowed for collecting information on various aspects. The other view-point was that the standard was based on actual research work carried out during the last 15–20 years and there was a pressing need for this specification with a view to providing a stimulus to the development of the compounded feed industry in the country. The latter view prevailed and IS:2052–1962 Specification for Balanced Feed Mixtures for Cattle was formulated. The main requirements of this standard are given in Table 5.

TABLE 5. *Requirements for balanced feed mixtures for cattle as prescribed in IS:2052–1962*

Characteristic	Requirement
i Moisture (% weight, max)	10.0
ii Crude protein (nitrogen x 6.25) (% weight, min)	20.0
iii Crude fat or ether extract (% weight, min)	3.5
iv Crude fibre (% weight, max)	12.0
v Acid insoluble ash (% weight, max)	3.0
iv *Phosphorus (% weight, min)	0.5

*The minimum amount of calcium has not been prescribed as it is presumed that the feeding of leguminous fodder or mineral mixture (*see* IS: 1644–1960 Specification for Mineral Mixture for Supplementing Cattle Feeds) would normally meet the calcium requirements.

Note: The values for requirements (ii) to (v) are on moisture-free basis.

Despite discussions over a period of five years, the industry continued to express the view that they were experiencing difficulties in the implementation of this standard. Accordingly, various issues were reopened in 1966 and comments and suggestions for revision of the standard were invited from all the manufacturers and scientific workers. During this series of discussions, the following points emerged:

(1) ISI:2052 was based on a sound scientific footing and the experience gathered from about 65 farms. Further, a sizeable quantity of compounded feeds being produced in the country conformed with this standard.

(2) The standard would be more implementable with the following amendments:

   (i) The title should be changed from 'balanced feed mixtures for cattle' to 'compounded feed for cattle'.

   (ii) The feed formulae should be re-checked and revised.

   (iii) A separate standard should be formulated for feeds meant for young stock (further details of this are given later).

   (iv) To minimise the cost of routine quality control, specifications for minimum content of phosphorus and fat were deleted (see Table 6).

   (v) The requirement for crude fat was deleted as a minimum of 2% crude fat was always expected to be available from the ingredients commonly used.

TABLE 6. *Requirements for compounded feeds for cattle as prescribed in IS:2052–1968*

Characteristic	Requirement
i Moisture (% weight, max)	10
ii Crude protein (nitrogen x 6.25) (% weight, min)	20
iii Crude fibre (% weight, max)	13
iv Acid insoluble ash (% weight, max)	4.0

Note: The values for requirements (ii) to (iv) are on moisture-free basis.

   (vi) TDN and DCP values given in the earlier version were retained for guidance, but it was emphasised that in view of limited testing facilities and long duration of test time it was not possible to conduct feeding trials. DCP values of raw materials should be prepared on a priority basis.

   (vii) Areas for desirable additional quality control measures by manufacturers were also identified (see Table 7). With these modifications the first revision of IS:2052 was issued in 1968.

TABLE 7. *Additional requirements (for quality control by manufacturer's) for compounded feeds for cattle prescribed in IS:2052–1968*

Characteristic	Requirement
Common salt, (% weight, max)	2
Calcium (as Ca) (% weight, min)	0.5
Phosphorus (as P), (% weight, min)	0.5
Vitamin A, IU/kg	5,000

Basic information relating to the preparation of standard reference tables giving TDN and DCP content of feed ingredients have already been compiled. These reference tables are expected to be ready within a year.

The 1968 version of IS:2052 has not posed any serious problems, and the only issue which has been discussed recently is the use of urea as an ingredient for compounding these feeds. The main reasons which had led to the exclusion of urea were non-availability of suitable supplementary carbohydrates and lack of data on deterioration during storage and transport and levels of incorporation. The committee examined these issues and it has recently been agreed that urea can be used at inclusion rates of up to 1% by weight, provided not less than 10% by weight of molasses is incorporated. The committee has also concluded that incorporation of urea as a partial replacement of protein even at this level would definitely be economical. IS:2052–1968 will be revised on these lines in the near future.

Feeds for young stock have been under consideration since the early sixties but work on the compilation of a standard gathered momentum during the revision of IS:2052 for cattle. An expert working for a private sector firm produced the draft proposals which were then thoroughly scrutinised by leading experts on cattle feed. Subsequently, on the basis of actual experience gained at Aarey Milk Colony and the Eastern Regional Station of the NDRI, a preliminary draft was compiled in April, 1969. Before this draft

TABLE 8. *Requirements for calf starter meal and calf growth meal as prescribed in IS:5560–1970*

	Requirement	
Characteristic	Calf starter meal	Calf grower meal
(2)	(3)	(4)
i Moisture, (% weight, max)	10	10
ii Crude protein (N x 6.25) (% weight)	23–26	22–25
iii Crude fat (% weight, min)	4	–
iv Crude fibre (% weight max)	7	10
v Total ash (% weight, max)	5.0	5.0
vi Acid insoluble ash (% weight, max)	2.5	3.5

Note: The values for requirements (ii) to (vi) are on moisture-free basis.

*While analysing for crude protein it should be ensured that the nitrogen has not been derived from urea or other ammonium salts.

was circulated the committee affirmed that adequate scientific work had been carried out in the country and a beginning should therefore be made to formulate this standard. Subsequently, IS:5560–1970 specification for feeds for young stock was agreed (see Table 8). It was also emphasised that in addition to these requirements manufacturers should exercise quality control on a common salt, calcium, phosphorus, vitamin A and TDN content.

Other standards

Mineral mixture for supplementing cattle feeds (IS:1664–1968)

A formula for a mineral mixture to supplement cattle feed, as developed at IVRI, NDRI and IARI, had been patented by the Ministry of Food and Agriculture. In 1958, it was decided to formulate an Indian standard and when this patent was released in 1959 a draft standard was compiled using it as a basis. The first version of this standard was finalised in 1960; it stipulated the requirements for moisture content, calcium, phosphorus, salt, iron, iodine, copper, manganese, cobalt, fluorine and also freedom from spores of pathogenic organisms, namely, *Bacillus anthracis* and *Clostridium* sp.

This standard was revised in 1968 to provide for the manufacture of mineral mixtures without salt.

Mineral mixture for supplementing poultry feeds (IS:5672–1970)

The proposal was received from a leading manufacturer who also sent considerable data on the subject in 1967. More data were collected by AFDC 15 and the proposed draft was discussed in 1969. The main discussion centred round the availability of minerals to poultry from various raw materials used in compounding mineral mixtures. The committee also considered whether manganese should be incorporated as it affected the utilisation of calcium and phosphorus in the mineral mixture. Finally, its maximum limit was prescribed as 0.27% by weight. Other requirements prescribed are moisture, calcium, phosphorus, iron, zinc, fluorine, copper, iodine and freedom from spores of the pathogenic micro-organisms mentioned earlier for

cattle. It also included a list of ingredients that may be used in compounding mineral mixtures.

Basic materials for mineral mixtures

For successful quality control of both these mineral mixtures, it was imperative that basic materials required for their manufacture should also be of uniform quality. In pursuit of this objective AFDC 15 has evolved the following:

(1) *Bone-meal as livestock feed supplement (IS:1942–1968)*
This standard was first issued in 1961 and then revised in 1968. The points which came up for special discussion were: protein content and the possible presence of pathogenic organisms, particle size, limits for fluorine, calcium and phosphorus on the basis of specific research investigations, and the method for the determination of moisture content.

(2) *Dicalcium phosphate, animal feed grade (IS:5470–1969)*
While preparing this specification, the initial discussion arose as to whether the fertiliser grade of dicalcium phosphate could be used for feeding animals. Finally, it was decided that for feed grade dicalcium phosphate additional requirements like purity, fluorine content, proper ratio of calcium to phosphorus and limits for copper, selenium and lead, were necessary and with these additional requirements, IS:5470–1969 was formulated. The toxic effect of fluorine was also discussed and its limit was fixed at 0.10%.

(3) *Calcined bone-meal (IS:7061–1973)*
The availability of phosphorus present in calcined bone-meal was discussed at length. After AFDC 15 was convinced that it would be available, a standard was prepared.

Specifications for feeds for laboratory animals

India has establishments at which many animals are kept for experimental purposes and it was therefore suggested that standards be formulated for feeds for laboratory animals. The suggestion was welcomed by the users of laboratory animals and draft specifications for guinea pigs and rats and mice were submitted by the experts from IVRI. The detailed specifications for feeds for these two categories of laboratory animals were prepared in close consultation with the users of laboratory animals and have been issued as IS:5654 Parts I and II (1970).

Feed ingredients

On the basis of considerable research experience and data from a number of scientific organisations and concerned industries, a series of standards have been formulated on oilcakes, foodgrain by-products, waste products of animal origin and other agricultural/industrial by-products. These standards along with the standard on methods of test for animal feeds are listed in Appendix A.

Review of standards

It will be seen that within a span of 10–12 years, it has been possible, with the co-operation of Indian scientists, to formulate Indian standards on practically all the important aspects of the animal feed industry. Despite this achievement, a critical evaluation in 1969 revealed that only an insignificant proportion of cattle and poultry feeds was covered under the ISI Certification Marks Scheme operated on a voluntary basis. A conference was therefore organised to discuss various issues relating to production, certification, availability and quality of ingredients and other supplements used by the compounded feed industry. This conference was held in November 1969 and written comments were invited from manufacturers on the acceptability of Indian standards and difficulties being faced by them in adopting Indian standards and procuring raw materials, if any. Some of the important comments which were received are given below:
(1) Manufacturers could product ISI certified feeds, provided there was a ready demand for them from farms and other large-scale purchasers.
(2) Except for solvent extracted oilcakes, it was not possible for compounded feed manufacturers to purchase good quality, tested and certified ingredients.
(3) The ISI certification scheme should be simplified and the certification fees should be reduced; further, the limits for acid insoluble ash and moisture in the feeds should be relaxed along with labelling requirements.
(4) The mushroom growth of organisations producing feeds without exercising quality control in their operations, enables them to market feeds at substantially lower prices. Such a situation creates problems in the field of competitive marketing particularly for those manufacturers who are conscious of quality.

During the conference (Indian Standards Institution, 1970) the following points were made:
(1) Some scientists expressed the opinion that one single standard could not be formulated for compounded feeds for cattle for meeting varying nutritional requirements for production and for work. They also felt that for animals being fed large quantities of good quality green forage or low quality roughages, the requirements would vary considerably. But the majority of scientists felt that IS:2052–1968 covered the average conditions of milk production and depending upon the nature and quantity of other feeds, the quantity of compounded feed could be altered to suit the age and type of the animal and local conditions.
(2) Raw materials of suitable quality were available in the country, particularly if some price incentive was paid for superior quality. In particular, solvent extracted oilcakes could be supplied in larger quantities if long term contracts were forthcoming.
(3) The FAO representative attending this conference remarked significantly that animal response was of course the best measure for judging the suitability and quality of feeds. Nevertheless, the importance of chemical analysis should not be under-estimated, for it helped in maintaining good supervision for some basic and important considerations involved in quality control.
(4) A strong plea was therefore made that all the Government Departments and local authorities should purchase ISI certified cattle and poultry feeds.
(5) The benefits of utilising the compounded feeds are not yet fully known to the cattle and poultry farmers, specially those who operate on a small scale and in rural areas. For the purpose, a documentary film exhibiting the benefits should be prepared.

ISI certification and implementation

Soon after the conference, a series of discussions were held with the industry to reduce the ISI certification fees and to simplify the ISI certification schemes as far as possible. Despite this only about 2,500 t of cattle feeds produced by eight licensees and 50 t of poultry feeds produced by three licensees have been brought under ISI certification. Together the eight licensees produce about 20,000 t of cattle feed, that is 10% of the total feed produced by members of the Compound Livestock Feed Manufacturers Association (CLFMA) (see Table 11). For poultry feeds, the figures are lower; in all they account for 150 t. It could thus be concluded that:
(1) Licensees cover only a small fraction of their production under ISI certification. Perhaps they do so when purchasers belonging to a particular region of the country, refuse to accept feeds without ISI certification.
(2) Only small scale units have come forward to join ISI certification.

(3) Whereas some impact has been made for cattle feeds, the impact for poultry feeds is practically nil, even though serious concern has been expressed about their quality in the highest circles.

Despite the fact that not many licenses have been issued under the ISI Certification Marketing Scheme, these Indian standards are quite popular. They are in great demand by the educational and research institutions and invariably they have been utilised as topics for high level discussions at various conferences and symposia. Some farmers who are making mixed feeds for their own use are also utilising these standards, particularly the feed formulae.

Plans for development of cattle and poultry in India

Before drawing any conclusions from this review of growth of standardisation activity, it would be desirable to have a wider view of India's plans for development of cattle and poultry, growth of compounded feed industry and the raw material situation. The current production of milk in the country is of the order of 23.5 mt. During the past five years the growth percentage has been rather low; in 1968–69 the milk production was of the order of 21.2 mt. During the Fifth 5-Year Plan a much higher growth rate, namely 5.5%, has been envisaged and it is hoped that by 1978–79 milk production will be of the order of 30 mt.

The production of eggs was of the order of 5,300 m in 1968–69 and the current production is estimated at 8,000 m. At a growth rate of 15%, the target for 1978–79 is 14,000 m.

The infra-structure for the cattle and poultry development is indicated in Table 9.

For cattle development, the programmes given in Table 9 would imply that during the Fifth 5-Year Plan period (April 1974 – March 1979) about 8m cows of breedable age (with an average lactation yield ranging from 700 to 1,000 kg) and about 5 m buffaloes (with lactation yields ranging from 1,000 to 1,500 kg) would be taken up. Of 8 m cows, 5.5 m would be drawn from the improved breeding stock under the existing Intensive Cattle Development Projects (ICDP) and Key Village Blocks. The remaining 2.5 m cows are intended to be taken up under the new ICD projects. Regarding buffaloes, the entire population of 3.5 m buffaloes already covered under the ICD projects and Key Village Blocks would be taken up for selective breeding or upgrading. The remaining 1.5 m buffaloes will have to be taken up under the new ICD projects. With these improved cattle and buffaloes it should be possible to increase total bovine milk production up to 30.5 mt (see Table 10).

In order to supplement the nutritional requirements it is estimated that the requirements of concentrated feeds (for 4 m cross bred cows and 2 m high-yielding buffaloes and followers that are expected to be in production during the Fifth Plan period) will be 6 m t/a. The requirements of concentrates for 2.2 m cross bred progeny which will be in different stages of growth in the areas to be covered under the integrated Cattle-cum-Dairy Development Projects are estimated at about 1.4 mt.

Poultry feeds

A 50% increase in egg production during the fourth 5-Year Plan period was possible as a large number of commercial poultry farms with larger numbers ranging from 500 to 50,000 birds were established in various parts of the country. In addition, generally superior chicks were available from government farms engaged in scientific poultry breeding programmes and private hatcheries that entered into collaboration with overseas farms for the introduction of hybrid stock on a continuous basis.

Because of the significant increase in egg production and existing market conditions, only a marginal increase in supply is now needed to any particular market to cause a substantial drop in wholesale egg prices. Therefore, the strategy for poultry development during the Fifth 5-Year Plan period will be to establish 'Co-operative Egg Marketing Federations' both at the National and State levels, to encourage the formation of 'Poultry Producers Co-operative Societies', and to create a 'Marketing Intelligence Cell'. In addition, scientific breeding programmes will be intensified in various regions and central poultry breeding farms will be set up in northern and hilly regions. The poultry breeding programmes in the states will also be effectively co-ordinated.

Quality control of inputs such as licensing of hatcheries and poultry feed manufacturing firms will receive increased attention and a number of feed analytical laboratories will be established to control the quality

TABLE 9. *Infra-structure of cattle and poultry development*

Numbers of	Likely level 1973–74 (base)	Additional	Level envisaged in 1978–79
Intensive cattle dev. projects	63	51	114
Key village blocks	621	92	713
Fodder seed production farms	38	33	71
Liquid milk plants	90	64	154
Milk product factories including creameries (in production)	18	34	52
Rural dairy centres	52	80	132
Frozen semen banks	8	10	18
Intensive egg & poultry production-cum marketing centres	81	150	142
Installed capacity of dairy plants in the organised sector (ml/d)	5.93	6.37	12.30

TABLE 10. *Projected milk production 1978–1979*

Bovine	Population (m)	Average lactation yield per bovine (t)	Total lactation yield (mt)	Expected annual milk production (mt)
(a) Cattle				
Dairy				
Cross-breed under present ICDP, key village blocks	1.00	2.00	2.00	1.60
Cross-breeds under the envisaged programme	3.00	2.50	7.50	6.00
Indigenous improved stock	7.00	0.70	4.90	3.27
Non-dairy				
Ordinary indigenous cows	37.60	0.15 (per annum)	–	5.64
Total	48.60			16.51
(b) Buffaloes				
High yielding	2.00	1.80	3.60	2.40
Medium				
(a) Better ones	2.00	1.20	2.40	1.80
(b) Moderate	4.00	1.00	4.00	2.66
Low yielders	18.40	0.60	11.04	7.36
Total buffalo milk	26.40			14.02
Total bovine milk production				30.53

of poultry feeds. Intensive broiler production projects will also be established.

Growth of compound feed industry

Cattle and poultry feeds are produced in large quantities by a number of plants in the public sector and by small-scale feed manufacturers. These units contribute approximately 50% of the total compounded feeds produced in the country. In addition, there are 23 manufacturers who have formed a Compound Livestock Feeds Manufacturer's Association (CLFMA). The production figures of CLFMA members during the last nine years are given in Table 11 (CLFMA, 1972), from which it will be seen that the total quantity of cattle and poultry feeds they produced in 1972 was 371,500 t. Thus, the total production of these feeds in India may be estimated at 750,000 t. Production figures of CLFMA members show that increase in cattle and other feeds over 1971 was 30.2%, whereas the increase in poultry feeds in 1971 was 59.5%. The total production of compounded livestock feeds was up by 43% in 1972, when compared to the total feeds produced in 1971.

TABLE 11. *Production of compound livestock feeds (excluding public sector and small scale units)*

Year	Cattle and other feeds	Poultry feeds	Total
1964	25.0	14.4	39.4
1965	50.3	28.4	78.7
1966	69.4	39.2	108.6
1967	91.5	42.6	134.1
1968	94.5	47.1	141.6
1969	131.3	57.7	189.0
1970	125.4	84.3	209.7
1971	144.8	114.7	259.5
1972	188.5	183.0	371.5

Other salient features of the industry are the following:

(1) The number of plants registered an increase from 42 in 1971 to 52 in 1972, and there now are only a few parts in the country where compounded feeds are not readily available. This is a particularly significant step forward, since the freight costs make it difficult for any one manufacturer to supply feeds outside the radius of approximately 300 km from the point of production at a reasonable cost to the farmer.

(2) During 1972 there was a tremendous shortage of raw materials and their prices showed an unprecented rise; the restrictions on state movements on cereals and cereal by-products especially affected and delayed supplies considerably.

(3) After detailed discussions with scientists, CLFMA members were convinced that de-oiled sal meal (*Shorea robusta*) could be safely utilised at a level of 20% in cattle and 5% in poultry feed.

Raw materials

Feed grains

The Food Corporation of India (FCI) is the main supplier of raw materials to various cattle/poultry feeding units falling under Co-operative and State Agro-industrial Corporations. Damaged foodgrains supplied by FCI to these units is of the order of about 8,000 t/a.

Generally, there is no shortage of cereal processing by-products like wheat bran, rice bran, husk, etc. However, their composition varies considerably as their suppliers cannot afford to carry out quality control on them. Nevertheless, some feed producing units have now set up their own quality control laboratories which test the ingredients with a view to either rejecting the supplies or adjusting the price on the basis of quality.

TABLE 12. *Export of deoiled oilcakes and rice bran during 1972–1973*

Oilcake	Quantity (000 t)	Value (000 Rupees)
De-oiled rice bran	123	26.826
Solvent extracted oilcake of copra	5	1.975
Solvent extracted groundnut oilcake	735	585.193
Solvent extracted linseed oilcake (decorticated)	26	20.282
Solvent extracted cottonseed oilcake (decorticated)	152	111.886
Solvent extracted cottonseed oilsake (undecorticated)	4	3.129
Solvent extracted kardi oilcake	65	15.597
Solvent extracted sesamum oilcake	1	512
De-oiled defatted oilcake meal	6	2.897
Solvent extracted niger oilcake	1	718
Total	1,118	796.015

Oilcakes

The total quantity of various oilcakes available in the country is over 4 m. t. Of this, about 25% is exported (see Table 12). The remaining 75% is available for animal feeding, and production of edible oilseed flours for human consumption (Rajan, 1973).

Fish meal

The current production of fishmeal in the country is of the order of 1,000 t although the installed capacity is much higher. There are about 10 fishmeal plants with daily raw material input capacities ranging from 5 to 40 t. The financial success of fishmeal plants depends on assured availability of fish in adequate quantity of over, say, 200 days in a year at a workable cost of less than Rs.200/t. However, such a situation hardly exists in and around any of those plants. To overcome this bottleneck the Central Institute of Fisheries Technology has designed small capacity fishmeal plants.

Blood meal

In a few slaughter houses in the country, blood is being collected suitably and transported to the production centre quickly enough for suitable processing. Since slaughter houses in the country (with the exception of a few) are non-mechanised, considerable difficulty is being experienced in obtaining the desired quality of blood. Realistic plans have been formulated for improving the slaughter houses in the country and this should enable increased production of blood meal.

Conclusions

(1) With the present knowledge available it has been possible to formulate Indian standards on all aspects of the compounded feeds industry. During this standardisation process, there was a meaningful dialogue among scientists from research and industry to their mutual benefit.

(2) Within a decade the compounded livestock feed industry in India has shown considerable development.

(3) Egg production has shown rapid progress and this tempo is likely to be maintained in the Fifth 5-Year Plan. For cattle development a strong infra-structure has now been built up and this would be expanded further and strengthened during the next few years.

(4) This would imply that there should be a faster growth of the compounded livestock feeds industry in India. The only vulnerable point is the assured supply of adequate quantities of raw materials.

(5) In the context of raw material shortage and development of improved livestock, quality control of feeds on the basis of Indian standards already formulated and revised to meet various viewpoints, will play a crucial role in the achievement of modest targets for milk and egg production.

(6) Complete quality control of feeds can be an elaborate process and lot to lot testing for certification can be possible only by examining a limited number of characteristics. For other important characteristics manufacturers have to adopt additional quality control measures. Other specific problems have to be dealt with on an individual basis.

(7) The importance of this limited quality control should not be underestimated for it would enable good supervision of some basic and important factors.

(8) Organisation of even this limited quality control effort for huge quantities of feeds involved presents formidable problems. But these must be overcome, for without this basic attention to quality control especially by smaller feed manufacturers improved livestock will not give the expected yields.

Thus effective quality control measures for livestock feeds with Indian standards as the base are a *sine qua non* for all animal husbandry development programmes in India. If this hypothesis is accepted, it is imperative that a small fraction of financial appropriations for the animal husbandry programmes should be allocated to quality control of animal feeds.

References

The Compounded Livestock Feeds Manufacturer's Association of India (1972) Annual Report, India.

Indian Standards Institution (1970) Quality control of animal feeds and feeding stuffs; *ISI Bulletin,* **22** No. 1, 3–8.

Rajan, S. S. (1973) Oilseeds availability and production, Souvenir, 38–40, New Delhi, India, Association of Food Scientists and Technologists (Northern Zone), 3.

Sen, S. K. (1969) Standardization and its principles, 4th article, Iran, Institute of Standards and Industrial Research of Iran, 3.

Sen, K. C. & Ray, S. N. (1964) Nutritive values of Indian cattle feeds and the feeding of animals *ICAR Bulletin* No. **25**, 22–133, New Delhi, Indian Council of Agricultural Research; 113.

Appendix A

Feed ingredients

1. *Oilcakes*

IS:1712–1970	Specification for cottonseed oilcake as livestock feed *(first revision)*
IS:1713–1970	Specification for decorticated groundnut oilcake as livestock feed *(first revision)*
IS:1932–1972	Specification for mustard and rape oilcake as livestock feed *(first revision)*
IS:1934–1961	Specification for sesamum (Til) oilcake as livestock feed
IS:1935–1961	Specification for linseed oilcake as livestock feed
IS:2151–1962	Specification for maize germ oilcake
IS:2154–1972	Specification for coconut oilcake as livestock feed *(first revision)*
IS:2053–1963	Specification for decorticated safflower (Kardi) oilcake as livestock feed
IS:3440–1966	Specification for solvent extracted linseed oilcake (meal) as livestock feed
IS:3441–1966	Specification for solvent extracted groundnut oilcake as livestock feed
IS:3591–1968	Specification for solvent extracted coconut oilcake (meal) as livestock feed *(first revision)*
IS:3592–1968	Specification for solvent extracted cottonseed oilcake (meal) as livestock feed *(first revision)*
IS:3593–1968	Specification for solvent extracted rice bran as livestock feed *(first revision)*
IS:5862–1970	Specification for solvent extracted nigerseed oilcake (meal) as livestock feed
IS:6242–1971	Specification for solvent extracted safflower oilcake (meal) as livestock feed
IS:7061–1973	Solvent extracted *Sal* seed meal
IS:1714–1960	Methods of sampling and test for oilcakes as livestock feed *(first revision)*

2. *Grain by-products*

IS:2152–1972	Specification for maize gluten feed *(first revision)*
IS:2153–1962	Specification for maize bran
IS:2239–1971	Specification for wheat bran *(first revision)*
IS:3160–1965	Specification for gram *Tur chuni*
IS:3161–1965	Specification for gram *Chuni*
IS:3162–1965	Specification for gram husk
IS:3163–1965	Specification for rice polish
IS:3648–1966	Specification for rice bran
IS:4193–1967	Specification for guar meal as livestock feed
IS:5063–1969	Specification for *Tur* husk

3. *Animal waste and other agricultural/industrial by-products*

IS:1162–1958	Specification for cane molasses
IS:1509–1972	Specification for tapioca as livestock feed *(first revision)*
IS:3336–1965	Specification for shark liver oil for veterinary use
IS:4307 1967	Specification for fishmeal as livestock feed
IS:5064–1969	Specification for tapioca spent pulp
IS:5005–1969	Specification for meat meal and meat-cum-bone meal as livestock feed
IS:6107–1971	Specification for dried silk worm pupae as livestock feed
IS:7060–1973	Specification for blood meal
IS:3198–1965	Specification for fodder yeast

Nutritional composition of Latin American feeds

L. R. McDowell, J. H. Conrad and Jenny E. Thomas
University of Florida, Gainesville, U. S. A.
and
L. E. Harris
Utah State University, Logan, U. S. A.

Summary

Information on the composition of feeds produced in Latin America has been collected and published in the form of feed tables. These are available as an atlas containing 3,390 feeds and an abridged version containing 650 feeds. The atlas is available only in English while the abridged version is also available in Spanish and Portuguese.

Résumé

La composition alimentaire des pâtures Latino-Américaines

Des informations sur la composition des pâtures produites en Amérique Latine ont été collectionnées et publiées sous la forme de tables d'alimentation. Ces-ci sont disponibles comme un atlas contenant 3,390 pâtures et une version raccourcie contenant 650 pâtures. L'atlas est procurable seulement en Anglais tandis que la version raccourcie est disponible aussi en Espagnol et Portugais.

Resumen

Composición nutritiva de los piensos de América Latina

Se ha recogido información sobre la composición de los piensos producidos en América Latina y se ha publicado en forma de tablas de piensos. Hay, también, una edición de estas tablas en forma de atlas que contiene 3.390 piensos y una versión abreviada que contiene 650 piensos. Del atlas existe la versión inglesa únicamente, mientras que de la edición abreviada hay también versiones en español y portugués.

Little information on the composition of Latin American Feeds is readily available. Although feed composition has been determined in many Latin American countries, little data has been widely published. Consequently, published North American and European feed tables are utilised in ration balancing. In 1969 a University of Florida/AID project was initiated to gather and compile Latin American feed composition data. Feeds entered on returned source forms were identified by international names and number with valid data being computerised and recorded on magnetic tapes.

Since the initiation of the project over 35,000 completed source forms have been received at the project centre from 69 laboratories in 21 Latin American countries. A preliminary edition of feed tables was published June 1972 based on 6,152 source forms. Less than 1½ years later a more complete edition of Latin American Feed Tables based on 27,196 source forms was available both in atlas (3,390 feeds) and abridged (650 feeds) editions. The abridged edition is published in English, Spanish and Portuguese, while the atlas is available only in English. In addition to Latin American Feed Tables, individual country and laboratory feed tables are also available to participating collaborators.

In the atlas edition the percentage of feed classes were as follows: forages 77.1; energy feeds 12.5; protein feeds 9.6, and others 1.0. Only 15.2% of the total feeds in the atlas edition were also found in the joint US–Canada international file versus 33.4% for the combined 713 energy and protein feeds. Of the total data published only 6% is combined with US–Canada feedstuffs. Because only 182 of the 3,390 feeds contained digestible protein and/or energy values, prediction equations were used to calculate additional values. The atlas contained the following nutritional information for the 713 energy and protein feeds as

254 a percentage: complete proximate analyses 94.5; Ca 49.4; P 48.5; Mg 9.1; K 7.2; Na 4.5; Cu 7.0; Fe 28.1; Mn 6.7; other minerals < 2.0; ascorbic acid 13.0; niacin 18.5; riboflavin 15.4; vitamin A equivalent 8.7; other vitamins < 1.0; lysine 9.1; methionine 9.0; tryptophan 7.0; and other amino acids 3.5–8.3. Of 713 energy and protein feeds the percentages containing 1 or more macromineral, trace mineral, vitamin or amino acid analyses were 50.5; 28.2, 18.5 and 9.1% respectively.

Abattoir by-products: potential for increased production in developing countries

S. K. Barat

Agricultural Services Division, FAO, Rome

Summary

The abattoir by-products resources of the developing world are objectively appraised as to their current status and scale of utilisation both at the international and domestic levels. There is a general lack of awareness of the potential economic worth of a wide range of these abattoir by-products in the developing countries, particularly of the inedible variety, leading to sustained waste of these valuable raw materials. These by-products, because of their essential linkage with meat production, are critically considered in the context of the meat economy of the developing world, the needs and requirements of which are assessed in relation to its constraints and constrictions, problems as well as promise.

Résumé

Les sous-produits d'abattoir: une possibilité d'augmentation de la production dans les pays en développement

Les ressources en sous-produits des abattoirs de monde en développement sont évaluées objectivement, concernant leur condition actuelle et l'échelle d'utilisation au niveau international et domestique. Il y a un manque total de conscience de la valeur du potentiel économique d'une grande série de ces sous-produits d'abattoir dans les pays en voie de développement, surtout de la variété immangeable, menant à des pertes subies de ces matières premières précieuses.

Ces sous-produits, à cause de leur connection essentielle avec la production de viande, sont examinés d'une manière critique au sujet de l'économie de la viande dans les pays en développement; les besoins et les exigences sont évaluées en rapport avec les contraintes et les diminutions, – des problèmes aussi bien que des promesses.

Resumen

Productos secundarios del matadero: potencial de aumento de producción en los países en vías de desarrollo

Se valoran objetivamente los recursos de productos secundarios del matadero del mundo en vías de desarrollo en cuanto a su estado actual y escala de utilización tanto a niveles n internacionales como nacionales. Hay una falta general de conocimiento del valor económico potencial de una variedad amplia de estos productos secundarios del matadero en los países en vías des desarrollo, particularmente de la variedad incomible, que conduce a un desperdicio sostenido de estas materias primas valiosas. A causa de su conexión esencial con la producción de la carne, estos productos secundarios se consideran críticamente en el contexto de la economía de la carne del mundo en vías de desarrollo, las necesidades y exigencias del cual se estiman en relación con sus limitaciones y estrecheces, sus problemas así como sus promesas.

Preface

There is a general lack of awareness in the developing world of the various economic possibilities in the field of animal by-product utilisation, of options open for commercial exploitation of some of these resources and of the more remunerative export outlets available for the same in the processed and semi-processed form. There are three major areas of optimum utilisation of inedible meat and meat by-products derived from both domestic and wild life sources which are of particular interest to developing countries (Barat, 1973).

1. As protein supplements in feed formulations

One of the major constraints that stands in the way of livestock/poultry/piggery development programmes today in the Third World is the lack of adequate supplies of feedstuffs especially feedstuffs at economic prices. The most costly component of these feed formulations is the protein supplement needed to ensure the optimum amino acid balance, and utilisation of inedible slaughter-house offals and wastes as a source of protein concentrates for animal feed therefore affords attractive economic possibilities.

2. As the raw stock for petfood industries

The fast-growing petfood industry in developed market economy countries offers a remunerative outlet for developing countries rich in domestic and wild life meat resources for such meat and meat by-products as are rendered unacceptable to their own human consumption needs. A high value sophisticated market for sub-human grade slaughter-house products which usually remain un/under-utilised in most developing countries, affords in fact a much better unit return on the relevant raw stock than may be at best possible in practice through the conventional rendering operation. It is estimated (Dale, 1973) that the current annual value of petfood sales is of the order of $1,000 m in North America, $200 m. in West Europe and $35—40 m. in Australia. The industry expects a doubling of the market by 1977 and a doubling again by 1985. A recent study by the New York Exchange (1973) shows that Americans are now spending more to feed pets than to feed babies. The survey reveals that 38% of US households own at least one dog and 22.6% have one or more cats. The present demand of the international petfood industry for abattoir by-products is estimated to be over 1 mt per year.

3. As phosphatic and nitrogenous concentrate for livestock and agricultural development

Many of the arid and semi-arid areas of the developing world where most of its livestock is born and bred have soil and pasture which to a large extent is chronically deficient in phosphorus and calcium, and because of the lack of these minerals, animals sustained on these pastures are unable to make full use of the food available. This phosphatic deficiency in livestock, however, can be easily remedied by feeding phosphates and the best form of such phosphates is sterilised bone-meal made from bones which are locally available. Ironically enough, countries which suffer from such chronic phosphorus deficiency frequently export vast quantities of bones instead of using them locally to improve their soil, pasture and livestock. As a source of nitrogen, concentrated organics made up of hoof and horn meal, tankage (meat and bone), dried blood etc and containing up to 14% nitrogen, are in great demand.

In most developing countries, however, this sector of the economy suffers from gross lack of imaginative planning leading to wasteful but nevertheless wholly avoidable losses due to non/mal-utilisation of a whole range of valuable animal by-products including hides, skins, blood, bones, intestines, hair, horns, hooves and a host of other carcass-derived items. This has resulted in a paradoxical situation in as much as the very countries in dire need of proteins and minerals both for human and animal nutrition as well as for soil enrichment are also the ones which make the least use of them from their own readily available potential sources. Thus it is not uncommon to find a developing country wasting valuable slaughter-house blood down the drain at one end and importing blood meal at the other at considerable expense of scarce foreign exchange, exporting crude bones at nominal price while importing expensive bone derivatives, cattle licks and rock phosphates; throwing slaughter-house offals to jackals and vultures and at the same time continuing to import processed proteins and other nitrogenous concentrates for feed formulations, selling hides and skins in the raw, uncured condition while repurchasing the same from abroad as finished leather and leather products.

In this connection it is to be emphasised that the need for optimum utilisation of these high protein by-products is the foremost within the developing countries themselves where human needs should receive top priority over all other considerations. In the international market, however, these basic needs of a subsistence-oriented economy have often to compete unfavourably with more powerfully backed 'extra-human' requirements of the developed world.

Edible and inedible by-products — a costly waste

In the context of this paper abattoir by-products stand for all those parts of the slaughtered stock which by their very nature are inedible or declared to be unsuitable for human consumption, constituting what is more popularly known as the 'fifth quarter' of the animal. However, the distinction between the edible and the inedible components of the carcass is rather arbitrary depending upon a variety of factors like the general economic and social condition of the people, their eating habits and dietary practices, religious beliefs and sentiments, the status of animal health in the area, degree of control and standards of veterinary and public health inspection as well as the structure and system of working of the abattoir industry. Since these parameters vary to a large extent from country to country it is patently obvious that what goes as an inedible by-product in one place under conditions of relative affluence need not necessarily be so in another with comparatively modest means. Increased affluence generates demand for meat which is more quality oriented and as a result certain parts of the animal are declared inedible and hence become unsaleable for human consumption. Indeed, the higher the income of the consumer, the higher is the percentage of potential edible offal which is used for inedible purposes. Based on the currently accepted norms of veterinary and public health inspection and control it is generally recognized that on an average 55% of the liveweight of cattle accounts for the edible portion, the inedible by-products constituting the rest of the 44% the shrinkage and loss being of the order of one percent only. The corresponding average dressed yields of hog and sheep carcass are 70 and 47% respectively. (See Tables 1 and 2).

TABLE 1. *Average yield of a 454 kg steer* (kg/head)*

Products	kg
Dressed beef	254.0
Hide	27.2
Rumen	24.9
Dry Tankage	5.9
Dry blood	3.6
Tallow	5.4
Grease	0.7
Oleo oil	18.1
Oleo stearine	4.5
Conc. Tankage	2.3
Hoof	0.9
Shin bone	0.9
Skull	1.8
Horn	0.3
Horn pith	0.3
Jaw bone	0.9
Tail	1.1
Neatsfoot oil	0.4
Sinews	0.9
Glue	0.9
Tripe	9.0
Heart	1.1
Kidney	0.8
Liver	5.4
Sweet bread	0.2
Bung gut	1 piece
Middle guts	6 m
Round guts	30.5 m
Weasand	1 piece
Bladder	1 piece
Sausage meat	2.7
Brains	0.3
Tongue	3.0

*Based on rendered yields in ANCO Cookers (Allbright − Nell Co., Chicago, Ill., USA).

TABLE 2. *Average yield of by-products from slaughtered animals (kg per head)**

By-products	Cattle	Hog	Sheep
Tallow (stock)	27.2		
Tallow	5.4		
Oleo fats	31.7		
Oleo oil	18.1		
Oleo stearine	4.5		
Fats − General		6.3	
Leaf lard		2.7	
Lard		11.8	
Offal	34.0	13.6	4.5
Tankage (cooked)	27.2	9.0	1.8
Tankage (pressed)	13.6	4.5	0.9
Tankage (dry)	5.9	2.3	0.4
Tankage (concentrated)	2.6	1.1	0.1
Blood (liquid)	18.1	3.2	1.8
Blood (dry)	3.6	0.8−1.0	0.2
Bone to tankage	6.8	1.8	
Grease	0.7	0.4	0.2

*Base on rendered yields in 'ANCO' Cookers (Allbright − Nell Co., Chicago, Ill., USA).

The magnitude of post-harvest losses in the plant industry caused by faulty handling, lack of storage facilities, insect damage, fungi, bacteria and rodents has often been justifiably highlighted. Receiving much less publicity and attention but of equal magnitude, however, are the wastage and losses involved in the handling and processing of animal by-products from the time the animal is slaughtered until the relevant products reach the end users.

The basic problem in the developing countries in this field is the centralised collection, salvaging and effective disposal of these animal by-products from widely scattered areas in terms of general considerations of economy, health, hygiene, and sanitation. In most of the primary producing countries the full poten-tialities for the development of these by-product resources are yet to be realistically worked out, the subsector often remaining the most neglected and technically the least efficient. This is mainly because in the majority of cases management resources and physical facilities are overwhelmingly deployed towards maximisation of prime carcass meat output and the optimum utilisation of by-products typically takes a very secondary place in the scale of priorities for pro-duction planning. As a general rule the less developed the economy the greater this disparity in emphasis appears to be. This is most unfortunate because it is precisely in a developing economy that such a need for economic utilisation of the by-products leading to increased return per animal as an effective incentive to primary producers remains paramount. Moreover, it is essentially a labour intensive industry oriented towards export promotion and diversification as well as import substitution affording maximum utilisation of in-digenous raw materials within the country. This in turn would stimulate development of a number of other vitally important ancillary processing industries creating additional employment opportunities in the small scale sector, particularly in the rural areas. Whereas destructive disposal of these valuable resources will constitute a downright waste at considerable recurring cost of labour, transport and conversion plant services, their optimum use, reuse and recycling as raw materials for a number of secondary ancillary industires will convert them into a positive source of profit revenue. Waste of this sort is therefore a costly luxury a developing country can ill afford and yet the tech-nology of the maximum exploitation of animal by-products on an economically viable scale has been neglected so much so that non/mal-utilisation of these potentially valuable raw materials still constitutes a veritable 'drain' on national wealth in most parts of the developing world. India, for instance, loses annually over US $70 m by not optimally utilising her available animal by-products (Barat, 1965). Besides, such waste has other implications as well, and indirectly leads to further losses. It ends to keep the cost of the primary product viz, meat, relatively high and to that extent restricts is com-petitive capacity, particularly in the international market, much to the detriment of the country's economy. In the domestic market too, failure to utilise optimally its by-products adds to the cost of meat production, thus pricing it out of the purchasing power of the majority of the country's most needy consumers. Moreover, such better utilisation of by-products will be immediately reflected in the increased net return per animal, thus ensuring the much needed additional incentive to the primary meat producer, ie the stock owner, for further improvement of his stock, the level and distribution of this personal cash income remaining the dominant factor in livestock production. Indeed, rational utilisation of the resultant by-products may well be a major factor in the overall profitability of the meat producing industry. In the case of small animals particularly, by-products provide as great an income as the sale of the dressed carcass.

However, a qualifying minimum throughput is generally required for ensuring the full benefit of the economics of scale in animal by-products utilisation. An annual throughput of at least 30–40,000 animals is required for setting up a mass production line in the abattoir as opposed to individual slaughtering which is mostly the case in developing countries. This is mainly the result of consumer preference for warm meat leading to the practice of daily slaughtering of a relatively small number of animals for consumption on the same day. Although it may not be quite possible to fully utilise these by-products in many of the slaughtering establishments in the developing countries mainly because of limited throughput, it is nevertheless felt that with the provision of elementary but basic facilities of handling, cleaning and washing and application of rather rudimentary technology, much of this potentially valuable raw material may be salvaged and, through preliminary preservation and preparative treatment in the place of production, could be rendered technically well worthwhile for subsequent transport and storage pending fuller systematic processing at convenient centres. In this connection it is to be noted, however, that in most developing countries abattoirs are generally managed by municipal councils with a view to avoiding monopoly exploitation by the private sector. These invariably lack investment capital, technical experience, marketing expertise and managerial talent to conduct the business efficiently. The authorities often take refuge under unrealistic taxation policy which results in evasion, corruption and nepotism.

Linkage with meat production

Since these by-products are mainly incidental to meat production, their supply by its very nature is inelastic and independent of the pressure of demand of their respective end uses. The volume of their actual output is thus to a large extent determined by the consumer demand for meat, an increase in which is generally reflected in an expansion of the corresponding flock and herd size. However, it is the relative emphasis on the particular aspect of the animal use as guided by contemporary economic conditions that substantially influence this output and availability of the relevant animal by-products. There are three main ways of expanding meat output through increase in:

(i) the number of animals in national herds and flocks;

(ii) the percentage offtake especially in areas of high stock density;

(iii) the live weight of animals at slaughter.

In some of the developing countries on the other hand, especially those with nomadic agriculture, livestock holding is reckoned as a symbol of prestige, wealth and status rather than a marketable resource putting undue premium on number instead of quality and as a result slaughtering is limited to the minimum with a view to maintaining or increasing the size of the herd or flock. In this way, large numbers of livestock are held under pastoral, tribal nomadism and transhumance systems. Under the impact of modern market oriented economy such traditional values are of course undergoing slow but steady change from livestock raising for prestige to livestock farming for profit.

The future and its needs

There is little doubt that the world demand for meat will go on increasing with a projected shortfall in supply at the end of the present decade by a little less than half of the current volume of meat trade (FAO, 1970). Europe is destined to become less self-sufficient in beef. These facts are of particular interest to the developing countries whose meat production potentials are not efficiently used as yet. The world meat market is a dynamic growth area. The value of meat trade rose by 30% over the preceding year to 6.1 billion in 1972. The world gross exports of beef and veal rose by $900 m. in 1972 (FAO, 1973). These are impressive figures and offer to the Third World substantial scope for export of meat. Apart from nutritional requirements in developing countries, meat consumption being highly income elastic in the initial stages of development, there appears to be good prospect of its increased demand in domestic markets particularly in the industrialising and semi-industrialised sectors. In this context pigs and poultry have great promise for increasing meat production due to their prolificity, early slaughtering maturity and remarkable ability to convert meal into meat. Besides, the production potential for many other animal resources in the developing countries needs to be systematically explored. In furthering the supplementation of national protein intake, utilisation of game resources obtainable through the cropping and culling programme of wildlife management in a number of developing countries merit due consideration.

A consistent and integrated framework of development has to be evolved. The long run nature of the basic interrelated problems outstanding in the field needs sustained efforts and the parameters of development must cover the entire spectrum of livestock improvement through disease control and animal health measures to improved breeding and feeding. The

TABLE 3. *Infrastructure for systematic development of meat and meat by-products*

1. Improvements in animal nutrition through supplementary feeds and concentrates; reserve of balanced feedstuffs, preferably based on indigenously available ingredients.

2. Range and pasture development through ecologically oriented optimum land use; utilisation of crop marginal uplands and fallow lands for pasture and rainfed or irrigated fodder production.

3. Progressive phasing out of the export of agricultural and agro-industrial by-products and feed ingredients for increasing use in indigenous feed formulations and grain-sparing animal rations.

4. Upgrading of local stock through cross-breeding where appropriate and selection programme based on recorded heritable characteristics of national herd; raising of quality livestock in numbers commensurate with improved feed supplies and environmental potentialities and capable of effectively utilising such available resources.

5. Control of major infections and epizotic and tick-borne diseases through sustained campaign of vaccination and dipping; setting up of immunised belts and disease free zones and effective zoosanitary regulations governing meat industry.

6. Transformation of traditional livestock raising from subsistence to commercial level; stratification of animal production through mixed farming system; establishment of ranches and feedlots modified to suit local conditions.

7. Modernisation of marketing system and rationalisation of price structure for live animals and meat at a remunerative level to ensure increased supply and offtake; quality control, grading and standardisation of all abattoir products.

8. Establishment of functional slaughterhouses under a viable system of operation with basic facilities for economic recovery of available by-products; development of organised stock routes, watering points, and road and rail transport; location of semi-intensive fattening centres close to slaughter points.

9. Centralisation of slaughtering operations at selected sites in rural areas and salvaging of all available animal by-products; provision of minimum basic amenities for their handling, collection and primary preservation.

10. Carrying out of exploratory animal by-products resource surveys and assessment of economic loss sustained due to their non/mal-utilisation; creating public awareness of their economic importance through pilot production, training, demonstration and factual publicity.

requisite infrastructure for an economically viable meat and meat by-products industry (Table 3) among others must include a workable system for grading and marketing of all commercially exploitable abattoir products. Problems have to be considered in their totality; attacking one or two on an *ad hoc* basis out of context of the whole would be counter-productive in the end.

The growth of abattoir by-products will follow the growth of the meat industry. The projected growth of meat production in developing countries for the decade 1970–1980 is estimated at 2.3% per year (7), which is not very different from the actual rate observed in the past. In the light of the present review of constraints and opportunities facing the livestock production in the developing world, it is most likely that given the minimum of requisite input a higher overall rate of growth will be achieved than was attained over the last decade.

References

Barat, S. K., (1965) *Leather Science,* Aug, **304.**

Barat, S. K., (1973) The profitable utilisation of by-products from the leather and allied industries. UNIDO, Vienna, ID/WG. 157/22, pp 1–13.

Dale, M. B., (1973) Commodities and Trade Division, FAO, Rome, Personal Communication.

FAO (1970) Agricultural Commodity Projections, 1970–80.

FAO (1972) Production Year Book.

FAO (1973) Report of the third session of the Inter-Governmental Group on Meat.

New York Stock Exchange (1973), Survey Report.

Lesser known oil seeds: a note on their nutritive value as determined by *in vitro* digestion

O. L. Oke

Department of Chemistry, University of Ife, Nigeria

and

I. B. Umoh

Department of Biochemistry, University of Benin, Nigeria

Summary

The *in vitro* digestibilities of the oven-dried chloroform extracted Nigerian oilseeds *Irvingia gabonensis*, *Parkia filicoidea* and *Citrullis vulgaris* were determined by both pepsin/trypsin and pepsin/pancreatin digestion. The results obtained showed good correlation with figures for true digestibility determined by rat feeding experiments. It is considered that *in vitro* enzymic digestion could be useful for screening potential protein foods.

Résumé

Des graines oléagineuses moins connues; une note sur leur valeur nutritive déterminée par la digestion *in vitro*

La digestibilité *in vitro* des extraits chloroformés, séchés au four, des graines oléagineuses de Nigeria, *Irvingia gabonensis*, *Parkia filicoidea* et *Citrullis vulgaris*, a été déterminée par la digestion par pepsine/ trypsine et pepsine/pancréatine. Les résultats obtenus ont montrés une bonne corrélation avec les chiffres concernant la vraie digestibilité, déterminée par des expériences d'alimentation des rats. On considère que la digestion enzymatique *in vitro* serait utile pour trier des éventuels aliments protéiniques.

Resumen

Las semillas oleaginosas menos conocidas: una nota sobre su valor nutritivo segun se ha determinado por la digestión *in vitro*

Las digestibilidades *in vitro* de las semillas oleaginosas de Nigeria *Irvingia gabonensis*, *Parkia filicoidea* y *Citrullis vulgaris* de las que se había extraído el cloroformo mediante el secado al horno, se determinaron por la digestión tanto pepsina/tripsina como pepsina/pancreatina. Los resultados obtenidos demostraron buena correlación con las cifras de la digestibilidad verdadera determinadas por experimentos alimentando ratas. Se considera que la digestión enzímica *in vitro* podría ser útil para examinar los alimentos con proteínas potenciales.

Introduction

Although animal experiments give the most correct results on the biological value of a protein, they are long, laborious and expensive. In cases where it is desired to compare several concentrated protein feeds, the animal method is not convenient and so there have been many attempts to introduce other methods. Also amino acid analysis does not tell how available the acids are especially in cases where heat is involved in the production of the protein concentrate which may lead to some loss in available lysine (and methionine).

Since digestibility may affect the nutritive value of a protein feed, the susceptability to enzymatic hydrolysis has provided a good method of estimating the digesti-bility or biological value of a protein. This method is quick and requires only a small amount of the sample. Akeson & Stahman (1965) have shown that the biological value (BV) of a feed protein could be predicted accurately from the amino acids released by a pepsin followed by pancreatic digestion. With 12 typical feed proteins they observed an excellent correlation ($r = 0.990$) with biological values obtained from rat feeding trials.

Using methods developed by Saunders & Kohler (1973) this work reports the results obtained with pepsin/ pancreatin and pepsin/trypsin *in vitro* hydrolysis on some of the lesser known oilseeds in Nigeria which are compared with results obtained by rat feeding trials. A good correlation between *in vitro* digestibility and BV would facilitate the subsequent work on the nutritive value of different protein fractions obtained from these oilseeds and other concentrated protein sources.

Materials and Methods

The oilseeds, *Irvingia gabonensis*, *Parkia filicoidea* and *Citrullis vulgaris* were brought from a local market.

(a) *Irvingia gabonensis* seeds were dried in the oven at 98°C for 24 hours and then ground up in small sample mill. The product was extracted continuously with chloroform and then air-dried.

(b) *Parkia filicoidea* seeds were boiled for 5 hours with water, the water was decanted, and cold water added. The pot was covered up and the material left to ferment for 3 days, after which it was washed with water and the softened testa removed. The seeds were then dried in the oven at 90°C for 24 hours, ground in a mill and subsequently extracted with chloroform, and the residue was then air-dried.

(c) *Citrullis vulgaris* seeds have very tough testa, they were cracked to release the seeds which were then dried in an oven at 90°C for 24 hours. They were then ground in a small sample mill and extracted continuously with chloroform and air-dried. The oilseeds were subjected to enzymatic digestion using the methods of Saunders and Kohler (1973) in which the undigested protein is analysed for nitrogen. For the pepsin/trypsin digestion 1 g of the oilseed was suspended in 1 ml of 0.01 N HCl and shaken gently for 48 hr at 95°C. After centrifuging, the residue was resuspended in 10 ml of water and 10 ml 0.1 M sodium phosphate, buffer, pH 8.0, and treated with 5 mg of trypsin. After shaking at 70°F for 16 hr it was again centrifuged, the residue washed, air-dried and analysed for nitrogen.

For pepsin/pancreatin digestion a sample of about 250 mg weight was suspended in 15 ml of 0.1N HCl containing 1.5 mg pepsin and gently shaken at 95°F for 3 hr. The suspension was neutralised with 0.5N NaOH, treated with 4 mg pancreatin in 7.5 ml of 02.M phosphate buffer (pH 8.0 containing 0.005M sodium azide) and shaken again for 24 hours at 95°F. The suspension was centrifuged and the residue was washed and analysed for total nitrogen content.

Results and discussion

The results obtained for both enzymatic hydrolyses using pepsin/trypsin and pepsin/pancreatin agreed very closely. *P. filicoidea* had the highest digestibility by enzymic hydrolysis, 97.3 and 97.8 followed by *C. vulgaris*, 96.4 and 97.0 and *I. gabonensis* 93.5 and 93.9. These figures were in accordance with the results obtained for the true digestibility (TD) of the oilseeds as shown in Table 1, 97.6, 96.2 and 93.5 respectively.

TABLE 1. *In vitro digestion of oilseeds*

	Digestibility (%)		
	pepsin/ trypsin	pepsin/ pancreatin	Rat assay
Citrullis vulgaris	96.42	97.03	96.15
Irvingia gabonensis	93.51	93.93	93.50
Parkia filicoidea	97.34	97.82	97.62

For these oilseeds therefore the *in vitro* method seemed to agree closely with the TD determined by animal experiments, but did not follow the same pattern as the BV unlike that found by Akeson and Stahman (1965) who determined the protein quality of 12 proteins using the amino acids released by an *in vitro* digestion with pepsin followed by pancreatin and obtained a correlation coefficient of 0.990 with BV.

Saunders and Kohler (1973) on the other hand using their simplified and quicker procedure obtained as good a correlation as with the original method when used on leaf protein samples dried under different conditions. They also found that their method correlated well with the TD as was found with the oilseeds used in this study.

Both BV and TD are important in deciding the nutritive value of a protein food as the net protein utilisation (NPU) is obtained as a product of their values.

It therefore looks as if the *in vitro* enzymatic digestion method could be used for screening potential protein foodstuffs because of its good correlation with TD.

No definitive conclusions could be arrived at yet since the number of samples was too small and not subjected to statistical analysis. This will be done later.

References

Akeson, W. R. & Stahman, M. A. (1965) Nutritive value of leaf protein concentrate, an *in vitro* digestion study. *J. Agri. Food Chem.*, **13** (2), 135–147.

Saunders, R. M. & Kohler, G. P. (1973) Pers. Comm.

Discussion

Dr Mayer: Could Professor Harris please expand on the efforts currently being made with regard to the recommendation to feed analysts of analytical methods of particular relevance from a nutritional point of view.

Professor Harris: With regard to the assessment of feeds by chemical analysis it is not normally necessary to determine ether extract or ash. Also crude fibre need only be determined when it is needed particularly to control this. With regard to forages, an *in.vitro* dry matter yields the most information and then nitrogen, cell walls digestible and metabolisable energy might be determined. In other words should we spend our time determining the ether extract, ash or crude fibre contents of forages or might it not be better to determine calcium and phosphorus or run some biological evaluations such as digestible and metabolisable energy?

Dr Osuji: Could Dr Bhagwan comment on the types and ages of the cattle to which the requirements given in Tables 5 and 6 of his paper refer. The minimum crude protein content of 20% DM seems to be particularly high and would seem to be wasteful when given to mature cattle.

Dr Bhagwan: It was decided to have only one standard for compounded cattle feeds so as to help towards possible mass production. Requirement of different classes of animals may be arrived at by adjusting the quantity of compounded feed which is fed.

Mr Woo: I would appreciate any comments Professor Harris may care to make on the relative merits of digital and analogue computers for feed formulation. In the example you have given it is presumed that a digital computer was used.

Professor Harris: I am afraid that I am not an expert on computers and cannot comment on the relative merits of digital and analogue computers. We were using a Burroughs 64000 digital computer but an IBM 360 or 370 or one of the ICL range could be used.

Dr Topps: I obtained the impression that in Professor Harris's attempt to establish international names for particular animal feeds, some very long names were being arrived at. This could cause the use of abbreviations which could in turn lead to confusion especially if more than one language is being dealt with.

Professor Harris: This has been considered and in addition to the long names presently used in the atlases we have published there is a short name on the computer. The long name is perhaps best used for teaching purposes and the short name for computer work. In addition to the name each feed has a five digit number assigned to it which is called the international feed reference number. When data is printed out from the computer this number as well as the name is given.

Professor Abou-Raya: Our experience in Egypt suggests that prediction equations of the type mentioned in Professor Harris's paper have only limited applicability, their use should be confined to the actual conditions under which the data is to be used.

Professor Harris: The work I described embodies two phases; the first phase is the accumulation of feed composition data which is being put into tables. My next point was that since the net energy for example varies in different conditions it might be better to use regression equations rather than constant tables. I do not believe that a regression equation used say in Utah would work in Egypt so each person in each area must have his own set of equations.

Seventh Session

Contamination of animal feeds by pests

Friday 5th April

Chairman
Dr W. Y. Magar
African Groundnut Council
Lagos
Nigeria

Pest control and storage problems in feed commodities in tropical and subtropical countries

F. Ashman

Tropical Products Institute, Tropical Stored Products Centre, Slough, England

Summary

Physical damage and quality deterioration is usually worse in the storage of non-processed raw materials.

Agricultural, technical and extension services should act to produce better quality raw materials by introducing improved harvesting, drying, storage and handling procedures.

Chemical changes occurring result in progressive deterioration of commodities as storage periods lengthen. Some control over this situation is essential and must be given special consideration during initial planning.

Storage structures at provender mills should be designed to prevent violent fluctuations of temperature and humidity, be easy to maintain, easy to clean and be suitable for the efficient application of pest control measures.

Segregation of raw, partially processed and processed feed from each other and from the processing plant (which in turn should be easy to clean) is basic in the initial planning of storage design and production flow. Second-hand bags are a major source of insects.

Dry mixing in particular will not eliminate insects and pests will be widely distributed and as a result cross infest clean produce on farms. This may be particularly dangerous if imported raw or partially processed commodities are responsible for the introduction and distribution of new pests.

Transport systems may be cross-infested so that processed feed and other high value commodities are contaminated, eg coffee, cocoa, tea, confectionery nuts, pulses, etc.

Treatment of animal feed commodities with pesticides may result in hazardous residues or reach levels unacceptable in some importing countries.

Efficiency of pest control depends on initial damage, moisture content, processing methods used, packaging techniques, etc.

Résumé

L'extermination des insectes nuisibles et les problèmes de stockage des pâtures dans les pays tropicaux et sous-tropicaux.

Les dégâts physiques et la détérioration de la qualité est d'habitude pire dans le stockage des matières premières non-traitées. Les services d'agriculture et les services techniques et auxiliaires devraient agir pour produire des matières premières de meilleure qualité, en introduisant des procédés améliorés de récolte, séchage, stockage et manipulation. Les changements chimiques qui se produisent, résultent dans la détérioration des produits à mesure que les périodes de stockage se prolongent. Une surveillance de cette situation est essentielle et il faudrait donner une considération spéciale pendant la planification initiale.

Les constructions pour le stockage des produits des moulins pour pâtures devraient être conçues pour prévenir les fluctuations violentes de température et d'humidité, être facile à maintenir et à nettoyer et convenables pour l'application efficiente des mesures pour l'extermination des insectes nuisibles. La séparation des pâtures brutes, partiellement traitées et complètement traitées et leur stockage loin de l'usine (qui à son tour devrait être facile à nettoyer) est fondamental dans la conception initiale du plan de stockage et l'écoulement de la production. Les sacs d'occasion sont une source majeure d'insectes.

Le mélange à sec en spécial, n'élimine pas les insectes qui ainsi seront largement distribuées et comme résultant vont infecter les produits purs dans les fermes. Ceci pourrait être particulièrement dangereux si des produits importés, bruts ou partiellement traités, sont responsables pour l'introduction et la distribution d'autres insectes.

Les systèmes de transport pourraient être infestés par infection croisée, de sorte que des aliments traités et des produits de grande valeur seront contaminés, par ex:

le café, le cacao, le thé, les noix pour la confiserie, les légumes à gousse, etc.

Le traitement des pâtures pour les animaux avec des pesticides pourrait produire des résidus hasardeux ou atteindre des niveaux qui ne sont pas acceptés dans certain pays importateurs.

L'efficacité de l'extermination des insectes nuisibles dépend de dégâts initiaux, du contenu en humidité, des méthodes de traitement et des techniques d'emballage.

Resumen

Control de pestes y problemas de almacenamiento en productos para piensos en los países trópicales y sub-trópicales

El daño físico y el deterioro de la calidad son generalmente peores en el almacenamiento de las materias primas no tratadas.

Los servicios agrícolas, técnicos y de extensión deben actuar para producir materias primas de mejor calidad por medio de la introducción de procedimientos perfeccionados de recolección, secado, almacenamiento y manejo.

Los cambios químicos que se producen tienen como resultado el deteriorio progresivo de los productos a medida que los períodos de almacenamiento se prolongan. Es esencial algún control de esta situación y se le debe dar consideración especial durante el planeamiento inicial.

Deben diseñarse las estructuras de almacenamiento y los molinos de forraje para evitar las fluctuaciones violentas de temperatura y humedad, deben ser fáciles de mantener, fáciles de limpiar y deben ser adecuados para la eficaz aplicación de medidas de control de peste.

La separación de piensos brutos, parcialmente tratados, y tratados, uno de otro, y de las instalaciones de tratamiento (que a su vez deben ser fáciles de limpiar) es básico en el planeamiento inicial del diseño del almacenamiento y salida de producción. Los sacos de segunda mano son una fuente principal de insectos.

La mezcla en seco en particular no eliminará los insectos y las pestes serán distribuídas ampliamente y como resultado se infestarán los productos limpios en las granjas. Esto puede ser particularmente peligroso si los productos importados en bruto o parcialmente tratados son responsables de la introducción y distribución de nuevas pestes.

Los sistemas de transportes pueden ser infestados de forma que los piensos tratados y otros productos de alto valor sean contaminados, por ejemplo, café, cacao, té, nueces para la fabricación de dulces, tubérculos, etc.

El tratamiento de los productos para piensos de animales con pesticidas pueden dar como resultado residuos peligrosos o alcanzar niveles inaceptables en algunos países importadores.

La eficacia del control de la peste depende del daño inicial, del contenido de humedad, métodos de preparación, técnicas de envasado, etc.

Introduction

Damage and deterioration in raw materials is often caused by storage pests and by poor threshing and handling techniques (Breese, 1964). Inadequate drying of cereals is an acute problem in many countries in the humid tropics. Traditional drying methods are often too slow and many permit serious insect damage, especially in maize and sorghum, before drying is completed. Artificial drying methods may be too advanced or expensive to introduce at farmer level (Hutchinson, 1971).

Oilseed cakes or cereal brans are particularly susceptible to attack by insects during storage. Transport systems and farms within a country may become infested by insects from these commodities and exports of such commodities frequently need to be fumigated before shipment and refumigated by importers. Other high value commodities such as dried fruit, confectionery nuts, coffee, cocoa and tea, etc. are at risk from direct cross-infestation or from infested transport systems.

Oilcakes, cereals and cereal brans are the principal culprits. As a typical example, Freeman (1965) compared imported rice brans from different sources and demonstrated how prevalent this problem was in these. It seems likely that if production of coarse grains increases in Southeast Asia, storage insects are likely to become a major problem unlike the comparatively insect-resistant paddy.

Methods of inspecting produce for insects and other general quality aspects have already been described by Ashman, (1966). Insects are not usually randomly distributed, they move in and out of commodities and populations of many species are dynamic and their numbers impossible to measure precisely. It is more sensible to follow the simple methods described (Ashman, 1973) for the early detection of insects than to attempt to sample using spears or probes which are inadequate.

Factors affecting quality in storage at all levels

Climatic aspects

Moisture contents of primary crops and other commodities used for animal feeds are related to climatic conditions during the drying and storage period (Cockerell et al. 1971). It is for example interesting to compare the moisture contents of maize in Zambia and the Western

TABLE 1. *Moisture content of maize in relation to ambient conditions*

| Country | Harvesting data | | Subsequent climatic conditions |
	Month	Moisture content (%)	Temperature relative rainfall humidity
Zambia*	May – June	11	Cool and dry up to October
W. Nigeria	August	20	Hot and wet until end November

* Data for Zambia from HMSO. Met. Office; for Nigeria from Cornes and Riley (1962).

State of Nigeria in relation to ambient conditions in Table 1.

It is known from work by TSPC in Zambia that no problems occur due to mould damage or insects until temperatures rise and the rains start in October.

In the Western State of Nigeria the supplies of dry maize will not be available until October – November, by which time it is known that considerable insect damage will have occurred if insecticides are not applied to the cobs in the drying cribs. Damage will continue during the storage of the dry maize at provender mills if fumigation or insecticide treatments are not applied because of continued high temperatures. The latter situation is typical of most wet tropical areas of the world.

Engineering aspects of insect and quality control

Storage structures should be constructed to reduce climatic fluctuations to a level significantly less than ambient conditions so as to reduce the hazard of condensation and the rate at which insects and moulds develop. Concrete structures are most suitable in this respect. The design of such structures should be simple, and raw materials and finished products at feed mills should be segregated in separate non-connected storage and all should be separated from the processing machinery. It should be emphasised that hygiene and segregation of stocks are two important factors in the control of pests and taints during storage.

Milling machinery is difficult to disinfest; if the mill is designed so that the building housing the mill can be sealed and successfully fumigated, disinfestation using fumigants will be simple, cheap and very effective. This fact is probably the most important consideration relative to the problem of pest control in provender mills and should be given special consideration in the initial design of new animal feed mills in climates where pests are likely to be a problem.

Floors and walls should be moisture proof, with controllable ventilation and design should incorporate rodent proofing.

Badly designed or cheap corrugated iron structures may be modified to improve performance with reflective paint on the roof to reduce heat radiation and by extending the roof to shade the walls.

Insecticides will persist longer on cool surfaces with a low pH – concrete is best painted with an emulsion to enhance persistence and activity of surface sprays; limewash should not be used for this reason.

Bulk storage

This permits the use of the most efficient and economic methods of pest control. Pest control using cool ambient aeration systems in bulk storage may be feasible; in the Middle East, for example, Navarro, *et al* (1969) have shown that cooling using low volumes of ambient air during the winter months will prevent pests from breeding in wheat and barley avoiding the need for insecticides. Many high altitude areas of the tropics could take advantage of this technique for the bulk storage of suitable commodities.

Hermetic storage of material bulks avoids the need for chemicals and will cause rapid mortality through insect metabolism, because insects present use up the oxygen. Plastic bags do not permit a completely hermetic effect unless a large number of insects are present, but Stirling (1971) has shown that at low populations levels *Sitophilus oryzae* is effectively controlled even though oxygen depletion is not sufficient to achieve 100% kill. There was no increase in numbers or damage at 27°C and 70% RH over a long storage period at oxygen and carbon dioxide levels of 8–10%. This observation may be of particular significance in the storage and packaging of animal feeds.

Physical barriers

Polythene sheets or cotton sheets may be left on stacks of dry produce in bags to prevent reinfestation physically by insects following fumigation. Condensation problems are most likely if polythene sheets are used in stores permitting violent temperature fluctuations. Certain packaging materials are also effective in preventing penetration of insects.

Chemical control of pests and fumigant residues

Fumigation techniques

Fumigants such as methyl bromide, phosphine, etc. may be used to control pests in stacks of produce, silos, or chambers using atmospheric, atmospheric recirculatory or vacuum systems to achieve a virtual 100% control. Reinfestation following fumigation is most likely in stacks or produce from chambers in cartons, bags or bales. It should be noted that all fumigants are dangerous and should only be handled by specially trained personnel.

Methyl bromide is likely to leave bromide residues which although not known to be hazardous to consumers, may exceed tolerances laid down by various countries. Hill and Thompson (1973) concluded that high residues are often found in exported animal feed from certain tropical countries where methyl bromide fumigations may need to be repeated on the same consignment several times. In shipments from some countries there has been an indication of an increase in bromide residues above 100 mg/kg during recent years. Thompson (1966) gives a comprehensive review of the use of methyl bromide.

Phosphine is least likely to leave chemical residues and may be more efficient for the fumigation of some commodities in stacks under gasproof sheets. Taylor (1974) compared the efficiency of methyl bromide for the fumigation of bags of cotton seed cake and maize germ meal under gas proof sheets at Mombasa, Kenya at a dosage of 64g/m^3 for 48 hours. Concentration time (ct) products ranged from 130 to 260 mgh/l in a stack of 105 t of cotton seed cake and in another stack of 50 t from 140 to 830 mgh/l. The concentration was very uneven, very high sorption took place near to jets reducing the availability of gas elsewhere and of course high residues were found in bags near to the jets; ambient temperatures averaged approximately 27°C. Brown (1959) quotes a ct product of 125 mgh/l for 99.9% kill of *Tribolium castaneum* (Herbst) pupae (the most resistent stage) using methyl bromide at 25°C. It is possible that levels lower than this occurred in both of Taylors' tests at other parts of the stack.

A 75-ton stack of cotton seed cake in bags was fumigated with Phostoxin tablets (each containing 3.0g aluminium phosphide) at a rate of 6 tablets per ton for 72 hours at approx. 27°C. Distribution of this gas was very even and ct products ranged between 40–42, averaging 41 mgh/l. Brown, *et al.* (1969) quote a ct product 10 mgh/l as very effective against *T. castaneum* even in a 2-day exposure. This suggests that phosphine would be more effective for the fumigation of cotton seed cake and because it is virtually non-sorptive would avoid the problem of chemical residues.

Fumigation of maize germ meal, (mean temperature 27°C) in bags under gas proof sheets with methyl bromide at 64g/m^3 for 48 hours was found by Taylor (1974) to be more satisfactory. Ct products ranged from 340–540 mgh/l and although there was some sorption, adequate ct products were achieved. For this reason no tests were conducted with phosphine.

It is widely believed (Thompson, 1966) that reaction with the protein components of a commodity is largely responsible for the degradation of adsorbed methyl bromide. The higher ct products obtained during the fumigation of maize germ meal were probably due to its containing 10% of protein compared with the 36–40% of protein in cotton seed cake. It would be anticipated that bromide residues would also be higher in the cotton seed cake.

Application of insecticide sprays and space treatment

Surface sprays

These are normally applied at wettable powder formulations diluted to give 1,000 mg of active ingredient/m^2 using malathion, iodofenphos, bromophos, etc. at 1% in a dispersion in water at 5 l/100m^2. This is efficient only if treated surfaces are clean, not too hot, at a pH approaching neutrality, free of cracks and crevices, etc. A communal effort is needed as reinfestation may occur continuously from other adjacent, badly run storage premises. Stacks of commodities are more difficult to protect as insects may penetrate so quickly that a surface spray may be of little value. Respraying of stores and fumigation sheets is recommended before a fumigated stack is unsheeted.

All infested stocks should be treated simultaneously and no entry of other infested material permitted into store unless a repeat fumigation and spraying is carried out.

Admixture of insecticides with raw or partially processed materials

This is the most efficient method of application. Most organo-phosphorus insecticides with a toxicity similar to malathion would be applied at rates of 8–12 ppm and would give a reasonable effect over several weeks provided commodities were dry. Stability is affected by enzyme activity in the commodity, this being most prevalent if fungi are actively developing (Rowlands, 1971).

In most tropical countries it would be wise to assume that incoming commodities are infested at least at population levels difficult to detect using conventional probe techniques.

Commodities flowing into or out of a bulk storage system can easily be treated if the flow rate is known and reasonably constant. Rate of treatment should not exceed 1.5 l/t; in practice the actual dosage achieved is often 60–80% of nominal. Insect mortality usually varies with particle size or surface area of the commodity (Wilkin 1967). This is of particular importance in pelletted or other semi-processed materials such as rice bran or carobs (locust beans *Ceratonia siliqua* L). Differential pick-up and biological efficiency of insecticides on a commodity of different particle size is best demonstrated in carobs.

TABLE 2. *Differential pick-up of insecticide sprayed at 11 ppm onto a carob mixture following kibbling*

Commodity	Particle size (cm)	Proportion of each in a mixture (%)	Differential pick up in the mixture (ppm)
Carobs:			
cubes	>0.10	62	2
medium	<0.10 >0.25	33	3
meal	<0.25	5	37

TABLE 3. *Mortality of oryzaephilus surinamensis adults on carobs of different particle sizes*

Applied malathion (ppm)	Carob fraction treated	kill of O.surinamensis adults. 25°C for 48 hr.
5.0	Cubes	100
5.0	Medium	90
5.0	Meal	15

Source: Wilkin (1966)

TABLE 4. *Mortality of adult T. castaneum on rice bran of different particle sizes*

Particle size (μ)	Mortality (%)*
8–180	8.5
375–500	24.6
850–1680	40.5
2700–12000 (small pellets)	55.9
12000–17000 (large pellets)	73.5

*After exposure for 10 days at 27°C and 70% RH insecticide dusts are difficult to handle and apply and do not adhere well to vertical surfaces.

In a pelletted commodity such as pelletted rice bran, similar effects are evident in that during handling, breakage of pellets occurs. Insecticides are likely to be more effective and more persistent if pellets do not crumble (Walker, 1971), as is shown in Table 4.

Residues of pesticides are likely to be highest in broken grains and dust and germ fractions of grains like wheat and maize after separation during milling.

Space treatment with insecticides

This is another method of insect control which is widely used. Emergence of adult 'Tropical Warehouse Moth' (*Ephestia cautella* Wlk), flight activity and mating were found to be highest at dusk in Ghana (Walker, 1971) and this behaviour appears to be applicable in most tropical and sub-tropical countries.

The writer has shown in Cyprus that applications of dichlorvos at dusk at a rate of $10\mu g/1$ ($10g/1000m^3$) to total store space, effectively controlled *Ephestia cautella* if continued each day until the onset of winter, ie the end of October. This insecticide effectively controls moths and some beetles in the free space but does not penetrate into commodities. It is used to prevent clean or fumigated material from becoming reinfested and to disinfest empty stores.

Concentrations of dichlorvos vapour found evenly distributed throughout stores one hour after application were approximately $1.0\mu g/l$ and after 3 hours had reduced to levels of about $0.2\mu g/l$. These concentrations are not hazardous to personnel but very toxic to *E. cautella* adults at time of peak flight activity. Residues in food in stacks of bags in Kenya were negligible, ie less than 1ppm in bags of maize nearest to the sprayer after repeated treatments with dichlorvos (Macfarlane, 1971).

Two time clocks control the dichlorvos sprayer unit developed by TSPC and the Pest Infestation Control Laboratory (PICL), one to switch the sprayer on at dusk and the second to control the quantity of dichlorvos metered into the store. One spraying unit was adequate to treat 6,000 t of carobs in a single store of $11,800m^3$ in Cyprus.

In the sub-tropics with cold winter periods, insecticide application may only be needed during hot weather in unheated storage systems. Pyrethrum fogging (Wheatley, 1970) using 0.5% pyrethrins in odourless white oil at 50 ml/$100m^3$ ($2.5g/1000m^3$) is equally effective but more labour intensive and expensive. These techniques are best integrated with fumigation, insecticide admixture and climatic control methods.

Some effects of processing on pests and storage quality of animal feed

Solvent extraction of rice bran and expeller cake kills insects, and the rice bran produced with an oil content of 1.0% or less is not suitable as a food for certain insect pest species. The tropical warehouse moth, *Ephestia cautella,* is unable to survive (Ashman, 1965) and the development of beetles is prolonged.

Temperatures and pressures during the expeller extraction of oilseeds are high enough to kill any insects present but reinfestation takes place during storage after cooling. Expeller cakes are the most frequently infested export commodities from many tropical countries and are a serious source of infestation in storage and transit.

Bulk storage and bulk shipments are now on the increase so that automatic insecticide spray application following cooling may be appropriate. Routine fumigation of cotton seed cake in railway trucks, using phosphine has a double advantage in that the exports can move direct to ships without further treatment and railway trucks are automatically disinfested. Many export commodities could be similarly treated.

Pelleting of raw materials or finished formulations has many advantages in that insecticides are easier to apply and are more effective and persistent. This application could be of value in the export or storage of large bulks of pelletted raw materials. Insects will not survive pelleting, so that insect-proof packaging could prove a useful technique to eliminate damage during storage and distribution of finished products, and help prevent the spread of pests to farms.

Dry mixing is not likely to eliminate all insects and although many are killed during the process, unless an Entoleter of similar centrifugal impact pest control technique is installed immediately prior to insect-proof packaging, no guarantee can be given that pests will be absent from formulations, so that fumigants or insecticides may have to be used.

This is important in many respects; second-hand bags, eg jute or other vegetable fibre bags, are a serious source of infestation. Provision should be made for cleaning, repair and disinfestation of these. Simple, cheap disinfestation procedures include fumigation using phosphine after placing rolls of sacks inside polythene bags or in gas-tight metal drums. Stacks of second-hand bags can be easily and cheaply fumigated under gas proof sheets using methyl bromide or phosphine.

Laboratory tests on penetration of insects through packaging materials do not provide reliable information on the behaviour of storage pests in a field situation where choice of movement is possible. Proctor and Ashman (1972) have shown that polythene film sack liners of $63.5\mu - 127.5\mu$ thickness gave some protection against reinvasion following in-sack fumigation with phosphine.

Conclusions

Correct design of storage structures and processing equipment in relation to climatic conditions, packaging, cleaning, maintenance, segregation of stocks and pest control procedures are essential considerations in preventing losses during storage.

Reference

Ashman, F. (1965) Experimental work; investigations in Britain Rice Bran. *Pest Infest. Res.,* 1964, 71–72.

Ashman, F. (1966) Inspection methods for detecting insects in stored produce. *Trop. Stored Prod. Inf.,* 12, 481–494.

Ashman, F. (1973) Methods and techniques of assessing quality in stored foods. *Trop. Stored Prod. Inf.,* 25, 33–35.

Breese, M. H. (1964) The infestability of paddy and rice. *Trop. Stored Prod. Inf.,* 8, 289–299.

Brown, W. B. (1959) Fumigation with methyl bromide under gas proof sheets. *Pest Infest. Res. Bul. No. 1,* Publ: HMSO,

Brown, W. B., Hole, B. D. & Goodship, G., (1969) Toxicity of phosphine to insects, *Pest Infest. Res.,* 1968, 59–62.

Cockerell, I., Francis, B. J. & Halliday, D. (1971) Changes in the nutritive value of concentrate feeding stuffs during storage *Proc. Conf. Development of Food Resource and Improvement of Animal Feeding Methods in the CENTO Region Countries, Ankara,* 1–7 June 1971, 181–192.

Cornes, M. A. & Riley, J. (1962) An investigation of drying rates and insect control in a maize crib with improved ventilation. *Tech. Rep. 12, W. Afr. stores Prod. Res. Unit. A. Rep. 1962.* 72–78.

Freeman, J. A. (1965) On the infestation of rice and rice products imported into Britain. *Proc. XIII Int. Congr. Ent., London, 1964,* 632–634.

Hill, E. G., Thompson, R. H. (1973) Pesticide residues in foodstuffs in Great Britain. A further report on the bromide contents of imported food and animal feeding stuffs. *Pestic Sci.,* 4, 41–49.

Hutchinson, M. T. (1971) A modified maize dryer for the small farmer. *Nigerian Agric. J.,* 8 (1), 20–25.

MacFarlane, J. A., (1971) Personal communication.

Navarro, S., Donahaye, E. & Calderon, M. (1969) Observations on prolonged grain storage with forced aeration in Israel. *J. stored Prod. Res.,* 5, 73–81.

Proctor, D. L., Ashman, F. (1972) The control of insects in exported Zambian groundnuts using phosphine and polythene lined sacks. *J. stored Prod. Red.,* 8 (2) 127–137.

Rawnsley, J. (1968) Biological studies in Ghana in the control of *Cadra cautells* (Wlk), the Tropical Warehouse Moth. *Ghana Jnl. Agric. Sci.,* 1 (2), 155–159.

Rowlands, D. G. (1971) The metabolism of contact insecticides in stored grains. Part II. *Residues Rev.,* 34, 91–161.

Stirling, H. (1971) Polythene sacks for small scale hermetic storage of grain *Trop. Stored Prod. Centre, TPI, Internal Rep.* 1–8.

Taylor, R. W. D. (1974) Personal communication.

Thompson, R. H. (1966) A review of the properties and the usage of methyl bromide as a fumigant. *J. stored Prod. Res.,* 1, 353–376.

Walker, D. J. (1971) Studies on the control of insects in rice bran. M.Sc. Thesis. Silwood Park, Imp. Coll., London Univ.

Wheatley, P. E. (1970) Insecticide treatment. Food Storage Manual III, 675–707. World Food Programme, Publ. Rome, FAO.

Wilkin, D. R. (1966) Personal communication.

Wilkin, D. R. (1967) Control of insects infesting carobs. *Pest Infest. Res.,* 1966, 32.

Aflatoxin in feedingstuffs – its incidence, significance and control

B. D. Jones
Tropical Products Institute

Summary

Moulds in foods and feedingstuffs have been a periodic problem for many years. However, the discovery of aflatoxin, a mycotoxin produced by fungi of the genus *Aspergillus,* in 1961, focused attention on the problem of fungal toxins and has resulted in large research and control efforts. Although it is recognised that there are other mycotoxins which may affect livestock and therefore should be avoided in feeds, this paper deals exclusively with the aflatoxins since the fungus *Aspergillus flavus,* from which they are derived, is widely distributed and aflatoxin has been reported as a contaminant of a number of feed materials.

The nature of the aflatoxin problem and the significance of the aflatoxin contamination of feeds is discussed from both the animal production and the public health aspects. Procedures which can be adopted to minimise aflatoxin contamination of products and the detoxification of aflatoxin contaminated material are reviewed. Current legislative and quality control measures which regulate the use of contaminated material and the current world situation regarding the incidence of aflatoxin in feeding-stuffs and feed materials are discussed.

Résumé

L'aflatoxine dans les produits alimentaires – son incidence, sa signification et son éradication

Les moisissures dans les aliments et les pâtures ont été un problème périodique pendant des nombreuses années. Cependant la découverte en 1961 de l'aflatoxine, une mycotoxine produite par les champignons de l'espèce *Aspergillus,* concentra l'attention sur le problème des toxines des champignons et résulta dans des vastes recherches et efforts d'éradication. Quoiqu'il soit reconnu qu'il y a d'autres mycotoxines qui peuvent affecter le bétail et qu'en conséquence elles devraient être évitées dans les aliments, le rapport s'occupe exclusivement des aflatoxines puisque le champignon *Aspergillus flavus,* desquels ils dérivent, est largement distribué et l'aflatoxine a été démontrée comme un contaminant d'un nombre de substances alimentaires.

On discute la nature du problème de l'aflatoxine et la signification de la contamination des aliments par aflatoxine sous l'aspect de la production animale et de la santé publique. On passe en revue les procédés qui peuvent être adoptés pour minimiser la contamination des produits et la détoxication des matériels contaminés avec aflatoxine. On discute les mesures législatives courantes et de vérification de la qualité, qui règlent l'utilisation des matériels contaminés et la situation mondiale actuelle concernant l'incidence de l'aflatoxine dans les substances alimentaires et les pâtures.

Resumen

La aflatoxina en los piensos – su incidencia, significación y control

Los mohos en los alimentos y en los piensos han sido un problema periódico durante muchos años. Sin embargo, el descubrimiento de la aflatoxina, una micotoxina producida por hongos del género *Aspergillus,* en 1961 enfocó la atención sobre el problema de las toxinas de los hongos, y ha dado como resultado una amplia investigación y esfuerzos de control. Aunque se reconoce que hay otras micotoxinas que pueden afectar al ganado y que, por consiguiente, deberían evitarse en los piensos, este documento trata exclusivamente de las aflatoxinas puesto que el hongo *Aspergillus flavus,* del cual se derivan, está distribuído ampliamente y se ha informado que la aflatoxina es un contaminante de cierto número de materiales para piensos.

Se discute la naturaleza del problema de la aflatoxina y la significación de la contaminación de la aflatoxina a los piensos tanto desde los aspectos de la producción animal como de la salud pública. Se revisan los procedimientos que pueden adoptarse para reducir al mínimo la contaminación de productos por la aflatoxina y la desintoxicación del material contaminado por la aflatoxina. Se discuten las medidas legislativas y de control de calidad actuales que regulan el uso del material contaminado y la situación mundial actual con respecto a la incidencia de la aflatoxina en piensos y en materiales para piensos.

In the last 30 years an increasing number of reports have appeared of diseases in animals of unknown aetiology which appeared to be associated with the ingestion of mouldy feed. Investigations have now shown that many of these diseases are caused by the presence in the feeds of mycotoxins, toxic metabolites produced by various species of fungi.

In some respects it is surprising that the significance of mycotoxins present in mouldy feed in the aetiology of animal disease should have escaped notice for so long, since ergotism, a mycotoxicosis caused by the ingestion of barley and rye infected with the fungus, *Claviceps purpurea* (Barger, 1931) has been known since the Middle Ages. Until recent times the subject of myco-toxins and mycotoxicoses, the disease syndromes which they produce, was studied by only a few veterinarians and mycologists, particularly in the USSR and Japan, where outbreaks of mycotoxicoses have been more frequent (Miyake *et al.*, 1940; Sarkisov, 1954). However, it was not until 1960 that this situation altered drastically with the developments related to 'Turkey X' disease which caused the deaths of some 100,000 turkey poults in Britain at that time (Blount, 1961). The cause of these deaths was traced to ground-nut meal which was included as a feed ingredient (Blount, 1961). Workers at the Tropical Products Institute and the Central Veterinary Laboratory, Weybridge succeeded in isolating a fungus from ground-nut meal which produced a fluorescent material, this material produced the characteristic symptoms associ-ated with Turkey 'X' disease (Sargeant *et al.*, 1961). This fungus was later identified as *Aspergillus flavus* Link ex Fries (Sargeant *et al.*, 1961) and the toxin was named 'aflatoxin'.

The demonstration that this toxin was extremely carcinogenic and its possible significance in problems of animal and human health has subsequently focused world wide attention on to the problems of mycotoxins in general and these have commanded large research and control efforts since then.

Chemistry of and production of the aflatoxins

The aflatoxins are a closely related group of secondary metabolites produced by certain strains of fungi of the *A. flavus* group. Whilst a number of other fungi have also been stated to produce these toxins (Hodges *et al.*, 1964; Kulik & Holiday, 1967; Basappa *et al.*, 1967; Scott *et al.*, 1967; van Walbeek *et al.*, 1968), several other groups have failed to substantiate these findings (Wilson *et al.*, 1967; Wilson *et al.*, 1968; Hesseltine *et al.*, 1966; Mislivec, 1968).

Four closely related toxins may commonly occur in crops infected with *A. flavus* viz aflatoxins B_1, B_2, G_1 and G_2. The distinguishing letters refer the colour of the fluorescence exhibited on thin-layer chromatograms (B = blue and G = green) and the suffixes to their respective positions on such chromatograms (Sargeant *et al.*, 1963). The structure of these aflatoxins, which

were determined by workers in the United States (Asao, *et al.*, 1963) are shown in Figure 1.

In some species eg cows, aflatoxins B_1 and B_2 are partially metabolised to give hydroxylated derivatives, which have been called aflatoxins M_1 and M_2 (Holzapfel *et al.*, 1966) or 'milk' toxins (see Figure 2). Two other hydroxy aflatoxins have also been isolated from cultures of *A. flavus*, these have been designated aflatoxin B_{2A} and G_{2A} (Dutton & Heathcote, 1968) (see Figure 2). Other related compounds aflatoxicol, parasiticol and aflatoxin GM_1, the structures of which are illustrated in Figure 3, are also produced by *A. flavus*, however aflatoxin P_1 (Dalezois *et al.*, 1971) which, primarily as its conjugate, is the principle urinary metabolite of aflatoxin B_1 in monkeys, has not been isolated from mould cultures.

A. flavus is an ubiquitous mould found throughout the world. However, isolates of *A. flavus* vary widely in their ability to produce aflatoxins and some strains do not produce aflatoxins at all (Boller & Schroeder, 1966; Wildman *et al.*, 1967) so that mould growth does not necessarily indicate the elaboration of aflatoxin. The presence of aflatoxin in a wide variety of agri-cultural products, including cotton seed (Loosmore *et al.*, 1964), maize (Van Walmelo *et al.*, 1965), and soya (Chong & Ponnamphalam, 1967) demonstrates the lack of a specific aflatoxin-substrate relationship. The conditions for growth of *A. flavus* and aflatoxin production on various substrates have been extensively studied. In general, in addition to the genetic require-ment already referred to, aflatoxin production depends on the moisture, the temperature, the nature of the substrate, and the available oxygen.

One of the most important of these conditions is the moisture content of or the relative humidity (RH) surrounding the substrate (Austwick & Ayerst, 1963). The optimum RH for aflatoxin production is about 85 to 90% (Diener & Davies, 1967) and the minimum RH for growth of *A. flavus* is 80% (Panassenko, 1941). The minimum substrate moisture content for *A. flavus* growth and aflatoxin production will depend on the substrate itself, however, this can be assessed as the moisture content of the substrate in equilibrium with 80% RH (Austwick & Ayerst, 1963). For groundnuts and groundnut cake this minimum moisture content is about 8% and 16% respectively (Austwick & Ayerst, 1963) so that provided groundnuts are dried to below this critical level and provided that the moisture levels in groundnuts and groundnut cake are kept below these critical levels during storage etc., then the risk of aflatoxin contamination is minimised.

As far as the temperature is concerned *A. flavus* shows maximum growth at about 18°C but grows well up to 42°C but aflatoxin production is maximal at 24°C and undetectable at 36°C (Rabie & Smalley, 1965).

In the case of groundnuts, which tend to be highly susceptible to contamination by aflatoxins and have therefore been extensively studied, it has been shown that the main factors contributing to high contamination levels are damage to the shell and splitting of the kernel which can be caused by insects, drought, or poor harvesting practices. Freshly harvested groundnuts are

Figure 1
Structures of the aflatoxins

B₁

B₂

G₁

G₂

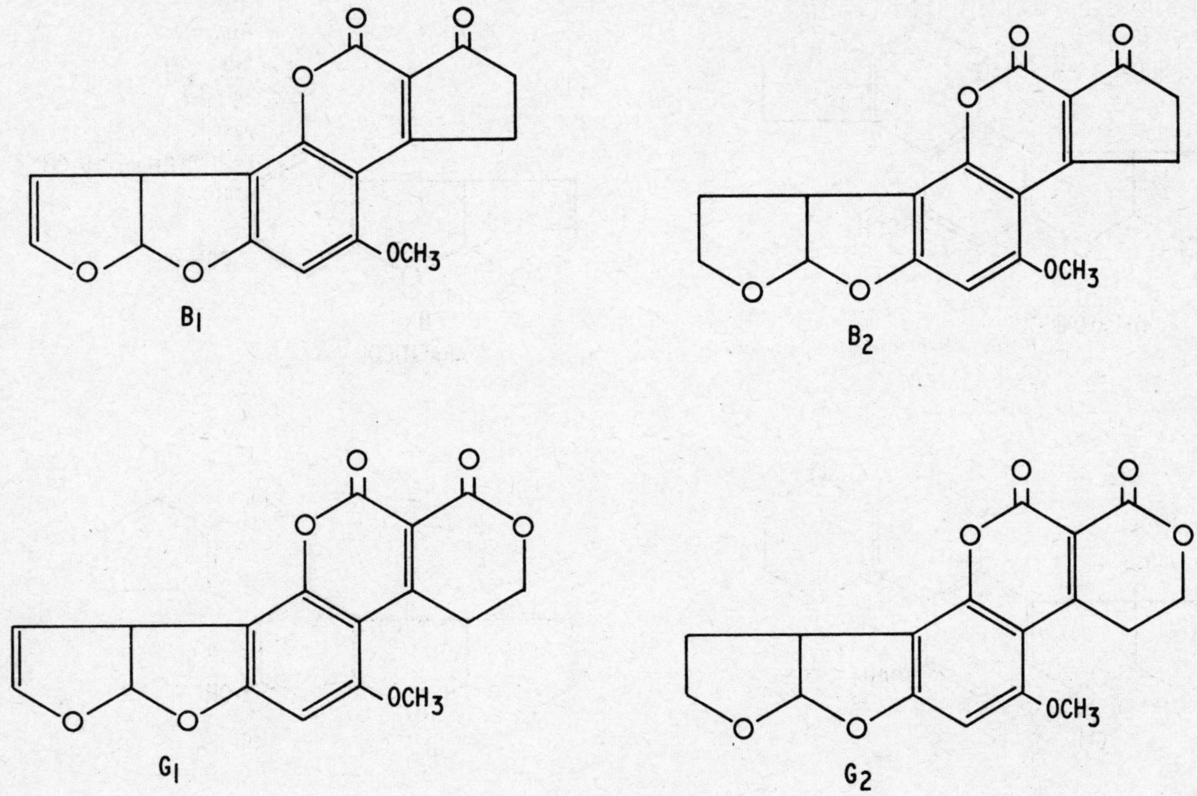

Figure 2
Structures of the aflatoxins

M₁

M₂

B₂A

G₂A

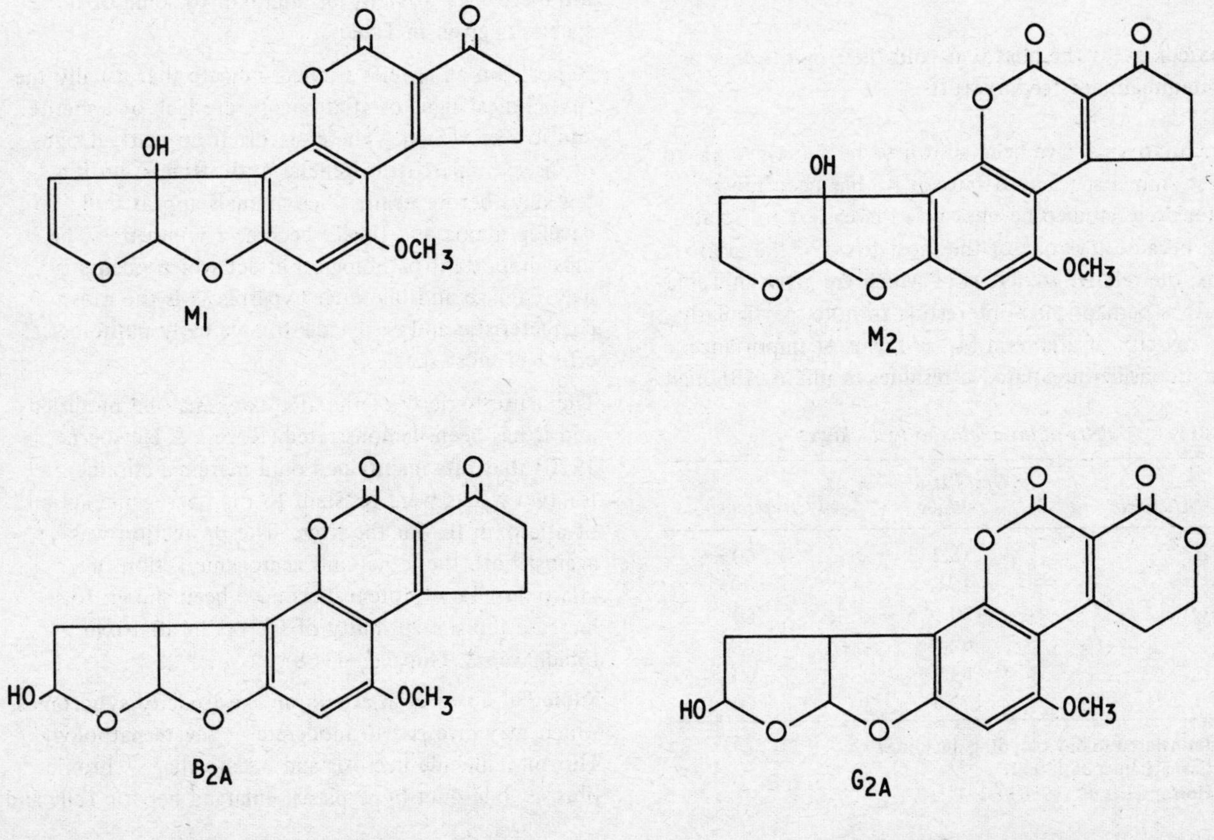

Figure 3
Structures of the aflatoxins

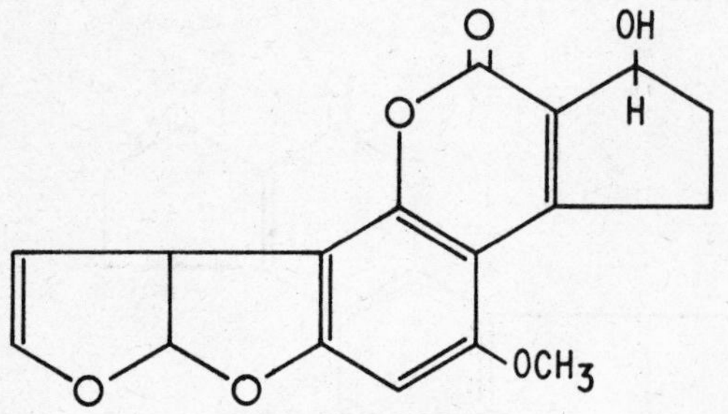

R₀
AFLATOXICOL

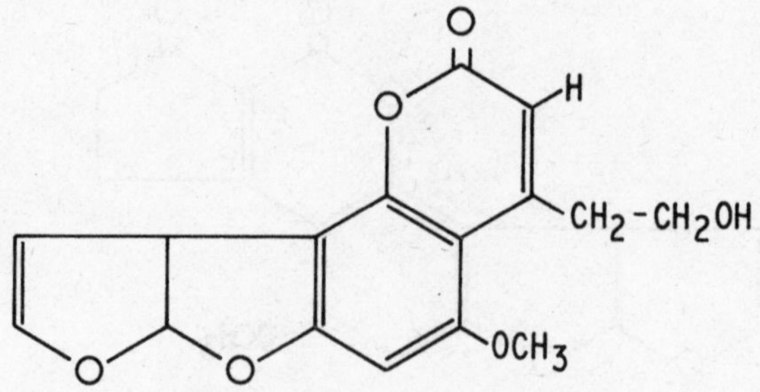

B₃
PARASITICOL

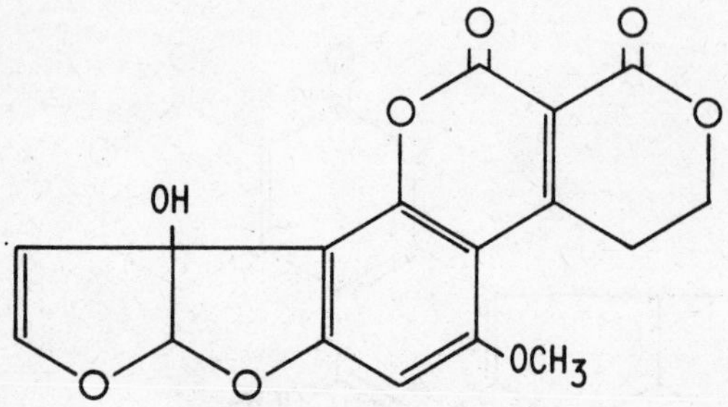

GM₁

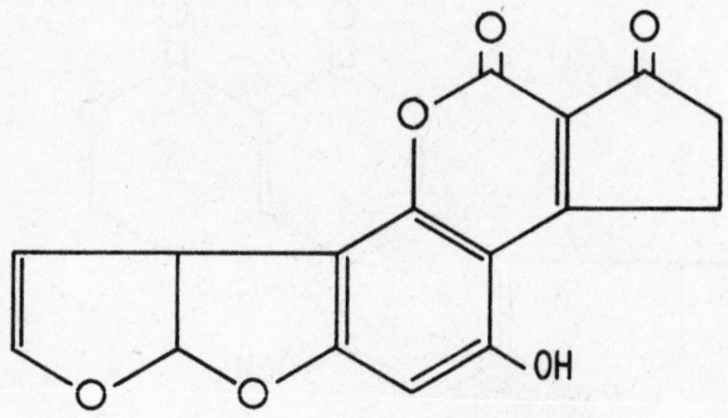

P₁

generally free from contamination with aflatoxin even when the pods and visible openings are infested with *A. flavus* (Ashworth *et al.*, 1969). Delays between harvesting and drying or slow drying encourage aflatoxin production. It has been recommended (McDonald & Harkness, 1965) that the groundnuts be picked by hand to reduce damage, followed by sun-drying in a thin layer with adequate protection from rain.

Toxicology of the aflatoxins and their significance as contaminants of feeding-stuffs

The aflatoxins have been shown to be acutely toxic to most animal species. Aflatoxin B_1 has been most extensively studied because it is the easiest to isolate and because it is one of the most toxic of the aflatoxins, the relative toxicities of which are given in Table 1. It is perhaps most interesting to note, particularly, the toxicity of aflatoxin M_1 as this is of importance when considering aflatoxin residues in milk. Although

early studies suggested that the duckling was the species most susceptible to acute poisoning, studies with this and other animals species have shown that the toxicity varies greatly depending upon the species tested, age, sex and nutritional status of the species, and the dose applied and the length of exposure. Of the various animal species tested the most highly susceptible appear to the duckling, rainbow trout, guinea pig, rabbit and turkey poult; the oral and intraperitoneal toxicity of aflatoxin to some of these species is given in Table 2.

Reports on all species studied indicate that usually the first clinical signs of aflatoxicosis are lack of appetite and loss of weight. There are often no marked signs of disease, apart from general unthriftiness, until a few days before death when animals appear dull, develop ataxia and finally become recumbent. The most important pathological effect which occurs is liver damage and bile duct hyperplasia is the most characteristic and easily identifiable early pathological effect in most species.

The acute toxicity of the aflatoxins may be modified and it has been demonstrated (Rogers & Newberne, 1970) that rats maintained on a marginal choline diet for two weeks were resistant to the necrogenic action of aflatoxin B_1 for the liver. The protection was against both the lethal and necrogenic action of aflatoxin. Low protein diets have been shown to increase the susceptibility of the rat to aflatoxin (Madhaven & Gopalan, 1968).

Sublethal doses of aflatoxin invoke toxicity syndromes, which may give rise to moderate or severe pathology. This may include necrosis and haemorrhage, chronic fibrosis, bile duct hyperplasia, enlarged hepatic cells and

TABLE 1. *Toxicity of the aflatoxins to ducklings*

Aflatoxin	Oral LD₅₀* mg/50g	Reference
B_1	18.2	(a)
M_1	16.6	(b)
G_1	39.2	(a)
M_2	62.0	(b)
B_2	84.8	(a)
G_2	172.5	(a)

*Adminstered to day old, 50 g ducklings.
(a) Carnaghan *et al.* (1963).
(b) Holzapfel *et al.* (1966).

TABLE 2. *Oral and intraperitoneal toxicity of aflatoxin to various animal species*

Species	Age (or weight)	Sex	LD50 in mg/kg body weight		Reference
			Oral	Intraperitoneal	
Duckling	1 day	M	0.37–0.56		(a), (b)
Trout	100 g	M–F	0.5	0.81	(c), (d)
Rat	1 day	M–F	1.0		(b)
	21 days	M	5.5		(b)
	21 days	F	7.4		(b)
Guinea pig	Adult	M		1.0	(e)
Rabbit	Weanling	M–F		0.5	(b)
Mouse			9.0		(f)
Sheep			2.0		(g)
Pig	Weanling	M–F	2.0		(b)

(a) Carnaghan *et al.* (1963)
(b) Wogan (1965)
(c) Bauer *et al.* (1969)
(d) Ashley *et al.* (1965)
(e) Butler (1965)
(f) Barnes (1967)
(g) Armbrecht *et al.* (1970)

finally liver tumours and/or hepatoma. There is a vast amount of information on the subacute toxic effects of the aflatoxins in various species and the subject has been extensively reviewed (Allcroft, 1969; Detroy *et al.* 1971; Butler, 1973).

The significance of the presence of aflatoxin in feedstuffs can be considered as twofold. Firstly, from what has already been stated it is apparent that either loss of animals may result, if they consume aflatoxin contaminated feed, or at best decreased weight gains, loss of egg production, fall in milk production etc. may occur. However, as a result of various feeding trials and other experimental and field observations, it is possible to obtain an indication of the effect on various animal species of various levels of dietary aflatoxin. From such data it should be possible to formulate maximum permissible levels of aflatoxin in feeds for different classes of livestock although it will be necessary to provide a sufficient margin of safety for individual variation in susceptibility and for reduced resistance which may occur in some cases due to inadequate nutrition and other factors.

Secondly, there is the possibility of ingested aflatoxins appearing, either as aflatoxins or their metabolic transformation products in animal tissues and/or milk. The evidence available on aflatoxin residues in foods of animal origin and the factors affecting such residues is discussed below.

The occurrence of aflatoxin in foods and feeds

A number of analytical surveys have been carried out to investigate the aflatoxin levels occurring naturally in a wide variety of agricultural commodities, which may be used as feedingstuffs. The commodities which have been shown to be contaminated include groundnuts and products (Allcroft & Lewis, 1963; Ling *et al.*, 1968; Allcroft *et al.*, 1961; Hiscocks, 1965), maize (Shank *et al.*, 1972), sorghum (Alpert *et al.*, 1971), sunflower (Lafont & Lafont, 1970), soya bean (Bean *et al.*, 1972), cotton seed (McMeans, *et al.*, 1968), cotton seed products (Whitten, 1968), copra (Arseculeratne & de Silva, 1971), barley (Nyiredy & Bodnor, 1966), oats (Lafont & Lafont, 1970) and cassava (Borker *et al.*, 1966).

Apart from the animal health and associated problems arising from the occurrence of alatoxin in feedingstuffs the problem of the transmission of aflatoxin through farm animals to the human food chain is also of great importance. Most animals metabolise aflatoxin rapidly but they may remain in relatively low concentrations in animal tissues (Allcroft *et al.*, 1966) and/or milk (Allcroft & Carnaghan, 1963).

A series of studies conducted in the USA (Keyl *et al.*, 1970) on the effect of feeding graded levels of aflatoxin to swine, cattle and poultry on possible transmission of aflatoxin into milk, meat and eggs confirmed the results of earlier work (Allcroft & Carnaghan, 1963; Abrams, 1965; Brown & Abrams, 1965; Platonow, 1965), which were that the chemical and biological testing methods available at the time could not establish the presence of detectable aflatoxin levels in the tissue and/or eggs. In a more recent study, Keyl and Booth (1971), using recognised chemical methods, were unable to detect aflatoxin in meat from swine and cattle fed rations containing 0.8 and 1 ppm of aflatoxin respectively. Further, eggs and meat from White Leghorn hens fed a ration of 2.7 ppm aflatoxin and the meat of broilers fed from one day to eight weeks of age on a ration containing 0.4 ppm aflatoxin contained no detectable aflatoxin. Aflatoxin has been detected in animal tissues when the aflatoxin levels in the animal's diet have been relatively high (van Zytvled *et al.*, 1970; Allcroft & Roberts, 1968).

The metabolites which have been detected in higher concentrations are the 'milk' toxins or aflatoxins M_1 and M_2. It has been shown (Allcroft & Roberts, 1968; Masri *et al.*, 1969) that the amount of aflatoxin M in the milk is proportional to the intake and levels of up to 50 mg/l have been reported. Surveys carried out in the USA (Brewington *et al.,* 1970) and UK (Allcroft & Carnaghan, 1963) on commercial milk samples indicated that no aflatoxin M was present, but it was found in retail milk from primary groundnut producing areas in South Africa (Purchase & Vorster, 1968).

Prevention of aflatoxin contamination and detoxification of contaminated materials

Although, in the long term, prevention would be the most effective and profitable means of minimising contamination, such measures are the most difficult to accomplish in practice because they necessitate general improvements in agricultural and storage methods that are traditional and commonplace in many of the areas where problems exist. Experiments with groundnuts carried out in West Africa (McDonald & Harkness, 1965; Bampton, 1963; McDonald & Harkness, 1963; McDonald & A'Brook, 1963; McDonald & Harkness, 1964; Burrell *et al.,* 1964; McDonald *et al.,* 1964) demonstrate the influence of agricultural practices on aflatoxin contamination.

One other possible way of preventing aflatoxin infestation is by the utilisation of aflatoxin resistant strains. This possibility has been examined for groundnuts and one report has appeared regarding a groundnut variety which is supposedly resistant to aflatoxin production (Suryanarayana & Tulpule, 1967). However, this claim has not been substantiated by other workers (Doupnik, 1969; Mixon & Rogers, 1973).

It has also been found possible, for some commodities, to remove contaminated portions without diminishing the usefulness of the material. The extent to which this is practicable depends on the characteristics of the commodity but it has been successful in the case of groundnuts (Patterson *et al.,* 1968) and maize (Anon, 1971). Such procedures have been successful because of the tendency for the aflatoxin to be localised in a relatively small proportion of the kernels in a batch and the association of the aflatoxin contamination with easily recognised defects in the size or colour of the kernels in the case of groundnuts and the fluorescence in ultra-violet light in the case of maize (Shotwell *et al.,* 1972). Although these procedures are used for edible products it is doubtful whether, for the present time at least, their use for animal feed material would be economic.

A variety of detoxification procedures have been applied to various aflatoxin-contaminated materials and these can be roughly classified under three headings viz solvent extraction procedures, chemical treatment and miscellaneous methods.

(1) *Solvent Extraction Procedures*

A variety of polar solvents or solvent mixtures including acetone: water (70:30 $^v/_v$) (Pons & Eaves, 1967), acetone: hexane: water azeotrope (Goldblatt, 1965), acetone: 25% aqueous ammonia (90:10) (Prevot *et al.* 1972), aqueous alcohols (Rayner & Dollear, 1968) and hexane: alcohol mixtures (Vorster, 1966) have been shown to be effective in removing aflatoxins from contaminated material, particularly oilseeds and oilseed residues. Such systems have the advantage that, under suitable conditions, they can essentially remove all the aflatoxins with little likelihood of forming from the aflatoxins products having adverse physiological activity and without appreciable reduction of the protein content or protein quality. On the other hand the cost of additional processing, the need for special extraction and solvent recovery equipment and the loss of some water soluble components from the material must be taken into account in assessing the feasibility of such procedures. It should also be noted that in the case of acetone containing solvent mixtures, adverse effects on flavour have sometimes been found due to the production of 'catty' odours caused by the presence of mesityl oxide.

(2) *Chemical Inactivation*

Numerous methods have been proposed for the destruction of aflatoxin by chemical means and many of these have been reviewed (Dollear, 1969).

Unfortunately many of these methods although they inactivate the aflatoxins they may decrease significantly the nutritive value of the processed material and may also produce products having undesirable side effects. However, some of the chemicals have proved useful and those such as sodium hypochlorite (Natarajan *et al.,* 1973) and hydrogen peroxide (Sreenivasamurthy *et al.,* 1967) have been shown to be applicable to the production of aflatoxin free protein isolates. The technique which, at the present time, would appear to have greatest application to animal feeds is the use of either ammonia or methylamine. For example a groundnut meal containing 0.07 ppm B_1, 0.03 ppm B_2 and 0.01 ppm G_1, ie a total of 0.11 ppm aflatoxins, was treated for 15 minutes at 15% moisture, 163°F, and 3 kg per/sq cm at an ammonia concentration of 6.7%. Such treatment resulted in the almost complete removal of the aflatoxins from the meal and the nitrogen content was increased by 0.46% as a result of this treatment (Dollear & Gardner, 1966).

In another series of experiments (Mann *et al.,* 1971) ammonia and methylamine treated material, which had negligible aflatoxin present, was fed to rats (28 days, 14% protein diet). These experiments indicated that there was significant lowering of the protein efficiency ratio for ammonia treated material, whilst the methylamine treated products had ratios equivalent to or higher than the untreated material. It has also been shown that the milk of cows, whose diet included aflatoxin contaminated cotton seed detoxified with ammonia, did not contain aflatoxin M (McKinney, 1973).

A number of processes for detoxification using ammonia or methylamine have been patented (Mann *et al.,* 1971; Masri *et al.,* 1969) and recently the Food and Drug Administration in the USA have decided to allow limited use of ammonia-detoxified material as ingredients in poultry layer rations at a maximum inclusion rate of 4% of the total ration, provided that the non-protein nitrogen content of the treated material does not exceed 1.0% (Anon, 1973).

(3) *Miscellaneous methods*

Aflatoxins are very stable to heat but they are destroyed to some extent by prolonged heating at 100°C at high moisture contents (Mann *et al.*, 1967) or by autoclaving at 1 kg/sq cm at 120°C again at high moisture content (Coombes *et al.*, 1966). However, it is unlikely that this could ever be used as the basis for a satisfactory method of aflatoxin removal, especially taking into account the adverse effect of such treatment on protein quality.

The possibility of destroying aflatoxin by ultra-violet (UV) irradiation has been the subject of frequent speculation. The instability of aflatoxin to UV light is well known and it has been shown that the principle photoproduct is significantly less toxic than aflatoxin (Andrellos *et al.*, 1967). However, no apparent change, as judged by the fluorescence test, was observed when groundnut meal was exposed to UV light for 8 hours (Feuell, 1966). The results of attempts to destroy aflatoxin in food by radiation, without also destroying the food have shown that this is very difficult or impossible (Frank & Grunewald, 1970). The doses required are so high that the radio-chemical alterations in the material are intolerable. The detoxification of some basic feed materials, however, especially protein concentrates, may be possible if irradiation is used in combination with other methods.

Over 1,000 organisms representing yeasts, moulds, bacteria, actinomycetes, and algae have been screened for their ability to destroy or transform the aflatoxins. Only a few of these viz *Flavobacterium aurantiacum* (Ciegler *et al.*, 1966), *Rhizopus spp.* (Cole *et al.*, 1972), *A. niger* (Burnett & Rambo, 1972) and steroid hydroxylating fungi (Detroy & Hesseltine, 1969) actually metabolised or transformed the aflatoxins. A process for the detoxification of aflatoxin containing edible products using *F. aurantiacum* has been patented (Ciegler & Lillehoj, 1969), but this sort of approach does not appear to offer commercial possibilities.

Analysis of aflatoxin residues in feedingstuffs

Two basic types of test have been developed for the detection of toxic material viz biological testing and chemical testing. In the original biological test, day-old ducklings were dosed with a concentrated extract of suspect material. The concentration of toxin present was assessed by examination of the damage to the livers after a period of seven days or after death (Carnaghan *et al.*, 1963). Other biological tests utilising various test systems have since been developed and these include fertile hens' eggs (Verrett *et al.*, 1964), brine shrimp eggs (Harwig & Scott, 1971), zebra fish eggs (Abedi & McKinley, 1968) and the organism *Bacillus megatarium*. However, such assays are clearly too time consuming for the routine analysis of materials such as required in a programme of quality control.

The chemical tests which have been developed rely on the characteristic fluorescence of the aflatoxins for quantitation. All of these methods depend on extraction of the toxin, either before or after defatting the sample, 'clean-up', if necessary, to remove oil and/or interfering fluorescent compounds and estimation of the amount of toxin present by the intensity of the fluorescence of chromatographically separated compounds. Some of the simpler methods, which follows the above basic procedure, which are particularly suitable for laboratories in less developed countries are described in a recent review (Jones, 1972) which also gives recommended methods, giving alternatives where appropriate, for a variety of agricultural commodities.

Since the publication of the above review a number of rapid methods for the detection of aflatoxin in agricultural products have been published and many of these utilise the so called 'milicolumns' (Liem & Belljaars, 1970; Pons *et al.*, 1973; Pons *et al.*, 1972; Cucullu *et al.*, 1970; Dantzman & Stoloff, 1972; Shannon *et al.*, 1973; Velasco & Whitten, 1973; Hesseltine & Shotwell, 1973). An instrument has also been developed which gives a direct reading of the aflatoxin level in parts per billion (ppb) of the fluorescence intensity of the aflatoxin band observed on one such milicolumn. Methods have also been published for the determination of aflatoxin in mixed feeds (Pons *et al.*, 1971) and one such method, which is at present being considered as the official method for the European Economic Community involves the use of two dimensional thin-layer chromatography (TLC) in order to resolve the aflatoxins in the final extract and to effectively remove any interfering fluorescent compounds.

In most laboratories the final estimation of the fluorescent intensity of the aflatoxin spots is done visually either by comparison with aflatoxin standards of known concentration or by the so called 'dilution-to-extinction' technique (Jones 1972) and therefore the precision of such methods is perhaps not as good as is desirable. The precision can be improved by utilising a thin-layer densitometer to estimate the fluorescence intensity and a number of studies have been carried out using a variety of available instruments (Belljaars & Fabry, 1972; Pons, 1971; Stubblefield *et al.*, 1967; Pons *et al.*, 1966).

The analytical method selected for any particular commodity will, to a large extent, depend on the nature of the commodity, although a number of collaborative studies have been undertaken with a view to comparing the merits of a particular or various methods for a number of commodities (Shotwell & Stubblefield, 1972; Shotwell & Stubblefield, 1973; Belljaars *et al.*, 1973; Jemmali, 1973; Anon, 1968). However, the results of international collaborative exercises carried out in 1971 and 1972 (Coon *et al.*, 1972; Coon *et al.*, 1973) indicate that there is still considerable scope for improvement in analytical methods, which do not approach the reproducibility that one might reasonably expect in modern food chemistry.

With regard to regulations governing aflatoxin levels in feeds and feedingstuffs, very few countries have legislation on this subject as far as is known. In Italy groundnut meal and cake with an aflatoxin content between 0.25 and 2 ppm must bear a label stating that the product contains aflatoxin and must not be used in feeds for ducks, turkey cocks, hens, pigs, lambs, calves or female animals in lactation. It is forbidden to use meal or cake containing more than 2 ppm of aflatoxin (Anon, 1967). The import of groundnuts and groundnut products into Denmark was banned in 1968 (Anon, 1968), however, it is understood that this complete ban has now been relaxed somewhat and that such products may be imported provided the aflatoxin content is zero. In Japan the maximum content of aflatoxin B_1 in imported groundnut extraction meal is limited to 1 ppm and the imported groundnut extraction meal must be used for mixed feed for livestock. The percentage of the imported groundnut extraction meal in mixed feed by weight is given in Table 3 (Anon, 1970).

TABLE 3. *Regulations on the use of imported groundnut extraction meal in feeds in Japan*

Species	Permitted groundnut meal in mixed feed (%)
Chickens, < 4 weeks	0
Calves, < 3 months	0
Pigs, < 2 months	0
Milk cows	3
Mixed feed for livestock other than referred to above	5

In the UK a code of practice recommended by the British Compound Animal Feedingstuffs Manufacturers National Association for the use of groundnut cakes and meals in mixed feeds (see Table 4) has been used.

No attempt has been made to legally enforce this code of practice which has been used successfully on a voluntary basis since 1965. The EEC countries have, for a number of years, been discussing maximum permitted levels for undesirable substances and products, which include Aflatoxin B_1, in feedingstuffs. These levels have now been agreed upon and they are given in Table 5 (EEC 1974).

TABLE 5. *Maximum permitted levels of aflatoxin B_1 in feedingstuffs in the EEC*

Substance	Feedingstuff	Maximum content in mg/kg(ppm) of unadulterated matter
Aflatoxin B_1	Straight feedingstuffs	0.05
	Complete feedingstuffs for cattle, sheep and goats (except dairy cattle, calves and lambs)	0.05
	Complete feedingstuffs for pigs and poultry (except young animals)	0.02
	Other complete feedingstuffs	0.01
	Complementary feedingstuffs for dairy cattle	0.02

Results of an international survey to assess the incidence of aflatoxin in feedingstuffs

A questionnaire was sent to as many countries as possible requesting information on the levels and incidence of aflatoxin contamination in feedingstuffs. However, unfortunately because of the nature of the results received and the different toxicity levels adopted in the various countries it has been difficult to collate the information as originally intended.

TABLE 4. *The UK code of practice for the utilisation of aflatoxin contaminated groundnut products in feedingstuffs as fed to livestock*

(A) *Where Groundnut Products have been tested*
  (a) *Where the reaction is negative* (ie below 0.1 ppm)
  Any percentage indicated by common usage.

  (b) *Where the reaction is weakly positive* (ie 0.1 to 1 ppm)

	Maximum per cent of Ground Nut to be included
Duck foods, Turkey starter food, Broiler starter food, Baby Chick food, Pig and Lamb creep feeds and Calf feeds	Nil
Turkey finisher and Broiler finisher foods	5
Pig and Laying foods (except those included in the previous categories) and Cattle rearing foods	7½
Adult Sheep and Cattle foods	15

  (c) *Where the reaction is moderately to strongly positive* (ie 1 ppm to 5 ppm)
  Not more than half the percentages given in (b) above.

  (d) *Where the reactions is very strongly positive* (ie over 5 ppm)
  Parcels should be restricted to rations for adult cattle and sheep, and be limited to not more than 2½ per cent for cattle and 5 per cent for sheep.

(B) *Where Ground Nut Products have not been tested*

Duck foods, Turkey starter food, broiler starter feed, Baby Chick food, Pig and Lamb creep feeds and Calf feeds	Nil
Turkey finisher and Broiler finisher foods	5
Pig and Layers foods (except those included in the previous categories), Cattle rearing food and Adult Cattle and Sheep foods.	7½

N.B. Concentrates and balancer foods can be adjusted accordingly.

A summary of the information received from the various areas in the world is given in Tables 6–11. Replies were not received from a number of countries and for others no information was available either because no analyses had been or were being carried out or because insufficient data on this subject had been accumulated. In some respects, therefore, the results may be somewhat biased against those countries for which details are available.

In considering this information it should be borne in mind that aflatoxin incidence may be seasonal depending to some extent on the climatic conditions prevailing in any particular year. In the case of groundnuts for example drought or too much rain at the wrong time is liable to increase the incidence of aflatoxin in the crop.

The data indicates that groundnuts and groundnut products are the commodities most likely to be contaminated and most likely to contain high levels of aflatoxin although as these products are the ones which are analysed for aflatoxin more than any other commodity the observed high incidence of contamination is not surprising. Other oilseeds and oilseed products from cotton seed, palm kernels, copra, sunflower, etc. may also be contaminated although the incidence and level of contamination would appear to be much lower than for groundnuts.

In general the incidence and levels of aflatoxin in grains and grain products appear to be low although maize is very often found to be highly contaminated. Samples of rice, sorghum and malt sprouts have also been found to be contaminated with over 1 ppm of aflatoxin (Scott, 1973).

Conclusions

It would appear likely that the problems associated with the aflatoxin contamination of feeds and feedingstuffs are going to be with us for some time, at least in the foreseeable future. Obviously improvements in agricultural and storage practices, particularly in the less developed countries, can help to reduce contamination levels and possibly completely eliminate aflatoxin from these commodities although this will be extremely difficult to realise in practice. The possibility of detoxifying contaminated material may also need to be considered, particularly as the need for protein sources for inclusion in animal feeds increases and the economic factors, which must be taken into account when considering detoxification, become more favourable. Strict quality control of feeds may also be necessary in order to prevent livestock losses and to prevent contamination of products of animal origin by aflatoxin.

However, before thinking about such control measures it is obviously very necessary to consider whether these are in fact needed. The dangers to certain animal species of ingested aflatoxin are well proven; but what are the effects of aflatoxin in man? There is a plethora of indirect information which suggests that aflatoxin may cause liver cancer in humans although it is, and will be, very difficult to obtain direct evidence on this

point. Therefore, in view of the uncertainty regarding such effects, until the point is proven conclusively one way or the other, there is perhaps a need to take precautions, as far as is practically possible, to eliminate aflatoxin from the human food chain. Obviously such a need must be balanced against what may be even a greater need, namely the need to provide, at a reasonable cost, a balanced diet for the world population in order to eliminate starvation and other nutritional disorders which are prevelant in many parts of the world.

Acknowledgements

I am grateful to my colleagues Mrs J. Bainton and Mr I. Cockerell for their assistance with the aflatoxin survey and to all those persons who supplied information in this regard.

References

Abedi, Z. H. & McKinley, W. P. (1968) Zebra fish eggs and larvae as aflatoxins bioassay test organisms. *J. Ass. Off. Anal. Chem.* **51** (4), 902–905.

Abrams, L. (1965) Mycotoxins in veterinary medicine. *Proc. Symp. Mycotoxins in Foodstuffs, Agr. Aspects,* Pretoria, South Africa, 103–114.

Allcroft, R. (1969) Aflatoxin-Scientific Background, Control and Implications (ed. L. A. Goldblatt), New York, Academic Press, 237–264.

Allcroft, R. & Carnaghan, R. B. A. (1963) Groundnut toxicity: an examination for toxin in human food products from animals fed toxic groundnut meal. *Vet. Record* **75**, 259–263.

Allcroft, R., Carnaghan, R. B. A., Sargeant, K. & O'Kelly, J. (1961) A toxic factor in Brazilian groundnut meal. *Vet. Record* **73**, 428.

Allcroft, R. & Lewis, G. (1963) Groundnut toxicity in Cattle: Experimental poisoning of calves and a report on clinical effects in older cattle. *Vet. Record* **75** 487–493.

Allcroft, R. & Roberts, B. A. (1968) Toxic groundnut meal: The relationship between aflatoxin B_1 intake by cows and excretion of aflatoxin M_1 in the milk. *Vet. Record* **82**, 116–118.

Allcroft, R., Rogers, H., Lewis, G., Nabney, J. & Best, P. E. (1966) Metabolism of aflatoxin in sheep: excretion of the 'milk toxin'. *Nature* **209**, 154–155.

Alpert, M. E., Hutt, M. S. R., Wogan, G. N. & Davidson, C. S. (1971) Association between aflatoxin content in food and hepatoma frequency in Uganda. *Cancer* **28**, 253–260.

Andrellos, P. J., Beckwith, A. C. & Eppley, R. M. (1967) Photochemical changes of aflatoxin B_1. *J. Ass. Off. Anal. Chem.* **50**, 346–350.

282 **Anon** (1967) *Boll. Laboratori, chim. prov.* **18** (3), 353—4.

Anon (1968a) Provisional IUPAC method for aflatoxin in peanuts, peanut butter and peanut meal. *IUPAC Inf Bull* No. 31, pp. 35—43.

Anon (1968b) *Danish Ministry of Agriculture Announcement No. 356,* Oct. 22.

Anon (1968c) Changes in Methods. *J. Ass. Off. Anal. Chem.* **51**, 485—489.

Anon (1970a) *Ann. Rep. Nat. Feeds and Fertilisers Inspection office,* Tokyo, Japan, 53—54.

Anon (1970b) Official Methods of Analysis of the Association of Official Analytical Chemists (ed. W. Horwitz) Association of Official Analytical Chemists, Washington, D.C., 11th Ed. 429—430.

Anon (1970c) Official Methods of Analysis of the Association of Official Analytical Chemists (ed. W. Horwitz), Association of Official Analytical Chemists, Washington D.C., 11th Ed. 431—433.

Anon (1970d) Official Methods of Analysis of the Association of Official Analytical Chemists (ed. W. Horwitz) Association of Official Analytical Chemists, Washington, D.C., 11th Ed. 431.

Anon (1971) U.V. light detects corn mold. *Agric. Res.* **20** (6), 15.

Anon (1973) FDA allowing feed use of ammoniated detoxified cotton seed meal. *Food Chem. News* (Feb. 12th), 8.

Anon (1974) E.E.C. Council Directive 74/63/EEC of 17 December 1973 on the fixing of maximum permitted levels for undesirable substances and products in feedingstuffs. *Official J. European Commun.* **17** (L38), 35.

Armbrecht, B. H., Shalkop, W. T., Rollins, L. D., Pohland, A. E. & Stoloff, L. (1970) Acute toxicity of aflaxtoxin B₁ in wethers. *Nature* **225**, 1062—1063.

Arseculeratne, S. N. & de Silva, L. M. (1971) Aflatoxin contamination of coconut products. *Ceylon J. Med. Sci.* **20** (2), 60—75.

Asao, T., Büchi, G., Abdel-Kader, M. M., Chang, S. B., Wick, E. L., & Wogan, G. N. (1963) Aflatoxins B and G. *J. Amer. Chem. Soc.* **85**, 1706.

Ashley, L. M., Halver, J. E., Gardner, K. W. & Wogan, G. N. (1965) Crystalline aflatoxins cause trout hepatoma. *Fed. Proc.* **24**, 627.

Ashworth, L. J. H., Schroeder, H. W., & Langley, B. C. (1965) Aflatoxins: Environmental factors governing occurrence in Spanish peanuts. *Science* **148**, 1228—1229.

Austwick, P. K. C. & Ayerst, G. (1963) Toxic products in groundnuts. Groundnut microflora and toxicity. *Chem. Ind. (London)* **2**, 55—61.

Bampton, S. S. (1963) Growth of *Aspergillus flavus* and production of aflatoxin in groundnuts. I. *Trop. Sci.* **5**, 74—81.

Barger, G. (1931) Ergot and Ergotism, London, Gurney and Jackson, p. 11.

Barnes, J. M. (1967) Toxic fungi with special reference to aflatoxin. *Trop. Sci.* **9**, 64—74.

Basappa, S. C., Jayaraman, A., Sreenivasamurthy, V. & Parpia, H. A. B. (1967) Effect of B-group vitamins and ethyl alcohol on aflatoxin production by *Aspergillus oryzae. Indian J. Expt. Biol.* **5**, 262—263.

Bauer, D. H., Lee, D. J. & Sinnhuber, R. O. (1969) Acute toxicity of aflatoxins B₁ and G₁ in rainbow trout (*Salmo gairdneri*). *Toxicol, App. Pharmac.* **15**, 415.

Baur, F. J. & Armstrong, J. C. (1971) Collaborative study of a modified method for the determination of aflatoxins in copra, copra meal and coconut. *J. Ass. Off. Anal. Chem.* **54** (4), 874—878.

Bean, G. A., Schillinger, J. A. & Klarman, W. L. (1972) Occurrence of aflatoxins and aflatoxin producing strains of *Aspergillus* spp. in soybeans. *Appl. Microbiol.* **24** (3), 437—439.

Beljaars, P. R., Cornelius, J. H., Verhülsdonk, A. H., Paulsch, W. E. & Liem, D. H. (1973) Collaborative study of two-dimensional thin-layer chromatographic analysis of aflatoxin B₁ in peanut butter extracts, using the antidiagonal spot application technique. *J. Ass. Off. Anal. Chem.* **56** (6), 1444—1451.

Beljaars, P. R. & Fabry, F. H. M. (1972) Quantitative fluorodensitometric measurements of aflatoxin B₁ with a flying-spot densitometer. I. Fluorodensitometric study of the behaviour of aflatoxin B₁ spots on different types of silica gel. *J. Ass. Off. Anal. Chem.* **55** (4), 775—780.

Blount, W. P. (1961) Turkey 'X' disease. *Turkeys* **9** (2), 52, 55—58, 61, 77.

Boller, R. A. & Schroeder, H. W. (1966) Aflatoxin producing potential of *Aspergillus flavus-oryzae* isolated from rice. *Cereal Sci. Today* **11**, 342—344.

Borker, E., Insalata, N. F., Levi, C. P. & Witzemann, J. S. (1966) Mycotoxins in feeds and foods. *Advan, Appl. Microbiol.* **8**, 315—351.

Brewington, C. R., Weihrauch, J. L. & Ogg, C. L. (1970) Survey of commercial milk samples for aflatoxin M. *J. Dairy Sci.* **53** 1509—1510.

Brown, J. M. M. & Abrams, L. (1965) Biochemical studies on aflatoxicosis. *Ondersterpoort J. Vet. Res.* **32** (1), 119—146.

Burnett, C. & Rambo, G. W. (1972) Aflatoxin inhibition and detoxification by a culture filtrate of *Aspergillus niger. Paper presented at 64th Meeting Am. Phytopath. Soc.,* Mexico City, August.

Burrell, N. J., Grundey, J. K. & Harkness, C. (1964) Growth of *Aspergillus flavus* and production of aflatoxin in groundnuts V. *Trop. Sci.* **6**, 74—90.

Butler, W. H. (1965) Mycotoxins in Foodstuffs (ed. G. N. Wogan), Cambridge, Mass., M.I.T. Press, 175—186.

Butler, W. H. (1973) Review of the toxicology of aflatoxin. *Pure Appl. Chem.* **35** (3), 217–222.

Carnaghan, R. B. A., Hartley, R. D. & O'Kelly, J. (1963) Toxicity and fluorescence properties of the aflatoxins. *Nature* **200**, 1101.

Chong, Y. H. & Ponnamphalam, J. T. (1967) The effect on ducklings of a diet containing moulded-soybeans. *Med. J. Malaya* **22** (2), 104–109.

Ciegler, A. & Lillehoj, E. B. (1969) *U.S. Pat. No. 3,428,458.*

Ciegler, A., Lillehoj, E. B., Peterson, R. E. & Hall, H. H. (1966) Microbial detoxification of aflatoxin. *Appl. Microbiol.* **14**, 934–939.

Clements, N. L. (1968) Rapid confirmatory test for aflatoxin B_1 using *Bacillus megatarium. J. Ass. Off. Anal. Chem.* **51** (6), 1192–1194.

Cole, R. J., Kirksey, J. W. & Blankenship, B. R. (1972) Conversion of aflatoxin B_1 to isomeric hydroxy compounds by *Rhizopus* spp. *J. Agr. Food Chem.* **20** (6), 1100–1102.

Coomes, T. J., Crowther, P. C., Feuell, A. J. & Francis, B. J. (1966) Experimental detoxification of groundnut meals containing aflatoxin. *Nature* **209**, 406–407.

Coon, F. B., Baur, F. J. & Symmes, L. R. L. (1972) International aflatoxin check sample programme: 1971 study. *J. Ass. Off. Anal. Chem.* **55** (2), 315–327.

Coon, F. B., Baur, F. J. & Symmes, L. R. L. (1973) International aflatoxin check sample programme: 1972 study. *J. Ass. Off. Anal. Chem.* **56** (2), 322–332.

Cucullu, A. F., Pons, W. A. & Goldblatt, L. A. (1972) Fast screening method for the detection of aflatoxin contamination in cottonseed products. *J. Ass. Off. Anal. Chem* **55** (5), 1114–1119.

Dalezios, J., Wogan, G. N. & Weinreb, S. M. (1971) Aflatoxin P_1: a new aflatoxin metabolite in monkeys. *Science* **171**, 584–585.

Dantzman, J. & Stoloff, L. (1972) Screening method for aflatoxin in corn and various corn products. *J. Ass. Off. Anal. Chem.* **55** (1), 139–141.

Detroy, R. W. & Hesseltine, C. W. (1969) Transformation of aflatoxin B_1 by steroid hydroxylating fungi. *Can. J. Microbiol.* **15**, 495–497.

Detroy, R. W., Lillehoj, E. B. & Ciegler, A. (1971) *Microbial Toxins – A Comprehensive Treatise, Vol VI – Fungal Toxins,* (ed. A. Ciegler, S. Kadis, & S. J. Ajl), New York, Academic Press, 90–126.

Diener, U. L. & Davies, N. D. (1967) Limiting temperature and relative humidity for growth and production of aflatoxin and free fatty acids by *Aspergillus flavus* in sterile peanuts. *J. Am. Oil Chem. Soc.* **44**, 259–263.

Dollear, F. G. (1969) Aflatoxin-Scientific Background, Control and Implications, New York, Academic Press, 359–391.

Dollear, F. G. & Gardner, H. K. (1966) Inactivation and removal of aflatoxin. *Proc. 4th Nat. Peanut Res. Conf.,* Tifton, Georgia, July, 72–81.

Doupnik, B. (1969) Aflatoxins produced on peanut varieties previously reported to inhibit production. *Phytopath.* **59**, 1554.

Dutton, M. F. & Heathcote, J. G. (1968) The structure, biochemical properties and origin of the aflatoxins B_{2A} and G_{2A}. *Chem. Ind.* 418–421.

Eppley, R. M. (1966) A versatile procedure for the assay and preparatory separation of aflatoxins from peanut products. *J. Ass. Off. Anal. Chem.* **49**, 1218–1223.

Feuell, A. J. (1966) Aflatoxin in groundnuts. IX. Problems of detoxification. *Trop. Sci.* **8**, 61–70.

Frank, H. K. & Grunewald, T. (1970) Radiation resistance of the aflatoxins. *Food Irradiation* **11**, 15–20.

Goldblatt, L. A. (1965) Mycotoxins in Foodstuffs (ed. G. N. Wogan), M. I. T. Press, Cambridge, Mass., 261–263,

Harwig, J. & Scott, P. M. (1971) Brine Shrimp *(Artemia salina* L.) larvae as a screening system for fungal toxins. *Appl. Microbiol.* **21** (6), 1011–1016.

Hesseltine, C. W. & Shotwell, O. (1973) New Methods for rapid detection of aflatoxin. *Pure Appl. Chem.* **35** (3), 259–266.

Hesseltine, C. W., Shotwell, O. L., Ellis, L. J. & Stubblefield, R. D. (1966) Aflatoxin formation by *Aspergillus flavus. Bacteriol. Rev.* **30**, 795–805.

Hiscocks, E. S. (1965) Mycotoxins in Foodstuffs (ed. G. N. Wogan) M.I.T. Press, Cambridge, Mass., 15–26.

Hodges, F. A., Zust, J. R., Smith, H. R., Nelson, A. A., Armbrecht, B. H. & Campbell, A. D. (1964) Mycotoxins: Aflatoxin isolated from *Penicillium puberulum. Science* **145**, 1439.

Holzapfel, C. W., Steyn, P. S. & Purchase, I. F. H. (1966) Isolation and structure of aflatoxins M_1 and M_2. *Tetrahedron Letters* **25** 2799.

Jemmali, M. (1973) Collaborative studies on the determination of aflatoxins in peanut products in France. *Pure Appl. Chem.* **35** (3), 267–270.

Jones, B. D. (1972) Methods of aflatoxin analysis. *Report No. G70,* London, Tropical Products Institute.

Jones, B. D. (1972) Methods of aflatoxin analysis. *Tropical Products Institute Report* No. G70, 11, 13 & 14.

Keyl, A. C. & Booth, A. N. (1971) Aflatoxin effects in livestock. *J. Am. Oil Chem. Soc.* **48**, 599–604.

Keyl, A. C., Booth, A. N., Masri, M. S., Gumbmann, M. R. & Gagne, W. E. (1970). *Proc. First U.S. – Japan Conference on Toxic Micro-organisms* (ed. M. Herzberg) UJNR Joint Panel on Toxic Micro-organisms and U.S. Dept. Interior, Washington, D.C., 72–75.

284   Krögh, P., Hald, B., Hasselager, E., Madsen, A., Mortensen, H. P., Larsen, A. E. & Campbell, A. D. (1973) Aflatoxin residues in bacon pigs. *Pure Appl. Chem.* **35**, (3), 275–281.

Kulik, M. M. & Holiday, C. E. (1967) Aflatoxin: A metabolic product of several fungi. *Mycopath. Mycol. Appl.* **30**, 137–140.

Lafont, P. & Lafont, J. (1970) Contamination of cereal products and animal feeds by aflatoxin. *Food. Cosmet. Toxicol.* **8**, 403–408 (French).

Lee, W. V. (1965) Quantitative determination of aflatoxin in groundnut products. *Analyst* **90**, 305–307.

Liem, D. H. & Beljaars, P. R. (1970) Note on a rapid determination of aflatoxin in peanut products. *J. Ass. Off. Anal. Chem.* **53** (5), 1064–1066.

Ling, K. H., Tung, C. M., Shek, I. & Wong, J. J. (1968) Aflatoxin B_1 in unrefined peanut oil and peanut products in Taiwan. *J. Formosan Med. Assoc.* **67** (7), 309–314.

Loosmore, R. M., Allcroft, R., Tutton, E. A. & Carnaghan, R. B. A. (1964) The presence of aflatoxin in a sample of cottonseed cake. *Vet. Record* **76**, 64–65.

Madhaven, T. A. & Gopalan, C. (1968) The effect of dietary protein on carcinogenesis of aflatoxin. *Arch. Path.* **85**, 133–137.

Mann, G. E., Codifer, L. P. & Dollear, F. G. (1967) Effect of heat on aflatoxins in oilseed meals, *J. Agr. Food Chem.* **15**, 1090–1092.

Mann, G. E., Codifer, L. P., Gardner, H. K. & Dollear, F. G. (1971) Process for lowering aflatoxin levels in aflatoxin-contaminated substances. *U.S. Pat No 3,585,041.*

Mann, G. E., Gardner, H. K., Booth, A. N. & Gumbmann, M. R. (1971) Aflatoxin inactivation: Chemical and biological properties of ammonia and methylamine treated cottonseed meal. *J. Agr. Food Chem.* **19** (6), 1155–1158.

Masri, M. S., Page, J. R. & Garcia, V. C. (1969) The aflatoxin M content of milk from cows fed known amounts of aflatoxin. *Vet. Record* **84**, 146.

Masri, M. S., Vix, H. L. E. & Goldblatt, L. A. (1969) Process for detoxifying substances contaminated with aflatoxin. *U.S. Pat. No. 3,429,709.*

McDonald, D. & A'Brook, J. (1963) Growth of *Aspergillus flavus* and production of aflatoxin in groundnuts. III. *Trop. Sci.* **5**, 208–214.

McDonald, D. & Harkness, C. (1963) Growth of *Aspergillus flavus* and production of aflatoxin in groundnuts. II. *Trop. Sci.* **5**, 143–154.

McDonald, D. & Harkness, C. (1964) Growth of *Aspergillus flavus* and production of aflatoxin in groundnuts. IV. *Trop. Sci.* **6**, 12–27.

McDonald, D. & Harkness, C. (1965) Growth of *Aspergillus flavus* and production of aflatoxin in groundnuts – Part VIII. *Trop. Sci.* **7** (3), 122–137.

McDonald, D., Harkness, C. & Stonebridge, W. C. (1964) Growth of *Aspergillus flavus* and production of aflatoxin in groundnuts. VI. *Trop, Sci.* **6**, 131–154.

McKinney, J. D., Cavanagh, G. C., Bell, J. T., Hoversland, A. S., Nelson, D. M., Pearson, J. & Selkirk, R. J. (1973) Effects of ammoniation on aflatoxins in rations fed lactating cows. *J. Am. Oil Chem. Soc.* **50**, 79–84.

McMeans, J. L., Ashworth, L. J. & Pons, W. A. (1968) Aflatoxins in hull and meat of cottonseed. *J. Am. Oil Chem. Soc.* **45**, 575–576.

Mislivec, P. B., Hunter, J. H. & Tuite, J. (1968) Assay for aflatoxin production by the genera *Aspergillus* and *Penicillium. Appl. Microbiol.* **16**, 1053–1055.

Miyake, I., Naito, H. & Tsunoda, H. (1940) Study on toxin production in stored rice by parasitic fungi. *Rice Utilisation Research Inst. Dept. Agric. and For.,* No. 1, 1.

Mixon, A. C. & Rogers, K. M. (1973) Peanuts resistant to seed invasion by *Aspergillus flavus. Oleagineux* **28** (2), 85–86.

Natarajan, K. R., Rhee, K. C., Carter, C. M. & Mattil, K. F. (1973) Detoxification of aflatoxins in raw peanuts and peanut meal by sodium hypochlorite. *Abst. Am. Chem. Soc. 166th Meeting,* Illinois, August, Abst. **79**;

Nyiredy, I. & Bodnor, M. (1966) Investigations on the occurrence of strains of *Aspergillus flavus* in home produced fodders and of their ability of producing aflatoxin. *Magyar Allat. Lap.* **21** (8), 352–354 (In Hungarian).

Panassenko, V. T. (1941) Mould fungi of confectionery goods and their control. *Microbiology (USSR)* **10**, 470–479.

Patterson, I., Crowther, P. C. & El Noubey, H. (1968) The separation of aflatoxin infected groundnut kernels. *Trop. Sci.* **10** (4), 212–221.

Platonow, N. (1965) Investigation of the possibility of the presence of aflatoxin in meat and liver of chickens fed toxic groundnut meal. *Vet. Record* **77**, 1028.

Pons, W. A. (1969) Collaborative study on the determination of aflatoxins in cottonseed products. *J. Ass. Off. Anal. Chem.* **52**, 61–72.

Pons, W. A. (1971) Evaluation of reflectance densitometry for measuring aflatoxins on thin layer plates. *J. Ass. Off. Anal. Chem.* **54** (4), 870–873.

Pons, W. A., Cucullu, A. F. & Franz, A. O. (1972) Rapid quantitative TLC method for determining aflatoxins in cottonseed products. *J. Ass. Off. Anal. Chem.* **55** (4), 768–774.

Pons, W. A., Cucullu, A. F., Franz, A. O., Lee, L. S. & Goldblatt, L. A. (1973) Rapid detection of aflatoxin contamination in agricultural products. *J. Ass. Off. Anal. Chem.* **56** (4), 803–807.

Pons, W. A., Cucullu, A. F. & Lee, L. S. (1971) Determination of aflatoxins in mixed feeds. *Proc. 3rd Intern. Congr. Food Sci. Technol.,* Washington, D.C., 1970, 705–711.

Pons, W. A., Cucullu, A. F., Lee, L. S., Robertson, J. A., Franz, A. O. & Goldblatt, L. A. (1966) Determination of aflatoxins in agricultural products; Use of aqueous acetone for extraction. *J. Ass. Off. Anal. Chem.* **45**, 694–699.

Pons, W. A. & Eaves, P. H. (1967) Aqueous acetone extraction of cottonseed. *J; Am. Oil Chem. Soc.* **44**, 460–464.

Pons, W. A., Robertson, J. A. & Goldblatt, L. A. (1966) Objective fluorometric measurement of aflatoxins on TLC plates. *J. Am. Oil Chem. Soc.* **43**, 665–669.

Prevot, A. F., Bloch, C. & Van derVoort, P. R. (1972) Pilot plant scale removal of aflatoxins. *Proc. Symp. on Control of Mycotoxins,* Göteborg, Sweden, August.

Purchase, I. F. H. & Vorster, L. J. (1968) Aflatoxin in commercial milk samples. *S. African Med. J.* **42** (2), 219.

Rabie, C. J. & Smalley, E. B. (1965) Influence of temperature on the production of aflatoxin by *Aspergillus flavus. Proc. Symp. Mycotoxins in Foodstuffs. Agr. Aspects.* Pretoria, South Africa. 18–29.

Ranfft, K. (1972) Ein kurzer beitrag zur aflatoxin analytik. *Z. Lebensmitt u. Forsch.* **150**, 129–133.

Rayner, E. T. & Dollear, F. G. (1968) Removal of aflatoxins from oilseed meals by extraction with aqueous isopropanol. *J. Am. Oil Chem. Soc.* **45**, 622–624.

Rogers, A. E. & Newberne, P. M. (1970) Nutrition and aflatoxin carcinogenesis. *Nature* **229**, 62–63.

Sargeant, K., Carnaghan, R. B. A. & Allcroft, R. (1963) Toxic products in groundnuts. Chemistry and origin. *Chem. Ind.* 53.

Sargeant, K., Sheridan, A., O'Kelly, J. & Carnaghan, R. B. A. (1961) Toxicity associated with certain samples of groundnuts. *Nature* **192**, 1096–1097.

Sarkisov, A. Kh. (1954) Mycotoxicoses. Goz. Izd. Selskohoziastvennoi Literatury, Moscow

Scott, P. M., Van Walbeek, W. & Forgacs, J. (1967) Formation of aflatoxin by *Aspergillus ostianus* Wehmer. *Appl. Microbiol.* **15**, 945.

Scott, P. M. (1973) Grain Storage (ed. R. N. Sinha & W. E. Muir) Avi Publishing Co. Inc., Westport, Connecticut, 343–365.

Shank, R. C., Wogan, G. N. Gibson, J. B. & Nondasutu, A. (1972) Dietary aflatoxins and human liver cancer. II. Aflatoxins in market food and foodstuffs in Thailand and Hong Kong. *Food Cosmet. Toxicol.* **10**, 61–69.

Shannon, G. M., Stubblefield, R. D. & Shotwell, O. L. (1973) Modified rapid screening method for aflatoxin in corn. *J. Ass. Off. Anal. Chem.* **56** (4), 1024–1025.

Shotwell, O. L. Goulden, M. L. & Hesseltine, C. W. (1972) Aflatoxin contamination: Association with foreign material and characteristic fluorescence in damaged corn kernels. *Cereal Chem.* **49**, 458–465.

Shotwell, O. L., Hesseltine, C. W., Burmeister, H. R., Kwolek, W. F., Shannon, G. M. & Hall, H. H. (1969) Survey of cereal grains and soybeans for the presence of aflatoxin. II. Corn and soybeans, *Cereal Chem.* **46**, 454–463.

Shotwell, O. L. & Stubblefield, R. D. (1972) Collaborative study on the determination of aflatoxin in corn and soybeans. *J. Ass. Off. Anal. Chem.* **55** (4), 781–788.

Shotwell, O. L. & Stubblefield, R. D. (1973) Collaborative study of three screening methods for aflatoxin in corn. *J. Ass. Off. Anal. Chem.* **56** (4), 808–812.

Stubblefield, R. D., Shotwell, O. L., Hesseltine, C. W., Smith, M. L. & Hall, H. H. (1967) Production of aflatoxin on wheat and oats: Measurement with a recording denisitometer. *Appl. Microbiol.* **15** (1), 186–190.

Suryanarayana Rao, K. & Tulpule, P. G. (1967) Varietal differences of groundnut in the production of aflatoxin. *Nature* **214** 738–739.

Sreenivasamurthy, V., Parpia, H. A. B., Srikanta, S. & Murti, S. A. (1967) Detoxification of aflatoxin in peanut meal by hydrogen peroxide. *J. Ass. Off. Anal. Chem.* **50**, 350–354.

Van Walbeek, W., Scott, P. M. & Thatcher, F. S. (1968) Mycotoxins from food-borne fungi. *Can. J. Microbiol.* **14**, 131–137.

Van Warmelo, K. T., Van der Westhuizen, G. C. A. & Milne, J. A. (1965) The production of aflatoxins in naturally infected high quality maize. *Tech. Commun. No.* 71. Dept. Agr. Tech. Serv., Pretoria, South Africa, 5pp.

Van Zytveld, W. A., Kelley, D. C. & Dennis, S. M. (1970) Aflatoxicosis: the presence of aflatoxins or their metabolites in livers and skeletal muscles of chickens. *Poult. Sci.* **49**, 1350.

Velasco, J. & Whitten, M. E. (1973) Evaluation of fluorisil tubes in detection of aflatoxin. *J. Am. Oil Chem. Soc.* **50**, 120–121.

Verrett, M. J., Marliac, J. P. & McLaughlin, J. (1964) Use of chicken embryo in the assay of aflatoxin toxicity. *J. Ass. Off. Anal. Chem.* **47**, 1003–1006.

Vorster, L. J. (1966) Studies on the detoxification of peanuts contaminated with aflatoxin. *Rev. Franc. Corps Gras* **13** (1), 7–12 (In French).

Whitten, M. E. (1966) A rapid screening method for detecting aflatoxins in cottonseed. *Cotton Gin Oil Mill Press* **67** (26), 7–8.

Wildman, J. D., Stoloff, L. & Jacobs, R. (1967) Aflatoxin production by a potent *Aspergillus flavus* Link isolate. *Biotechnol. Bioeng.* 9, 429–437.

Wilson, B. J., Campbell, T. C., Hayes, A. W. & Hanlin, R. T. (1968) Investigation of reported aflatoxin production by fungi outside the *Aspergillus flavus* group. *Appl. Micribiol.* 16, 819–821.

Wilson, B. J., Harris, T. M. & Hayes, A. W. (1967) Mycotoxin from *Penicillium puberulum. J. Bacteriol.* 93, 1737–1738.

Wogan, G. N. (1965) Mycotoxins in Foodstuffs (ed. G. N. Wogan), Cambridge, Mass. M.I.T. Press, 163–173.

TABLE 6. *Aflatoxin levels in feeds and feedingstuffs in Africa*

Country	Commodity	Year	No. of samples	Aflatoxin levels in ppm <0.05	0.05–0.1	0.1–0.5	0.5–1	>1	Average aflatoxin level ppm	Analytical method ref.
Angola	Groundnut cake	NS	1						0.1	NS
	Groundnut cake	NS	1						0.22	134
Gambia	Groundnuts (FAQ)	1970	24	16	2	2	2	2		135
	Groundnuts	1972	46	33	4	2	2	5		135
	Groundnut cake	1972	1			1				135
Ghana	Groundnuts	NS	20		6	14			0.1	136
Madagascar	Groundnut cake	NS	1						0.6	NS
Malawi	Groundnut meal	1968	NS	Range <0.025–0.25					0.05–0.01	135
	Groundnut meal	1969	NS	Range <0.025–0.25					0.05–0.01	135
	Groundnut meal	1970	NS	Range <0.025–0.25					0.05–0.01	135
	Groundnut meal	1971	NS	Range <0.025–0.25					0.025–0.05	135
	Groundnut meal	1972	NS	Range <0.025–0.1					0.05–0.1	135
Mali	Groundnut cake	1972	7						1.5	135
Mauritius	Groundnuts	NS	19	1 sample >0.02						135
Mozambique	Groundnut cake	NS	2			2			0.01	NS
	Groundnut meal	NS	214	Range 0–0.50					0.01	NS
Nigeria	Groundnut cake	NS	59	Range 0.01–1.8					0.45	NS
	Groundnut cake	NS	6	Range 0.08–0.17					0.13	134
	Groundnut cake	1972	8		4	4				
	Groundnut cake	1973	6		2	4				
Senegal	Groundnut cake	NS	4	Range 1.3–2.9					2.2	NS
	Groundnut cake	NS	7	Range 0.11–0.18					0.15	134
South Africa	Groundnut cake	NS	4			4			0.1	NS
	Groundnut cake	1972	1			1				
	Groundnut meal	1964 1974	NS		most samples			few		NS
Sudan	Groundnuts (FAQ)	NS	6	3	3					135
	Groundnut cake	NS	48	24	21	3				135
	Cottonseed cake	NS	38	10	15	13				135

N.S. not stated.

TABLE 7. *Aflatoxin levels in feeds and feedingstuffs in North America*

Country	Commodity	Year	No. of samples	Aflatoxin levels in ppm <0.01	0.01–0.07	0.07–0.5	0.5–1.5	>1.5	Average aflatoxin level ppm	Analytical method ref.
Canada	Maize	NS	NS	All samples <0.005 ppm						137
USA	Groundnut meal	1970	7	3	(0.05–0.25, 3 samples; 0.25–1, 1 sample)					135
	Whole milk	NS	220	All samples <0.005						
	Soya beans	NS	8						0.02	134
	Maize	NS	7	Range 0–0.16					0.02	134
	White maize	NS	11	Range 0–0.14						134
	Yellow maize	NS	49	Range 0–0.31						134
	Maize grade 1	NS	3	3						134
	Maize grade 2	NS	67	67						134
	Maize grade 3	NS	126	126	Range <0.005–0.025					134
	Maize grade 4	NS	65	65						134
	Maize grade 5	NS	7	7						134
	Maize subgrade	NS	25	25						134
	Cotton seed	1964–5	928	882	22	18	5	1		138
	Cotton seed	1965–6	1319	1253	48	18	0	0		138
	Cotton seed	1966–7	943	888	27	20	8	0		138

N.S. not stated.

TABLE 8. *Aflatoxin levels in feeds and feedingstuffs in South and Central America*

Country	Commodity	Year	No. of samples	Aflatoxin content in ppm				Average aflatoxin content ppm	Analytical method ref.
				<0.05	0.05–0.25	0.25–1	>1		
El Salvador	Cotton seed	1970–1971	13		2 samples contained >0.02			0.04	137
Nicaragua	Cotton seed	1970–1971	17		1 sample contained 0.03			0.03	137
Guatamela	Cotton seed	1970–1971	69		1 sample contained 0.06			0.06	137
Argentina	Groundnut cake	NS	6		Range 0.1–0.25			0.2	NS
	Groundnut cake	NS	2		Range 0.2–0.25			0.23	NS
	Groundnut cake	NS	6		Range 0.03–0.96			0.31	138
Brazil	Groundnut meal	NS	7		Range 0.8–2.3			1.6	NS
	Groundnut cake	NS	38		Range 0.1–2.93			0.79	134
	Groundnut meal	1970	20			9	11		NS
	Maize	NS	18	18				<0.005	109
	Cotton seed	NS	17	17				<0.005	109
	Soya bean meal	NS	16	16				<0.005	109
	Wheat offals	NS	14	14				<0.005	109
	Rough rice	NS	2			2		0.4	109
	Hay (alfalfa or Rhodes grass)	NS	NS		Range 0.5–2				137

TABLE 9. *Aflatoxin levels in feeds and feedingstuffs in Asia*

Country	Commodity	Year	No. of samples	Aflatoxin levels in ppm					Average aflatoxin content ppm	Analytical method ref.
				<0.05	0.05–0.1	0.1–0.5	0.5–1.0	>1.0		
Bangladesh	Groundnut cake	NS	3	Aflatoxin range 0.3–1.7					1.0	NS
Burma	Groundnut cake	1972	17	11		6				135
Ceylon	Copra	NS	38	22	0.05–0.25 (10)	0.25–1 (5)		1		136, 139
India	Rice bran	NS	1	1		·			<0.005	134
	Maize	NS	1		1					134
	Sesame cake	NS	1	1					0.03	134
	Tapioca	NS	1	1					<0.005	134
	Guar meal	1972	1			1				NS
	Rice bran cake	1972	2			2				134
	Turbuj meal	1972	1			1				NS
	Safflower meal	NS		Range 0.2–0.4						NS
	Cotton seed	NS	388	285	29	28	18	28		138
	Cotton seed meal	NS	NS	Range 0.2–0.4						NS
	Groundnut cake	1970	178	4	4	165	5			135
	Groundnut cake	1971	2200	406	230	709	553		302	135
	Groundnut cake	1972	2340	279	166	864	646		385	135
	Groundnut cake	1973	2850	12	13	585	1244		996	135
	Groundnut cake	1970	580						0.35	NS
	Groundnut cake	1971	602						0.29	NS
	Groundnut cake	1972	911						0.32	NS
	Groundnut cake	1973	405						0.77	NS
	Groundnuts	NS	874	832	10	32				135
Indonesia	Groundnut cake	NS	3	Range 0.6–0.8					0.65	NS
	Groundnut cake	NS	1						0.52	135
	Cassava chips	NS	2						0.12	NS
Philippines	Maize white	NS	46	Range 0–0.19					0.02	NS
	Maize yellow	NS	86	Range 0–0.35					0.08	NS
	Maize	NS	18	Range 0–0.40					0.10	NS
	Bran	NS	3	Range 0.05–0.20					0.12	NS
	Copra	NS	171	Range 0–0.39					0.01	140
	Copra meal	NS	2	Range 0–0.006					0.006	140
	Copra pellets	NS	3	Range 0–0.006					<0.005	140
	Poultry feed	NS	66	Range 0–0.17					0.015	NS
	Hog feed	NS	11	Range 0–0.2					0.04	NS
	Rice bran	NS	8	Range 0–0.05					0.008	NS
Taiwan	Groundnut cake	NS	12	Range 0–0.29						134
	Groundnuts	NS	8	Range 0–0.43						134

TABLE 10. *Aflatoxin levels in feeds and feedingstuffs in Australasia*

Country	Commodity	No. of samples	Aflatoxin level in ppm	Average aflatoxin content in ppm	Analytical method ref.
Australia	Groundnut meal	3	0.25−0.5		135
	Sorghum (good)	10	0.005		135
	Sorghum (mouldy)	1	1		135
	Maize (mouldy)	6	0.25−2.0	1.4	135
	Maize (good)	12	<0.005		135
	Poultry feed	16	<0.005		135
	Sunflower meal	8	<0.005−0.5	0.1	135
	Safflower meal	6	<0.005		135
	Rapeseed meal	6	<0.005		135
	Groundnuts	2	<0.005		141
	Millet	1		0.13	134
	Groundnut meal supplement	1		2.0	134
	Pig feed	2	0.06−1.00		134
	Layers mark	1		0.05	134
	Cracked maize	1		0.1	134
	Wheat (rain damaged)	7	0.4−5	2.2	142
	Barley	NS	<0.005		142
	Cotton seed	NS	<0.005		142
	Fishmeal	NS	<0.005		142
Gilbert and Ellice Islands	Copra	20	<0.005−0.14		109
	Babai	1		0.5	109
	Copra	7	<0.005−7.0	1.4	109
Marshall Islands	Copra	4	<0.005−0.1	0.03	109
Caroline Islands	Copra	2	<0.005		109
Western Samoa	Copra	3	<0.005		109

TABLE 11. *Aflatoxin levels in feeds and feedingstuffs in Europe*

Country	Commodity	Year	No. of samples	Aflatoxin content in ppm					Average aflatoxin content ppm	Analytical method ref.
				<0.05	0.05−0.1	0.1−0.5	0.5−1.0	>1.0		
Czechoslovakia	Feedingstuffs	1969	102	68% + v$_e$.						134
	Feedingstuffs	1970	155	32% + v$_e$.						134
	Feedingstuffs	1971	50	19% + v$_e$. 1 sample >0.1						134
	Feedingstuffs	1972	161	85% + v$_e$. 53 samples >0.1						134
	Feedingstuffs	1973	107	24% + v$_e$.						134
	Dried milk	1971	2						<0.005	134
	Dried milk	1972	3						<0.005	134
	Dried milk	1973	40						<0.005	134
Denmark	Groundnut meal	NS	NS	Range 0.075−1.25						143
	Soya meal	NS	NS	Range 0−0.002						143
Eire	Irish oats	NS	26						<0.005	144
	Imported oats	NS	13						<0.005	144
	Groundnut meal	NS	16	<0.05−1.0					NS	144
	Pig Feed	NS	12						<0.005	144
	Poultry Feed	NS	11	2 samples + v$_e$.					<0.005	144
	Grains	NS	>200						<0.005	134
	Calf feed	NS	1						<0.005	135
	Groundnut cake	NS	43	<0.05−0.15					NS	NS
	Cottonseed cake	NS	1						0.1	NS
France	Soya bean cake	NS	51	36	8	7				139
	Groundnut cake	NS	12	6	4	2				139
	Maize	NS	34	29	4	1				139
	Wheat	NS	32	24	6	2				139
	Sunflower cake	NS	15	12	3					139
	Cocoa cake	NS	7	6	1					139
	Sorghum	NS	12	10	2					139
	Oatmeal	NS	19	15	4					139
	Rye	NS	21	20	1					139
	Barley	NS	14	14						139
	Pig feed	NS	30	18	8	4				139
	Poultry feed	NS	60	43	13	4				139
	Cattle feed	NS	20	12	7	1				139

Trial of the elimination of aflatoxin in groundnuts by physical methods

A. Bockelee-Morvan and P. Gillier
Groundnut Department, I.R.H.O., Paris, France

Summary

Aflatoxin contamination of groundnut kernels from Francophone West Africa is largely confined to those from pods which become physically damaged by pests before harvesting, or during harvesting. Trials are described in which it is shown that a significant reduction in overall aflatoxin contamination of unshelled groundnuts can be achieved by the removal of defective pods by hand or pneumatic sorting. The removal of abnormal kernels from shelled groundnuts by mechanical or hand sorting also greatly reduces overall aflatoxin contamination.

Résumé

Essais d'élimination de l'aflatoxine des graines d'arachide par des méthodes physiques
La contamination par l'aflatoxine des graines d'arachides de l'Afrique de l'Ouest Francophone est largement limitée aux gousses qui deviennent physiquement endommagées par les insectes, avant ou pendant la récolte. On décrit des essais dans lesquels on a montré qu'on peut accomplir une réduction significa-tive de la contamination globale avec aflatoxine des graines écorcées, en enlevant les gousses défectueuses manuellement ou par triage pneumatique. L'enlève-ment des noyaux anormaux des graines écalées par triage mécanique ou manuel, réduit grandement la contamination globale avec aflatoxine.

Resumen

Pruebas de eliminación de aflatoxina en los cacahuetes por métodos físicos
La contaminación de las semillas de cacahuetes por la aflatoxina en el Africa Occidental de habla francesa se confina en gran parte a aquéllos cuya vaina sufre daño físico a consecuencia de pestes antes de la recolección, o durante ella. Se describen pruebas en las que se demeustra que puede lograrse una reducción significa-tiva en la contaminación de aflatoxina total de los cacahuetes no descascarados, quitándoles la vaina defectuosa a mano o con separadoras neumáticas. La extracción de las semillas anormales de cacahuetes descascarados mediante separación mecánica o manual reduce también en gran manera la contaminación de aflatoxina total.

Introduction

The value of groundnuts from African producer countries is reduced by the presence of aflatoxin in quantities estimated at an average of 0.1–0.2 parts per million (ppm) and regulations in countries importing press-cake are getting more and more strict.

If it can be achieved, the detoxification of press-cake will be costly, and there would be every advantage in limiting it to the fraction of the harvest which is contaminated. In this connection, the I.R.H.O. is studying practical means of processing separately in the oil mill the largest possible aflatoxin-free portion of the harvest, and the smallest possible contaminated portion in the detoxified press-cake. This paper describes two trials of the separation of contaminated groundnuts by hand sorting of the pods or seeds, carried out in Senegal, Mali and Niger in 1973.

Pod sorting trial

The trial concerned 9 lots of 50 to 400 kg of ground-nuts harvested in very different situations in Mali, Niger and Senegal. In Mali and Niger each lot was drawn from 1,000 kg of groundnuts bought in the markets and representing a sample of harvests in the zone. In Senegal, each lot was drawn from the harvest of 100 farmers spread throughout the area in which the various varieties were grown.

The pods which were apparently intact were separated from those which were split, broken, pierced by insects or millipedes (*Diplopoda*), gnawed by termites or with dried ends (black end). Table 1 gives the main characteristics of each category of pods, expressed as percentages of the original lot before sorting (random).

The intact pods represented between 56 and 92% of the lots (mean 73%), and their aflatoxin content was an average 24% of that of the lot before sorting. An analysis of the results shows that only lots with aflatoxin contents which were not very high before sorting yielded lots which were free from or contained less than 0.55 ppm aflatoxin afterwards.

In certain parts of these countries, the 1972 harvest suffered from the harmful effects of drought. Because of this, the lots of groundnuts used in the experimentation had very diverse characteristics with regard to ripeness, insect damage, and aflatoxin content, the quality being sometimes much lower than that obtained in a normal year. For this reason it seemed preferable

not to mention actual flatoxin contents in ppm but to express them as percentages.

However, in all cases an appreciable diminution of the aflatoxin content was noted. In the pods set aside, it was those which were pierced (generally by millipedes) which had the highest contents, on an average nine times more than a random lot. Although they represented only a small percentage (4.3%) they contained an average 40% of all the aflatoxin in the lot. Next came the split, broken, termite-gnawed and black-ended pods.

The following important lessons can be drawn at the cultivation level:
(1) The advantage of careful harvesting and threshing to avoid split or broken pods.
(2) The role of millipedes in contamination by *Aspergillus flavus*, the pods pierced by these having very high aflatoxin levels.

An industrial firm is putting the finishing touches to a pneumatic shelling apparatus, the principle of which is the separate shelling of whole pods, pierced pods and those which are broken, split, gnawed by termites and have black ends. Hand sorting by the producers is at present employed in Senegal for unshelled edible groundnuts.

TABLE 1. *Pod sorting trial*

Category of pods		(1) Good pods	(2) Split pods	(3) Broken pods	(4) Pierced pods	(5) Gnawed by termites	(6) Black ends	(2+3+6)	Random
Lot 1	a	16.5	17.0	–	63.9	2.6	–	17.0	100
	b	(76.0)	(5.5)		(8.8)	(9.7)		(5.5)	(100)
	c	22	309		726	27		309	100
Lot 2	a	0	0	–	100	0	–	0	100
	b	(92.0)	(1.3)		(2.4)	(4.3)		(1.3)	(100)
	c	0	0		4,166	0		0	100
Lot 3	a	38.7	8.1	–	31.0	22.2	–	8.1	100
	b	(90.0)	(2.8)		(8.0)	(9.2)		(2.8)	(100)
	c	48	289		587	241		190	100
Lot 4	a	2.8	56.5	–	10.5	29.2	1.0	57.5	100
	b	(61.6)	(29.6)		(3.8)	(4.2)	(0.2)	(29.8)	(100)
	c	4	191		276	695	500	193	100
Lot 5	a	1.5	75.3	–	14.2	8.9	0.1	75.4	100
	b	(56.1)	(33.2)		(3.2)	(6.5)	(0.1)	(33.3)	(100)
	c	3	227		364	137	100	226	100
Lot 6	a	9.3	9.1	10.7	39.8	30.5	0.6	20.4	100
	b	(84.0)	(3.8)	(2.7)	(2.5)	(6.4)	(0.6)	(7.1)	(100)
	c	11	239	396	1,592	476	106	287	100
Lot 7	a	42.2	10.7	18.1	8.4	5.3	15.3	44.1	100
	b	(70.0)	(3.4)	(5.0)	(3.5)	(5.5)	(12.6)	(21.0)	(100)
	c	60	214	362	240	96	121	210	100
Lot 8	a	15.6	2.8	10.9	49.4	3.9	17.4	31.1	100
	b	(64.4)	(3.3)	(3.6)	(3.6)	(1.3)	(24.0)	(30.9)	(100)
	c	24	85	302	1,453	300	72	100	100
Lot 9	a	31.5	5.6	15.4	28.0	1.3	18.2	39.2	100
	b	(73.0)	(2.6)	(5.1)	(2.5)	(0.6)	(16.2)	(23.9)	(100)
	c	43	215	302	1,120	217	112	164	100
	a	17.5			38.4	11.5		32.6	100
	b	(73.0)			(4.3)	(5.3)		(17.3)	(100)
	c	24			890	217		188	100

a – 1st figure. Percentage of total aflatoxin content in the fraction under consideration.
b – 2nd figure. Percentage in weight of the fraction considered in the initial unsorted lot (random)
c – 3rd figure. Aflatoxin content of the fraction considered in relation to the total for the random lot = 100.

Seed sorting trial

This trial was carried out in Senegal on two lots of groundnuts for oil milling and two lots of edible groundnuts. The object was to determine which categories of seeds were the most contaminated in order to work out a method of sorting seeds applicable to oil milling groundnuts in countries where the product is marketed as seeds. For edible groundnuts, sorting is always necessary, whether done at producer level or, more often, at industrial level.

First of all perfect seeds were separated from those presenting some form of anomaly; the latter, representing 30.7% of the total, were divided into four categories:

A. Unripe, wrinkled seeds, otherwise normal;
B. Seeds showing signs of mould attack, otherwise normal;
C. Broken or skinned seeds, otherwise normal;
D. Seeds of abnormal colour, without visible attack by moulds.

The characteristics of each category (Table 2) are expressed as a percentage of the lot of defective seeds before sorting (30.7% of the initial lot). Provided that they had no abnormal colouring, the broken or skinned seeds were little contaminated. The seeds with abnormal colouring were highly contaminated, whilst those which showed visible signs of mould attack had very high contents of aflatoxin.

The sorting of categories A, B and C, which made up about 18% of all the seeds, gave lots which were almost or entirely aflatoxin-free.

These criteria are now being applied to the sorting of edible groundnuts and give every satisfaction, the wrinkled and broken seeds being eliminated mechanically. Seeds attacked by moulds or with abnormal colouring are removed by hand, as well as the skinned seeds, which can be exported.

Conclusions

The sorting of unshelled groundnuts and seeds enables a considerable improvement in the quality as regards aflatoxin. The sorting criteria defined can be used for hand sorting, and their application to industrial methods of sorting contaminated groundnuts can be envisaged. This would involve pneumatic shelling, grading, and the removal of seeds with abnormal colouring by electronic sorting.

These studies were carried out by the I.R.H.O. within the more general framework of the separation, before industrial processing, of contaminated groundnuts, which also includes the detection at the time of purchase of contaminated lots by means of rapid visual (Dickens method) or chemical (minicolumn method) tests and the development of methods of cultivation, harvesting, threshing and drying likely to limit the risks of aflatoxin production.

TABLE 2. *Visual criteria for seed sorting*

Category of seeds		Lot 1	Lot 2	Lot 3	Lot 4	Mean
(a) Wrinkled, unripe	a	4.5	4.3	0.9	2.3	
	b	(37.4)	(37.8)	(37.6)	(37.8)	(37.6)
	c	12	11	2.5	6.2	7.9
(b) Visible mould damage	a	86.4	84.7	91.6	90.3	88.2
	b	(12.1)	(11.9)	(11.9)	(12.1)	(12.0)
	c	715	715	770	745	740
(c) Broken, skinned	a	0.3	0.6	0.7	0.9	0.6
	b	(41.3)	(41.2)	(41.3)	(41.5)	(41.3)
	c	0.8	1.4	1.6	2.1	1.5
(d) Abnormal colour (no visible mould attack	a	8.3	10.4	6.8	6.5	8.2
	b	(9.2)	(9.1)	(9.2)	(8.6)	(9.1)
	c	95	114	77	74	90

Discussion

Miss Carey: My question is directed to Mr Ashman. I wonder if the use of irradation techniques to disinfest crops rather than by the use of pesticides which may leave residues has been considered?

Mr Ashman: Quite a lot of work has been carried out in this field. The biggest problem is that the insects have to be brought into close contact with the energy source and this necessitates a very sophisticated and expensive system of operation. A plant was established in Turkey with American assistance fairly recently, but I think the use of radiation techniques to control insects is far too sophisticated especially for the developing countries.

Dr Babatunde: I would be grateful if Dr Jones could supply some information on the work carried out in West Africa on the prevention of aflatoxin contamination of groundnuts by improved farming practices. Also what progress has been made towards the development of varieties of groundnuts which are resistant to contamination by aflatoxin?

Dr Jones: The work to which I referred was carried out by MacDonald and Harkness during the period 1962–65. They showed that damage to the shell allowed access to the kernels by *Aspergillus flavus* while the crop was still in the ground, and that delayed harvesting after cessation of the rains could then lead to the kernels drying out to moisture contents beneficial to *A. flavus* growth. Delayed drying after harvesting or storage at too high moisture contents could also subsequently result in aflatoxin contamination. With regard to the question on resistant varieties no groundnut variety has yet been found to resist *A. flavus* growth and aflatoxin production.

Dr Traore: I would first like to thank the government of the United Kingdom for my participation at this conference, and the assistance given to Mali in the establishment of our aflatoxin control laboratory. In fact it was Dr Jones who visited Mali in 1972 to assist us in this latter regard. We have now been actively involved in aflatoxin control for about one year and have found sampling to be our main difficulty. Dr Jones rather glossed over sampling in his paper and I would appreciate his further comments on this. I should mention that all samples of groundnuts or groundnut parts used for edible purposes which we have looked at so far had aflatoxin contents below 0.05 ppm. Also we have not found any evidence of aflatoxin contamination of maize or sorghum in Mali. Finally we are currently using groundnut cake containing on average 1.5 ppm of aflatoxin at inclusion rates of 10% in pig and poultry rations and the autopsies showed no hepatic lesions. Perhaps susceptibility to aflatoxin varies not only with the species of animals but also between individual animals, and perhaps it is also dependent on the conditions under which the animals are kept.

Dr Jones: It is a very difficult problem to obtain a representative sample for aflatoxin analysis from many commodities. This is especially so for maize and groundnut kernels as aflatoxin contamination may be confined to only a few kernels, and if a sample is taken contaminated kernels may be missed or picked out and the aflatoxin content will be either lower or higher than the average for the whole lot. Workers in the USA and Canada have carried out statistical analyses on this problem and their countries operate fairly vigorous sampling procedures, but even so it is likely that their sampling is far from perfect. If aflatoxin incidence is to be investigated in Mali a very elaborate sampling scheme will be necessary which will entail a great deal of work. Toxicity certainly varies between animal species. It will also vary within a species depending on the age of the animal and its diet. Low protein diets tend to enhance susceptibility to aflatoxin. Pigs and poultry are not particularly susceptible to aflatoxin. However, it should be mentioned that pigs fed a diet containing 0.3 ppm aflatoxin produced bacon containing 0.05 ppm of aflatoxin, so there is a possibility of contamination of

296 animal products entering the food chain under these circumstances.

Mr Redfern: I would like to enquire whether the African Groundnut Council has considered taking any steps to establish a common standard for aflatoxin levels of groundnuts or groundnut products exported from its member countries?

Chairman: This is being considered and will be discussed at a meeting we are soon to hold in Khartoum. We are of course studying the legislation on aflatoxin which has been introduced by some of our customers and the results of research on the effect of aflatoxin on animals. However, we feel that much of this legislation is unrealistic and is not soundly based in scientific terms. We certainly accept an ethical obligation to market groundnuts and groundnut products of the highest quality but we face many problems in trying to effect improvements. There is certainly a lot of scope for reduction in aflatoxin contamination but if we are to accept legislation we must be assured that it is soundly based and realistic.

Eighth Session

Use of concentrate feeds in developing countries

Friday 5th April
Morning and Afternoon

Chairman
Professor K. A. Alim
University of Alexandria
Arab Republic of Egypt

Concentrate feeds for livestock in Trinidad and Tobago: an economic view of developments since 1960

J. Cropper
University of the West Indies, Trinidad

Summary

As a result of Government measures in the form of land settlement schemes, import controls and tax incentives, the livestock industry in Trinidad and Tobago has expanded rapidly in the past 10 years — principally broilers, eggs, milk and pork. All these enterprises are based on concentrate feeding.

This growth has occurred without any parallel effort to expand supplies of local feed ingredients, and consequently imports have risen sharply. Although self sufficiency has been attained for several livestock products, an analysis of the pig industry has shown that this achievement has brought little benefit to the country in terms of saving of foreign exchange; however, some employment has been created.

The vulnerability of the livestock industry to movements in the prices and supplies of feeds on the world market, which is presently being experienced, was recognised rather late. The approaches adopted to try to insulate the industry from external forces include combinations of both short and longer term strategies. These can be grouped under two heads, which are respectively attempts to reduce the cost of imported feed ingredients and attempts to replace imports by locally available feeds.

The currency used throughout this paper is the Trinidad and Tobago dollar: at present rates of exchange T & T $1.00 equals approx. £0.21 and US $0.45.

Résumé

Les pâtures concentrées pour le bétail en Trinidad et Tobago; un aperçu économique des développements depuis 1960

Comme résultat des mesures prises par le gouvernement, concernant les projets de mise en possession des terres, des limitations des importations et de réduction des taxes, l'industrie du bétail en Trinidad et Tobago s'est développée dans les dernières 10 années — surtout les poulets à rôtir, les oeufs, le lait et les cochons. Toutes ces entreprises sont basées sur des pâtures de concentrés. La croissance s'est produite sans un effort parallèle pour amplifier les provisions locales d'ingrédients alimentaires et par conséquent les importations ont montées vivement. Même en atteignant une production suffisante de certains produits du bétail, une analyse de l'industrie des cochons a montrée que cette réalisation a apportée peu de bénéfice pour le pays en termes d'économie de devises étrangères; cependant on a crée un nombre d'emplois.

La vulnérabilité de l'industrie du bétail, concernant les fluctuations des prix et l'approvisionnement sur le marché mondial qu'on éprouve à présent a été reconnue un peu tard. Les essais faits d'isoler l'industrie des forces externes incluent les combinaisons des stratégies à longue et à court terme. Ces-ci peuvent être groupés en deux catégories, qui sont respectivement des tentatives de réduire le coût des ingrédients pour les provisions importées et des tentatives de remplacer les importations par des produits disponibles localement.

Resumen

Piensos concentrados para el ganado en Trinidad y Tobago: una visión económica de los desarrollos desde 1960

Como resultado de las medidas del gobierno en forma de proyectos de asignación de tierras, controles de importación y exención de impuestos, la industria de la ganadería en Trinidad y Tobago se ha expandido rápidamente durante los últimos 10 años — principalmente aves para asar, huevos, leche y cerdo. Todas estas empresas están basadas en la alimentación con concentrados.

Esta expansión se ha producido sin un esfuerzo paralelo para aumentar los suministros de ingredientes de piensos locales y, como consecuencia, han aumentado las importaciones extraordinariamente. Aunque se ha conseguido la autosuficiencia para varios productos para el ganado, un análisis de la industria del cerdo ha demostrado que este logro ha supuesto poco beneficio para el país en cuanto a ahorro de divisas extranjeras; sin embargo, se han creado algunos puestos de trabajo.

Se reconoció más bien tarde la vulnerabilidad de la industria ganadera a los movimientos de los precios y suministros de piensos en el mercado mundial, que se están experimentando actualmente. Las medidas adoptadas para tratar de aislar la industria de las fuerzas externas, incluyen combinaciones de estrategias tanto a corto como a largo plazo. Estas pueden agruparse en dos apartados que son, respectivamente, los intentos para reducir el coste de los ingredientes de piensos importados, y los intentos de sustitución de importaciones por piensos disponibles en la localidad.

Introduction

The sister islands of Trinidad and Tobago are located at the southern end of the chain of Caribbean Islands, off the northern eastern tip of the coast of Venezuela: that is 11°N and 62°W.

The climate can be classified as wet, lowland tropics. Rainfall averages approximately 160 cm annually, with the major part falling between the months of June and December. The average daily temperature is 27°C with a diurnal variation of 5°C. The highest point in the island is approximately 1,000 m although most of the land is less than 300 m.

The population in 1972 was 1.05 m having an average annual per capita income of $1260.

The total land area is 5128 km^2, of which 35% is under agriculture, 45% under forests and 20% is built-on and waste land. The main agricultural enterprises are sugar and cocoa (each approximately 40,000 ha) while smaller areas are planted with citrus, coconuts, coffee and food and vegetable crops. The contribution of the various sectors of agriculture, including livestock, to the Gross Domestic Product is listed in Table 1.

TABLE 1. *The contribution of selected sectors of agriculture to the gross domestic product: Trinidad & Tobago, 1970*

Sector	Gross Domestic Product	
	$m	Proportion of total (%)
Cocoa & coffee	11.4	22.1
Citrus	1.8	3.5
Coconuts	4.0	7.7
Vegetables	5.0	9.7
Other food crops	6.3	12.2
Forestry	2.4	4.6
Fisheries	5.8	11.2
Poultry meat & eggs	11.4	22.1
Milk and beef	2.2	4.3
Pork	1.4	2.7
Total agriculture (except sugar)	51.7	100.0

The author is grateful for the comments on an early draft of the paper by Professor D. T. Edwards and E. F. Unsworth of the Departments of Agricultural Economics and Livestock Science, U.W.I., respectively.

The country has been a traditional exporter of primary agricultural produce, sugar in particular, as well as cocoa, coffee and citrus, and export agriculture has been for many years a principal economic activity. However, during this century, the discovery of oil and natural gas and the development of refining facilities for local as well as imported oil have led to the emergency of the oil industry as the major economic activity in the country.

In 1962, the country became fully independent of Britain. Since this time economic policy has aimed at moving the country away from its dependence on export agriculture. This policy has involved the further expansion of the oil industry, the promotion of manufacturing industries and attempts to diversify the country's agriculture, principally by the expansion of livestock production.

The expansion of livestock production

(a) *The means of promoting expansion*

In Trinidad and Tobago as in many developing countries which have owed their existence to the export of primary products to the metropole, the basic resources of production were geared to the production of export products (Beckford, 1972). As a result, the Government has to become actively involved in measures to promote the production of alternative products, either by diverting existing resources or by providing new ones. The policy of the Government of Trinidad and Tobago has been to provide new resources for the expansion of livestock production, leaving the production of sugar and other export crops unaltered.

Eggs and poultry meat were the first products to be promoted in the early 1960's, while since the mid 1960's efforts have also been made to increase pork and milk output; there has been little effort to expand beef and sheep production.

Livestock production has been promoted by such means as subsidies, fiscal incentives, tax rebates, distribution of Government 'Crown' land, control of imports, price control and the research and extension of information.

Fiscal incentives and tax rebates have been given mainly to industries serving agriculture such as factories for

During the period, quantities of coconut and citrus meal have remained at similar levels, since there has been no expansion in the output of copra and citrus (see Figure 6). In 1972 they provided only 5% of all concentrate ingredients used in the country. No information is available on the use of molasses, although usage is thought to have increased somewhat.

(d) *Economic effects of expanded livestock production*

The overall effect of the livestock expansion programme on the balance of payments situation has been to reduce imports of livestock products somewhat, until 1970 when prices of imports as well as quantities of milk imports increased significantly (see Figure 9). At the same time the total cost of imported livestock feeds has also increased due to greatly expanded quantities being imported. As a result the combined cost of animal feeds and livestock products increased over the period. This does not mean that the import substitution policy failed, for during this time population increased by approximately 25% and per capita consumption of several livestock products also increased significantly. However, since a high proportion of the production costs of eggs, poultry meat, pork and milk production is comprised of feed, imports of these products have been substituted largely by imports of feedingstuffs. The effects of the import substitution policy on the pork industry have been examined in detail by Exeter and Edwards (1972).

Figure 9

Value of imports of livestock products and animal feed ingredients: Trinidad and Tobago 1960—1972

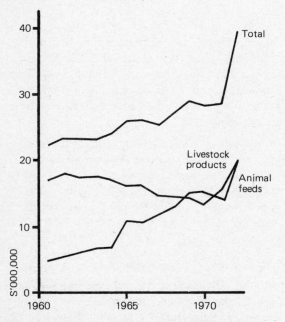

Exeter showed that, in so far as import substitution means a greater degree of self-sufficiency, the policy had met with success in the pork industry. On the other hand, examined on the basis of the net import savings, the policy was seen to be only moderately successful, because local production was based almost entirely on imported feeds while at the same time the average feed conversion ratio of 1:4.8 was poor. In addition local production of fresh pork had not contributed significantly to the generation of income and

employment in the economy, since such a high proportion of the inputs (feed in particular) were imported. 303

The situation changed somewhat when a pork processing factory was opened but this improvement was short-lived since poor management of the factory led to the disruption of pork production. Thus, as is shown in Figure 3, since 1969 local production has fallen and imports of pork and pork products have increased while consumption has fallen. In addition there were considerable costs to the economy as a result of implementing the programme; these arose from Government investment in establishing and subsidising new pig farms, while at the same time consumers had to pay higher prices for reduced quantities of pork products.

Exeter concludes that 'The conditions for a successful import substitution policy might therefore be summarised by saying that the industry selected should be one for which the market is large and where there is an opportunity to produce a considerable proportion of the inputs from domestic sources. The measures to encourage expansion should concentrate on removing restraints to production, so that the policy can be implemented at a low opportunity cost in terms of the absorption of resources, and secondly on keeping imported inputs to a minimum, thus increasing the amount of net import savings made.

'Applying these conditions to the pig industry in Trinidad, it can be concluded that although import substitution was relevant to the economy between 1965—69, the pig industry was not an appropriate industry to be developed under such a policy in view of the limited opportunities for reducing the import content of feeds'.

The same conclusions may be drawn also with respect to the broiler and egg industries since these are similar in most respects to the pig industry, although for the broiler production the feed conversion ratio is much more efficient (1:2.5). A similar situation exists also in milk production where the feeding system relies heavily on concentrates.

The basic weakness in the import substitution policy for livestock products was recognised at a policy level for the first time in the Third 5-Year Plan, 1969—1973 (Government of Trinidad and Tobago, 1969) which states: 'It is Government's stated intention to examine every possible method of reducing the foreign exchange of animal feeds, including the use of locally produced ingredients, so as to improve the economic position of the nation's livestock farmers, and to maximise the benefits which can accrue from a viable livestock sector'. The further problem, namely the susceptibility of the local livestock industries to increased world prices for feed ingredients and to shortfalls in supply, has been recognised only recently at the time of the worldwide shortage of feed grains.

The government has embarked simultaneously on two strategies aimed at combatting these problems:
(a) attempts to reduce the cost of imported feed ingredients, and,
(b) attempts to replace imports by locally available feeds.

Attempts to reduce the cost of imported feed ingredients

(a) Price control

A survey of the feed manufacturing industry in 1970 (Ali *et al* 1970) found that manufacturing capacity was grossly underutilised (45% excess capacity) and recommended that the Government reduce animal feed prices in order to squeeze out less efficient manufacturers. While the Government has continued to control feed prices they have been unwilling to take such positive action as was recommended. In fact during 1973, feed manufacturers withheld supplies of feed from farmers as a protest against the Government's unwillingness to raise the prices for animal feeds in the face of increasing import prices. While some price increases were undoubtedly in order at this time, there has been no reduction in the number of feed manufacturers, and it can be concluded that the price increases awarded were sufficient to allow even the least efficient firm to remain in operation.

(b) Local ownership of feed manufacturing firms

Of the seven manufacturing firms in the country, four are foreign owned, while two others are closely associated with foreign companies: only the smallest firm is truly local. The same survey referred to above found that much of the feed purchasing and formulation for these six firms was controlled by their principals in the USA and Canada. The result of this situation was that the cheapest source of ingredients was not always used, for example, ingredients originating in South America were imported via North America, thus adding to the cost. In addition many ingredients were imported as a result of inter-company transactions between parent and subsidiary companies, thus raising questions about the pricing of commodities.

The Government has recently stated its intention to take control of two feed manufacturing companies, but has so far not moved to implement this intention. It is assumed that the purpose of this policy will be to allow greater flexibility in purchasing and feed formulation than in the past.

(c) Bulk purchasing of ingredients

Limited quantities of corn and soya bean meal have been imported in bulk through the country's flour mill. In addition, it was proposed that a grain terminal be established to import feed grains in bulk. It was learnt however that the principal shareholder in the flour mill, the proposed terminal and one of the larger feed manufacturers was the same North American company, and it was feared that this concentration of investment would not be in the best interests of the livestock industry and the country in general. The Government has, therefore, stated its intention to acquire a controlling interest in the grain terminal, which began operation late in 1973. It is not possible to determine whether Government ownership *per se*

will be able to influence significantly the price of imported ingredients, although the bulk handling facilities in themselves should have a positive effect.

Attempts to replace imports by locally available feeds

While the attempts referred to above can have an immediate effect on reducing the cost of feeds, the major benefit to the country will arise if locally available feeds can replace imports. In the discussion which follows, emphasis is placed on net foreign exchange savings as the criterion for recommending a particular policy. Attention is not paid to the profitability of production, since in the first instance it should be determined whether a policy is justifiable for the country on macro-economic and social grounds before considering private returns.

(a) Existing by-products

A survey by Gohl (n.d.) in 1968–1970 revealed the availability of many different by-products available for animal feeding, for example, filterpress mud, brewers grain, lime skins, pea shells, coffee grounds, cocoa pods and shrimp offal. The more conventional of these, such as coconut and citrus meal, are already fully utilised in livestock feeding. However, the coconut and citrus industries have suffered a number of problems in recent years and additional quantities of by-products are unlikely to be forthcoming in the future.

The use of other products has been limited by their irregular availability, unstandardised nature, unsuitable form for feeding and in some cases by lack of information about their suitability as animal feeds. It seems unlikely that these products could have a significant impact on the livestock industry even if these problems were resolved because available quantities are generally small.

Only wheat middlings (a by-product of the local milling of imported wheat) are likely to increase in availability. This will occur when the milling capacity is expanded from presently supplying 60–70% of the country's flour to supplying all of the flour: at present the remainder of the flour is imported in milled form.

(b) Specially produced products

(i) *Corn and soya bean:* corn (maize) and soya bean meal are the two major feed ingredients imported at the present time. In Trinidad and Tobago, corn is a traditional small farmer crop, but for human rather than animal food, while soya bean was grown successfully on an experimental scale for the first time in the mid 1960's. Since 1971, a 200 ha pilot project (Chaguaramas Agricultural Development Project, 1973) has been underway to examine the feasibility of commercial production of these crops. The yield and cost results have so far been encouraging, although it has been questioned if these results could be maintained on very large areas under continuous cultivation.

intensive conditions tend still to be semi-luxury items for which the end market is limited. Unfortunately we have to deal in what is practical rather than what is desirable in human nutrition terms and the scale of the feed industry must be related to the end product market. Perhaps the classic case of inadequate market research concerned the establishment of a broiler industry in one South-East Asian country where the initial planning went along the lines of 'in America we eat over 30lbs of chicken meat per head per year; if in this country the population (of which 80% lived in rural areas) ate only one broiler chicken per year then . . . . .'. The consequences in terms of unused capacity from the broiler breeding flock were discouraging. Fortunately in this particular instance the planning of the feed industry was ahead of the planning of the breeding industry.

The second aspect of the background survey to the establishment of a feed industry should concern the availability of resources upon which the viability of the industry depends. Although the prime concern of this survey should be to examine raw material resources it is important to recognise that other resources vitally affect the economics of the feed industry. The work of the nutrition scientist in utilising the available raw materials is of no use if other management skills are not available to ensure that the right quality of product is available to the livestock farmer at the right time, with regularity of supply and at the right price to ensure that he can operate profitably. The various management disciplines of buying, production, marketing, finance and distribution all have an important contribution to make in the operation of a feed industry. It is possible to train managers in these disciplines but the training requirement must be recognised during the planning stage of a project.

It is vital that the background survey establishes precisely how the products of the feed industry will move from the factory to the livestock producer. A distributor network is normally the most efficient form of contact between the manufacturer and the livestock producer. Newly established feed industries tend to have a very much wider distribution area than would normally be regarded as economic in more developed countries. The feed manufacturer would find it totally uneconomic to have direct contact with the widespread consumers of his products. The role of the distributor in the organisation of the physical distribution and in the provision of services such as credit and management advice is vital. A distributor network operating under the supervision of the feed manufacturer is a simple way of multiplying up his skills and making them available to a wider cross-section of the community. A well trained feed distributor tends often also to become a distributor of other agricultural inputs: a need that is not always well supplied in many countries.

The key area of the background survey of resources will however be the raw material situation. Raw material costs are approximately 75/80% of the on-farm cost of compound feeds. The availability of adequate supplies of raw materials must be a key factor therefore in the success of a feed project. The feed manufacturer should as far as possible depend upon local resources but there are few countries in the world where all the necessary ingredients for a feed industry are available. In many cases feed industries have been established on the basis of using imported proteins until locally available supplies can be developed. Problems of availability of foreign exchange must lead the feed manufacturer to try and develop local sources of proteins even if this involves going into the production of for example fishmeal or soya beans or maize on his own account. Nevertheless there will be some items which may necessarily have to be imported on a long term basis such as vitamins and other micro-ingredients. It is important that this factor is recognised at an early stage and that long term provision is made for the availability of foreign exchange for these items.

If the feed manufacturer is obliged to go into the production of raw materials then he should take a similar view to that recommended for the feed manufacturers approach to breeding. He should see his role as a catalyst in developing local production and he should where possible withdraw at an early stage from direct production of raw materials although he may have to retain an indirect involvement through the provision of capital to cover the carrying of stocks from one harvest to another. If the feed manufacturer withdraws from direct production then the benefits of the development of the feed industry are spread more widely through the agricultural community.

Production

Having established the background in which the industry will operate and having planned his marketing strategy the feed manufacturer must consider the actual mechanics of production. There is always a temptation to translate into developing countries the sophistications of more developed ones. The feed manufacturer should consider carefully what is appropriate for the market in which he is operating. The priorities in the early stages are to establish the supplies of raw material to specified quality standards and to ensure that adequate supplies of standard quality products are available to the livestock producer. If capital is a limiting factor then it should be used at the introducing sophistications into the manufacturing process. The old computer dictum of 'rubbish in — rubbish out' applies equally well in the feed industry.

All the initial marketing effort of the feed manufacturer must be designed to create confidence in himself and his products. The livestock producer must be taught to judge the product not in terms of price per ton but in terms of its economic value. The feed manufacturer should assist the livestock producer to develop recording systems to assess the efficiency of his livestock production enterprise. Without records the livestock producer can assess a compound feed only in terms of its price per ton. With records he can assess that feed in terms of cost per unit of output. The cost of feed per dozen eggs produced or per lb of chicken meat is a much more important figure than the cost per ton of the feed itself. The feed manufacturer has a most important role to play in assisting the livestock producer to assess the value of his products and in so doing he helps the livestock producer to know where to improve the efficiency of his production systems.

Relationships between Government and the feed industry are a potential source of problems for the individual feed manufacturer. However these problems can be kept to a minimum if there is a conscious effort on the part of the feed industry to put over to the Government an understanding of its methods of operation. These problems tend to arise in four areas:

(1) Raw materials
(2) Quality control
(3) Relations between the feed industry and academic institutions
(4) Relations with Government extension services.

In the initial background survey for a feedingstuffs project it is important to establish Government policies in relation to raw materials. There may be potential areas of conflict between needs of the feed industry and the human population particularly in the supply of food grains. It is important to establish with the appropriate Government department the need of the feed industry for continuity of supply of raw materials. There have been occasions when comprehensive measures introduced to ensure the fair availability of supplies of grain in times of shortage have overlooked the problems of the feed industry. The whole livestock industry must put over to Government planners the importance of continuity of supply of all raw materials. It may not be possible always to guarantee the full requirements of the industry but at least the minimum needs of the industry should feature in Government planning.

The Government and the feed industry must always work closely together in the area of quality control. The feed industry must accept the need for some form of control but it is important that industry and the Government work closely together to ensure a degree of control appropriate to the particular environment. There must be an understanding on the part of the Government of the problems and of the objectives of the feed industry. The feed manufacturer may be the easiest point at which to introduce controls but there is no point in laying down tight standards for finished products if the feed manufacturer cannot obtain the appropriate quality of raw materials.

The feed industry must endeavour to convince the Government of its own philosophies. If the feed industry projects an image of attempting to make a quick dollar at the expense of the livestock producer then it must expect a high degree of control. But the real control of the feed manufacturer lies in the ability of the livestock producer to assess the economic value of his products. The necessity of the feed manufacturer to gain repeat orders is a far better control than any comprehensive list of restrictive regulations.

There must be a mutual understanding and willingness to co-operate to work out new problems. For example the normal distribution radius of the feed plant in developed countries is 130/160 km but in developing countries much greater distances can be the norm. Problems of weight loss in transit are substantial particularly in hotter climates. This type of problem is not to be solved by an arbitrary declaration that the feed manufacturer is responsible for bag weights at the ultimate point of consumption. It is a problem which requires careful evaluation on both sides.

The complexity of feed control measures should also be related to the available policing facilities. There is no point in applying sohpisticated control procedures which can be enforced only with the major manufacturing companies. It is important that control measures be simple enough to be enforced at all levels in the feed industry.

The feed manufacturer must of necessity maintain the closest possible links with Government extension services. If the feed industry has done its planning well then there should be complete identity between the objectives of the Government and the feed industry. Indeed in some cases the feed industry may well be Government operated. Nevertheless it requires an enormous effort on the part of the feed manufacturer to communicate his views through to the furthest points of a Government extension service. Without this effort a feed industry cannot play its full role in a co-ordinated livestock development programme.

Similarly it is important that the feed industry should maintain close links with the research scientists in academic institutions. Only in this way can the feed industry keep the research scientist fully informed about the problems which face him and encourage the research scientist in both his fundamental and applied research to look at problems specific to the agriculture of the country concerned. It seems occasionally that the research scientist tends to work over ground that has been well tilled elsewhere and that he does not have an adequate understanding of the very real problems of the local feed manufacturer. This is a difficult area in which to generalise. Nevertheless in the last resort the onus must be on the feed manufacturer to ensure that the academic world understands his problems and has an opportunity to contribute to their solution.

The potential problems outlined in this paper are applicable to all types of feed manufacturing project whether they be in the private, co-operative or public sector. However all problems are specific to the individual country concerned. The best method of dealing with problems is to anticipate them at the planning stage of the feed manufacturing projects. The key features of this planning will be to identify the precise role of the feed industry in the livestock production cycle, to identify the resources available to the feed industry and to establish in association with the Government an understanding of the problems, objectives and requirements of the feed industry.

In every problem lies an opportunity for the feed manufacturer to do his job more effectively.

Experience of animal feed production in Barbados and the Eastern Caribbean

E. G. B. Gooding and V. A. L. Sargeant
Caribbean Development Bank, Bridgetown, Barbados

Summary

Before the Second World War the countries of the Eastern Caribbean were almost self-sufficient in meat, and the animal population was fed almost entirely from locally produced feed. During the war, however, animal production declined and draught animals on the sugar estates were replaced by tractors. Also local production of maize and legumes for animal feeding declined, and new more sophisticated animal production industries based on the importation of cheap concentrates emerged.

The recent rise in international prices for feedingstuffs is posing a serious threat to the viability of animal production enterprises in the Eastern Caribbean, and there is now an urgent need to base them on locally produced rather than imported feedingstuffs. There are possibilities for the adoption of the new livestock production systems based on sugar cane, while more maize and soya could possibly be grown in the Caribbean area.

Résumé

Essais de production des pâtures pour les animaux en Barbade et Les Petites Antilles

Avant la deuxième guerre mondiale les territoires des Petites Antilles se suffisaient à eux-mêmes pour la viande et le bétail était nourri presque entièrement avec du fourrage produit localement. Pendant la guerre cependant, la production d'animaux diminua et les animaux utilisés dans les plantations de sucre à canne furent remplacés par des tracteurs. La production locale de mais et de légumes déclina également et des nouvelles industries, plus sophistiquées, pour la production d'animaux, surgirent basées sur l'importation des produits concentrés peu coûteux.

L'augmentation récente des prix internationaux des pâtures pose un danger sérieux à la viabilité des entreprises de production d'animaux dans les Petites Antilles et il y a maintenant un besoin urgent de les baser sur des matières nutritives produites localement, plutôt que de les importer. Il y a des possibilités d'adoption des systèmes de production du nouveau bétail basées sur la canne à sucre, tandis que plus de mais et soja pourrait être planté dans la région des Antilles.

Resumen

Experiencia de la producción de piensos para animales en Barbados y en el Caribe del Este

Antes de la Segunda Guerra Mundial los países del Caribe del Este eran casi autosuficientes en carne, y la población animal se alimentaba casi completamente con piensos producidos localmente. Sin embargo, durante la guerra la producción animal bajó y los animales de tiro fueron sustituídos por tractores en los ingenios azucareros. También bajaron la producción de maíz y de legumbres para alimento de los animales y surgieron nuevas industrias de producción animal más complejas, basadas en la importación de concentrados baratos.

La reciente subida de los precios internacionales de los piensos supone una seria amenaza a la viabilidad de las empresas de producción animal en el Caribe del Este y hay ahora una necesidad urgente de basarlas en piensos producidos localmente más bien que en piensos importados. Hay posibilidades de adopción de los nuevos sistemas de producción de ganado basados en la caña de azúcar, al tiempo que se podría cultivar, posiblemente, más maíz y soja en el área del Caribe.

The Eastern Caribbean islands are small and agricultural land values are high, usually £250 to £1,500 per hectare. Almost every available and safely cultivable acre is planted in export crops such as sugar cane or cotton, or citrus, bananas or spices, and food crops such as yams, sweet potatoes and green vegetables. Consequently there is neither room nor economic sense in extensive animal husbandry systems.

Before the Second World War large numbers of draught animals (oxen) were kept by sugar plantations for hauling carts, ploughing and other work; in Barbados alone the population of cattle was estimated at 25,000. They were integrated into a system of mixed farming. Dry lot feeding was practised using grass grown on land too poor for cane, while corn (used in chopped green form or as meal), legumes, and cotton seed meal came from crops grown in rotation with cane; coconut meal was a by-product of coconut oil production and molasses of sugar production. Relatively small amounts of imported feeds were sometimes used, notably pollard (wheat middlings) and linseed or soya bean meal. Not very much beef was imported; the local product was derived from old animals and sterile heifers, and was of very poor quality; veal was obtained from female calves. There was some production of fresh milk on estates from a few of the breeding cows and there were a number of dairies, using mainly the same concentrates along with some cut forage, selling unpasteurised milk.

Oddly, perhaps, mechanisation of agriculture came in during the war: several factors interacted such as the decline of cotton production and rising prices for pollard, the necessity for using all molasses for alcohol production, labour difficulties, and so on. By the end of the war the working animals on estates were virtually nil though there may have been as many as 10,000 cattle in Barbados which were mainly owned by peasants. The import of meat and dairy products was soaring and meat has become the major single item on the food bill of most Caribbean territories. Although there has been a slow rise in production of beef, consumption has fast outstripped it. Mayers (1970) has shown that the total production of meats in the Commonwealth Caribbean was static at 60 m kg per year from 1956 to 1961, it then rose, rising to 72 m kg by the end of 1967, but consumption rose from 62 m kg in 1956 to over 130 m kg in 1967.

Milk production declined after the war, largely in the face of competition from the cheaper processed milks from abroad, but recently there has been a resurgence of interest, especially in Barbados, where today 25 registered dairies with about 800 producing cows (mainly imported Holsteins or their descendents) sell about 2.3 m kg per year to a central pasteurising depot. This probably represents about two thirds of the island's total milk production. Large quantities of processed milk are, however, still imported to meet consumption demands. Pangola grass (*Digitaria decumbens* Stent.) is the main source of fodder, plus concentrate feeds. It should be noted in passing that the carrying capacity of tropical grasses is lower, they are relatively poor in digestible material and high in

fibre (Butterworth, 1967), a condition which may be related to the tropical environment (Gooding, 1972). This underlines the proposition that, where land is scarce, and perhaps even where it is not, concentrate feeding is extremely important for economic and efficient production of livestock and milk in the tropics, indeed, that is what this conference is about.

Feeding of concentrates in Eastern Caribbean

The history of concentrate feeding in the Eastern Caribbean, its present status, and what perhaps can be done for the future, should be examined a little more closely.

The pre-war situation has already been noted. During the war even more attention was given to locally available materials notably corn meal, cotton seed meal and molasses and in Barbados a 'Balanced Animal' (BA) feed was devised which was suitable for all livestock except poultry. Formulae were changed according to the exigencies of supply but were maintained around 75% TDN and 20% crude protein. Similar exercises were undertaken in other territories. In Barbados a BA feed is still being manufactured though most of the ingredients are now imported.

In the mid 1950's imports of branded feeds started, and proved competitive in price as they were largely based on the then relatively cheap American corn and soya bean meal. Some of the local mixing plants, (two in Barbados, and one each in Grenada, St. Vincent, St. Lucia, Dominica and Antigua), expanded their lines from the simple BA type feed into a whole range of such products as special dairy mixes, feeds for all stages of pig, poultry and egg production, and in the process some became associated with North American suppliers, with formulations based on US practice. However, a few years ago rising costs and the feeling that too much money was being exported led to a more intensive search for local ingredients. Work at the University of the West Indies (Gohl, 1969; Gohl, 1970; Devendra & Gohl, 1970; Gohl & Devendra, 1970) and in Barbados (Spalding, 1968) led to the formulation of a variety of concentrate feeds, cheaper than those from imported ingredients, and a number of farms started (as before the war) to make up their own rations. As an example, pig rations with the formulations shown in Table 1 were used on one large farm in Barbados.

A problem with such rations, (especially for pigs), was that although they may have been 20% or 30% cheaper than the commercial product, the conversion ratio was about 4:1 instead of 3 or 3.5:1. The difference in cost, therefore, per kg of pig was minimal and the animal occupied pen space for just so much longer. Also some ingredients (eg brewers' grain) were not generally obtainable.

In the rations for cattle a wider range of substitutes for imported grain and protein meals were included from time to time by the major commercial manufacturers. These included citrus pulp, wheat middlings (from flour mills in Trinidad and Jamaica), coconut meal, molasses, urea, locally grown corn, etc. However, with the increase in dairy farming, pig production and poultry production, most of these components are now in short supply and the main suppliers are still outside the region.

In the last year or two there has been a terrifying rise in the prices of many ingredients. Table 3 shows comparative prices for 1971 and 1973. Largely as a result of this the price of meat has also risen sharply.

TABLE 1. *Farm compounded rations for pigs*

Growing ration	
Brewers grain	25 kg
Coconut meal	50 kg
Fishmeal	4 kg
Molasses	20 kg
Minerals	3 kg
Finishing rations	
Coconut meal	20 kg
Wheat middlings	25 kg
Soya bean meal or fishmeal	20 kg
Molasses	20 kg
Minerals	2 kg

When wheat middlings and soya bean meal started to rise sharply in price the formula for the finishing ration was changed to that in Table 2.

TABLE 2. *Pig finishing ration with brewers' grains substituted for soya bean and wheat middlings*

Coconut meal	40 kg
Brewers grain	25 kg
Fish meal	4 kg
Molasses	20 kg
Minerals	3 kg

TABLE 3. *Prices of feed ingredients c.i.f. Barbados*

	EC$ per tonne	
	1971	1973
Corn (maize)	165	359
Sorghum	149	176
Soya bean meal	286	686
Linseed meal	216	490
Wheat middlings	131	274

Note: Currency equivalents: EC$1 approximately US$.50= £0.21

Prices of the formulated products have, of course, had to show corresponding rises. Table 2 shows a few examples.

TABLE 4. *Prices EC$ per 45.5 kg bag of manufactured feeds (ex factory in Barbados)*

	1971	1973
BA feed	9.64	17.80
Dairy mix	9.20	15.32
Cage layer	13.60	20.40
Broiler starter	15.92	25.82
Broiler finisher	15.64	24.88

In Barbados in the past three years the legally controlled price of pigs has risen from EC$1.23/kg to EC$1.98/kg liveweight and farmers say that this is still an uneconomic price, while that of whole chickens (again legally controlled) have risen from EC$2.31 to EC$3.30/kg dressed. Similar rises have occurred throughout the region. The ex-farm price of milk in Barbados (the only Commonwealth Caribbean island with a well developed dairy industry), has risen from approximately 33 cents/kg ex farm to 50 cents, surely some of the most costly milk in the world. These are prices obtaining in relatively poor developing countries.

Table 5 shows imports of feeds and feedingstuffs for the 4 years, 1957, 1965, 1972 and 1973; it will be seen that for Barbados alone the quantities involved run into several million kilograms costing millions of dollars, and that price rises in 1973 doubled the outflow of money. If we are to stop importing inflation we must make better use of local raw materials. Even if they do not bring down prices to any great extent this will at least keep the money in the region.

TABLE 5. *Imports and costs of feeds and feedingstuffs into Barbados*

Year	Weight (m kg)	Value c.i.f. Barbados (m EC$)
1957	11.6	2.2
1965	17.5	2.9
1972	34.1	6.7
1973 (est.)	34.1	13.0

This is no easy matter. So far no very satisfactory solution has been found and attempts to produce better forages and the ingredients for concentrates which have been made in most of these territories, has so far been rather half-hearted.

Growing crops for concentrates

Maize

Low yields of about 1,700 kg/ha with associated low or even nil financial return have discouraged interest in this crop in the Eastern Caribbean, even though the demand as an ingredient for animal feeds alone is about 10,000t. However, recent experiments have shown that yields could be greatly increased simply by improved cultural practices (Gooding & Hoad, 1966), while certain hybrids specially bred for the tropics by a US company in Jamaica appeared very promising, with commercial yields of about 3,400 kg/ha in Barbados (Gooding & Johnson, 1970) and up to 6,800 kg/ha in St. Vincent (Baynes, 1972) under experimental conditions. However, this variety was extremely susceptible to insect pests and when in one year almost the whole crop in Barbados failed because of ear worm attack despite spraying, enthusiasm for growing maize collapsed. In any case the return per hectare was small about EC$175 at the best when grown in rotation with sugar cane and when all overheads and land preparation costs were absorbed by the cane. However, with current prices and the possibility of mechanisation, maize should be seriously studied once more, though it should be noted that even the breeder of these new cultivars believes that under tropical conditions, with short days and relatively long nights, yields of over 6,000 kg/ha will not be possible (Seghal, 1968), though in the USA twice that amount can be obtained.

Even so, locally grown cheap maize remains the key to compounding a low cost animal feed in this region, and is essential for poultry production, so serious consideration must be given to extensive, mechanised production. The Eastern Caribbean islands do not have suitable terrain and land values are too high. However there are areas in Belize and Guyana which might be well suitable and the Caribbean Development Bank has already been approached in connection with a 200 ha scheme in the former territory. This is only a fraction of the region's needs, but a step in the right direction.

A problem to be borne in mind is transport from such territories as Guyana and Belize to the Eastern Caribbean; there seems little doubt that the only economic way would be by bulk carrier with appropriate bulk installations at the point of receipt.

Sorghum (for grain).

Sorghum has not been extensively tested in the Caribbean so far, though some forage sorghums seemed promising (Gooding & Hoad, 1966). Certain varieties of grain sorghum may well be suitable for relatively low rainfall areas where corn will not do well (eg parts of Antigua or Barbados) but information is still sadly lacking.

Soya.

Soya bean meal has for many years been the main source of protein in feeds compounded in the region, but the commercial varieties of the USA, Colombia and Peru have not been successful in the Caribbean. However, recent work at the University of the West Indies in Trinidad has suggested that yields comparable with commercial yields in other countries (1,700 to 2,250 kg/ha) may be attainable, and commercial trials in Guyana have confirmed this. At the prices prevailing until recently, however, the value of the crop was too low for it to be competitive with sugar or other local crops but the current price situation could alter this and the possibility of growing maize and soya in rotation might now prove feasible.

Root crops.

Yams, sweet potatoes and cassava have frequently been considered for use as animal feeds but their cost must rule them out except under quite extraordinary circumstances. Current prices for such crops are about 22 cents (EC) per kg. Assuming 25% dry matter this makes the basic cost of the carbohydrate 88 cents without any further processing and this is higher than the cost of refined sugar!

Sugar cane production

While small quantities of molasses have been used for a long time in many formulations, the recognition of the possible importance of sugar cane as a primary source of feed for cattle has been long delayed, along with what is almost a corollary, the use of urea as a substantial part of the nitrogen source (for ruminants).

Recently 'Comfith' has appeared on the scene, a Canadian development in which the sugar-containing pith of the cane is removed from the rind, and, together with the chopped green cane tops, urea, some organic protein such as fishmeal, minerals and vitamins, provides a complete ration for fattening cattle or for dairying. Lengthy experiments have been carried out in Barbados with this feed and results have been described in several papers (Donefer et. al. 1973). Average weight gains of about 0.9 kg per day have been achieved, from 36 to 545 kg, and, on the information so far available, the system should be economic. One of the great virtues is that full use is made of the sugar cane's high productivity of carbohydrate per hectare (it is reported to be the world's most effective crop in this respect) and a hectare of cane at the moderate yield of 75 t can provide about 5.5 animals for slaughter per year at 550 kg liveweight, in contrast to about one per hectare on improved Pangola grass, ie the productivity is over five times greater. Obviously this is a most promising development. The biggest drawback is that it is a fattening system, and a calf (or a little more than one calf) must be put into the system to replace every animal taken out for slaughter, so an ancillary cow-calf operation is necessary. If there is an existing dairy industry the bull calves can supply the Comfith operation, and as it now seems that Comfith can be used as a dairy ration, this could very well be the basis of a joint dairy-fattening operation.

The Canadian International Development Agency has asked the Caribbean Development Bank to undertake a survey of the potentialities for this technology in the lesser developed countries of the Commonwealth Caribbean, in the hope that viable projects will be found that can be funded by the CDB. This survey is to ascertain to what extent this system can be developed, bearing in mind constraints imposed by the supply of livestock, cane, management, etc., and the price obtainable for beef.

Agricultural by-products

Bananas

Banana rejects are, in effect, a by-product of the banana packing operation throughout the Eastern Caribbean (except Barbados). The raw material can be very cheap and the Caribbean Development Bank has approved a small livestock scheme in which these play a part. However, in their fresh form they are perishable and bulky; to process them for concentrates would be costly. Solar drying systems have been proposed (McDowell, 1973) but seem scarcely practicable for large-scale operations. Nevertheless, here is a high-energy raw material, cheap, often going to waste. Clearly further investigation is necessary to devise systems whereby such a product could be used in ration formulations. In the meantime reject bananas are being used for feeding animals in mixed farming systems, where bananas and animals are grown on the same farm, or where the source of the bananas is close to an animal production unit.

Citrus pulp meal.

The dried pulp from citrus canning operations is a good energy source and is used, when available, in rations in Trinidad and Jamaica, and has been used in commercial rations mixed in Barbados. However, quantities are limited. Further, the economics of production are such that it cannot be produced by small-scale operators: it is unlikely ever to be an important ingredient in concentrates throughout the region.

Oilseed meals

Copra oilcake

This is the most abundant local source of protein, containing 21% of protein and still being available for about EC$181/t (compare with other protein materials in Table 3). However, the annual production in the Caribbean region is only about 15,000t against an estimated requirement of 20-25,000 t for the Eastern Caribbean alone. This is, of course, a by-product of coconut oil production and there is no question of growing coconut trees specifically for its production.

Cotton seed meal

As a result of the great decline in the cotton industry the supply of cotton seed meal has dwindled to almost nothing, but for 1974 Barbados is expected to produce about 300 t of cotton seed and Antigua perhaps twice as much. There is hope for continued expansion of the cotton industry but even so this meal is likely to become of only local importance.

Sugar cane by-products

Comfith has been already dealt with as a primary product but molasses, however is a by-product, and one that has been sadly neglected. Preston's recent work in Cuba is now well known (Preston 1969; Muñoz *et al.,* 1970) but as long ago as 1946 cattle in Iowa, USA were being wintered on corn cobs and molasses-urea. However, the idea of using molasses as a basic energy food, and urea to be replace much of the organic nitrogen, has taken a long time to be adopted in the Eastern Caribbean and only two farms (both in Barbados) are operating this system but using some concentrates as well, while one or two other farmers are experimenting rather tentatively. Molasses as an animal feed could have had a real future, as the Commonwalth Caribbean *exports* about 182ml/a (enough to maintain over 100,000 head of cattle on the molasses-urea system), and was until recently quite cheap at 8 cents (EC) per litre the equivalent of about 11 cents (EC) per kg of digestible carbohydrate, but within the past year the export value has more than doubled; at the new price the economics are doubtful. (The farmers who

are currently using this are getting it locally at the old price which is in effect, a subsidy). In terms of beef production the value of the animals at EC$1.65/kg on the hoof just about matches the export value of the molasses necessary to raise them, so at current prices a switch to molasses would have to be justified in terms of social rather than economic benefits. However, if the value of beef increases, or the export prices of molasses falls, the situation would be quite different.

Systems of animal production

There has been a tendency in the Caribbean to try to follow systems devised for temperate climates, viz., grazing or zero grazing, supplemented by concentrates in the case of cattle, or complete concentrate feeding for pigs, poultry, etc. By and large these systems have been adapted reasonably well. However, it is felt that more thought should perhaps be given to systems of animal production in relation to local conditions and available foods, possibly to integrated mixed farming in which maize, for example, is not necessarily regarded as a crop to earn cash in its own right but in which the final crop from the field may be milk or beef. The beginning of such a system is seen in Comfith in which the beef, not the cane, is the crop.

Even in more conventional systems there is much to be learned. The Ministry of Agriculture in Barbados has put forward suggestions for systems of raising beef (Nurse, 1973) but the fact remains that the ideal combination of grazing or forage and concentrates (and if so what concentrates) to provide the required ends with greatest economy is not known. Studies are being made in Barbados at present to compare the performance and economics of several dairy farmers and there is a great measure of disagreement among farmers. One farmer, for example, feeds over 1 kg of concentrates per kg of milk produced; at the other extreme another, who relies heavily on irrigated Pangola, feeds only about 0.3 kg per kg of milk, (these are herd averages, not per cow in milk). One has 77 ha with 200 animals, viz, 2.6 animals per ha of improved Pangola grass, another has 140 head on 16 ha of rough grazing, but uses molasses-urea very heavily and, in view of the necessity for intensive production mentioned early in this paper, this approach seems highly commendable. The fact remains, however, that the concept of growing ingredients for animal feeds has not been seriously pursued in the Eastern Caribbean, and it is not known whether an industry can be built on such crops. It is strongly believed by the authors that if the region is to escape from the serious problems of wild price fluctuations on the international market, and the drain on local currencies is to be minimised much more work and experience is required on intensive or semi-intensive systems of animal production in the region, with emphasis being placed on the system, as well as on the production with the use of local ingredients with the concentrate rations.

Much of this paper has been condensed from an unpublished Caribbean Development Bank report entitled *The Production and Utilisation of Locally Grown Animal Feeds'*, prepared by V.A.L. Sargeant, 1972.

References

Baynes, R. A. (1972) Yields of maize (*Zea mays* L.) in four Eastern Caribbean islands as influenced by variety and plant density. *Trop. Agric. (Trinidad)* **49**, 1, 37.

Butterworth, M. H. (1967) The digestibility of tropical grasses. *Nutr. Abst. Rev.* **37**, 349.

Devendra, C. & Gohl, B. I. (1970) The chemical composition of Caribbean feedingstuffs. *Trop. Agric.* **47**, 335.

Donefer, E., James, L. A. & Lawrie, C. K. (1973) Use of a sugar cane derived feedstuff for livestock. *III World Conference on Animal Production,* Melbourne, Australia.

Gohl, B. I. (1969) The availability and possible uses of by-products and crops produced locally as sources of animal feeds. University of the West Indies.

Gohl B. I. (1970) Animal feeds from local products and by-products in the British Caribbean. Rome: FAO.

Gohl, B. I. & Devendra, C. (1970) The use of urea in ruminant feeding. *Caribbean Farming* **2**, 26.

Gooding, E. G. B. & Hoad, R. M. (1966) Barbados Sugar Producers' Association Diversification Section, Annual Report.

Gooding, E. G. B. & Johnson, C. A. (1970). Experiments on corn, 1968-1969. Barbados Sugar Producers' Association.

Gooding, E. G. B. (1972) Plant responses to tropical conditions. *Proc. Caribbean Food Crops Society.* In the press.

Mayers, J. M. (1970) The marketing and demand for meat in the Commonwealth Caribbean. University of the West Indies.

McDowell J. (1973) Solar drying of crops and foods in humid tropical climates. Caribbean Food and Nutrition Institute.

Muñoz F., Morciego, F. & Preston, T. R. (1970) *Revista Cubana de Ciencia Agricola,* **4**, 91.

Nurse, J. O. J. (1973) Beef Production in Barbados, 1972. Ministry of Agriculture, Science and Technology, Barbados.

Preston, T. R. (1969) *Revista Cubana de Ciencia Agricola,* **3**, 141.

Seghal, S. (1968) Factors affecting yield of tropical maize. *Proc. Carib. Food Crops Soc.* **6**, 102.

Spalding, R. W. (1968) Feeding the dairy herd. Ministry of Agriculture, Barbados.

Reports by session chairmen

Dr P. C. Spensley

The first session of the Conference started with two introductory papers which gave an overall perspective of many of the topics dealt with in detail later on in the Conference.

In the first of these Dr Nestel gave an account of the present position world-wide in animal production and and feed supplies. He drew attention to the fact that although the less developed countries contained the majority of the world's cattle, sheep and pigs, because of their relatively low productivity, they produced no more than one-third of the world's meat and dairy products. This indicated that the opportunity for genetic and nutritional improvement was much greater in the less, than in the more developed areas.

The point was also made that before grain, which is the major animal feed in the more developed countries, could be fed to livestock in less developed countries, supplies for use directly as human food would have to be increased.

The importance of animal draught power and the sophisticated equilibrium which exists in some subsistence farming systems was also mentioned. In these circumstances low animal productivity might not be out of accord with optimum utilisation of the farmers' resources.

Dr Nestel went on to discuss the need to develop new systems of ruminant meat production, with emphasis being placed on rearing stock on land unsuited for food production and finishing them either on by-products or on crops such as sugar cane or cassava, which are efficient convertors of solar energy and which can produce annual yields of dry matter which are multiples of those possible in temperate climates.

After referring to various protein sources such as oilseed cake, single cell protein, fishmeal and tropical pasture legumes, Dr Nestel described experiments which had shown that native cattle and pigs digested nutrients more efficiently than exotic breeds when on a low protein diet. He underlined the importance of this to peasant farmers, particularly as these animals are also well adapted to their environment.

In the second paper on 'Factors Limiting the Production of Animal Products in the Tropics' which was presented by Dr A. J. Smith, the major factors affecting the production of domestic animals and hence animal products in the tropics were discussed. These included the direct and indirect effects of climatic environment, the genetic merit of the available livestock, the system of animal feeding, the incidence of animal diseases, management factors, the efficiency of local research and extension services, the availability of credit, the existence of processing and marketing facilities, price structure and policy, and the priority given to the industry by local governments.

The production of animal feeds was also governed by many factors of which the two most important ones were soil type and climate. Dr Smith's paper directed attention particularly to the situation in the humid tropics.

Within such regions, food for animals could be obtained from natural and planted forage, planted field and tree crops, by-products of field and tree crops and by-products from non-crop materials. The factors which limited the production of animal feeds from these sources were discussed.

In a short paper on animal production in the Philippines, Professor Alim discussed the main types of livestock in the Philippines, the carabao, cattle, pigs and poultry. The major problem with carabao was the long calving period and research into the reproductive rate and fertility of the carabao could be of great value. The improvement of pastures had the highest potential for increasing cattle production. With regard to the feed milling industry the major problem at present was ensuring adequate year-round supplies of local raw materials of appropriate quality.

In reply to a question from Dr Babatunde, Dr Smith agreed that management was as important as the genetic potential of stock. He knew of many examples of badly managed exotic breeds having been out-performed by local cattle.

Mr. Blair-Rains observed that in the past the developed countries had imported oilseed protein from the less developed countries to produce meat and now they were importing meet itself. In view of the low consumption of animal protein in less developed countries were these exports desirable? Sympathy was expressed for this point of view but it was emphasised that it was up to the Governments of the less developed countries to decide the relative priorities of foreign exchange requirements for economic development and better nutrition for the people. It was also pointed out that a similar situation often existed within the less developed countries in that farmers sold animal protein whilst they and their families were not eating enough themselves.

Mr. Sunkwa-Mills mentioned the importance of education, training of extension personnel and land tenure in the development of animal production. Dr Smith agreed that these were important as were also credit facilities.

Mr Bird

Four papers were presented during the second session of the conference. They covered world supplies, trade and utilisation of cereals, oilcakes, fishmeal and the potential of the SCP in animal feeds.

Mr Low's paper noted increasing production of cereals both in developed and developing countries, the rapid rise in demand for human consumption in developing and for animal feeds in developed countries, the dramatic price increases and the world potential for further expansion of production. It was however, emphasised, that unless a rational attitude can be developed and effective policies implemented to bring about changes in population growth and in the world's political and economic regimes, serious shortages in grain supplies could develop.

Mr. Breslin emphasised the new status of oilcakes and meal in the seed crushing industry, the expansion of world production of oilcakes and the great rise in prices in recent years. The dominating position of soya bean in the oilcake industry was explained in terms of its higher cake/oil ratio. After giving some consideration to the various factors which will tend to influence supply and demand in the future, the speaker mentioned that FAO had predicted an anticipated surplus of oilcakes in 1980.

Mr. Bellido discussed the production, supplies, trade and technical aspects of using fishmeal as an ingredient in compound animal feeds.

Dr. Tolan discussed the technical aspects and economics of using single cell protein rather than conventional protein concentrates such as oilseed cakes and meals in compounded feeds. It would appear that the large-scale production of single cell protein might well be economic in certain areas of the world in future.

Among the matters discussed in the discussion periods were the likely pattern of cereal prices during 1975, the economics of single cell protein production, the need to increase oilseed production in the less developed countries, the production of low solubility fishmeal for ruminant feeding, the economics of extracting oil from fishmeal and the future role of single cell protein in animal feeding.

Conclusive answers were not obtained for all the questions which were posed particularly those relating to prices. However, it was considered that any demand for additional food by the third world to raise standards of nutrition to an acceptable level could lead to shortages rather than the surpluses of cereals and oilseeds which were presently anticipated by many. It was therefore imperative to increase production of these commodities. It was mentioned that single cell protein was unlikely to contribute very greatly to world protein supplies for the foreseeable future, but could have certain specialised uses, eg in milk-replacers for calves.

Dr Devendra

In the developed countries widespread applications of the results of scientific research has led to a quantum jump in livestock production efficiency. Until recently efforts to improve efficiency in developing countries have been based largely on similar applications of the same results. There have been difficulties, of course, in translating techniques from temperate to tropical environments: and much of the research carried out in developing countries has been addressed to these difficulties. Outstanding successes have been achieved in some enterprises but the overall contribution to development has been small, and all too often the success has been central to the enterprise. The macro economic consequences have been generally unfavourable; in some instances, for example the development of livestock production has created a high demand for scarce foreign exchange resources.

This unfavourable effect is particularly evident when one considers energy sources. The developing countries have followed the developed in accepting grain as the desirable main energy source, and where grain has not been available, they have tended to resort in the case of ruminant nutrition to exclusive use of pasture, thereby placing a constraint on productivity.

In the search of maximum utilisation of the available feed resources it is perhaps important to bear in mind the sources and availability of potentially useful feeding-stuffs together with extent to which they are presently utilised, the scope for import substitution, and research into their utilisation and effective guidelines for application.

There is now a general realisation of the need for research into and reappraisal of other potential locally available sources of energy.

It was timely to devote an entire session of this Conference to this important topic, and I congratulate the organisers on a choice of paper subjects which ensured that coverage would be as comprehensive as possible in the time available. On the commodity side we were brought up to date on the research on use of sugar cane — which occupies a position of unique importance in this field — of a root crop and of two very different types of waste product. We had coverage of ruminant and non-ruminant nutrition; we were made aware of toxicological implications and we were given insight into the need for development of appropriate processing technologies.

I also congratulate again the four speakers on their excellent presentation of their papers.

Dr Preston left us in no doubt as to the potential of sugar cane to contribute to livestock development and he gave us an up to date account of the latest findings of research on its utilisation. Paradoxically, the discoveries relating to the role of added starch and protein in sugar cane based diets proved to be the most thought-provoking — indeed the most surprising. This underlines the importance of thorough research on all aspects of utilisation of these potential new energy sources. This research must go hand-in-hand with economic evaluation. Dr Preston showed as a graph of production costs for a particular situation which was extremely favourable. This is most encouraging, but I would emphasise the need for detailed economic appraisal of the applicability of the new techniques in each situation. I regret that more time was not available for formal discussion of this most interesting topic.

Dr Müller's paper reminded us that cassava provides leaf meal as well as root meal, and I would endorse the views expressed by one speaker from the floor who suggested that there is a need for increased attention to use of leafy materials in animal nutrition in the tropics. Dr Müller's conclusion that pelleting greatly improves the nutritive value of cassava underlined again the importance of processing technology. Animal nutritionists must work closely with processing specialists in developing the new energy sources. One aspect of cassava utilisation that merits more attention is its feeding to ruminants and this is worthy of attention.

The point pertaining to processing was also developed by Dr Bressani in his paper on the use of coffee processing waste. Like Dr Müller, he was concerned with processing as a means both of preservation and detoxification, though in this instance the position on the latter topic is not as clear cut as in the case of cassava. The findings on adaptation by animals to diets containing coffee pulp are of particular interest and worthy of considerable further research. There are of course other examples of instances in which an industrial waste produce which currently creates an environmental problem could be put to use in animal nutrition.

The paper delivered by Dr Gomez on behalf of Dr Clavijo and Dr Maner elucidated the main nutritional problems arising in the use of waste bananas in swine feeding. Here processing proved necessary in some instances not on grounds of preservation but on those of nutrition. We were thus reminded that special attention must be paid to the problems of drying — and, of course, to the economics of the drying operation.

I feel that this session has successfully underlined the need for developing countries of the tropics to depart from the traditional and previous dependence on cereal grains in favour of other sources of energy for animal feeding. Tremendous opportunities exist for these new sources of energy, the effective utilisation of which could create a new dimension in animal feeding, leading to increased productivity from tropical animals and indeed the productive capacity of the developing countries as a whole.

Professor Fuller discussed the use of fats and oils as a source of energy in mixed feeds for pigs and poultry. He referred to the possibilities for increasing densities of mixed feeds, reducing the total weight and volume of feed that must be transported, mixed, stored and handled; and he showed that nutrient-energy ratios being constant, efficiency is usually improved more than proportionally by the use of fats. The animal meets its energy requirements more economically, requiring less energy expenditure in eating; because of the lower heat increment, and metabolisable energy values underestimate the net or useable energy value. The heat increment of the entire diet is reduced as a result of an associated dynamic effect of fats in a synergistic effect far greater than that expected for summation of individual energetic values.

These concepts were extended by Professor Fuller to the amelioration of heat stress. Increased fat content in the diet resulted in increased energy and nutrient intakes, protein requirement also being lowered resulting in a further diminution of heat increment. Growth rate was greatly improved by such dietary manipulation; and it seems clear that the control of heat increment is an important factor in combating the reduced growth rate in poultry and pigs associated with high temperature environments.

Following a report of adverse Nigerian experience of feeding high fat levels Professor Fuller described antoxidant requirements, and expanded on his findings that certainly 10% and possibly up to 30% fat could be fed without adverse effect. His experiments were essentially orientated towards practical environmental conditions rather than refined laboratory temperature control.

The paper presented by Mr Clarke discussed the use of low quality forage in ruminant diets. It made the point that domestic ruminants need not compete with man for nutrients in so far as some 64% of the world's agricultural land may be considered to be non-arable and suitable for grazing only, so that the ruminant's ability to digest cellulose and hemicellulose may well represent optimal production. A better utilisation of roughages in developing countries could be of considerable value when surplus grains are unavailable.

The paper discussed the effects of reducing particle size on the utilisation of roughages, particularly in the context of the feeding of straws. For instance in the utilisation of barley straw it was shown that there is an optimal particle size of about 6 mm. Alternatively chemical treatment, for instance with sodium hydroxide, can be employed.

It is noted that roughage has a beneficial effect in the feeding of concentrates under tropical conditions, being related to the basic biochemistry of ruminant fermentation and the utilisation of metabolites. Salivary secretion and associated flow rates that the duodenum also appear to be implicated.

The utilisation of tropical roughages was discussed in the context of supplementation of nitrogen intake, for instance with urea and naturally occurring protein supplements.

The addition of low quality roughage may affect growth rate but does not lead to inferior carcase composition. Dry matter intake is frequently greater than with concentrate alone though liveweight gain may decrease above 10% or sometimes as high as 30% inclusion. Killing out percentages can be affected however, the proportion of gut fill being greater.

In general discussion, it was reported that roughages vary widely in their response to alkali treatment for ruminant nutrition. It also appeared that the economics of the alternative milling possibilities remain to be evaluated. In comment on the importance of genetic considerations in nutritional work it was pointed out by a delegate that related stock improvement began to show returns only after an extended period, certainly not until the second generation. There was some discussion of the effect of fat intake on ruminant performance from which it emerged that fat appears to depress the digestion of roughage.

A number of short papers was presented, which were of considerable interest. Professor Darwish discussed the feeding and replacement value of lime treated corn cobs for dairy cows. Digestion coefficients increased; and, in use, it was shown that there was only an insignificant decrease in milk yield.

The paper presented by Dr Babatunde referred to earlier work which raised doubts about the adequacy of meat and bonemeal in sustaining good growth when fed as the only source of protein supplement in the rations of pigs. The product can readily exhibit a wide degree of variability in quality depending on such factors as the amounts of tissue and bone included during processing. In the experiments discussed in the paper it was shown that the various meat and bonemeal diets were all inferior to control diets or groundnut meal diets for feeding pigs. In subsequent discussion it was pointed out that the calcium levels fed in the study approximated to those known to be lethal even with zinc supplementation.

Dr Were discussed shortages of human and animal foods which had led to a study of alternative supplies of feed for stock. He reported on the use of flower residues after the extraction of pyrethrum, some 5,000 t/a being available in Kenya. It has been proved possible to replace up to 50% of maize silage in rations for fattening cattle on feedlots with pyrethrum residues.

Dr Castello presented a paper on the use of tuber meals as carbohydrate resources for broiler rations. In the work reported upon in the paper it was concluded that sweet potato and cassava meals can partially but not totally replace maize in broiler rations. Birds fed other roots (*Colocasia esculentum, Dioscorea alata* and *Amorphophallus campanulatus*) performed poorly.

Professor Abou Raya reported on trials of by-product feeding (particularly of maize and rice) in mixtures with concentrates, supplemented with urea and molasses. He stressed particularly, as Dr Clarke had done earlier, the value of pelleting. Professor Oyenuga discussed the nutritional value of conofor seed, which is eaten by man but also has possibilities for animal feed.

The session as a whole brought together a considerable body of experience on the feeding of a range of materials of varying degrees of promise. It was apparent that while some areas of work, for instance on the feeding of roughages and lipid nutrition, have been extensively worked over, much remains to be done and that there remain many other newer areas of nutritional evaluation of tropical produce to be explored.

Dr Bressani

The session concerned itself with five papers dealing with the nutritive value of oilseed meals and cakes, the presence in them of physiologically adverse factors and their elimination, and some applications in animal feeding as carried out in Turkey, the Sudan and Nigeria.

The residues remaining after removal of oil by pressure or solvent extraction are concentrated sources of protein, protein content varying from between 17 and 50%.

Hydraulic press oilseed cakes have variable residual oil contents depending on the efficiency of processing. The residual oil may be of variable composition, from drying oils to hard vegetable fat. The residual oils in the meals are nutritionally valuable as energy sources, particularly in areas where conventional energy sources, such as cereal grains, are costly and mainly used by the human populations.

The meals are mainly used as a source of protein and their essential amino acid make-up is of great importance. The oilseed cakes are mostly deficient in lysine or sulphur containing amino acids. However, some are also important sources of these two amino acids, as for example soya with its high lysine content and sesame with its high sulphur amino acid content. Processing conditions are important in determining the protein quality of the oilseed cakes, since extreme conditions may increase essential amino acid deficiencies, eg lysine may be rendered nutritionally unavailable. On the other hand toxic and anti-physiological factors can often be eliminated or inactivated by the adjustment of processing conditions, while other toxic compounds can be eliminated by chemical treatment. Even though oilseeds contain toxic compounds they can be eliminated to a large extent, but it is important to be aware of their possible presence.

Oilseed cake production in Turkey, Sudan and Nigeria is increasing, with cotton seed being most abundant and cheap. However, the quality of the meals leave much to be desired because of poor technology from inadequate storage facilities, processing techniques, marketing and transportation problems. Good quality meals can be obtained only if these problems are solved. Other oilseeds meals of importance are sesame and groundnuts.

Cashewnut scrap kernel meal, although not as good as cashewnut meal could be an important protein source in Nigeria, even though it is lysine deficient. In feeding trials with swine, cashewnut scrap meal proved to be superior to groundnut meal.

Several questions were raised. Among these, concern was expressed about the stability of oil seed meals containing residual oil. However, if the seed was un-damaged and properly dried before processing, the residual oil should be quite stable. Furthermore, most of the oilseed meals contain natural antioxidants. Solvent extracted meals contain less oil, but they may be nutritionally inferior when both oil and protein contents are taken into consideration. Solvent extracted meals have a higher protein content.

There is no doubt that production of oilseed meals should be increased to meet protein needs. At the same time their potential protein supplementary value should be protected by improving harvesting and storage facilities, processing conditions and the ways in which they are used.

Professor Oyenuga

This session dealt with methods of assessment and standardisation of animal feeds including current efforts being made in major centres towards the standardisation of feed analytical data.

In the paper presented by Dr Carpenter methods of assessing the nutrient quality of concentrate feeds were discussed. The nutrients commonly identified in these feeds include vitamins, minerals, energy and the Weende or proximate analysis which identifies proteins, fat, crude fibre and nitrogen-free extract (NFE).

Although vitamins and minerals, including these required in traces, are indispensable for efficient body utilisation of feed nutrients and therefore for animal productivity, Dr Carpenter felt that a feed formulator may be indifferent to their content, since these essential micro nutrients could be more economically added from commercially available pre-mixes, which generally cover all contingencies.

Other imprecisely known factors like the UGF (un-identified growth factors) particularly in poultry pro-duction are better treated in the same way as anti-biotics.

Energy, the most important nutrient, is now assessed in the metabolisable form (ME) rather than as starch equivalent (SE), making the more cumbersome and expensive use of animal calorimeters and respiratory chambers unnecessary. The results should, however, be expressed on the same basis eg on dry matter or on the basis of a stated moisture content.

Proximate analysis which is the basis of existing legisla-tion on feedstuff composition still provides a crude kind of 'finger print' or profile of a sample. However, care must be taken in the interpretation of relative availability or digestibility of crude fibre and NFE. Due note should be taken of the newer methods of fractionating the crude fibre components particularly of forage as recently spear-headed by Dr Van Soest. Dr Carpenter also referred to the need to evaluate the sugar and starch constituents of NFE and referred to the use of the anthrone reagent which gives a simple colori-metric measurement that includes both sugar and starch.

Those interested in the prediction formula for cal-culating the classical ME values for poultry and other details are requested to refer to Dr Carpenter's excellent paper on this subject. Dr Carpenter dealt further with the analytical methods for estimating fats and proteins of feeds and the component amino acids and their nutritional availability. In particular he dealt with the determination of fluorodinitro-benzene (FDNB) reactive lysine as a method for estimating available lysine.

In the paper presented by Mr Bindloss the position with regard to feed standards in relation to commercial transactions was discussed. Mr Bindloss emphasised the need to confine legislation on non-nutrient con-stituents of feeds to that which was absolutely necessary to safeguard public health and safety, as unnecessarily rigorous standards in this regard would inevitably lead to higher consumer prices for animal products.

The paper presented by Professor Harris outlined the efforts which he and his associates are making towards the compilation of feed composition tables which can be used on a world-wide basis. His appeal for co-operation by additional laboratories engaged in feed analysis should be noted.

Dr Bhagwan's paper described the efforts being made in India on the formulation of standards for feed materials and mixed concentrate feeds. The difficulties experienced in implementing these standards were discussed and an outline was also given of the Govern-ment's plans for livestock development in India.

In the discussion periods Professor Fuller made some very important points with regard to fat rancidity while Professor Abou-Raya mentioned that he had found proximate analysis for crude fibre and NFE content to be as useful as the more detailed breakdown of car-bohydrate constitutes for most practical purposes.

There was no discussion of the papers presented in summary form by Dr Oke, Professor Harris (on behalf of Dr MacDowell) and Dr Barat.

Dr Magar

In the paper presented by Mr Ashman the need for an integrated approach to pest control starting at harvest was emphasised. Drying of grains to a safe moisture content for storage as quickly as possible and the need for adequate storage facilities suited to the local environment were mentioned as being of particular importance if losses were to be avoided. Governments should also consider possible storage problems when formulating policies on the use of insectides and the introduction of new varieties of crops.

Insect control is equally important at the processing and marketing phases where large volumes of com-modities are gathered together and the risks of in-festation are very much increased. It should be remembered that high protein products such as cakes are particularly susceptible to insect attack. Processors should note also that pelleted feeds are more easily and efficiently protected with insecticides than meals or flours. Mr. Ashman emphasised the need for careful planning of central stores and milling facilities to facilitate pest control and the need for separating the raw material from the milling machinery and the final products.

It was emphasised that it was necessary to review each situation separately and to select the method of chemical control considered most suitable and to use this in conjunction with good hygiene. The application of pesticides alone is certainly not sufficient and will be all the more costly if hygiene is ignored.

Finally it would appear that the application of control measures to materials in transit might have good possibilities.

Dr Jones gave a short history of the discovery of mycotoxins indicating that storage moulds have only relatively recently received attention. This was following the discovery of aflatoxin, a metabolite of the ubiquitous mould *Aspergillus flavus*. A range of aflatoxins is now known of which the most toxic is the M1 type found in milk. This is formed from B1 which is again the most toxic of the group found in other foodstuffs.

Moisture content, temperature, availability of oxygen and the nature of the substrate all influence the production of aflatoxin. It appears that groundnuts are particularly susceptible to contamination and that high levels are evident in pods with damaged shells or split kernels and where slow drying has been encouraged by poor handling techniques at the farmer level. If aflatoxin formation is ever to be completely controlled more attention is needed on the farm.

Dr Jones in his paper then discussed the toxicology of aflatoxins and pointed out that the species, age, sex and nutritional status of the animal all influence the response to aflatoxin in the diet. This is why it is particularly difficult to establish safe intake levels and to legislate accordingly.

There are two possible effects from the ingestion of toxic meal; one is a direct effect on the animals resulting in death or a reduction of liveweight, the other is the possibility that aflatoxins or their by-products may be transmitted into the human food chain by the consumption of animal products. So far milk and bacon are two products where aflatoxin or its derivatives have been detected.

Detoxification of animal feeds by solvent or chemical means would seem a distinct possibility. The advantages are obvious, but at present the cost and the difficulty of maintaining the foodstuffs in an otherwise unaltered form are two problems that continued research will hopefully overcome.

In certain crops such as groundnuts, sorting the commodity to remove the physically damaged material would seem to be a good method of reducing aflatoxin levels in food stuffs. This, coupled with improved harvesting and drying practices could in the long term do much to eliminate the problem.

Sampling and analytical methods were referred to. Although a lot of progress has been made in the development of quick methods of analysis, there is room for continued improvements in their reliability and accuracy. This should be pursued, because a quick and reliable method would be a considerable benefit to developing countries. Sampling emerges as a major problem, and some guidance aimed at improving methods currently used would also be of considerable benefit.

In the paper presented by Mr Bockelee-Morvan the point was made that the chemical detoxification of aflatoxin contaminated materials is at present very costly. Ideally this should only be necessary on a very small portion of the total bulk. To achieve this, contaminated produce must be separated from clean produce prior to processing. In the groundnut crop, this can either be done before shelling or before the kernels are crushed. Mr Morvan went on to describe two trials recently undertaken in West Africa.

The first involved pod sorting and it was established that removal of split, broken or pest damaged pods reduced the aflatoxin content by an average of 76%. Pods pierced by millipedes were the most highly contaminated and although they represented a very small percentage of the total, they contained an average 40% of all the aflatoxin. Work is in progress on a pneumatic shelling apparatus which distinguishes and separates undamaged pods from the remainder.

The second trial on kernels showed that if, wrinkled, mouldy, or discoloured seeds were removed, the remainder were almost entirely aflatoxin free. Broken and skinned seeds were not usually toxic unless discoloured as well, and in any case this group together with the shrivelled seeds is removed by mechanical grading.

In conclusion it was suggested that pneumatic shelling followed by grading and hand sorting would give a product largely free from aflatoxin.

During the discussion period the desire was expressed that legislation on aflatoxin should be more realistic in order to avoid unnecessary wastage of valuable high protein animal feeds which would increase costs and deprive the world of important sources of food and feed. This is fully realised and is indeed an important problem.

Mr Halliday (in place of Professor Alim)

The first paper by Mr Cropper emphasised the need for a clear idea of the overall economic implications of livestock development programmes. The substitution of livestock products by local production is of little or no economic benefit with regard to the saving of foreign exchange if most of the feedstuffs have to be imported. Sugar products, such as sugar cane itself and molasses had probably the most potential for use as animal feeds in Trinidad, and the production systems developed by Preston and others could well be of great benefit to Trinidad in this regard. The paper by Mr Wood, also illustrated well the necessity for very careful planning in the establishment of feed production enterprises in developing countries. Co-operation by feed manufacturers with government and the livestock producers was of vital importance. It was emphasised that the manufacturer

Summaries of papers circulated at the conference for information only

The use of brewers' dried grains in poultry and pig rations in Nigeria.

A. A. Ademosun
Department of Animal Science, University of Ife, Nigeria.

Brewers' dried grains (BDG), a by-product of the brewing industry, has been investigated as a feed ingredient in poultry and pig rations. Although it is high in protein, its high fibre content limits its use as a replacement for feed grain. Experiments have indicated that it can be incorporated into poultry starter and grower mash at levels of up to 10 and 15% respectively without any adverse effect on the performance of the birds and results in saving on the cost of the feed. BDG can also be used as an energy diluent in an effort to produce lean pork.

Résumé

L'utilisation des graines séchées des brasseurs dans les rations pour la volaille et les cochons en Nigeria

On a investigué les graines séchées des brasseurs (GSB), un sousproduit de l'industrie de la brasserie, comme ingrédient dans les rations pour la volaille et les cochons. Quoique riche en protéines, la grande teneur en fibres limite son usage comme un remplacement des graines comme nourriture. Des expérimentations ont indiquées qu'elles peuvent être incorporées dans les mélanges pour les jeunes poussins et pour la volaille en développement, dans des proportions respectives de 10–15%, sans aucun effet défavorable sur le rendement de la volaille et avec des résultats d'économie du coût de la nourriture. GSB peuvent être utilisées aussi comme un diluant de l'énergie pour produire de la viande maigre de porc.

Resumen

El uso de los granos desecados de la industria cervecera en las raciónes de las aves de corral y de los cerdos en Nigeria

Los granos desecados de la industria cervecera (BDG), un producto secundario de la industria de la fabricación de la cerveza, se ha investigado como un ingrediente de los piensos en las raciones de las aves de corral y de los cerdos. Aunque es rico en proteína, su alto contenido de fibra limita su uso como una sustitución del pienso de granos. Los experimentos han indicado que puede incorporarse en la masa iniciadora y estimuladora del crecimiento de las aves de corral a niveles de hasta el 10 y el 15%, respectivamente, sin ningún efecto adverso en el desarrollo de las aves, y produciendo ahorro en el coste del pienso. Puede usarse BDG también como un disolvente de energía en un esfuerzo para producir cerdo magro.

Feed Materials for use in rations of aquatic animals in French Polynesia

G. Cuzon
National Centre for Ocean Exploration, Tahiti

There is a scarcity of locally produced animals feeds in French Polynesia which is traditionally met by importation. Recent rises in prices of internationally traded feedingstuffs have provided a stimulus to utilise more local products.

The meats from a gasteropod *Trochus niloticus* which is collected for its shells can be used with good results as an ingredient of prawn (*Penaeus merguiensis*) and fish foods after chopping and grinding. It can also be used for pig feeding. Fish meal can also be produced by sun drying and milling trash fish and fish offal.

Coconut meal is produced in Tahiti to the extent of 400–500 t per month, and can be used as an ingredient of feed for fresh water prawns (*Macrobrachium rosenbergii*). There are also prospects for producing

Spirulina (presently cultured in Mexico) and *Leucana leucocephela* and preliminary trials with *L. leucocephela* in Tahiti have indicated that it can be grown with high yields and that it has good prospects as a fish food. Cassava leaves, brewers' grains and brewers' yeast are also products with potential for use as animal feeds and fish foods.

It is concluded that fishmeal, coconut meal and *L. leucocephela* meal offer the best prospects for wide use as animal feeds in the Pacific area.

Résumé

Les produits alimentaires pour l'usage dans les rations des animaux aquatiques dans la Polynésie Française

Dans la Polynésie Française il y a un manque de pâtures pour les animaux, produites localement, qui par tradition est satisfaite par l'importation. Les augmentations récentes des prix des nourritures négociées internationalement, ont été un stimulant pour utiliser d'avantage les produits locaux.

La chair d'un gastropode *Trochus niloticus* qui est recueilli pour sa coquille, peut être utilisée avec des bons résultats comme un ingrédient des nourritures pour les crevettes (*Penaeus merguiensis*) et des poissons, après broiement et hachement. Cette nourriture peut être utilisée également comme pâture pour les cochons. Les pâtures de poisson peuvent être produites aussi par séchage au soleil et en moulant les rebuts et les viscères de poisson.

La pâture de noix de coco est produite à Tahiti en quantité de 400–500 t par mois est peut être utilisée comme un ingrédient des nourritures pour les crevettes d'eau douce (*Macrobrachium rosenbergii*). Il y a aussi des perspectives de production de la *Spirulina* (cultivée à présent au Mexique) et la *Leucana leucocephela,* et les essais préliminaires avec *L. leucocephela* à Tahiti ont indiqués qu'elles peuvent être cultivées avec des grands rendements et qu'il y a des bonnes perspectives de les utiliser comme nourriture pour les poissons. Les feuilles de manioc, les graines des brasseurs et la levure des brasseurs sont aussi des produits importants à être utilisés comme pâtures pour les animaux et comme nourriture pour les poissons.

On conclut que les pâtures de poisson, de noix de coco et de *L. leucocephela* offrent les meilleures perspectives pour l'usage étendu comme nourritures pour les animaux dans la région du Pacifique.

Resumen

Materiales para piensos para uso en las raciónes de animales acuáticos en la Polinesia Francesa

Hay una escasez de piensos para animales producidos localmente en la Polinesia francesa que se satisface tradicionalmente mediante la importación. Las alzas recientes en los precios de los piensos del comercio internacional han servido de estímulo para utilizar más productos locales.

La carne de un gasterópodo *Trochus niloticus* que se recoge por sus conchas puede usarse con buenos resultados como un ingrediente de alimento para los camarones (*Penaeus merguiensis*) y los peces, después de cortada y molida. Puede usarse también para alimento de los cerdos. Puede producirse también harina de pescado, secando al sol y moliendo peces de desecho y despojos de peces.

La harina de coco se produce en Tahití en cantidad de 400–500 t. por mes, y puede usarse como un ingrediente de piensos para camarones de aguas frescas (*Macrobrachium rosenbergii*). Hay también perspectivas para producir *Spirulina* (actualmente cultivada en Méjico) y *Leucana leucocephela,* y las pruebas preliminares con *L. leucocephela* en Tahití han indicado que puede cultivarse con altos rendimientos y que tiene buenas perspectives como un alimento para peces. Las hojas de cazabe, los granos y la levadura de la industria de la cerveza son también productos potenciales para usar como piensos para los animales y alimentos para los peces.

La conclusión es que la harina de pescado, la harina de coco y la harina de *L. leucocephela* ofrecen las mejores perspectivas para uso amplio como piensos para animales en el área del Pacífico.

Dried rumen contents as feeds for poultry in Egypt

A. K. Deek, A. R. Abou Akkada, A. Khaleel and K. El-Shazly
University of Alexandria, Arab Republic of Egypt.

Rumen contents from cows and water buffaloes slaughtered in Alexandria abattoirs, were strained, mixed with hydrochloric acid, the supernatant decanted and the bottom layers in the large containers dried at 40–50°C. The dried preparation contained, crude protein 38 to 40%, crude fat 12%, total ash 25%, soluble carbohydrates 21%, fibre 4% and calcium 3%. In a series of feeding experiments, this preparation was included in the rations of day-old chickens of Alexandria breed in order to replace 50, 75 and 100% of fishmeal. It was shown that dried rumen contents sucessfully replaced the fishmeal. Furthermore, the body weight gain, efficiency of feed utilisation and dressing percentage in 8 weeks of chickens fed on rations containing proportions of dried rumen contents plus fishmeal were superior to those of birds fed on diets containing only fishmeal or dried rumen contents. Further experiments have indicated that the dried rumen contents can serve as an effective substitute for fish or blood meals but inferiorly replaced a combination of both proteins in the diets of young chickens. It is therefore concluded that if the rumen contents available at commercial abbatoirs in Egypt were dried for use as poultry feeds, the amounts of fish or blood meals required for the broiler industry would be reasonably reduced.

Résumé

Le contenu du rumen séché comme pâture pour la volaille en Égypte

Les contenus du rumen des vaches et des buffles d'eau abbatus dans les abattoirs d'Alexandria ont été filtrés, mélangés avec de l'acide chlorhydrique, la couche surnageante décantée et les couches profondes séchées dans des grands réservoirs, à 40–50°C. La préparation séchée contenait des protéines brutes 38–40%, de la graisse brute 12%, cendres totales 25%, hydrate de carbone soluble 21%, fibre 4% et calcium 3%. Dans une série d'expériences d'alimentation faites à Alexandria, cette préparation a été incluse dans les rations des poussins âgés de 1 jour, afin de remplacer 50, 75 et 100% des pâtures de poisson. On a demontré que le contenu séché de rumen remplaçait avec succès la pâture de poisson. D'ailleurs, le gain en poids corporel, l'efficacité de l'usage de la nourriture et le pourcentage d'assaisonnement, en 8 semaines d'alimentation de la volaille avec des rations contenant des proportions de contenu de rumen séché et en plus de la farine de poisson, était supérieur à celui de la volaille nourrie avec des diètes contenant seulement de la farine de poisson ou le contenue séché de rumen. D'autres expériences ont indiquées que le contenu séché du rumen peut servir comme un substitut efficace pour les pâtures de poisson ou de sang, mais remplaçait moins bien une combinaison des deux protéines dans la diète des jeunes poussins. Par conséquent, on conclut que, si les contenus de rumen, disponibles dans les abattoirs commerciaux en Egypte, seraient séchés pour l'utilisation comme nourriture pour la volaille, les quantités de pâtures de poisson ou de sang nécessaires pour l'industrie du poulet rôti seraient raisonnablement réduites.

Resumen

Los contenidos desecados del herbario como piensos para las aves de corral en Egipto

Se colaron los contenidos de los herbarios de vacas y búfalos de agua sacrificados en los mataderos de Alejandría, se mezclaron con ácido clorhídrico, la parte superior se decantó, y las capas del fondo en los grandes recipientes se secaron a 40–50°C. La preparación seca contenía proteína bruta 38 al 40%, grassa bruta 12%, total de cenizas 25%, hidratos de carbono solubles 21%, fibra 4% y calcio 3%. En una serie de experimentos de alimentación, se incluyó esta preparación en las raciones de pollos de un día de raza Alejandría a fin de sustituir el 50, 75 y 100% de harina de pescado. Se comprobó que los contenidos desecados del herbario sustituían con éxito a la harina de pescado. Además, el aumento de peso, la eficacia de la utilización del pienso y porcentaje de "dressing" en ocho semanas, de pollos alimentados con raciones que tenían proporciones de contenido de herbario desecado más harina de pescado, eran superiores a los de las aves alimentadas con dietas que contenían únicamente harina de pescado o contenido de herbario desecado. Experimentos posteriores han indicado que el contenido del herbario desecado puede servir como un sustituto eficaz de las harinas de pescado y de sangre, pero era inferior cuando reemplazaba a una combinación de ambas proteínas en las dietas de los pollos jóvenes. La conclusión, por consiguiente, es que si se desecase el contenido de los herbarios disponibles en los mataderos comerciales en Egipto para usar como piensos de las aves de corral, las cantidades de harinas de pescado o de sangre exigidas por la industria de los pollos de asar se reduciría razonablemente.

Problems of feeding concentrates to livestock in Zambia

E. A. Gihad
University of Zambia, Zambia.

During the wet season Zambian livestock are entirely dependent on grazing (mainly natural ranges), except for high milk producing cows which are supplied with some supplementary protein concentrates. In the dry season livestock are largely maintained on the standing hay and/or browsing.

The resources of concentrates in Zambia are limited, the estimated requirement being 120,000 t of which 70% would be energy or basal feeds which could be provided as maize or its milling by-products, and 30% would be protein concentrates.

The potential for producing protein concentrates within Zambia is very limited at present. Oilseed production is too low to justify the establishment of oil mills, but there are prospects for feeding heat-treated whole soya beans to pigs and poultry and unprocessed sunflower and cotton seed to ruminants. Dehydrated alfafa meal could also be produced.

Résumé

Les problèmes concernant les pâtures concentrées pour le bétail en Zambie

Pendant la saison humide, le bétail en Zambie est entièrement dépendant des pâturages (surtout les étendues naturelles), à l'exception des vaches produisant des grandes quantités de lait, qui sont nourries avec certains suppléments de concentrés de protéines. Dans la saison sèche, le bétail se nourrit largement avec du foin stocké et/ou en butinant.

Les ressources de concentrés en Zambie sont limitées, les nécessités sont évaluées à 120.000 t. desquelles 70% sont composées de nourriture, ou de base ou d'énergie, qui peut être fournie comme mais ou ses sousproduits moulus et 30% concentrés protéiniques.

Les possibilités de produire des concentrés protéiniques en Zambie à présent sont très limitées. La production de graines oléagineuses est trop réduite pour justifier l'établissement de moulins, mais il y a des perspectives de nourrir les cochons et la volaille avec des graines de soja traitées par la chaleur et les ruminants avec du tournesol non traité et des graines de coton. Des pâtures d'alfalfa pourraient être produites.

Resumen

Los problemas de los peinsos concentrados para el ganado en Zambia

Durante la estación de las lluvias, el ganado de Zambia depende completamente de los pastos (principalmente pastos naturales), excepto las vacas de producción de leche alta, a las que se les suministran algunos concentrados de proteínas suplementarios. En la estación seca el ganado se mantiene en gran parte del heno existente y/o ramoneando.

Los recursos de concentrados en Zambia son limitados, estimándose las necesidades en 120.000 t. de las cuales el 70% serían energéticos o básicos, que podrían suministrarse del maíz o de los productos secundarios de su molienda, y el 30% serían concentrados de proteínas.

El potencial para producir concentrados de proteína en Zambia es muy limitado al presente. La producción de semillas oleaginosas es demasiado baja para justificar el establecimiento de molinos de aceite, pero hay perspectivas para alimentar a los cerdos y aves de corral con semillas de soja completas tratadas térmicamente, y a los rumiantes con semillas de girasol y de algodón no tratadas. También se podría producir harina de alfafa deshidratada.

De-oiled neem fruit (*Azadirachta indica*) cake for use as a cattle feed in times of famine

R. E. Patil, D. V. Ranganekar, K. L. Meher and C. M. Ketkar
Ahmednager, Maharashtra State, India

The shortage of concentrates for cattle feed in India particularly in times of drought has made it necessary to utilise materials such as neem oilcake which are not conventionally used as cattle feed. It is estimated that 352,000 t of neemcake might be available in India as a by-product from oil expression for soap production.

Neemcake has a very bitter taste but cattle will eat it when they become accustomed to it. It contains around 14% of crude protein, 13.7% of ether extract and 26.5% of crude fibre. Palatability can be increased by the inclusion of molasses in rations containing neemcake.

Experiments have shown that neemcake can be used successfully as an ingredient of concentrate mixtures for dairy cattle at inclusion rates of up to 10%.

Résumé

Les tourteaux de fruits deshuilés de margosa (*Azadirachta indica*) utilisés comme pâture pour le bétail en temps de famine

Le manque de produits concentrés pour nourrir le bétail en l'Inde, surtout en temps de sécheresse, a rendu nécessaire d'utiliser des tels matériels comme les tourteaux d'huile de margosa, qui conventionnellement ne sont pas utilisés comme pâture pour le bétail. On évalue que 352.000 t. de tourteaux de margosa sont disponibles en l'Inde comme un sous-produit de l'huile utilisée pour la fabrication du savon.

Les tourteaux de margosa sont très amères, mais le bétail les mangent quand il s'habitue. Les tourteaux contiennent environ 14% protéine brute, 13% extrait éthéré et 26.5% fibre brute. La saveur peut être améliorée par l'addition de mélasses dans les rations contenant des tourteaux de margosa.

Des expériences ont démontrées que les tourteaux de margosa peuvent être utilisés avec succès comme un ingrédient dans les mélanges de concentrés pour le bétail de ferme, dans des proportions jusqu'à 10%.

Resumen

La torta de orujo del fruto desaceitado de la margosa (*Azadirachta indica*) para usar como pienso del ganado en tiempos de carestía

La escasez de concentrados de piensos para el ganado en la India, particularmente en tiempo de sequía, ha hecho necesario utilizar materiales tales como la torta de orujo de la margosa, que no se usa corrientemente como pienso para el ganado. Se estima que podría disponerse en la India de 352.000 t. de torta de orujo de margosa como producto secundario de la extracción del aceite para la producción de jabón.

La torta de orujo de la margosa tiene un gusto muy amargo, pero el ganado la comerá cuando se vaya acostumbrando a ella. Contiene alrededor del 14% de proteína bruta, el 13,7% de extracto de éter y el 26,5% de fibra bruta. El gusto puede mejorarse mediante la inclusión de melazas en las raciones que contengan torta de orujo de margosa.

Los experimentos han demostrado que la torta de orujo de margosa puede usarse con éxito como un ingrediente de mezclas concentradas para las vacas de leche incluyéndose en proporción de hasta el 10%.

A pilot study on the use of sugar cane by-products for growing cross-bred heifers.

D. V. Rangnekar, V. P. Nigadikar, B. N. Soble,
A. L. Joshi, and R. E. Patil
Bharatiya Agro-Industries Foundation, Uruli-Kanchan,
Dist. Poona, India

The performance of Gir x Holstein crossbred heifers on an experimental ration based on sugar cane bagasse, molasses and urea was compared with that of those fed a conventional ration of cereal straw supplemented with concentrates. The performance on the experimental ration was as good as that on the conventional ration, while feed costs were appreciably lower. It was concluded that these were good prospects of using complete feeds based on sugar cane by-products in India on an economic basis for dairy cattle, particularly the growing animals.

Résumé

Une étude pilote sur l'utilisation des sous-produits de la canne à sucre pour l'elevage des races croisées de génisses

Les performances des races croisées de génisses Gir x Holstein avec une ration expérimentale de bagasse de la canne à sucre, mélasses et urée ont été comparées avec celles de mêmes animaux nourris avec une ration conventionnelle de paille de céréales complétée avec des concentrés. Les performances avec les rations expérimentales étaient aussi bonnes que celles avec la ration conventionnelle, tandis que le coût des pâtures était appréciablement plus bas. On est arrivé à la conclusion qu'il y avait des bonnes perspectives d'utiliser en l'Inde, sur une base économique, des pâtures complètes basées sur les sousproduits de la canne à sucre pour les animaux de ferme, surtout ceux en voie de croissance.

Resumen

Estudio piloto sobre el uso de los productos secundarios de la caña de azúcar para la cría de terneras híbridas

El comportamiento de las terneras híbridas de Gir x Holstein con una ración experimental basada en bagazo de caña de azúcar, melazas y urea se comparó con la de las alimentadas con una ración convencional de paja suplementada con concentrados. El resultado con la ración experimental fué tan bueno como el obtenido con la ración convencional, mientras que los costes de los piensos fueron apreciablemente más bajos. La conclusión fué que había buenas perspectivas de usar piensos completos basados en productos secundarios de la caña de azúcar en la India en una base económica para vacas de leche, particularmente para los animales de cría.

List of registered delegates

Dr A. K. Abou-Raya,
Animal Nutrition Section,
Animal Production Department,
Faculty of Agriculture,
Cairo University,
Cairo,
ARAB REPUBLIC OF EGYPT

Mr I. Acolatse,
Direction de l'Élevage,
Lomé
TOGO

Mr D. Adair,
Tropical Products Institute,
Industrial Development Department,
Culham,
Abingdon,
Berks,
UK

Chief S. A. Adeniyi,
Ministry of Agriculture & Natural Resources
P.M.B. 5007,
Secretariat,
Ibadan,
NIGERIA

Mr E. Ajulo,
Nigerian Stored Products Research Institute,
Kano,
NIGERIA

Mr M. M. Ali,
University of Aberdeen,
School of Agriculture,
Division of Agricultural Chemistry,
581 King Street,
Aberdeen AB9 1UD
UK

Prof. K. A. Alim,
Alexandria University,
Faculty of Agriculture,
Shatby,
Alexandria,
ARAB REPUBLIC OF EGYPT

Dr N. Anand,
Tropical Products Institute,
56–62 Grays Inn Road,
London WC1X 8LU
UK

Dr R. J. Andrews,
RHM Animal Feed Services Limited,
Feedingstuffs Marketing Department,
Dean Grove House,
Cole Hill,
Wimborne,
Dorset BH21 7AE
UK

Chief E. O. Ashamu,
NIGERIA

Mr F. Ashman,
Tropical Products Institute,
56–62 Gray's Inn Road,
London WC1X 8LU
UK

Dr G. M. Babatunde,
Department of Animal Science,
University of Ibadan,
Ibadan,
Western State,
NIGERIA

Mr E. F. Bajunirwe-Butsya,
Ministry of Agriculture & Animal Resources,
P.O. Box 7003,
Kampala,
UGANDA

Mr B. P. Baker,
United Molasses Trading Company Limited,
Bowater House East,
68 Knightsbridge,
London SW1
UK

338　Dr S. K. Barat,
Agricultural Services Division,
FAO,
Via delle Terme di Caracalla,
Roma,
ITALY

Mr A. Barranco,
Tropical Products Institute,
56–62 Gray's Inn Road,
London WC1X 8LU
UK

Mr K. W. Bean,
Editor 'World Crops',
Riverside House,
Hough Street,
London S.E.18
UK

Ingo A. D. Bellido,
Pesca,
P.O. Box 2881,
Lima,
PERU

Mr C. M. Bentley,
Food & Allied Industries Limited,
10 Intendance St.,
Port Louis,
MAURITIUS

Dr H. Bhagwan,
Indian Standards Institution,
Manak Bhavan,
9 B.S. Zafar Marg,
New Delhi 11001,
INDIA

Mr A. A. Bindloss,
Unilever Ltd,
P.O. Box 91,
St. Bridget's House,
Bridewell Place,
London EC4P 4BP
UK

Mr A. D. Bird,
CAFMNA,
58 Southwark Bridge Road,
London SE1 0AS
UK

Mr P. R. Bird,
Faure-Fairclough,
14–18 Holborn,
London EC4
UK

Mr A. B. Blair,
Société de Gestion,
Excomm S.A.,
8 Avenue Calas,
Geneva,
SWITZERLAND

Dr R. Blair,
Agricultural Research Council,
Poultry Research Centre,
King's Buildings,
West Mains Road,
Edinburgh EH9 3JS
UK

Mr A. Blair Rains,
Land Resources Division,
Tolworth Tower,
Surbiton,
Surrey
UK

Mr A. Bockelee-Morvan,
Institut de Recherches pour les Huiles et Oléagineux,
8–13 Square Petrarque,
75016 Paris,
FRANCE

Mr A. Bogale,
Livestock and Meat Board,
Shola Poultry Centre,
P.O. Box 2449,
Addis Ababa,
ETHIOPIA

Mr K. A. Bouchard,
Cooper Nutrition Products Limited,
Stepfield,
Witham,
Essex CM8 3AB
UK

Mr G. R. Breag,
Tropical Products Institute,
Industrial Development Department,
Culham,
Abingdon,
Berks,
UK

Dr Brenes,
Piensos Hens S.A.,
Avda. Infanta Carlota, 123
Barcelona – 15,
SPAIN

Mr P. J. R. Breslin,
Tropical Products Institute,
56–62 Gray's Inn Road,
London WC1X 8LU
UK

Dr R. Bressani,
Institute of Nutrition of Central America & Panama
　　(INCAP),
P.O. Box 1188,
Guatemala,
GUATEMALA

Mr E. M. R. Britton,
Barclays Overseas Development Corporation Limited,
54 Lombard Street,
London EC3P 3AH
UK

Dr A. W. A. Burt,
Burt Research Limited,
23 Stow Road,
Kimbolton,
Huntingdon PE18 0HU
UK

Mr E. Caen,
Société Anonyme Louis Dreyfus,
6 Rue Rabelais,
75361 Paris Cedex 08,
FRANCE

Miss H. M. Carey,
International Project in the Field of Food Irradiation,
Institut Für Strahlentechologie,
75 Karlsruhe 1,
Postfach 3640,
FEDERAL REPUBLIC OF GERMANY

Dr K. J. Carpenter,
Department of Applied Biology,
University of Cambridge,
Downing Street,
Cambridge CB2 3DX
UK

Dr L. S. Castillo,
Philippine Council for Agricultural Research,
University of the Philippines,
Los Baños College,
Laguna,
PHILIPPINES

Mr R. J. Clarke,
General Foods Limited,
Banbury,
Oxfordshire,
UK

Mr V. J. Clarke,
School of Agriculture,
Sutton Bonnington,
Loughborough,
Leicestershire,
UK

Dr H. A. V. Clavijo,
Santa Domingo Experimental Station,
Instituto Nacional de Investigaciónes Agropecuarias,
Apartado No. 2600,
Quito,
ECUADOR

Mr A. Clayton,
Commonwealth Development Corporation,
33 Hill Street,
London W1A 3AR
UK

Mr W. A. Cleaver,
Crosfields International Ltd.,
18 Buckingham Gate,
London SW1
UK

Mr F. Clermont Scott,
Beecham Research Laboratories,
Manor Royal,
Crawley,
Sussex
UK

Mr E. C. Clyde Parris,
Animal Nutritionist,
Government of Trinidad,
TRINIDAD

Mr S. Conrad-Mason,
Ministry of Agriculture,
Tanteen,
St. Georges,
GRENADA

Mr D. G. Coursey,
Tropical Products Institute,
56–62 Garys Inn Road,
London WC1X 8LU
UK

Mr J. G. Cridlan,
Tradax England Limited,
Kempson House,
Camomile Street,
London EC3
UK

Mr J. Cropper,
University of the West Indies,
Department of Agricultural Economics,
St. Augustine,
TRINIDAD

Mr. G. R. Cruickshank,
Poultry World,
161 Fleet Street,
London EC4P 4AA
UK

Mr Cuzon,
CNEXO-COP,
Vairao,
Tahiti,
FRENCH POLYNESIA

Prof. A. E-M. Darwish,
Faculty of Agriculture,
Assiut University,
ARAB REPUBLIC OF EGYPT

Mr J. Davie,
Ministry of Overseas Development,
Eland House,
Stag Place,
London SW1E 5DH
UK

Mr C. J. A. Delaitre,
Ministry of Agriculture & Natural Resources,
Port Louis,
MAURITIUS

Dr D. A. V. Dendy,
Tropical Products Institute,
56—62 Grays Inn Road,
London WC1X 8LU
UK

Mr L. Desvaux de Marigny,
Mauritius Chamber of Agriculture,
29 Brown - Sequard St,
Curepipe,
MAURITIUS

Dr C. Devendra,
Malaysian Agricultural Research and Development
    Institute,
P.O. Box 208,
Sungei Besi,
Selangor,
MALAYSIA

Mr A. Dewandre,
38 Avenue Woine,
Yvoir 5190,
BELGIUM

Mr J. Djoukam,
Agricultural Engineering School of Cameroon,
B.P. 138,
Yaoundi,
CAMEROON

Mr N. Ellis,
Scottish Agricultural Industries Limited,
25 Ravelston Terrace,
Edinburgh EH4 3ET
UK

Mr J. L. U. Eme,
Agricultural Development Authority,
East Central State,
Enugu,
NIGERIA

Mr K. Eslami,
Range & Pasture Development Fund of Iran,
31 Maykadeh Avenue,
Tehran,
IRAN

Dr B. L. Fetuga,
Department of Animal Science,
University of Ibadan,
Ibadan,
NIGERIA

Mr G. N. Finch,
Roger Williams Technical and Economic Services Inc.,
37/41 Bedford Row,
London WC1R 4JH
UK

Dr P. J. Findlen,
Foreign Agricultural Service,
422 West Administration Building,
U.S. Department of Agriculture,
Washington DC 20250,
USA

Mr R. Fitzhenry,
Department of Agriculture & Fisheries,
Dublin 2,
IRELAND

Mr P. F. Flanagan,
Guthrie Estates Ltd,
120 Lane End Road,
Sands,
High Wycombe,
Bucks HP12 44X
UK

Mr B. N. Fox,
Shell International Chemical Company Limited,
Ref. CLS/1 Shell Centre,
London SE1
UK

Mr B. J. Frentzel,
E. D. MacLeod & Company Limited,
Corn Exchange Chambers,
2 Seething Lane,
London EC3
UK

Prof. H. L. Fuller,
Department of Poultry Science,
University of Georgia,
Athens,
Georgia 30602,
USA

Mr P. J. Gallimore,
BOCM/Silcock Limited,
Basing View,
Basingstoke,
Hampshire RG21 2E2
UK

Miss E. Gatumel,
Société Française des Pétroles,
BP,
10 Quai Paul Doumer,
92401 — Courbevoie,
FRANCE

Mr D. George,
Oloagun Enterprises,
NIGERIA

Prof. A. K. Göğüs,
University of Ankara,
Faculty of Agriculture,
Department of Animal Nutrition,
University of Ankara,
Ankara,
TURKEY

Dr G. Gomez,
CIATT (Centro Internacional de Agricultura Tropical),
Apartado-Aereo 67–13,
Cali,
COLOMBIA

Mr M. J. Gollin,
United Molasses Trading Company Limited,
Bowater House East,
68 Knightsbridge,
London SW1
UK

Mr E. G. B. Gooding,
Caribbean Development Bank,
Bridgetown,
BARBADOS

Mr H. C. H. Graves,
President,
National Council of Concentrate Manufacturers,
7 Parkside,
London SW1X 7JW
UK

Mr R. F. Hancock,
Commodities & Trade Division,
c/o FAO Regional Office for Europe,
Palais des Nations,
Geneva,
SWITZERLAND

Prof. L. E. Harris,
Department of Animal Science,
Utah State University,
Logan,
Utah 84321
USA

Mr H. R. Harrison,
Agricultural Planning Associates Limited,
12 de Walden Court,
85 New Cavendish St,
London W1N 7RA,
UK

Mr A. W. Hartley,
Spillers Limited,
Research and Technology Centre,
Station Road,
Cambridge CB1 2JN
UK

Mr R. Harzallah,
A/D Office de l'Élevage et des Pâturages,
TUNISIA

Dr M. J. Head,
Lecturer in Animal Nutrition,
University of Surrey,
Guildford,
Surrey,
UK

Mr G. C. Hoetink,
Koninklijke Wessanen N.V.,
Zaanweg 51,
Wormerveer,
THE NETHERLANDS

Prof. M. Hoffmann,
Karl Marx Universität,
Sektion Tierproduktion und Verterinärmedizior,
Fachgruppe Tierfütterung,
701 Leipzig,
Marg. Blank-Strasse 8,
GERMAN DEMOCRATIC REPUBLIC

Mr T. D. Holderness-Roddam,
United Molasses Trading Company,
Bowater House East,
68 Knightsbridge,
London SW1
UK

Mr W. B. Holmes,
8 Reyger Street,
Worcester,
Cape Province,
REPUBLIC OF SOUTH AFRICA

Mr A. Hone,
Institute of Commonwealth Studies,
Queen Elizabeth House,
21 St. Giles,
Oxford OX1 3LA
UK

Mr J. R. Hopkins,
Ministry of Agriculture, Fisheries & Food,
Government Building,
Lawnswood,
Leeds 16
UK

Mr B. W. Howells,
Beecham Agricultural Products,
Beecham House,
Brentford,
Middlesex,
UK

Mr M. J. Hoxey,
Colborn International Limited,
Barton Mills,
Canterbury,
Kent,
UK

Mr C. I. Iheanitu,
Livestock Feeds Unit,
Agricultural Development Authority,
East Central State,
PMB 1024,
Enugu,
NIGERIA

Mr J. Jacobsen,
Elias B. Muus Odense Limited,
2—4 Frederiksgade,
5100 Odense,
DENMARK

Dr S. Jani,
The National Milling Corporation,
P.O. Box No. 9502,
Dar-es-Salaam,
TANZANIA

Dr K. Jewers,
Tropical Products Institute,
56—62 Gray's Inn Road,
London WC1X 8LU
UK

Dr B. D. Jones,
Tropical Products Institute,
56—62 Gray's Inn Road,
London WC1X 8LU
UK

Dr N. R. Jones,
Tropical Products Institute,
56—62 Gray's Inn Road,
London WC1X 8LU
UK

Mr M. Jones,
Continental Foods Corporation,
Apartado 735,
Panama 1,
REPUBLIC OF PANAMA

Mr V. Jones,
Beecham Research Laboratories,
Nutritional Research Centre,
Walton Oaks,
Dorking Road,
Tadworth,
Surrey,
UK

Mr A. Kaboash,
Department of Poultry Research,
Wye College,
Wye,
Kent,
UK

Mr B. D. S. Kapita,
National Milling Company,
Lusaka,
ZAMBIA

Dr M. C. Keith,
Unilever Research Laboratory,
Greyhope Road,
Aberdeen AB9 2JA
UK

Mr J. Keys,
J. E. Hemmings & Son Limited,
Barford Mills,
Barford,
Nr Warwick,
UK

Mr G. Kidd,
Imperial Chemicals Industries Limited,
Agricultural Division,
P.O. Box 1,
Billingham,
Teesside,
UK

Mr K. M. Kinani,
Ministry of Agriculture and Animal Resources,
P.O. Box 7003,
Kampala,
UGANDA

Prof. J. L. R. Kirkaldy,
Protein Tumboh Tumbohan Malaysia Sendirian Bhd,
62 Main Road,
Petaling,
Selangor,
MALAYSIA

Mr D. W. Knight,
Hebden Water Feeds Limited,
Lee Mill,
Hebden Bridge,
Nr Halifax HX7 7AE
UK

Mr D. L. Knights,
BP Proteins Limited,
Britannic House,
Moor Lane,
London EC2 9BU
UK

Mr H. Koenig,
UNIDO,
Lerchenfelderstrasse 1,
A-1011 Vienna,
AUSTRIA

Mr Kuffuor,
Production Manager,
Tema Food Complex Corporation,
P.O. Box 282,
Tema,
GHANA

Mr H. Kumar,
Plenty and Son Limited,
Bescon Division,
Hambridge Rd,
Newbury,
Berkshire,
UK

Mr P. W. G. Lake,
Beecham Agricultural Products Division,
Beecham House, B/01,
Great West Road,
Brentford,
Middlesex,
UK

Dr B. M. Laws,
Pauls and Whites Foods Limited,
Research and Advisory Department,
New Cut West,
Ipswich,
Suffolk IP2 8HP
UK

Dr M. Lengelle,
Organisation for Economic Co-operation and
    Development,
Directorate for Agriculture and Food,
2 Rue Andre — Pascal,
Paris — 16^e,
FRANCE

Professor D. Lewis,
School of Agriculture,
University of Nottingham,
Sutton Bonnington,
Loughborough,
Leics LE1 5RD
UK

Mr G. L. Lewis,
BOCM Silcock Limited,
Unilever Limited,
St. Bridget's House,
Bridewell Place,
London EC4
UK

Miss L G Lewis,
Barclays Bank International Development Fund,
Barclays Bank International Limited,
54 Lombard Street,
London EC3
UK

Prof. I. E. Liener,
Department of Biochemistry,
University of Minnesota,
College of Biological Sciences,
140 Gortner Laboratory,
St. Paul,
Minnesota, 55101,
USA

Mr E. M. Low,
Home Grown Cereals Authority,
Haymarket House,
Oxendon Street,
London SW1Y 4EF
UK

Dr W. Y. Magar,
Director, Scientific & Technical Department,
African Groundnut Council,
P.O. Box 3025,
Lagos,
NIGERIA

Miss E. C. Marchant,
Shell International Chemical Company Limited,
Shell Centre,
London SE1 7PG
UK

Dr L. Mayer,
FAO,
Via delle Terme di Caracalla,
00100 Roma
ITALY

Mr P. N. Mbugwa,
National Agricultural Research Station,
P.O. Box 450,
Kitale,
KENYA

Mr H. M. McMichael,
Guthrie Corporation,
Guthrie Diversification Project,
c/o Labu Estate,
Labu, Negri Sembilan
MALAYSIA

Dr James M. McNab,
Agricultural Research Council's Poultry Research
    Centre,
King's Buildings,
West Mains Road,
Edinburgh EH9 3JS
UK

Mr R. B. Miller,
Chairman,
Cane Commodities Caribbean Limited,
1st Avenue,
George Street,
Bridgetown,
BARBADOS

Miss D. Montague,
Agricultural Supply Industry Limited,
53 Beak St.,
London W1R 3LF
UK

Dr J. Montilla,
Instituto Investigaciónes Veterinarias,
Apartado 70 — Maracay,
Estado Aragua,
VENEZUELA

Mr D. R. Moore,
Croda Agricultural Limited,
90 Kingwood Close,
Leigh-on-Sea,
Essex,
UK

Mr F. B. Moors,
Department of Agriculture,
P.O. Box 206,
Apia,
WESTERN SAMOA

Mr A. M. Morgan-Rees,
Tropical Products Institute,
56–62 Gray's Inn Road,
London WC1X 8LU
UK

Dr G. G. Morris,
Tate and Lyle Limited,
Group Research and Development,
Philip Lyle Memorial Research Laboratory,
P.O. Box 68,
Reading,
Berks RG6 2BX
UK

Dr A. M. S. Mukhtar,
Department of Animal Husbandry,
Faculty of Veterinary Science,
University of Khartoum,
P.O. Box 32,
Khartoum North,
DEMOCRATIC REPUBLIC OF THE SUDAN

Mr Z. Müller,
FAO,
UNDP/BF Project SIN 67/505,
P.O. Box 6 Nee Soon,
SINGAPORE 26

Mr G. Mwaiswaga,
Ministry of Agriculture,
Central Veterinary Laboratory,
P.O. Box 9254,
Dar-es-Salaam,
TANZANIA

Dr P. Nababsing,
Mauritius Sugar Industry Research Institute,
Reduit,
MAURITIUS

Dr J. Nabney,
Tropical Products Institute,
56–62 Gray's Inn Road,
London WC1X 8LU
UK

Mr K. C. Nah,
Pig & Poultry Research & Training Institute,
10½ M.S. Sembawang Road,
SINGAPORE 26

Mr H. Najar,
Alain Savary Street,
No. 30,
Tunis,
TUNISIA

Dr B. L. Nestel,
(IDRC),
Centro Internacional de Investigaciónes para el
    Desarrollo,
Apartado Aereo 53016,
Bogota, D.E.,
COLOMBIA

Dr T. H. Nguyen,
Ets Guyomarc'h,
BP 234,
56006 Vannes,
Morbihan,
FRANCE

Mr M. Norsworthy,
Booker McConnell Ltd,
Bucklersbury House,
83 Cannon Street,
London EC4N 8EJ
UK

Dr O. L. Oke,
Department of Chemistry,
University of Ife,
Ile-Ife,
NIGERIA

Mr V. O. Ola,
Vegetable Oils (Nigeria) Limited,
P.O. Box 121,
Ikeja,
NIGERIA

Dr P. O. Osuji,
University of the West Indies,
Department of Livestock Science,
St. Augustine,
TRINIDAD

Prof. V. A. Oyenuga,
Department of Animal Science,
University of Ibadan,
Ibadan,
NIGERIA

Mr C. H. Peeler,
International Federation of Agricultural Producers,
c/o National Farmers Union,
25/31 Knightsbridge,
London SW1X 7NJ
UK

Miss E. Phillipo,
Livestock International,
24 West Bar,
Banbury,
Oxfordshire,
UK

Mr J. R. Pickford,
Cooper Nutrition Products Limited,
Stepfield,
Witham,
Essex CM8 3AB
UK

Mr G. Pinson,
Tropical Products Institute,
56–62 Gray's Inn Road,
London WC1X 8LU
UK

Mr B. Pitt,
Ministry of Agriculture,
St. Georges,
GRENADA

Mr B. Platt,
Farming World,
BBC Bush House,
Strand,
London WC2
UK

Dr T. R. Preston,
Comision Nacional de la Industria Azucarera,
Humboldt 56, 3° − Piso,
MEXICO, DF 1

Mr R. G. Pringle,
United Molasses Trading Company Limited,
Bowater House East,
68 Knightsbridge,
London SW1
UK

Mr C. R. J. Pritchard,
BP Proteins Limited,
Brittanic House,
Moor Lane,
London EC2Y 9BU
UK

Dr K. Polzhofer,
Unilever Research Laboratory,
D-2000 Hamburg 50,
Behringstrasse 154
FEDERAL REPUBLIC OF GERMANY

Mr R. E. Pye,
W. and J. Pye Limited,
Fleet Square,
Lancaster,
UK

Mr M. P. Read,
Barkers and Lee Smith Limited,
Barkers Mills,
Wymorrdham,
Norfolk,
UK

Mr J. Redfern,
Agricultural Marketing & Export Services,
The Malthouse,
Donhead St. Mary,
Shaftesbury,
Dorset SP7 9DN
UK

Mr D. Ritchie,
Pedigree Petfoods Limited,
Melton Mowbray,
Leicestershire LE13 1BB
UK

Dr J. Robb,
Unilever Research Laboratory,
Colworth House,
Sharnbrook,
Bedfordshire,
UK

Dr R. Roberts,
J. Bibby Agriculture Limited,
Richmond House,
1 Rumford Place,
Liverpool L3 9QQ
UK

Mr E. F. Roscoe,
Faure Fairclough Limited,
14/18 Holborn,
London EC1N 2PR
UK

Mr J. B. Rose,
BOCM Silcock Limited,
Unilever Limited,
St. Bridget's House,
Bridewell Place,
London EC4
UK

Dr G. D. Rosen,
Birchwood Lodge,
The Birches,
Orpington,
Kent,
UK

Mr B. McC Rutherford,
BOCM Silcock Limited,
Unilever Limited,
St. Bridget's House,
Bridewell Place,
London EC4,
UK

Mr E. Sama,
Chef de la Région d'Élevage des-Savanes,
Dapango,
TOGO

Mr J. A. Sandys
Commonwealth Development Corporation,
33 Hill Street,
London W1A 3AR
UK

Mr J. Secher,
Muus International Limited,
2–4 Frederiksgade,
5100, Odense,
DENMARK

Mr J. Sheehan,
Department of Agriculture & Fisheries,
Dublin 2,
IRELAND

Dr B. J. Shreeve,
Central Veterinary Laboratory,
New Haw,
Weybridge,
Surrey KT15 3NB
UK

Mr M. Skottke,
Federal Agency for Economic Assistance,
6236 Eschborn 1,
Stuttgarter Strasse 10,
Postfach 5180,
FEDERAL REPUBLIC OF GERMANY

Dr A. J. Smith,
Centre for Tropical Veterinary Medicine,
Easter Bush,
Roslin,
Midlothian,
UK

Mr D. Smith,
Spillers Farm Feeds Limited,
Remington House,
65 Holborn Viaduct,
London EC1A 2FH
UK

Mr K. G. Smith,
Ministry of Agriculture, Fisheries & Food,
Pest Infestation Control Laboratory,
London Road,
Slough,
UK

Dr P. C. Spensley,
Tropical Products Institute,
56–62 Gray's Inn Road,
London WC1X 8LU
UK

Mr M. de Speville,
Food & Allied Industries Limited,
10 Intendance St.,
Port Louis,
MAURITIUS

Dr H. Swan,
School of Agriculture,
Sutton Bonnington,
Loughborough,
Leics LE12 5RD
UK

Mr T. O. Sunkwa-Mills,
Ministry of Agriculture,
P.O. Box 5779,
Accra North,
GHANA

Ir Tan Hong Tong,
Chemara Research Station (KGSB),
c/o Chemara Research Station,
1½ Mile Labu Road,
Seremban,
N. Sembila,
MALAYSIA

Mr W. C. Tang,
Protein Tumboh Tumbohan Malaysia
Sendirian Bhd.
62 Main Road,
Petaling,
Selangor,
MALAYSIA

Mr J. G. Thomas,
University of Lancaster,
Biological Sciences,
Bailrigg,
Lancaster,
UK

Mr Thompson,
Beecham Agricultural Products Division,
Beecham House,
Great West Road,
Brentford,
Middlesex,
UK

Mr W. H. Timmins,
Tropical Products Institute,
Industrial Development Department,
Culham,
Abingdon,
Berks,
UK

Mr A. Tolan,
Ministry of Agriculture, Fisheries & Food,
Great Westminster House,
Horseferry Road,
London SW1
UK

Dr J. H. Topps,
University of Aberdeen,
School of Agriculture,
581 King Street,
Aberdeen AB9 1UD
UK

Dr Traore,
Centre National de Recherches Zootechniques,
Sotuba,
Bamako,
MALI

Mr L. Tulloo,
c/o Mauritius Co-operative Agricultural Federation,
Dumas St.,
Port Louis,
MAURITIUS

Mr J. H. Urwin,
United Molasses Trading Company,
Bowater House East,
68 Knightsbridge,
London SW1
UK

Mr R. Valdez,
University of Aberdeen,
School of Agriculture,
581 King Street,
Aberdeen AB9 1UD
UK

Mr F. Vallée,
Fédération Nationale des Coopératives de Production et
    d'Alimentation Animals,
129 Bd Saint Germain,
75008 Paris,
FRANCE

Mr D. M. Veira,
University of Aberdeen,
School of Agriculture,
Division of Agricultural Chemistry & Biochemistry,
581 King Street,
Aberdeen AB9 1UD
UK

Dr M. C. Walsh,
State Laboratory,
Upper Merrion Street,
Dublin, 2,
IRELAND

Mr H. R. Were,
Ministry of Agriculture − Animal Production,
P.O. Box 30028,
Nairobi,
KENYA

Dr G. D. Wesoloski,
Central Soya Company Inc.,
1200 North Second Street,
Decatur,
Indiana, 46733,
USA

Mr P. E. Wheatley,
Tropical Stored Products Centre (TPI),
London Road,
Slough SL3 7HL,
Berkshire,
UK

Miss B. Wheeler,
An Foras Taluntais,
(The Agricultural Institute),
Dunsinea,
Castleknock,
Co. Dublin,
IRELAND

Dr D. R. Williams,
BOCM/Silcock Limited,
Basing View,
Basingstoke,
Hants RG21 2EQ
UK

Mr P. Williams,
Anutrol Limited,
23 West Smithfield,
London EC1A 9PE
UK

Mr R. A. Wilkins,
Director, Field Operations,
Bookers Sugar Estates Limited,
22 Church Street,
Georgetown,
GUYANA

Mr K. K. Woo,
Zuellig (Gold Coin Mills) Private Limited,
14 Jalan Tepong,
Jurong,
SINGAPORE 22

Mr J. A. Wood,
United Agricultural Merchants Limited,
Basingstoke,
UK

Mr A. E. Wright,
United Molasses Trading Company Limited,
Bowater House East,
68 Knightsbridge,
London SW1
UK

Mrs T. A. Zablan,
Xavier University,
College of Agriculture,
Cagayan de Oro City,
PHILIPPINES

Printed for Her Majesty's Stationery Office by Hobbs the Printers Ltd.,
(2148) Dd817500 2M 2/75 G3313